DIGITAL TECHNOLOGY WITH MOS INTEGRATED CIRCUITS

RONALD J. WEBB
Elgin Community College
Elgin, Illinois

McGRAW-HILL BOOK COMPANY

New York Atlanta Dallas St. Louis San Francisco Auckland Bogotá Guatemala Hamburg Lisbon London Madrid Mexico Milan Montreal New Delhi Panama Paris San Juan São Paulo Singapore Sydney Tokyo Toronto

McGraw-Hill Publications in Electronic Computer Technology

Gordon Silverman, Project Editor

Books in This Series:

Small Computer Theory and Applications by Denton J. Dailey
Mathematics for Computers by Arthur D. Kramer
Integrated Circuits for Computers: Principles and Applications by William L. Schweber
Computers and Computer Languages by Gordon Silverman and David Turkiew
Digital Technology with MOS Integrated Circuits by Ronald J. Webb

Sponsoring Editor: John J. Beck
Editing Supervisor: Mitsy Kovacs
Design and Art Supervisor: Nancy Axelrod
Production Supervisor: Kathleen Donnelly

Dedicated to my wife, Camille

Library of Congress Cataloging-in-Publication Data

Webb, Ronald J.
Digital technology with MOS integrated circuits.

(Electronic computer technology)
Includes index.
1. Digital integrated circuits. 2. Metal oxide semiconductors. I. Title.
TK7874.W39 1988 621.3815′3042 87-29738
ISBN 0-07-068792-7

Digital Technology with MOS Integrated Circuits

1 2 3 4 5 6 7 8 9 0 DOCDOC 8 9 4 3 2 1 0 9 8 7

ISBN 0-07-068792-7

CONTENTS

PREFACE

Digital Technology with MOS Integrated Circuits is written for technical students preparing for careers in the field of electronics. The preparation of this work was approached with two primary goals: first, to introduce students to the various types of digital technology that are currently used in industry; and second, to explain the underlying principles that make this technology work. Understanding how things work not only makes electronics more interesting and easier to learn but also provides some protection against premature obsolescence. The best preparation for a career in electronics in a changing world is a good grounding in the basic principles. Without this background it is difficult to read the literature of the trade and to keep up with new developments. Study of the basic concepts also provides employment flexibility. In electronics, it is not always easy to predict what the demands of the job market will be in two years. The diversity and flux of the electronics industry make for a variety of employment opportunities, and the basic tools will provide maximum flexibility in filling the available jobs. It is for these reasons that practice and craft have been blended in this text with the underlying principles.

To accomplish these goals a technique for reinforcing the learning process has been introduced. Because the study of digital technology involves a great deal of new terminology as well as an extensive use of acronyms, students are given the opportunity to think over what they are learning without truly testing themselves. A list of questions under the heading "Review" follows most sections, and immediately after each question (introduced by the letter **Q**), the answer (signaled by the letter **A**) has been given. This form of constant review and reinforcement will, it is hoped, aid in the learning process. These questions supplement the detailed examples that appear throughout the chapter. Each chapter concludes with a generous selection of problems.

The organizational plan of the book is in keeping with the goals of providing the "why" and "how" of modern electronics. Chapter 1 provides an introduction to the physical concepts which are important to the study of digital logic. Digital logic may, of course, be studied as an abstract system, using graphics symbols and Boolean equations, but without an understanding of the physical realities of current, voltage, resistance, and capacitance, actual circuit implementation is fraught with hazards. Particularly important in Chapter 1 are the concepts of logic signal strengths, ac and dc loading, and fanout. Chapter 2 extends these electrical fundamentals into the realm of active devices. The physical characteristics of MOS transistors and the process of fabricating integrated circuits are explained. The discussion builds up to a key concept in the understanding of VLSI, how MOS switches are used to produce logic signals. Chapter 3 relates these simple switches to combinational logic functions and circuits. Switching algebra and Karnaugh mapping are also introduced. These are tools that will be used throughout the book. Chapter 4 completes the basic theory portion of the text with an introduction to sequential logic. The tools needed to analyze and design clocked sequential networks are presented with examples using D, JK, and T flip-flops. The state machine concept is also introduced, since it provides an easy and reliable method of design and clarifies the operation of control logic in VLSI. The first four chapters emphasize MOS digital technology, but not to the exclusion of bipolar implementation. The remainder of the text applies equally to MOS and bipolar technology. Emphasis is placed on using standard 54/74 series parts.

Chapter 5 discusses the storing and shifting of data. These two operations are common to most digital systems. Storing, in this chapter, refers to working storage, using registers. The chapter begins by defining *digital data* and *digital data organization* and proceeds to showing how data is handled. Serial and parallel bus structures are also introduced along with the standard packaged parts commonly used to implement data handling. Chapter 6 deals with counters. Counters are the most commonly used of sequential state machines and deserve separate treatment. All the basic types of counters are discussed, and the popular 74xx160 family of programmable counter is highlighted. Chapter 7 discusses arithmetic circuitry. Digital computers were originally designed to perform arithmetic computation, and arithmetic logic remains an important part of control applications. Video games, avionics, digital modems, and industrial control systems are among the many applications requiring computational circuitry. Moreover, Chapter 7, together with Chapters 5 and 6, will prepare the reader for an understanding of microprocessors.

Chapter 8 discusses one of the fastest-growing segments of digital technology, semiconductor memory. Mass storage of data for extended periods of time has always been a requirement of computers. While magnetic storage has traditionally been favored because of volatility issues, recent advances have made semiconductor memory more attractive and a focus of intense development. Understanding this technology may become increasingly important in the future. RAM, ROM, EPROM, and PLA are described both in terms of data storage and in terms of replacements for combinational logic. Memory mapping and bus representation are also discussed.

Chapter 9 provides something which is missing from most digital technology texts—a thorough discussion of control logic and the types of circuits used to implement control logic. Chapter 9 depends on the groundwork laid in Chapters 5, 6, 7, and 8 to develop a basic understanding of microprocessors and control sequencers. Stored program control is explained in terms of standard logic parts. A simple 4-bit microprocessor example provides insight into the operation of larger, more complex machines. In addition, user programmable control hardware is introduced as an alternative to microprocessors in applications requiring greater speed and/or density.

Chapter 10 introduces analog-to-digital and digital-to-analog conversion. This is an old topic, but it assumes new significance as advances are made in the speed and power of digital computational circuitry. Digital signal processing (DSP) has invaded all areas of the analog world, including broadcast television, motion pictures, medical instrumentation, and process control in manufacturing. General-purpose digital signal processing is even becoming accessible to owners of personal computers. Soon, the TV repairer may need to understand the fundamentals of data conversion.

Chapter 11 exposes the painful realities involved in implementing real digital systems. Transmitting data across a printed circuit board or between printed circuit boards is becoming increasingly difficult as data rates become faster. Transmission line problems, once associated only with long cables, are now appearing on 12-inch printed circuit boards. These problems cannot be prevented or solved unless they are properly understood. Very often they are ignored in the design phase of a project and remain to haunt the technician or service person who must make the system work. This is where theory meets reality, and it is important that both designer and technician understand the inescapable limitations which nature has placed on us. Chapter 12 further emphasizes physical reality with a survey of basic troubleshooting equipment and procedures. It is hoped that this final chapter will provide the reader with some basic practical knowledge and a general sense of direction.

I would like to thank several of my colleagues for their help and advice in preparing this text. Acknowledgments are made to Mike McClaughry and Mike Rebeschini of Motorola's Applied Digital Research Laboratories for their expert review of Chapter 2. Thanks are extended to John Swan, also of Motorola, for sharing his idea of a simplified 4-bit microprocessor and to Amin Varis and Steve DeLong of Rockwell's Switching Systems Division for their helpful suggestions which have been incorporated into the middle and end portions of this text. Special thanks go to Carol Brade, Rockwell SSD Librarian, for her support. Finally, I would like to thank the numerous students and fellow teachers at Elgin Community College who have taught me the purpose and meaning of education.

Ronald J. Webb

1

BASIC ELECTRICAL CONCEPTS

This chapter is a review and expansion of concepts which are important to understanding digital circuitry. It may be used as a review for those who have completed the prerequisites or as an introduction for those who have not.

There are two sides to the study of digital electronics, the logical side and the physical side. The logical aspects are just *ideas* which can be implemented in many different ways (electrically, optically, with fluids under pressure). These ideas are somewhat abstract and could be taught without reference to electronics. The physical side, however, requires some understanding of how those ideas are implemented electrically and thus involves some knowledge of basic electrical principles. The two concepts which are most important to this understanding are *resistance* and *capacitance*. These two properties of matter are relatively easy to explain but are often misunderstood. When they are not fully understood, perceptions are formed which may result in confusion when real physical circuits are built and tested.

Even when constrained to electrical implementations of digital functions, there are various technologies to choose from. Older designs were implemented with relays, tubes, and resistor-transistor logic (RTL) and diode-transistor logic (DTL). Current designs use mostly transistor-transistor logic (TTL), complementary metal-oxide semiconductors (CMOS), and emitter-coupled logic (ECL). Within these newer technologies there are *families* of parts which may be used in the construction of digital circuits. Each family provides more or less the same logical functions, but the parts themselves do not behave identically in physical circuits. The differences are important and can best be explained in terms of resistance and capacitance. With this in mind, the objective of Chapter 1 is to provide the background needed to understand the physical differences between current logic implementations. Specifically, we will:

1. Explain the material basis of resistance and capacitance
2. Show how resistance and capacitance are related to voltage and current
3. Use the above information to explain three logic signal strengths—active, resistive, and high-impedance

1-1 PHYSICAL BASIS OF CONDUCTION

Any discussion of resistance must begin with an explanation of conduction. The question we should ask is: Why do some materials conduct electricity while others do not? The answer can be found by examining the structure of the atom. There is a model of the atom which explains conduction and has proven useful in predicting observable events. This model was proposed by Niels Bohr in 1913. The Bohr model suggests that electrons whirl around a core, or nucleus, like planets around the sun. Recent theories maintain that the electrons do not have actual positions in the atom, but form *clouds*. A cloud represents the *probability* of the electron's position. The denser the cloud, the more probable it is that an electron is in the area. Since we cannot describe with absolute precision the location of individual particles within the atom, we are forced to use statistical methods to predict their behavior. For our purposes, however, these complications may be ignored, and we may assume that the electrons do occupy definite orbits.

The orbits within the atom are sometimes called *energy levels* or *energy states*. Electrons within these levels have a dual nature: they sometimes appear to behave like particles and other times like waves. The wavelike characteristic helps to explain why electrons can occupy only certain prescribed orbits. An electron can fit only into an orbit whose circumference is an integer multiple of the electron's wavelength. Figure 1-1 illustrates this phenomenon. Each allowable orbit is associated with a fixed quantity, or *quantum*, of energy. In order to advance to a higher level, an electron

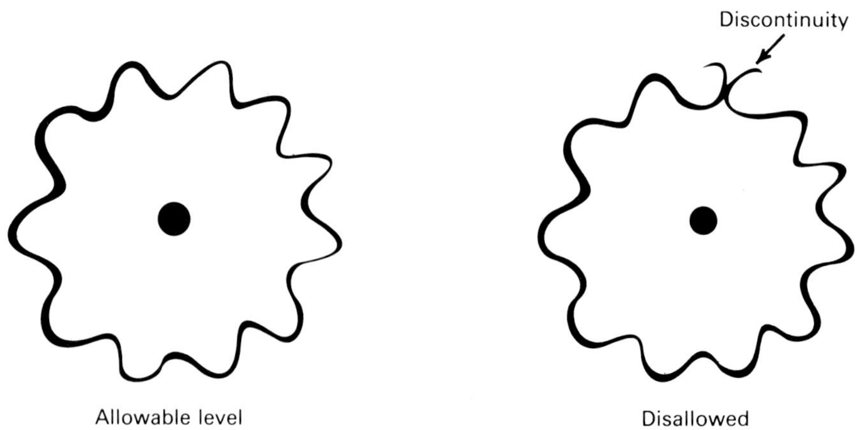

Fig. 1-1 An electron can fit only into an orbit whose circumference is an integer multiple of the electron's wavelength.

must gain the exact amount of energy needed to advance to an allowable orbit. A very definite and precise amount of energy is required to raise an electron from one level to the next. Figure 1-2 illustrates the energy levels for the hydrogen atom. Hydrogen is the simplest of elements and is composed of only one proton and one electron. The unit of energy used in Fig. 1-2 is the *electron volt* (*eV*). The meaning of this unit will become clear later in this chapter.

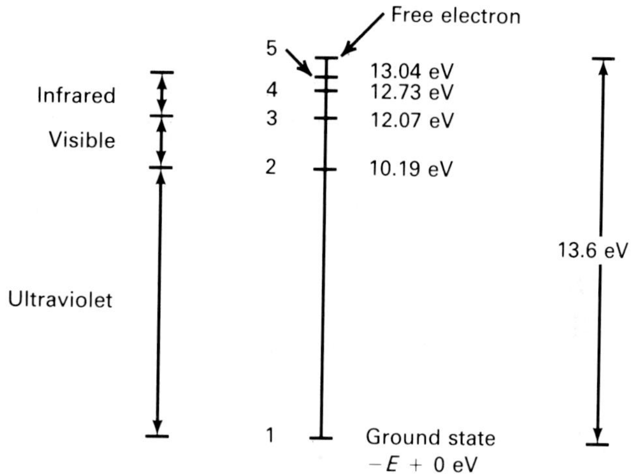

Fig. 1-2 Energy level diagram for hydrogen. When an electron loses energy, it radiates with a definite wavelength. Energy losses between the second and third level are visible to the human eye.

The lowest energy state is called the *normal*, or *ground*, state. The other states are called *excited*, or *radiating*, states. When an electron gains energy, it is said to be "excited." When an electron loses energy, it is said to "radiate." When an electron radiates, the energy lost is sometimes absorbed by another part of the material, and sometimes it leaves the material altogether. If it leaves *and* its wavelength is within the visible spectrum, the energy will appear to us as light. Energy from outside may enter the atom and be absorbed, or it may just pass through. Whether it is absorbed or allowed through depends on the wavelength of the energy. For example, some wavelengths of visible light will bump electrons in the hydrogen atom from state 2 to state 3. These wavelengths would likely be absorbed rather than allowed through the atom. Wavelengths which cannot bump electrons to higher levels will pass right through. This explains why materials are transparent to certain wavelengths of energy. Clear, uncolored glass is transparent to all visible energy.

As an electron acquires new energy, it moves farther away from the nucleus. When its energy is great enough to move out of the influence of the nucleus, the electron becomes *free*. In the normal, or unexcited, state, the atom is electrically neutral. It contains an equal number of electrons in orbit and protons in the nucleus. The charge associated with an electron is negative, and the charge associated with a proton is positive. When an electron becomes free, it leaves behind a positive *ion*. The positive ion is an atom which contains more protons than electrons. In all but the simplest case, the nucleus will also contain neutrons. Neutrons have no charge and do not influence the electrical characteristics of the atom. Figure 1-3 illustrates the Bohr model using the silicon atom as an example.

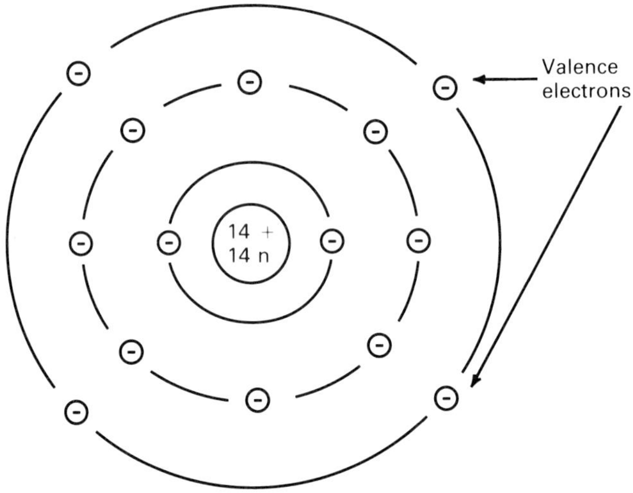

Fig. 1-3 Atoms in their normal state are electrically neutral. The 14 units of positive charge in the silicon nucleus are balanced by 14 electrons in orbit. The 4 electrons in the outer level are called *valence electrons*.

The outer energy levels of an atom are collectively referred to as the *valence band*. If any of the outer electrons in the valence band receive enough energy to jump, they will leave the atom and become free. Beyond the valence band is the *conduction band*. Electrons in the conduction band are free to move around in the material. These electrons are *charge carriers* for electric current. Between the valence band and the conduction band there is a gap in which no electron may reside. In order for an electron to become free, it must receive sufficient energy to jump this gap. There are a number of ways an electron may acquire this energy. *Heat*, or *thermal energy*, can excite an electron into the conduction band. *Illumination*, or *light energy*, may be absorbed by electrons and cause them to move up into the conduction band. *Bombardment with subatomic particles* and *electric discharge* (sparks) can also excite electrons into conduction.

Types of Materials

All materials can be placed into one of three broad categories: *conductors*, *insulators*, and *semiconductors*. Figure 1-4 illustrates how these groups differ. Insulators have large gaps between the valence and conduction bands. At room temperature, very few electrons have enough energy to jump across these gaps. As a result, there are virtually no free electrons available for conduction. Metals, on the other hand, have gaps which are

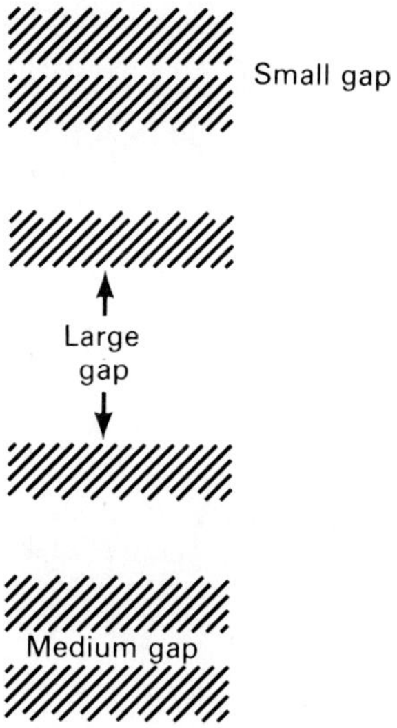

Conductors: Electrons in the outer level can easily move into the conduction band.

Insulators: The large gap makes movement to the conduction band almost impossible at room temperature.

Semiconductor: The gap is such that, at room temperature, some electrons have enough energy to jump.

Fig. 1-4 Energy gaps between valence and conduction bands.

extremely small or even nonexistent. Metals are rich in free electrons. Copper, for instance, at room temperature, has approximately one free electron per atom. This wealth of charge carriers permits very good conduction. Between these two extremes are the materials called *semiconductors,* which have gaps that are moderately difficult to jump. This means that at room temperature there will be enough electrons in the conduction band to permit some conduction. Insulators do not conduct electricity under normal circumstances. Conductors conduct freely. Semiconductors *may* conduct in a potentially *controllable* way. In insulators there are too few charge carriers to support conduction, in conductors there are so many charge carriers that conduction occurs uncontrollably, but in semiconductors, the number of charge carriers falls between the two extremes, which affords some measure of control.

1-2 ELECTRIC CURRENT

The *drift velocity* of free electrons in the conduction band determines the amount of electric current flowing in a conductor. Drift velocity is the result of an applied electric field and is not the same as random *thermal velocity*. The average speed of free electrons at room temperature in copper, in the absence of an electric field, is about 1.6×10^8 cm/s. This is thermal velocity, and it depends on the temperature of the material. Drift velocity, on the other hand, depends on the strength of an external electric field. An applied electric field will cause a drift pattern similar to that shown in Fig. 1-5. The drift movement caused by an external electric field is normally much slower than the thermal movement. The thermal speed of an electron is the total distance *covered* moving from point A to point B divided by the time it took to get there. The drift speed, on the other hand, is the distance d between points A′ and B divided by the same

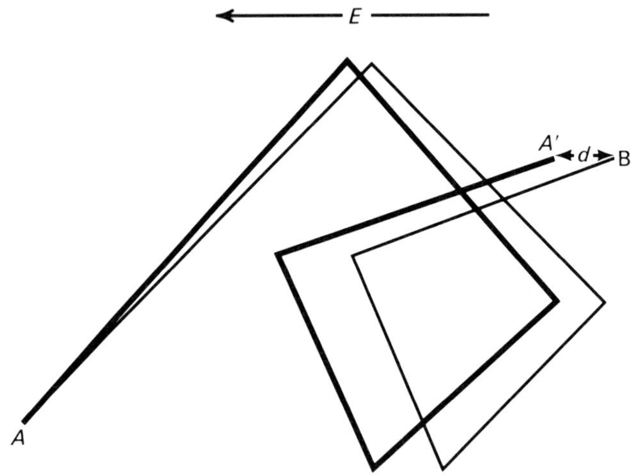

Fig. 1-5 The darker line represents the path a single conduction electron might take without the influence of an electric field. The lighter line represents the same path under the influence of an electric field $\overleftarrow{E}$. The drift movement is represented by distance d.

elapsed time. The average thermal velocity for copper at room temperature is about 1.6×10^8 cm/s, but the drift velocity V_d is only on the order of 1×10^{-2} cm/s.

Drift velocity is the net speed at which electric charge flows in a conductor. It requires the presence of an electric field and is defined to be in the direction opposite to the applied field. *Drift current* is the movement which we commonly call *electric current*. Drift current always has direction, occurs only when an electric field is present, and is relatively slow. For example, if we provided a steady current of 1 A in a copper wire with a diameter of 0.171 in (common 14-gauge house wiring), we could expect that it would take about 33 min for the electrons in this wire to drift 1.0 cm. In contrast, the electric field itself sets up across the material at close to the speed of light. What does this mean? It means that if we connect a piece of conductive material between the terminals of a battery, electrons will begin leaving the material at the positive junction and entering the material at the negative junction *immediately* (at near the speed of light). Inside the material, however, conduction electrons will drift in the direction of the positive junction at velocity V_d. If we could look at a cross-section of the conductor, we would see that electrons cross back and forth randomly. When the electric field is applied, they continue to cross back and forth randomly, but more electrons cross in the direction of the positive terminal than the other way.

Net electron flow, due to the electric field, is *electric current*. It can be measured as a net amount of charge Δq passing through a cross-section within some interval of time Δt. We do, in fact, define electric current in this way:

$$i = \text{electric current} = \frac{\Delta q}{\Delta t} \tag{1-1}$$

where the unit of current is one *coulomb of charge per second,* or one *ampere*. Another way to say this is that 1 C is the *net* number of electrons transferred through a cross-section in 1 s by a steady current of 1 A. The charge carriers need not always be electrons—they could be ions in solution (which may be positive or negative) or *holes* (which are always positive and will be discussed in Chapter 2). One coulomb represents 6.3×10^{18} charge carriers, either positive or negative. Positive charge carriers always move in the direction opposite to negative charge carriers when an electric field is present. By convention, current is usually defined in terms of positive charge carriers even when none are present.*

1-3 RESISTANCE

If conduction electrons were not impeded in any way, drift velocity would increase without bound. However, *collisions* occur between the free electrons and the rest of the material. There is a continual interchange of energy between the moving electrons and the stationary ions. Energy is lost by the electrons and absorbed by the ions upon impact. The *resistivity* of a material is *inversely* related to the average distance between these collisions and *inversely* related to the number of free electrons available for conduction. This means that a material is more resistive if collisions occur more often and less resistive if there are more electrons available for conduction. The resistivity of matter varies greatly in magnitude. Figure 1-6 shows the enormous differences between the resistivity of various

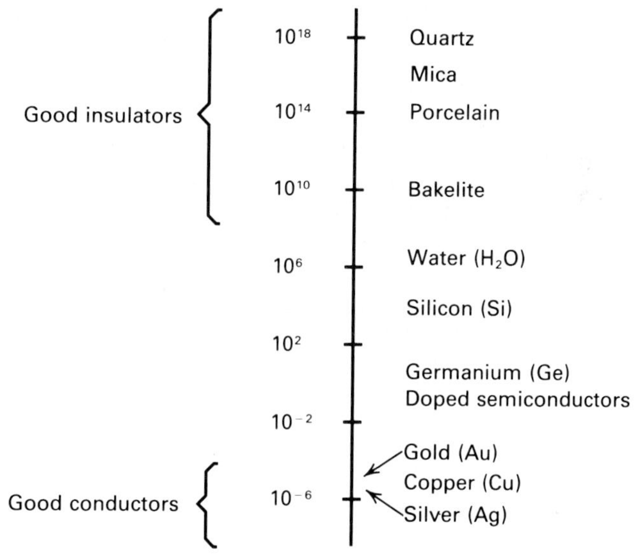

Fig. 1-6 Resistivity (in ohm-centimeters) of some common materials.

*Nineteenth-century physicists defined current in terms of positive charge carriers. When electrons were discovered, there was some movement toward redefining conventional current in terms of negative charge carriers. However, it was soon discovered that both types of carriers exist, and calculations are tidier when negative signs do not have to be routinely carried around.

materials. Probably no other characteristic of matter has a greater range of values.

Resistivity is a characteristic of all materials but does not, by itself, tell us how a given piece of material will impede electron flow. To know this, we must also know the physical size of the specimen. The unit of resistivity is the *ohm-centimeter* (Ω-*cm*), and the unit of actual resistance is the *ohm*. Resistance is related to resistivity in the following way:

$$R = \rho \frac{l}{A} \tag{1-2}$$

where ρ is resistivity (in ohm-centimeters), l is the length (in centimeters), and A is the area (in square centimeters). See Fig. 1-7*a*.

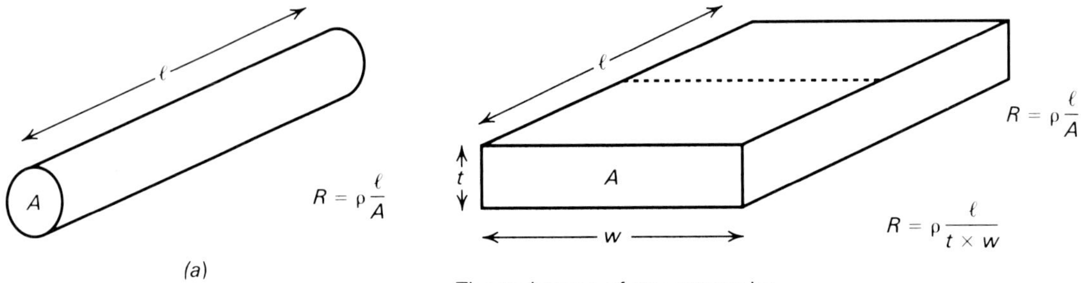

Fig. 1-7 Resistivity and resistance. *Sheet resistance* is the resistance of one square of any conductive sheet. The use of sheet resistance assumes uniform thickness.

EXAMPLE 1-1

The resistivity of copper, at room temperature, is 1.7×10^{-6} Ω-cm. What is the resistance of a 50-ft length of 14-gauge copper wire (diameter of 0.171 in)? One inch equals 2.54 cm.

SOLUTION

The diameter of the wire, in centimeters, is 0.171 in × 2.54 cm/in = 0.434 cm. The length, in centimeters, is 50 ft × 12 in/ft × 2.54 cm/in = 1524 cm. The cross-sectional area of the wire is:

$$A = \pi \times \left(\frac{0.434}{2}\right)^2 = 0.148 \text{ cm}^2$$

The total resistance of the wire is:

$$R = \rho\frac{l}{A} = 1.7 \times 10^{-6}\ \Omega\text{-cm} \times \frac{1524\ \text{cm}}{0.148\ \text{cm}^2} = 0.0175\ \Omega$$

A concept which is useful in the design of printed circuit boards and integrated circuits is *sheet resistance*. Figure 1-7*b* shows a section of conductive material which might be part of a metal run on a printed circuit board or integrated circuit. The area A in Eq. (1-2) is a product of thickness t and width w. If the thickness of a conductive sheet can be considered uniform, then the resistivity of the material can be specified in ohms per square ($\Omega/\square$). In one square, the length-to-width ratio will always be unity.

Loss

In the previous section, electric current was defined as the net number of charge carriers passing through a cross-sectional area in a given period of time. If there were no collisions, this number would continually increase until the *source* supplying the electrons failed to keep up. In real matter, however, collisions limit the drift velocity. Collisions also cause the material to become warmer when electric current passes through it. The greater the electric field, the greater the velocity of the free electrons *between collisions* (in the direction of drift), and the greater the energy absorbed on impact. Stronger electric fields mean more net drift and warmer material.

The situation is similar to marbles rolling down a flight of stairs. Some amount of energy is required to raise the marbles, in the first place, to the top of the stairs. Once the marbles have reached the top, that energy is stored in the marbles as *potential energy*. If the marbles were to fall freely from the top to the bottom level, their velocity would increase continuously until they struck the bottom. However, as they roll down the stairs, they give up energy with each bump and thereby achieve a fairly constant velocity on the way down. If the risers were higher (meaning fewer steps for the same total distance), they would fall farther between the steps and achieve a greater average velocity. Thus higher risers would be less resistive, the marbles would move down the stairs faster, and more marbles could fall in the same amount of time. Regardless of speed, however, each marble will have lost one-half of its initial potential energy when it falls down one-half the stairs and all its initial potential energy when it reaches the bottom.

1-4 CURRENT AND VOLTAGE IN A RESISTIVE MATERIAL

If a resistive material is connected across a battery, it will support a constant current. If there are longer paths between collisions, greater drift velocity will be achieved, and more current will flow. This is another way of saying that the material is less resistive. Potential energy is developed in the battery because electrons are torn away from their atoms by chemical action and moved, against their natural inclination, to a higher potential. The potential energy which they acquire is commonly measured in *joules* (J), and the electrical potential is measured in *volts* (V). The relationship between the electrical potential energy and the electric potential is as follows:

Electrical potential *energy* = quantity of charge × electric potential

or

$$K = qV \tag{1-3}$$

where K is the electric potential energy in joules, q is the quantity of charge in *coulombs* (C), and V is the electric potential measured in volts. A *volt* is defined as the difference in potential between two points A and B when it takes 1 J of energy to move 1 C of charge from position A to position B. A volt is one joule per coulomb. If, instead of using 1 C as the unit of charge, we used e (the charge associated with one electron), then K would be measured in another unit of energy, the *electron volt (eV)*. The electron volt is used when very small quantities of energy are being discussed.

It should be mentioned here that a battery is a *voltage source*. This means that as long as a battery is properly charged and is not overloaded, it will supply a fixed voltage. A *current source* is another type of power source. It will, with proper usage, provide a fixed amount of current. A voltage source is a stand-alone unit while a current source is made of transistors powered by a voltage source. An *ideal* voltage source will provide any amount of current with a fixed amount of voltage, and an *ideal* current source will provide a constant amount of current regardless of the voltage. Internal resistance between the source of electrons and the terminal causes a voltage source to be nonideal. A current source must operate within a specified voltage range; therefore, it can never really be ideal. Voltage and current sources are commonly called *power sources*, or *power supplies*.

EXAMPLE 1-2

Suppose an ideal 1-V battery is connected across a resistor as shown in Fig. 1-8. One ampere of current flows through the resistor. How much energy, per second, is being dissipated in the resistor? Assume that the connecting wires have negligible resistance.

SOLUTION

It takes 1 J of energy to raise 1 C of charge to a potential of 1 V. We know that 1 A is equal to 1 C of charge passing per second. This means that every second, 1 C of charge flows through the resistor. Since 1 J was required to raise each coulomb to 1 V, 1 J/s must be lost in the resistor when the charge is returned to the ground potential. Therefore, 1 J/s is being dissipated in the resistor in the form of heat. The joule per second is such a commonly used unit that it has its own name, the *watt* (W). The watt is a unit of electric power.

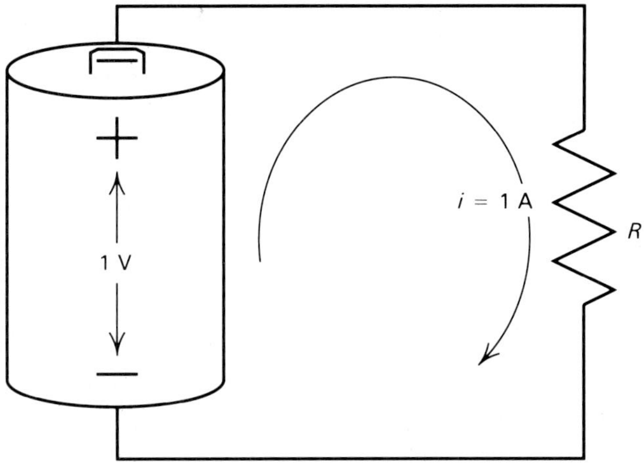

Fig. 1-8 Every second 1 C of charge flows through the resistor and gives up 1 J of energy. One joule per second is called *one watt*.

Review 1-4

1. **Q.** What sources of energy can cause valence electrons to move into the conduction band? **A.** thermal energy, illumination, particle bombardment, and electric discharge

2. **Q.** What is the fundamental difference between a conductor and an insulator? **A.** the size of the energy gap between the valence band and the conduction band in the material

3. **Q.** Explain the differences between thermal velocity, drift velocity, and the speed of an electric field. **A.** Thermal velocity is random and dependent upon temperature. Drift velocity is dependent upon an external electric field and is normally much slower than thermal speed. An electric field may set up across a material at near the speed of light, which simply means that movement at both ends of the material starts immediately.

4. **Q.** What is electric current? **A.** the *net* movement of charge through a cross-section of material within some interval of time

5. **Q.** What determines the resistivity of a material? **A.** two things: the number of available charge carriers, which is a function of the energy gap between valence band and conduction band, and the mean free path between collisions, which is a function of the internal structure of the material

6. **Q.** What is a volt? **A.** It is the difference in electric potential between two points when it takes 1 J of energy to move 1 C of charge from one point to the other

7. **Q.** What is a watt? **A.** It is a measure of power, equal to 1 J/s lost or used.

1-5 USEFUL LAWS

We have explored the physical basis of four electrical quantities, *resistance, voltage, current,* and *power*. We related *resistance* to the mean free path between collisions and the availability of charge carriers. We said that *current* is the net flow of charge carriers through a conductor when an electric field is present. We defined *voltage* in terms of potential energy per unit charge. And we described *power* as energy used or lost per second. These explanations help to establish the physical basis of electricity but do not provide the tools needed to predict electrical behavior. We would like to define each of these quantities in terms of the others so that we can make calculations and analyze circuits.

The expression which relates resistance to voltage and current is:

$$r = \frac{v}{i} \tag{1-4}$$

It is important to realize that this is only a definition of resistance. A very similar statement, which is often confused with Eq. (1-4), is *Ohm's law*. Ohm's law was first published in a pamphlet in 1827 by George Ohm, a German physicist. Ohm said that the resistance of many types of conductive materials is independent of either the voltage across the material or the current through the material. Initially it was thought that all conductive materials had this property, and the relationship was stated as:

$$v = Ri \tag{1-5}$$

where R is a constant.

The difference between Eqs. (1-4) and (1-5) is that in (1-4) we do not assume that the resistance is constant. We now know that, in many cases, R is not a constant. The relationship $r = v/i$ is the definition of resistance in a material regardless of whether that material obeys Ohm's law. Several important devices do not obey Ohm's law, and yet this definition of resistance remains useful. It can be used, for example, to derive an equation for electric power. Electric power was represented earlier as energy lost in a resistor. In the example, 1 A of current flowed through the resistor with 1 V of potential. Using Eq. (1-4), it is easy to see that the resistor must have been a 1-Ω resistor. One joule per second, 1 W, of power was dissipated in that resistor, and, with a little thought, you could prove to yourself that electric power is related to resistance, current, and voltage in the following way:

$$\text{Power (in watts)} = \frac{\text{joules}}{\text{coutombs}} \times \frac{\text{coulombs}}{\text{seconds}} = \frac{\text{joules}}{\text{seconds}} = \text{watts}$$

that is,

$$\text{Power (in watts)} = \text{voltage} \times \text{current} = v \times i \tag{1-6}$$

and, combining (1-6) with (1-4),

$$P = i \times v = \frac{v}{R} \times v = \frac{v^2}{R} \tag{1-7}$$

Kirchoff's Laws

Another important contribution to electrical science was Gustav Kirchoff's *current law*. Kirchoff's current law states that the sum of the currents entering and leaving a node must be 0. A node is a connection between two or more conductors. This law follows from the assumption that any amount of charge entering a node must be balanced by an equal amount of charge leaving that node; otherwise charge would either be created or destroyed in the node. We can, in fact, store charge in a node, but not an infinite amount. If charge is being stored in a node, the capacity for storage will eventually be exhausted, and the node will return to compliance with Kirchoff's law.

Kirchoff also proposed that the sum of voltages around any closed path is zero. This is Kirchoff's *voltage law*. For example, in a circuit such as the one shown in Fig. 1-9, any voltage developed by the battery will be

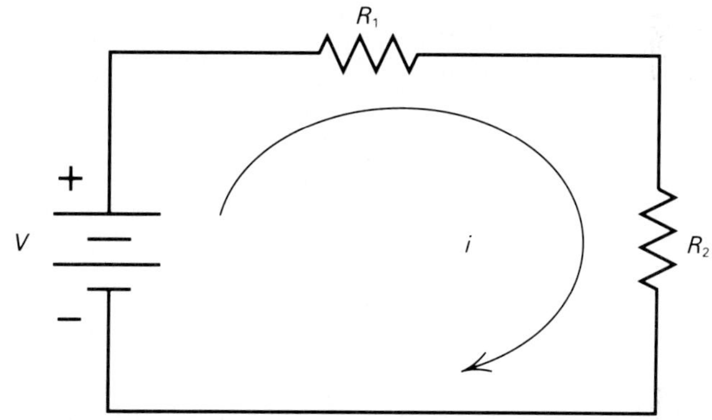

Fig. 1-9 According to Kirchhoff's law, all the voltage developed by the battery must drop across the two resistors. Moreover, the current through R_1 must equal the current through R_2.

dropped across the resistors. The sum of those voltages dropped across the resistors will always equal the voltage produced by the battery. This is useful for finding the voltage which appears across any given resistor. In Fig. 1-9, we know that each resistor conducts the same amount of current i (this follows from Kirchoff's current law) and that the voltage dropped across a conductor is equal to the resistance of the conductor *times* the current through the conductor (Eq. 1-5). With these two facts, we can derive the following equation:

$$V = iR_1 + iR_2 \tag{1-8}$$

If we know the values of the battery voltage V and the values of the resistors R_1 and R_2, then we can solve for the current i. When i is known, we can solve for the voltage across each resistor.

There is, however, a shortcut which derives from Kirchoff's laws, called *voltage division*. This is best explained by example. Using the circuit in Fig. 1-9, the voltage v_2 (across R_2), can be calculated with the voltage divider equation as follows:

$$v_2 = iR_2 = \frac{R_2}{R_1 + R_2} V \qquad (1\text{-}9)$$

Likewise, the voltage across R_1 is:

$$v_1 = iR_1 = \frac{R_1}{R_1 + R_2} V \qquad (1\text{-}10)$$

These relationships will be useful later when digital output circuits are discussed.

Review 1-5

1. **Q.** What is an ideal voltage source? **A.** An ideal voltage source is a device which provides a fixed amount of voltage regardless of the number of electrons drawn from it.
2. **Q.** What is an ideal current source? **A.** An ideal current source is a device which provides a fixed amount of current regardless of how the voltage across it changes.
3. **Q.** What is a power supply? **A.** A power supply is a voltage or current source.
4. **Q.** What does it mean to say that a material "obeys Ohm's law"? **A.** This means that the material has a resistance which is independent of the voltage applied across it.
5. **Q.** Can Kirchoff's current law ever be violated? **A.** Yes, for some amount of time, charge can be stored in a node. Once the capacity for charge storage is exhausted, however, the node will return to compliance with Kirchoff's law.
6. **Q.** Kirchoff's laws lead to what circuit analysis technique? **A.** the voltage division technique

1-6 CAPACITANCE

Without going into a lot of detail, we can say that, in general, power supplies use some type of external energy source to pull (or strip) electrons away from an electrically neutral material. This operation causes excess negative charge to accumulate on one terminal while the other terminal is left with a net positive charge. In a battery, the energy for this stripping process is provided by chemical action. The same is true in the case of power stations which burn fossil fuels. Other energy sources such as nuclear fission, solar energy (nuclear fusion), wind (sun and gravity), flowing water (gravity), and geothermal power (stored heat) can also be used to generate excess charge. In every case, some source of stored energy is used to strip electrons away from a neutral material and deposit them in a handy place. That energy is transferred to the electrons and is measured in *joules per coulomb*, or *volts*.

If two wires are connected to the terminals of a power source, the action of stripping away electrons and concentrating electrons extends to the wires. Some electrons will be pulled out of the positive wire, and other electrons will be pushed into the negative wire. This creates an electric *pressure* between the wires which is called *voltage* or *potential*. If the tips of the two wires are brought closely together, a concentrated *electric field* develops between them. Figure 1-10 shows the invisible lines of force associated with such a field. Electric energy is stored in this field, and the field is most concentrated where the lines are the shortest. If we could

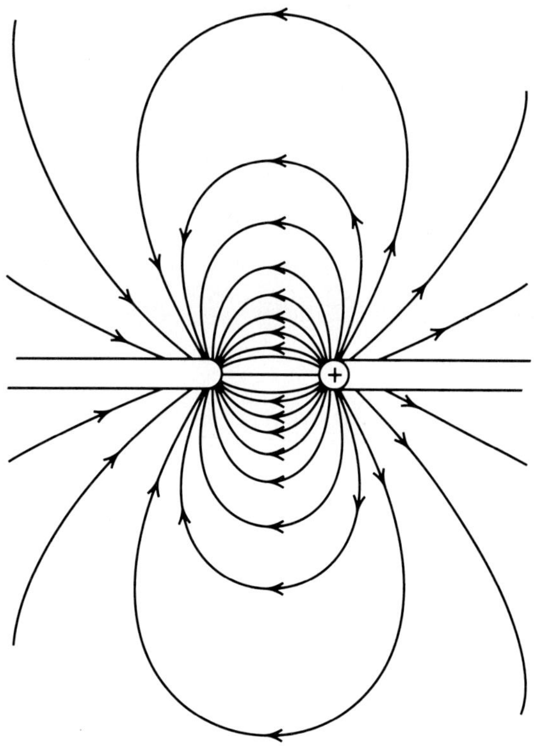

Fig. 1-10 When two oppositely charged wires are brought close together, an electric field sets up between them. Electric energy is stored in this field, and the field is most concentrated where the lines are shortest.

shorten the distance between the wires and straighten out more of the force lines, we would increase the amount of energy stored. This is the reasoning behind the parallel plate *capacitor*.

Figure 1-11 shows a capacitor constructed with two parallel plates. The larger the plates (area wise), the greater the number of short force lines,

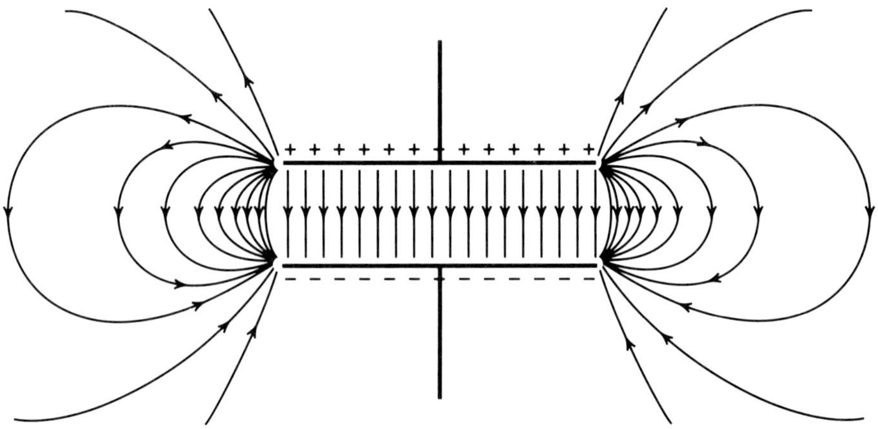

Fig. 1-11 A parallel plate capacitor increases the number of force lines by adding more surface area. The force lines away from the edges are straighter and therefore shorter. The field is most concentrated where the lines are the shortest. Greater field concentration means more stored energy.

and the greater the energy stored between them. The *capacity* to store energy also depends on the distance between the plates and the permittivity constant of the insulating material between the plates. An insulator, when used in a capacitor, is called a *dielectric*. For the parallel plate capacitor, the total capacitance is:

$$C = \epsilon \frac{A}{D} \tag{1-11}$$

where A is the area of the plates, D is the distance between them, and ϵ is the permittivity constant of the material between the plates.

Most capacitors are not constructed with parallel plates because of physical size limitations. In *metal-oxide semiconductor (MOS)* digital technology, however, our concerns are not with the techniques of capacitor construction, but rather with techniques for avoiding unwanted capacitance. Little capacitors will form wherever unlike charge accumulations can pile up but not touch. We will see that this is a major problem in digital electronics because it limits the speed of digital logic. The formula for parallel plate capacitance will appear again with reference to diode and transistor fabrication. Another equation defines capacitance in terms of charge and voltage:

$$C = \frac{q}{v} \tag{1-12}$$

where C is the capacitance in *farads* (F), q is the charge in coulombs, and v is the voltage in volts. One farad is equivalent to one coulomb per volt. Equations (1-11) and (1-12) together indicate that larger parallel plates can store more coulombs per volt. In digital logic circuits tiny parallel plate capacitors appear everywhere in logic gate inputs, in printed circuit board traces, and between wires in cables. They limit the speed of digital logic and sometimes corrupt the waveforms. When fast logic families are used, these little capacitors assume great importance, and understanding their function becomes crucial.

Digital Output Circuits

In order to appreciate the effect of capacitance in digital circuits, we must look at some basic circuits containing resistors and capacitors. Figure 1-12*a* illustrates a simple digital circuit. This circuit is digital because its output voltage can have only one of two *discrete* values, either 0 or 5 V. When the switch is closed, terminals A and B are shorted (connected directly together), and V_{out}, the voltage difference between the output terminals, is 0. Meanwhile, current flows through resistor R, and 5 V drop across the resistor as we would expect from Kirchhoff's voltage law. If the switch is opened, current ceases to flow and voltage ceases to drop

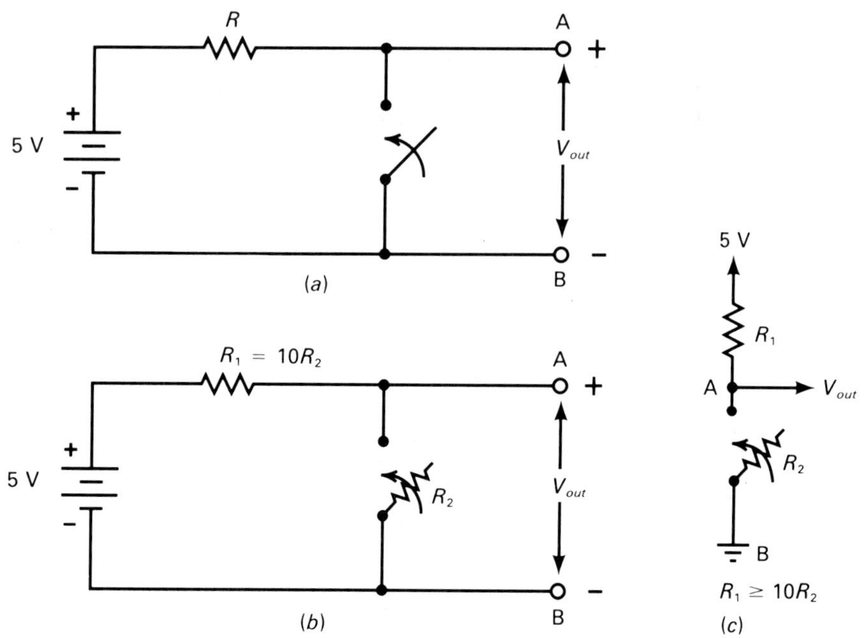

Fig. 1-12 A simple digital cricuit. (*a*) Oversimplified; (*b*) with switch resitance; (*c*) equivalent representation.

across the resistor. The instant after the switch opens, electrons continue to be pulled out of the resistor by the battery, but no new electrons replace them because the connection to the negative terminal is broken. This leaves the entire resistor and terminal A with a positive potential. When the switch opens, terminal A rises to 5 V above terminal B, and V_{out} equals 5 V.

Digital circuits no longer use mechanical switches (relays), but the concept is the same. Transistor switches turn on (close) and turn off (open) like mechanical switches. However, while mechanical switches have almost zero resistance, the transistor switches commonly used in digital logic have resistances in the range of 25 to 1000 Ω. In Fig. 1-12*b* the mechanical switch is replaced with a transistor switch having resistance R_2. In order to obtain a low logic level output (near 0 V), the original circuit resistance must be at least 10 times larger than R_2 as shown. When the switch is opened, the output voltage rises to 5 V as before. But when the switch is closed, V_{out} does not return to 0 V. We can use the voltage division shortcut from Sec. 1–5 to see why.

$$V_{out} = \frac{R_2}{R_2 + 10R_2} 5 = \frac{1}{11} 5$$

$$= 0.45 \text{ V}$$

The output will not reach 0 V when the switch is closed because a voltage divider is formed with the resistor and closed transistor switch. Actually, transistors do not obey Ohm's law. That is, the resistance of a transistor switch is not constant for all voltages. But for approximation purposes we may consider the resistance of the transistor switch to be constant.

Another reality complicates the picture further. What if the simple digital circuit must charge a capacitor? This is a reasonable suggestion since the output of any logic circuit must connect to the input of at least one other circuit in order to be useful, and all digital inputs have some capacitance. Figure 1-13 shows the simple digital circuit connected to a capacitive load. Previously, the opening and closing of the switch resulted in nearly

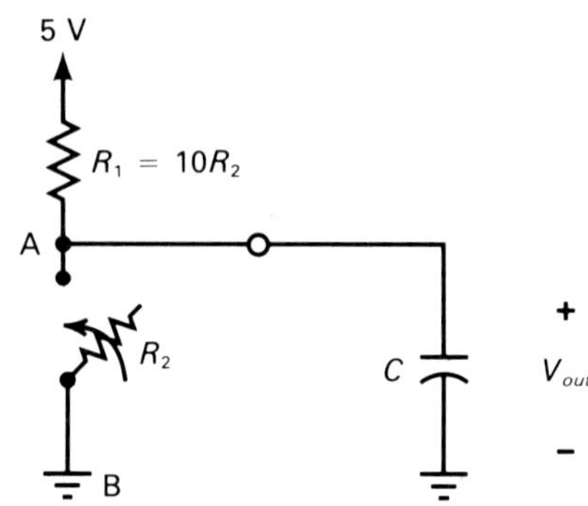

Fig. 1-13 Digital circuit driving a capacitive load.

instantaneous transitions. The capacitor will slow these changes down. If the switch has been opened for a long time, the voltage difference between A and B will be 5 V as before. Negative charge is built up on the bottom plate, and an equal amount of positive charge is left on the top plate. When the switch is closed, the electrons trapped on the negative plate will flow to the positive plate through resistance R_2. Ultimately, the voltage across resistance R_2 will reach 0.45 V as before, but the extra charge stored on the capacitor will slow down the transition. It will take some amount of time for the capacitor to *discharge* to 0.45 V. The exact time required depends on the amount of resistance and capacitance in the circuit.

Larger values of resistance or capacitance will cause greater delays. Figure 1-14 shows the charging and discharging curves for the circuit in Fig. 1-13. The equation for the discharge curve is:

$$v = V_F + V_0 e^{-t/RC} \tag{1-13}$$

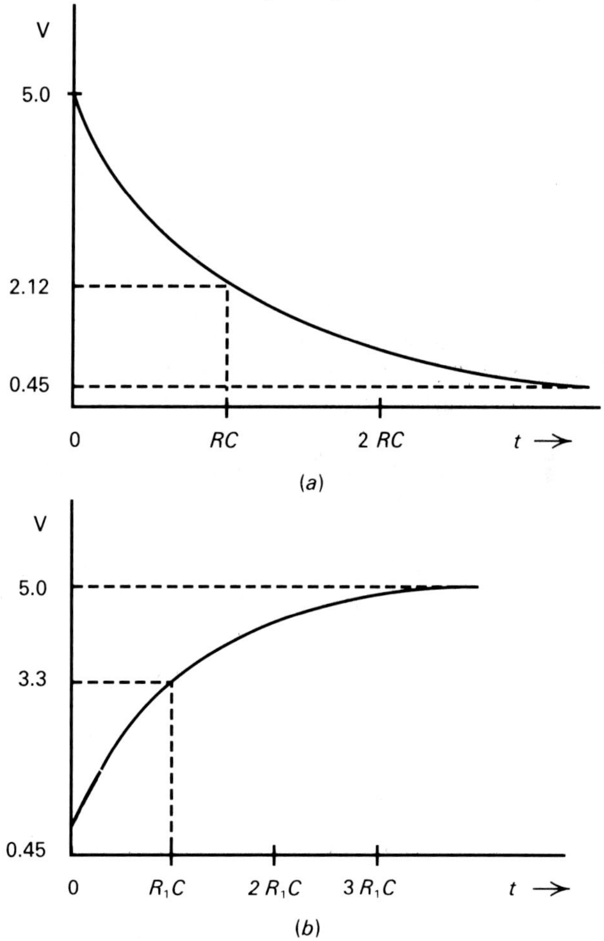

Fig. 1-14 Charging and discharging curves for the simple digital circuit in Fig. 1-13. (*a*) Discharging to 0.45 V; (*b*) charging to 5.0 V.

where V_0 is the difference between the initial and final voltage on the capacitor, V_F is the final voltage, t is the elapsed time, C is the capacitance being discharged, and R is the resistance through which the charge is lost. The lowercase v represents the voltage on the capacitor at any given time. The product RC is called a *time constant*. Roughly 63 percent of the voltage change is accomplished in one time constant. After five time constants, the voltage on the capacitor will be within 1 percent of its final value. In the example, most of the charge is lost through R_2 because R_2 is so much smaller than R_1, but the correct value for R in the equation must be calculated to include R_1. This will be explained later. For now, consider the value of R in the equation to be approximately equal to R_2.

After several time constants, the circuit will reach a *steady state*, with V_{out} equal to 0.45 V. If the switch is opened, the capacitor will *charge*. The battery will supply charge to the capacitor sufficient to make the output equal 5 V. Once again, this will take several time constants, except now the charge must flow entirely through R_1. This will take about 10 times longer than the discharge process because the R_1 is 10 times larger than R_2. The equation for charging the capacitor is:

$$v = V_i + V_0(1 - e^{-t/RC}) \qquad (1\text{-}14)$$

where V_i is the initial voltage. Ideally, the charging time should equal the discharge time in order to obtain symmetrical waveforms and maximum speed of operation. However, this is not possible in this example because R_1 must be at least 10 times larger than R_2 in order to obtain low logic levels near 0 V.

The capacitance in the above example was *resistively* driven high and *actively* driven low. To *drive* a load means to supply charge to or remove charge from the load. Digital circuit outputs must drive the inputs of other digital circuits. Logic device inputs behave like small capacitors in the range of 4 to 6 picofarads (pF) each (1 pF $= 1 \times 10^{-12}$ F). Printed circuit board traces contribute additional capacitance at roughly 30 pF/ft. When digital inputs are purely capacitive, current flows only until the capacitance is either fully charged or discharged. Some inputs are resistive as well as capacitive. This means that current continues to flow after the capacitance has charged or discharged.

When inputs are driven *high*, charge flows from the positive terminal of the power supply through the logic output into the load. When inputs are driven *low*, charge flows out of the load back through the logic output into the ground terminal of the power supply. The manner in which these inputs are driven defines the *signal strength* of the logic. *Resistive signal strength* means that the connection between the power supply and the load is made through a fixed resistor. *Active signal strength* means that the connection is made through a transistor switch.

Resistive-high, active-low outputs are quite common. Active-high, active-low outputs are also common. In the latter case, transistor switches are used in both the upper and lower connections to the power supply. Figure

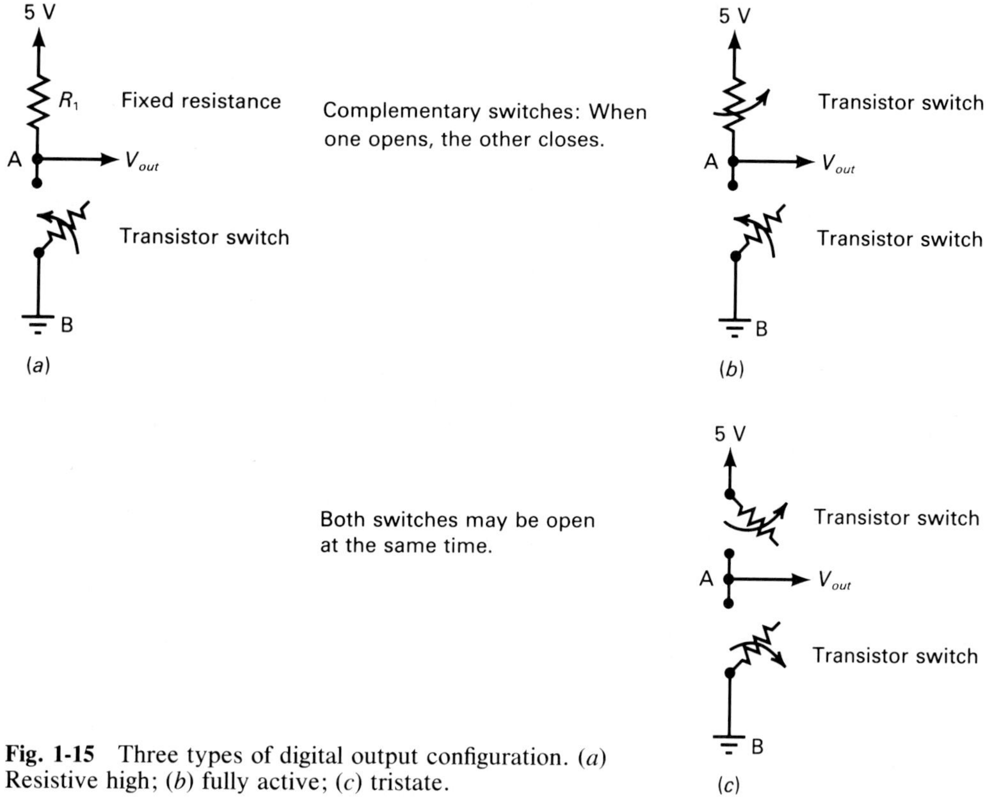

Fig. 1-15 Three types of digital output configuration. (*a*) Resistive high; (*b*) fully active; (*c*) tristate.

1-15*b* illustrates this type of output configuration. Fully active output circuits use complementary switches. This means that when one switch is closed, the other is open, which assures that the load is always driven either high or low. This type of output has two advantages. First, when a purely capacitive load is driven, the load will discharge fully to 0 V. Second, the transistor switches can be chosen so that they provide equal amounts of resistance. This allows symmetrical output waveforms and permits the logic to operate at maximum speed. In some cases, the output is configured so that both switches may be opened at the same time. The output configuration is then called *high-impedance*, or tristate. High-impedance signal strength indicates that no connection is made to either terminal of the power supply.

Thévenin's Theorem

Sometimes it's hard to tell where the charge is going when there is more than one resistor in a circuit. M. L. Thévenin, a French engineer, wrestled with this problem and published in 1883 a method for simplifying resistive circuits. Using Thévenin's theorem, any portion of a resistive network can be reduced to an equivalent circuit consisting only of a voltage source and a resistor. The equivalent circuit is referred to as a *Thévenin-equivalent circuit*.

Thévenin's theorem is best illustrated by example. Suppose we have a network such as the one shown in Fig. 1-13. Suppose also that we are interested only in the current through, or voltage across, capacitor C when the switch is closed. First pretend that C is not in the circuit and find the voltage drop that would develop between nodes A and B given that C has been removed. This is called the *open-circuit voltage*, or V_{oc}, because the circuit has been "opened" between the output and the capacitor. Using the voltage division rule which was discussed earlier, we find that $V_{oc} = 0.45$ V. Now with C still removed, short-circuit *all* voltage sources (there is only one in this case). This is equivalent to replacing each voltage source with a wire to ground. In addition, remove *all* current sources (there are none in this case). Now find the Thévenin-equivalent resistance R_{th} by determining the resistance which exists between node A and node B, given that all sources have been disabled. This resistance would be the resistance measured with an ohmmeter between nodes A and B if the sources were appropriately disabled.

We find in the above example that shorting the voltage source leaves us with two resistors in *parallel*. This means that a charge may flow through *either* resistor to ground. We represent this situation with the following notation: $R_1 \| R_2$. This indicates that R_1 is in parallel with R_2. If you have trouble knowing when resistors are in *parallel* and when they are in *series*, turn to Appendix A for a short tutorial.

When two or more resistors are connected in series, their values are *added* to find the equivalent resistance. When two or more resistors are connected in parallel, the following equation applies:

$$\frac{1}{R_{eq}} = \frac{1}{R_1} + \frac{1}{R_2} + \frac{1}{R_3} + \cdots + \frac{1}{R_n} \tag{1-15}$$

For only two resistors, this simplifies to:

$$R_{eq} = R_1 \| R_2 = \frac{R_1 R_2}{R_1 + R_2} \tag{1-16}$$

In the above example, R_{eq} is the Thévenin resistance. Using Eq. (1-16), we find that $R_{th} = (10/11)R_2\ \Omega$. Knowing V_{oc} and R_{th}, we can replace the entire circuit, except for capacitor C, with the equivalent circuit shown in Fig. 1-16*c*.

EXAMPLE 1-3

Show how digital output types *a* and *b* in Fig. 1-15 drive a 50-pF capacitive load. Assume that the transistor switches each have 100 Ω of resistance and R_1 is 1000 Ω. The outputs are initially off, then turned on for 0.1 μs (1 μs = 10^{-6} s), and then turned off again. Use eqs. (1-13) and (1-14) to obtain the rise and fall times. Sketch the resultant waveforms.

SOLUTION

Figure 1-17*a* shows how the capacitive load is driven with an active-low, resistive-high output. It takes roughly 10 times longer to charge than to

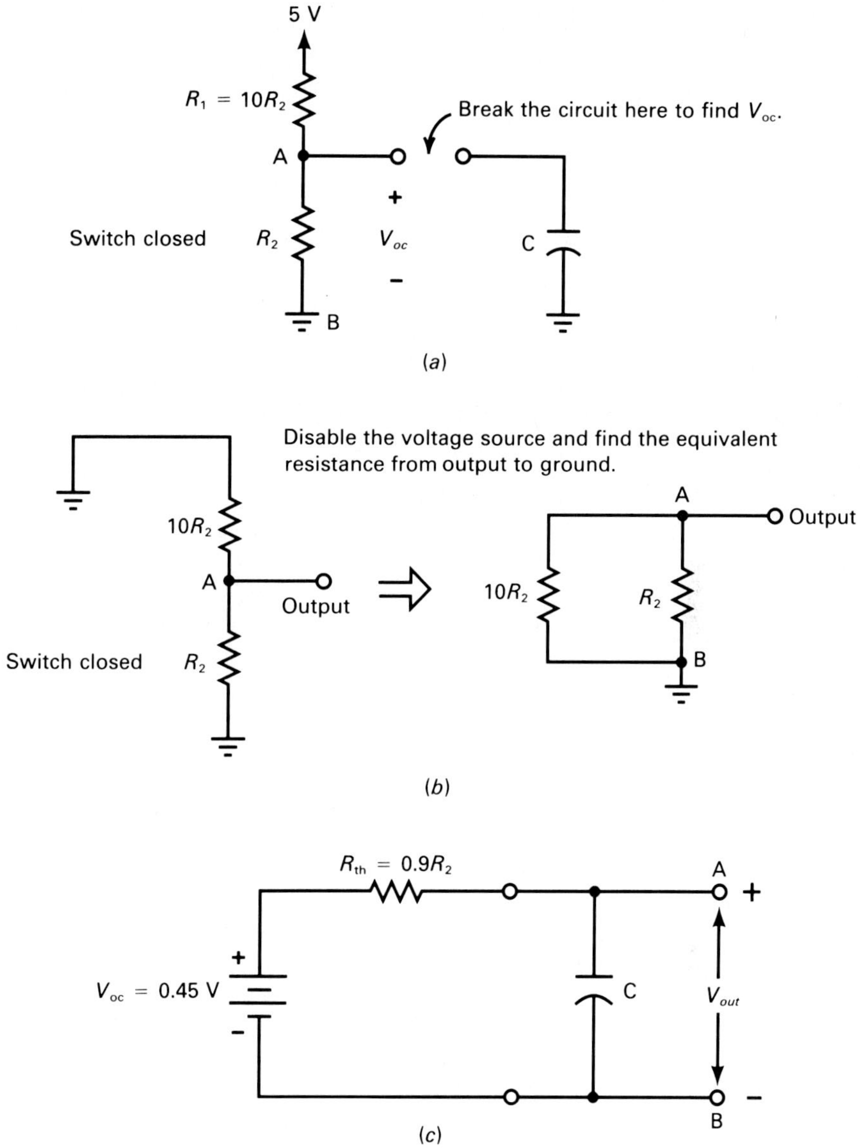

Fig. 1-16 Finding the Thévenin equivalent of the circuit in Fig. 1-13 when the switch is closed. First find V_{oc} (*a*); then find R_{th} (*b*). Use V_{oc} and R_{th} to construct the equivalent circuit (*c*).

discharge. The output, in this case, never fully charges because 0.1 μs represents only two time constants. The output never achieves a 0 V either because of the resistor divider. Figure 1-17*b* shows the fully active output reaching full value in roughly 25 ns (1 ns $= 10^{-9}$ s), or five time constants. The fully active output achieves both 0- and 5-V levels and produces a symmetrical waveform.

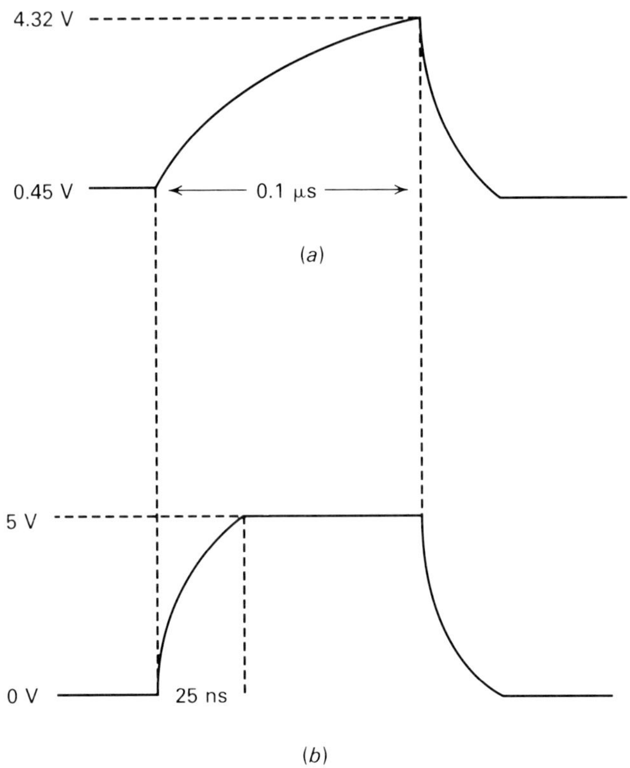

Fig. 1-17 Waveforms for Example 1-3.

Digital Inputs

Some digital inputs are purely capacitive. This means that, with the exception of small leakage currents, no charge will flow after the input has been fully charged or discharged. Other digital inputs require that a certain amount of current *sink* to ground in order to keep the input at its low level. When fully active outputs drive purely capacitive loads, they are able to achieve output low voltages of nearly 0 V. But when fully active outputs drive mixed inputs (resistive and capacitive), the final output low value will depend on how much current the transistor switch must sink in the steady state. A typical sink current for mixed inputs is 0.4 mA (1 mA $= 10^{-3}$ A). This means that the transistor switch must sink about 0.4 mA of steady-state current in order to maintain a low value at the input. When more than one input is driven, the currents add, and the output low voltage rises. If the output low voltage rises too high, the input will not interpret it as a low value. This reality limits the number of mixed inputs that any output can drive. The number of inputs that a digital output can drive is referred to as the allowable *fanout*. When purely capacitive inputs are driven, the fanout is limited only by time constraints. That is, any number of inputs can be driven, as long as there is a sufficient amount of time before the next transition. When mixed inputs are driven, the fanout is limited by the resistance of the transistor switch.

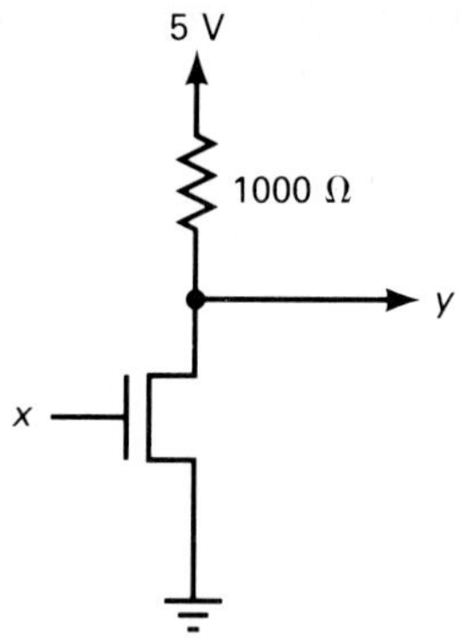

Fig. 1-18 Transistor circuit for Prob. 12 and Example 1-4.

EXAMPLE 1-4

Suppose that the output circuit in Fig. 1-18 has a switch resistance of 200 Ω. If it must sink 0.4 mA when pulling an input low, what will the output low voltage be? If the logic specification states that a logic low voltage can be no higher than 0.8 V, what is the fanout of this digital output?

SOLUTION

Driving one load low involves sinking 0.4 mA through a 200 Ω resistance. This results in an output voltage of 0.08 V. Each additional load will add 0.4 mA, and the total output voltage will be:

$$V_{out} = n(0.4\ \text{mA})(200\ \Omega)$$

where n is the number of inputs driven. If V_{out} must be no greater than 0.8 V, solving for n yields a maximum fanout of 10.

Review 1-6

1. **Q.** What is capacitance? **A.** It is the capacity to store charge carriers.
2. **Q.** What is the time constant in a simple *RC* circuit? **A.** It is the product *RC*.
3. **Q.** What effect does capacitance have in digital circuits? **A.** It slows down the transitions from one output voltage to the other.
4. **Q.** How many time constants are required to achieve a 99 percent transition? **A.** five
5. **Q.** How does Thévenin's theorem help in analyzing digital circuits? **A.** It helps to simplify circuits with more than one resistor so that a single *RC* time constant for the circuit can be found.
6. **Q.** If the resistive-high and fully active output structures use the same type of transistor switch, which would you expect to be faster when driving a purely capacitive load? **A.** the fully active output
7. **Q.** What is fanout? **A.** It is the number of inputs a digital output can drive.
8. **Q.** How is fanout determined? **A.** When driving purely capacitive loads, fanout is limited only by speed requirements. Fanout for mixed loads is limited by the amount of current an output can sink while maintaining an output low voltage below some specification.

REVIEW QUESTIONS

1. Review electrons and energy levels. Try to explain, in general terms, why a 1-in-thick sheet of plate glass is transparent while a piece of paper only 0.01 in thick is opaque.

2. Fluorescent tubes contain mercury which becomes vaporized when the light is turned on. Electrons in the vapor are excited by the electricity but then fall back to lower energy states, radiating energy in the ultraviolet range. If the tube were clear glass, an insignificant amount of visible light would be produced. However, the inside of the tube is coated with a layer of chemicals which radiate visible light when exposed to invisible ultraviolet light. Based on what you know about energy levels within the atom, explain how this might occur. (Think about how different materials require different-sized packets of energy to excite their electrons and how a large packet might be spread out over several levels.)

3. A *thermistor* is a semiconductor whose resistance changes dramatically with temperature. With proper calibration, it can be used as a "thermometer." With your knowledge of energy levels, explain why this is possible.

4. Cadmium sulfide (CdS) *photodetectors* change their resistance when exposed to light. They can be used in the home to control automatic night-lights which turn on in the evening and off at dawn. Explain this phenomenon.

5. Explain the relationship between *thermal motion*, *drift current*, and *electric field*.

6. How much energy is required to move 1 C of electrons from the positive terminal to the negative terminal of a 1-V battery?

7. What is the effect of resistance in a circuit? What is the effect of capacitance?

8. What makes the output of a circuit digital?

9. How do transistor switches differ from mechanical switches?

10. Explain the difference between resistive and active outputs.

REVIEW PROBLEMS

1. If we were able to count electrons as they passed through a cross-section of wire, and we counted 18.9×10^{18} electrons in 2 s, how much current, in amps, is the wire conducting?

2. The resistivity of aluminum at room temperature is 2.8×10^{-6} Ω-cm. What is the resistance of a 50-ft length of 14-gauge aluminum wire?

3. An aluminum block measures 1 cm × 1 cm × 7 cm. What is the resistance between the two square ends?

4. The resistance of AWG 18 solid wire is 6.51 Ω/1000 ft for copper and 10.7 Ω/1000 ft for aluminum. What diameter is size AWG 18 wire?

5. If we measured the temperature of a length of house wire and determined that 1 J of energy was being wasted every 10 s while the wire was conducting 6 A of current, how many volts are dropping across that length of wire?

6. The resistivity of tungsten increases as the material becomes warmer. When we first close the switch, the tungsten filament in a light bulb presents only about 0.1 Ω to a dc equivalent of 120 V. After a very short time, the temperature of the filament has risen to the point where the resistance is about 140 Ω. What size, in watts, is the bulb?

7. A 5-in length of metal foil runs straight across a printed circuit board. If the foil is 12.5 mils wide and has a sheet resistance of 0.03 Ω/□, then what is the resistance of the foil (1 mil = 10^{-3} in)?

8. The car battery cables to the starter motor have about the same resistance as the starter motor itself. If the car battery is measured at 14 V, how much voltage drops across the starter when we turn the key?

9. In the above problem, if the starter is drawing 250 A (not unusual on a cold morning), how much power is being lost in the connecting wires? Where is that energy going?

10. If the plates of a parallel plate capacitor are separated by a distance of 0.8 mm and the permittivity constant of the air between them is 8.859×10^{-12} F/m, how large must the plates be to achieve 1 F of capacitance?

11. A capacitor has been charged to 1 V. If 1 C of charge has been stored on the plates of the capacitor, how large, in farads, is the capacitor? If the capacitor is completely discharged through a resistor, how much energy is dissipated in the resistor?

12. The transistor circuit shown in Fig. 1-18 works in the following way. When 5 V is applied to the input x, the transistor turns on, and the output y goes to 0 V. When 0 V is applied to the input x, the transistor turns off, and the output y rises to 5 V. If the on-resistance of the transistor is 80 Ω, what is the actual operation of the circuit?

13. The capacitor in Fig. 1-19a has been fully charged by the battery. Assume that the switch has no resistance. If the switch is thrown to the 1kΩ side, how long will it take to discharge the capacitor to within 1 percent of its final value? How many time constants is this? After

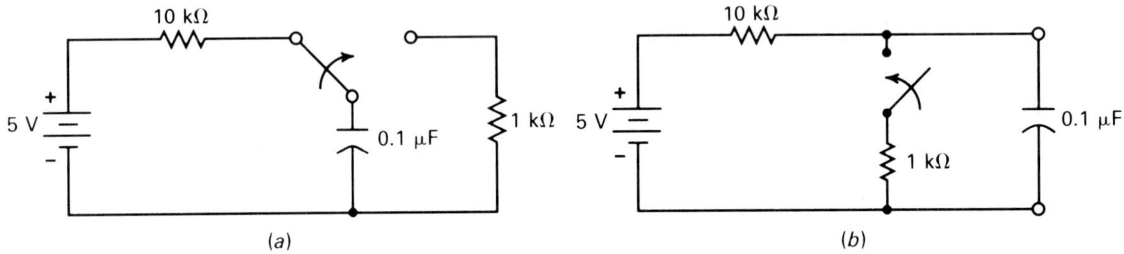

Fig. 1-19 Charging and discharging a capacitor.

the capacitor is fully discharged, the switch is thrown back to the 10-kΩ side. How many seconds are required to *charge* the capacitor to within 1 percent of its final value? How many time constants is this?

14. Sketch charging and discharging curves for the circuit in Fig. 1-19*a*. Use eqs. (1-13) and (1-14).

15. Redo Probs. 13 and 14 with Fig. 1-19*b* circuit. How is this different?

16. The circuit in Prob. 12 is called an *inverter*. Figure 1-20 shows two inverters, one driving the other. Draw a Thévenin-equivalent circuit

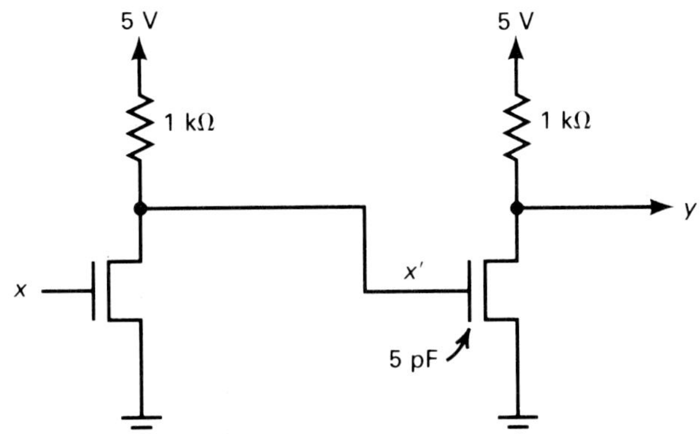

Fig. 1-20 Cascaded inverters.

for the first inverter when the switch is closed ($x = 5$ V). Note that the input of the second inverter has 5 pF (1 pF $= 10^{-12}$ F) of capacitance, and must be included as part of the first circuit. If the on-resistance of the transistors is 80 Ω, how long will it take to charge and discharge the input of the second inverter to within 1 percent of its final value?

17. If the output circuit in Fig. 1-18 has a switch resistance of 16 Ω and must sink 0.4 mA when pulling one input low, what will the output low voltage be? If the logic specification states that an output low can be no greater than 0.4 V (and each input load contributes 0.4 mA), what is the output low fanout of the circuit in Fig. 1-18?

2 BASIC SEMICONDUCTOR DEVICES

This chapter expands the physical concepts introduced in Chapter 1 to include the semiconductor devices commonly used in digital logic. The semiconductors are a special group of materials which conduct poorly in their pure form but can be modified for better conduction. In metals, conduction depends only upon the movement of electrons through the solid, but in semiconductors, the solid itself can contribute to charge conduction. This chapter explores the basic characteristics of semiconductor materials and explains how they can be modified to improve and control conduction. Design and fabrication processes are also discussed, and a bridge is made between the theory and the physical reality of the devices used in digital integrated circuits.

2.1 SEMICONDUCTOR CRYSTALS

Semiconductors are crystalline in nature. Crystals are familiar items in everyday life. Table salt, snowflakes, and diamonds all have crystalline patterns. The patterns are regular and repetitive. Snowflakes, although formed in an almost infinite number of geometric patterns, contain only 60° angles and are, therefore, essentially all alike. Table salt normally forms cubes. Other materials form long needles, rhomboids, and hexagonal patterns. Each crystalline material has a characteristic pattern. A crystalline solid is distinguished from other solids by the fact that its atoms are arranged in a *periodic* fashion. True crystalline solids have a single continuous periodic pattern throughout the entire material. This periodic arrangement is referred to as a crystal *lattice*. Most crystalline solids have random regions of disjointed lattice with no overall continuous pattern. These solids are called *polycrystalline*. Solids which have no periodic structure at all are called *amorphous* solids.

There are various ways in which crystal atoms are bound together, and these determine the lattice arrangement and the physical properties of the material. For example, pure graphite and diamond both contain only atoms of carbon, yet graphite is soft and diamonds are very hard. Carbon atoms have four electrons in the outer shell. In the diamond form of carbon, each outer electron is shared with a neighboring atom in an electron-pair bond. This is called *covalent bonding* and is illustrated in Fig. 2-1, Covalent

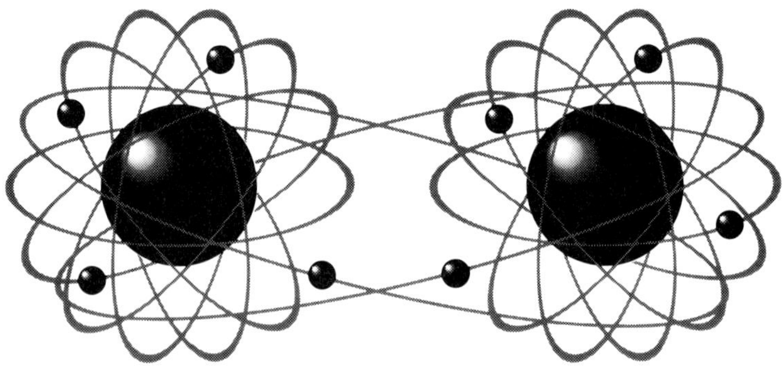

Fig. 2-1 Covalent bonding between two carbon atoms.

bonding makes the diamond crystal both hard and strong. In the graphite form of carbon, only three of the outer electrons are shared with neighboring atoms in covalent bonds, the fourth is bound to a neighboring atom by a much weaker bond. This difference in crystal lattice explains why the two substances, which are otherwise chemically identical, have such different physical properties.

Many materials bond covalently, but not all can assume the diamond crystal lattice. If you look at the portion of the periodic table shown in Fig. 2-2, you will see that silicon (Si) and germanium (Ge) appear just

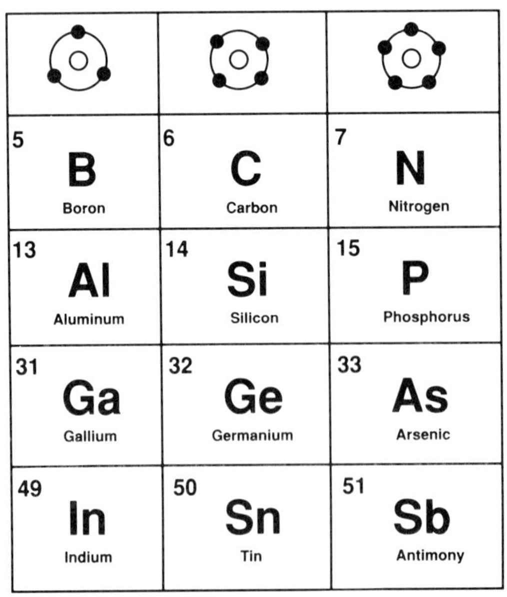

Fig. 2-2 Relevant portion of the periodic table of elements. Columns contain elements with the same number of valence electrons.

below carbon and therefore have chemical properties similar to carbon. Both can assume the diamond lattice. In order to more easily discuss this structure, we will adopt a convention in drawing atoms. The silicon atom was illustrated in Chapter 1, Fig. 1-3. We will simplify that drawing by showing only the core and the valence electrons. Fig. 2-2 uses this convention at the top of each column. Valence electrons will always be represented by darkened circles, and the inner electrons will not be drawn. Like carbon, silicon has four electrons in the outer band and can be made to covalently bond with adjacent atoms in a diamond lattice. A portion of pure, single crystal silicon is represented in the three-dimensional drawing of the diamond lattice in Fig. 2-3*a*, and a two-dimensional representation is shown in Fig. 2-3*b*. We will always use two-dimensional representations while remembering that the lattice is actually three-dimensional and extends with the same regular pattern into the paper and out toward the viewer.

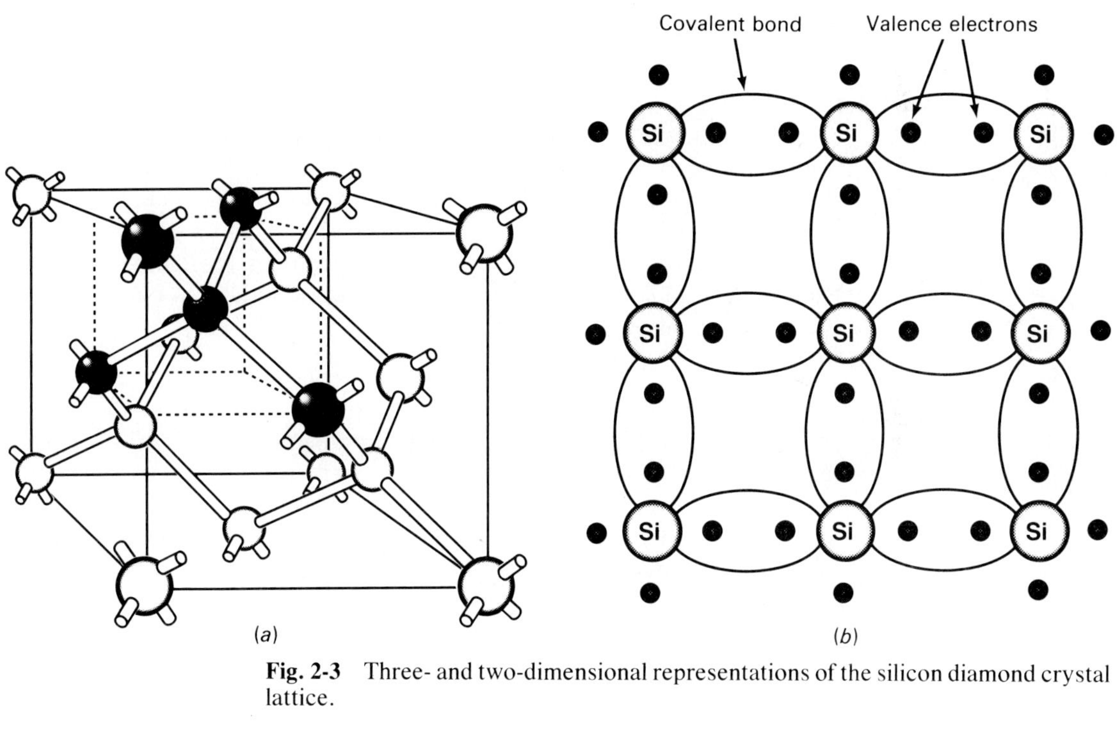

Fig. 2-3 Three- and two-dimensional representations of the silicon diamond crystal lattice.

Electrical Characteristics

Covalent bonds are quite strong, and electrons which are bound covalently have little chance of becoming free. At very low temperatures crystals of this type behave as insulators. At room temperature, however, some of the bonds are broken because of thermal energy. This permits a small amount of conduction. In semiconductors, the energy required to jump the gap between valence band and conduction band is essentially the same amount of energy required to break the covalent bond. At room temperature this is about 0.72 eV for germanium, 1.1 eV for silicon, and 5 eV for diamond. At room temperature, the *average* valence electron has only about 0.025 eV above the valence band and cannot break away. However, in materials with a band gap of around 1 eV (Ge and Si), a significant number of electrons will be excited thermally into the conduction band to become charge carriers. When an electron breaks away, it leaves behind a *hole*. The importance of the hole is that it may serve as a charge carrier comparable in effectiveness to the free electron. When a hole exists, it is *relatively easy* for a valence band electron in a neighboring atom to leave its covalent bond to fill that hole. The valence band electron moving from a bond to fill a hole leaves another hole behind, and this hole may then be filled by another valence band electron. In the absence of an external electric field, this movement is random. However, when an external electric field is applied, the movement of these holes may constitute a positive flow of current.

In a pure semiconductor, there are an equal number of free conduction band electrons and valence band holes. An external electric field will cause *the free electrons to move in one direction and the holes to move in the other*. Conduction can be accounted for by either treating the holes as positive charge carriers within the valence band or treating the free electrons as negative charge carriers in the conduction band. In processed semiconductors, however, regions are created in which conduction occurs *either* via free electrons in the conduction band *or* via holes in the valence band. The fundamental difference between metals and semiconductors is that metals conduct only by means of electrons in the conduction band while semiconductors may conduct with either conduction band electrons or valence band holes.

Doping

In pure semiconductor material, conduction is minimal because relatively few charge carriers are available. However, certain impurities can be added to improve conductivity, and the addition of these impurities to pure, or *intrinsic*, semiconductor material is called *doping*. These impurities fall into two groups which flank the semiconductors in the periodic table. Referring back to the portion of the periodic table shown in Fig. 2-2, the important fact to notice is that semiconductor elements have four electrons in their valence band and the elements used for doping have either three or five valence electrons. *Donor* impurities such as phosphorus (P), arsenic (As), and antimony (Sb), have five valence electrons and are called *pentavalent*. *Acceptor* impurities such as boron (B), aluminum (Al), gallium (Ga), and indium (In) have three valence electrons and are called *trivalent*.

Consider the silicon crystal lattice in Fig. 2-3. If a donor atom replaces a silicon atom, the donor atom will bond with the silicon atoms without disturbing the lattice. As an example, Fig. 2-4*a* shows a phosphorus atom embedded in a silicon lattice. Four electron-pair bonds are formed just as if the phosphorus atom were a silicon atom. However, there is one important difference. Phosphorus, in its normal, electrically neutral state,

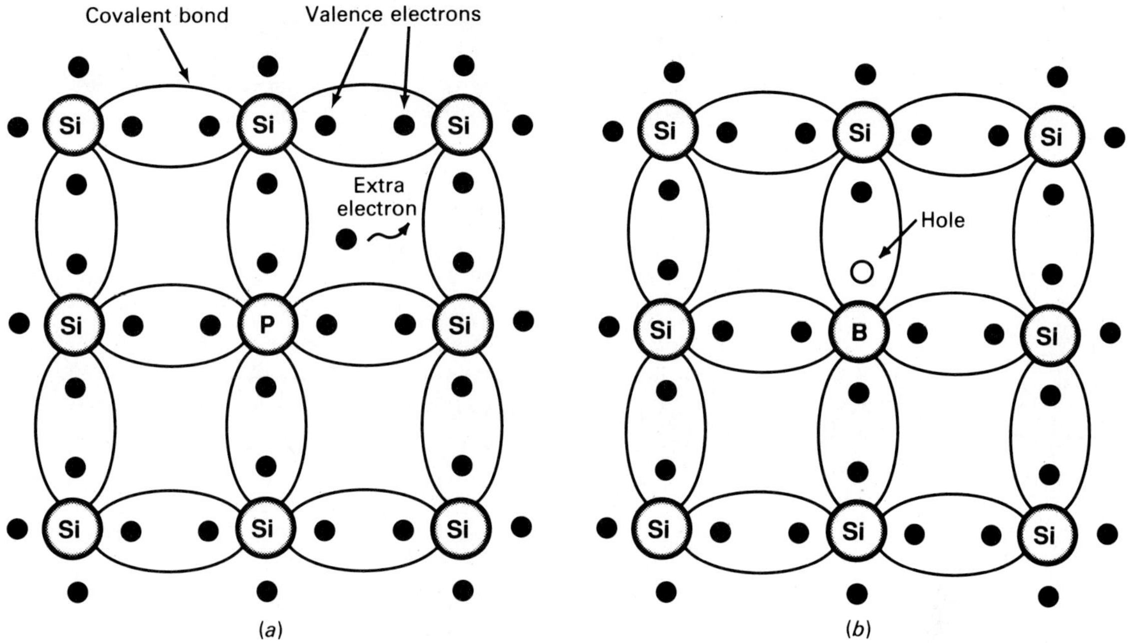

Fig. 2-4 Impurity atoms replace silicon atoms in a silicon diamond lattice.

has five electrons in the valence band, and only four of these electrons will covalently bond with adjacent silicon atoms. There will always be an extra electron which forms no bond. The energy required to excite this extra electron into the conduction band is quite small (≈ 0.01 eV) because no covalent bond needs to be broken. And since the average valence band electron at room temperature has approximately 0.025 eV, most of these leftover electrons become charge carriers. Germanium or silicon which is doped with donor impurities is called *n-type* material. *The n means that there are extra negative charge carriers in the crystal.*

If an acceptor atom replaces a germanium or silicon atom in a crystal lattice, the valence electrons will *try* to form four covalent bonds but will be short one electron. Figure 2-4*b* shows that, in this case, there is a hole. A hole is an unstable condition in the lattice, and it is quite easy for a neighboring atom to give up an electron to fill it. Germanium or silicon which is doped with acceptor elements is called *p-type* material. *The p indicates an excess of positive charge carriers.* The addition of acceptor, or *p*-type, impurities to the intrinsic semiconductor makes it easy for valence band electrons to move since very little energy is needed for an electron to leave its covalent bond to occupy a hole.

In pure semiconductor material the number of holes equals the number of free electrons. Thermal agitation always produces new holes with corresponding free electrons, and these hole-electron pairs continually disappear as a result of *recombination*. The hole concentration will always equal the free electron concentration in a pure semiconductor. But if the semiconductor is doped with donor, or *n*-type, impurities, there will be more free electrons than holes, and electrons will be the *majority carriers* in that material. If, on the other hand, the semiconductor is doped with acceptor, or *p*-type, impurities, there will be more holes than free electrons, and holes will be the majority carriers in the material. In the first case, conductivity is improved by increasing the movement of charge within the conduction band, and in the second case conductivity is improved by increasing the movement of charge within the valence band.

Now if you doubt that holes behave as legitimate positive charge carriers, consider the following experimental evidence. Fig. 2-5*a* shows a piece of

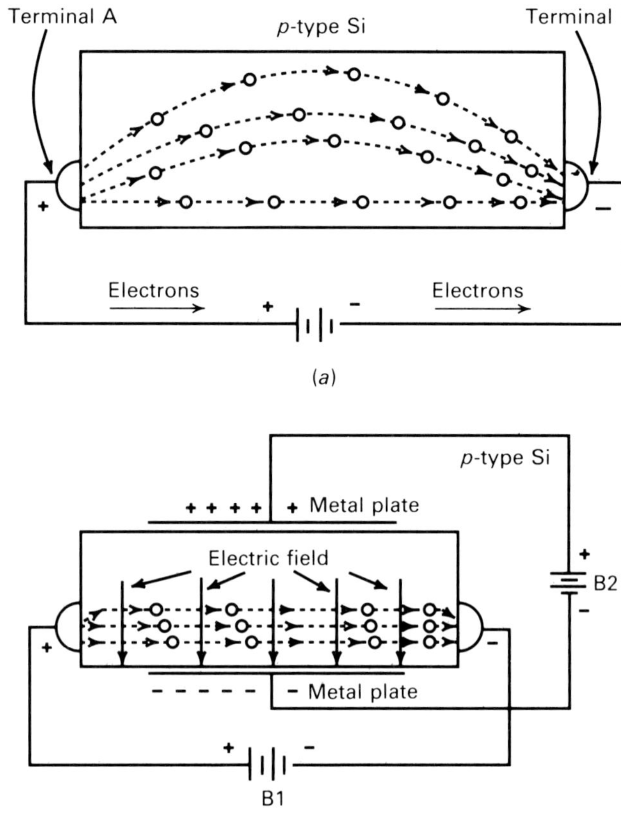

Fig. 2-5 Evidence for the existence of holes.

p-type silicon with battery connections at either end. Note that the connections to the *p*-type material are made at the low end of the block. The electric field will produce a current which arcs toward the upper side of the block as shown. This means that the charge carriers at the low side travel a shorter distance and will arrive to the opposite end sooner. Because of the curved paths, however, the overall current, that is, the charge moving from one end to the other, will be slower than if all the paths were straight. Electrons do enter the material from the negative terminal of the battery, and we might be tempted to view this as electron current. However, if plates are connected to a second battery and placed above and below the semiconductor as shown in Fig. 2-5*b*, the time required for the charge carriers to travel from one terminal to the other is shortened, suggesting that the paths were straightened by the external electric field. This straightening could occur only if the charge carriers were positive. If the charge carriers had been negative, they would have been pulled toward the positive plate, causing longer paths. It appears then that the holes do behave like a true positive charge carriers in *p*-type semiconductors.

There is an important difference between the movement of majority carrier holes and majority carrier electrons in semiconductor material. Free electrons have a higher mobility than holes. *Mobility* is a measure of the ease with which charge carriers drift through the material. While electrons and holes will always move randomly due to thermal motion, they also exhibit a net drift in the presence of an external electric field. The drift velocity is normally much smaller than the random thermal velocity. Mobility is the *average particle drift velocity per unit electric field* and is expressed in units of centimeters per second per volts per centimeter, or $cm^2/(V \cdot s)$.

$$\mu = \frac{V_{d\,avg}}{E} \tag{2-1}$$

The mu symbol μ normally carries a *p* or *n* subscript to indicate either hole or electron mobility. The mu symbol (μm) may also be used to represent a unit of distance, one micrometer, which is 10^{-6} m or 10^{-4} cm. The context will make clear the intended usage.

Crystal Growth

The crystal growing process is the first step in fabricating an integrated circuit. The very existence of integrated circuits today depends on breakthroughs made in the growth of pure, single crystal silicon. The process uses *quartzite*, a pure form of sand, which is placed in a furnace with various forms of carbon. Reactions take place which produce a very pure polycrystalline form of silicon. The impurity concentrations of this silicon are in the parts-per-billion range because the presence of foreign atoms of different sizes and chemical properties will disrupt the regular lattice. The pure silicon is then melted in a fused-silica (SiO_2) crucible with a very small amount of dopant. Boron is commonly added for *p*-type silicon, and phosphorus is commonly used for *n*-type silicon.

A seed crystal of pure silicon is suspended above the crucible and is lowered until it just enters the melt. Part of the seed melts and joins the rest. The seed is then pulled slowly upward at a rate of a few millimeters per minute. As it is pulled, it solidifies into an ingot of single-crystal silicon. This process, called the *Czochralski technique*, is the most common method for producing single-crystal silicon. Another technique, called the *float-zone process*, starts with a preformed, polysilicon ingot which has a seed crystal attached. The ingot is slowly turned into single-crystal silicon by melting a small band and moving it upward so that a molten *float zone* moves up from the seed crystal to the top of the ingot. One advantage of the float-zone method is that no crucible is used, and so one source of contamination is eliminated. Another advantage is that multiple passes can be easily made. Foreign atoms will tend to move along with the float and concentrate at the end of the ingot. This results in a substantial reduction of impurities with each pass. The main problem with the float-zone technique is gravity. The molten silicon must be contained by its own surface tension, and this limits the floating zone to very short segments. Gravity also causes convection currents in the melt. Convection, caused by differences in weight due to uneven heating, is a major source of defects. Gravity problems may be solved in outer space, where successful crystal growing experiments have been conducted under gravity-free conditions.

When the ingot is complete, the seed crystal is removed, and the ingot is shaped with grinders to produce a uniform diameter. The resulting cylinder is also ground along the length of the ingot to produce one or more flat regions which identify the crystal orientation of the ingot and the type of ingot produced. Finally, the ingot is sliced into *wafers* with a diamond saw. The wafers are *lapped* on both sides to obtain uniform flatness and polished on one side to obtain a mirror finish.

The crystal growing, grinding, and slicing process produces a semiconductor *substrate* with a fairly well controlled doping concentration, either $p-$, ("p minus"), or $n-$ ("n minus"). The minus signs indicate that the doping is light ("minus" will be spelled out to avoid confusing the minus signs with hyphens). Near the surface of the wafer, a more exact doping profile is needed, and a further step, called *epitaxy*, is required. *Epitaxy* refers to the growth of layers in the surface of the semiconductor substrate. It is done either with vapors or liquid, in high-temperature ovens or in ultrahigh-vacuum conditions with thermal beams of atoms or molecules. Molecular-beam epitaxy is very precise but requires near-perfect vacuum conditions. This is not conducive to volume production. Chemical vapor epitaxy allows high-volume production but cannot be as tightly controlled as molecular beam techniques. In either case, epitaxy improves the uniformity of the doping concentration near the surface of the wafer.

Other Crystals

While *very large scale integration (VLSI)* technology rests firmly on the shoulders of single-crystal silicon, other types of crystals are also grown. Germanium preceded silicon as the material of choice. Germanium has a mobility that is roughly three times higher than silicon, which is good. But intrinsic germanium also has a much larger concentration of free electrons at room temperature, and this is not desirable. Transistor fabrication involves the careful control of carrier concentration in the substrate, and a more neutral starting material has advantages. Compound semiconductors, such as gallium arsenide, indium phosphide, indium gallium arsenide, and indium antimonide have also been grown into single crystals. These synthetic materials require special processing. Epitaxial reactors are used to build crystal lattices one atomic layer at a time. Alternate layers of indium arsenide and gallium arsenide have been grown on indium phosphide substrates, and the fastest transistor, to date, was fabricated with indium gallium arsenide layered with aluminum gallium arsenide. The motivation for these experiments is to improve processing techniques so that high-mobility substances can be grown on more stable substrate materials. Ultimately it may be feasible to grow compound lattices on silicon.

Gallium arsenide has long been labeled "the technology of the future" because of its high charge carrier mobility and self-insulating properties. Higher mobilities mean greater speeds and faster logic. Self-insulation means that large separate areas are not needed to isolated one transistor from another in the substrate and transistors can be spaced more closely. Gallium arsenide, however, is not found in nature. Gallium and arsenic do not crystallize at the same temperature and pressure and therefore do not easily combine into a solid. The first gallium (GaAs) field-effect transistor was demonstrated in 1966, but widespread use of this material in digital integrated circuits (ICs) is not expected until packaging and processing techniques have been improved so that the price of GaAs approaches that of silicon. Silicon is simple, plentiful, and has a natural oxide which forms easily. Presently, GaAs is used in light-emitting diodes, microwave amplifiers and oscillators, and some specialized digital applications.

Review 2-1

1. **Q.** What lattice pattern does silicon assume when it is grown into a single crystal? **A.** diamond
2. **Q.** What chemical bond determines this pattern? **A.** covalent
3. **Q.** Name three descriptive terms for phosphorus (P) atoms which relate to semiconductor doping. **A.** donor, *n*-type, pentavalent
4. **Q.** What is the advantage of the float-zone technique in crystal growing? What are the problems? **A.** Greater purification can be achieved. Containment of the molten material is difficult and there are problems with convection currents
5. **Q.** What semiconductor material may challenge silicon? Why? What is the single most important factor in silicon's dominance? **A.** gallium arsenide higher mobilities economics

2-2 PN JUNCTIONS

The creation of diffusion areas is the second major step in the fabrication of an integrated circuit. A *diffusion region* is a layer within the substrate of opposite polarity. The process of creating such a layer involves several repetitive operations. Initially, a thin oxide layer (silicon dioxide) is grown on the surface of the wafer by exposing it to pure oxygen at high temperatures. Then a thin film of *nitride* (silicon nitride) is deposited on top of the oxide, and a film of light-sensitive *photoresist* is spun over the nitride. Figure 2-6*a*. shows a cross-section of the substrate at this point.

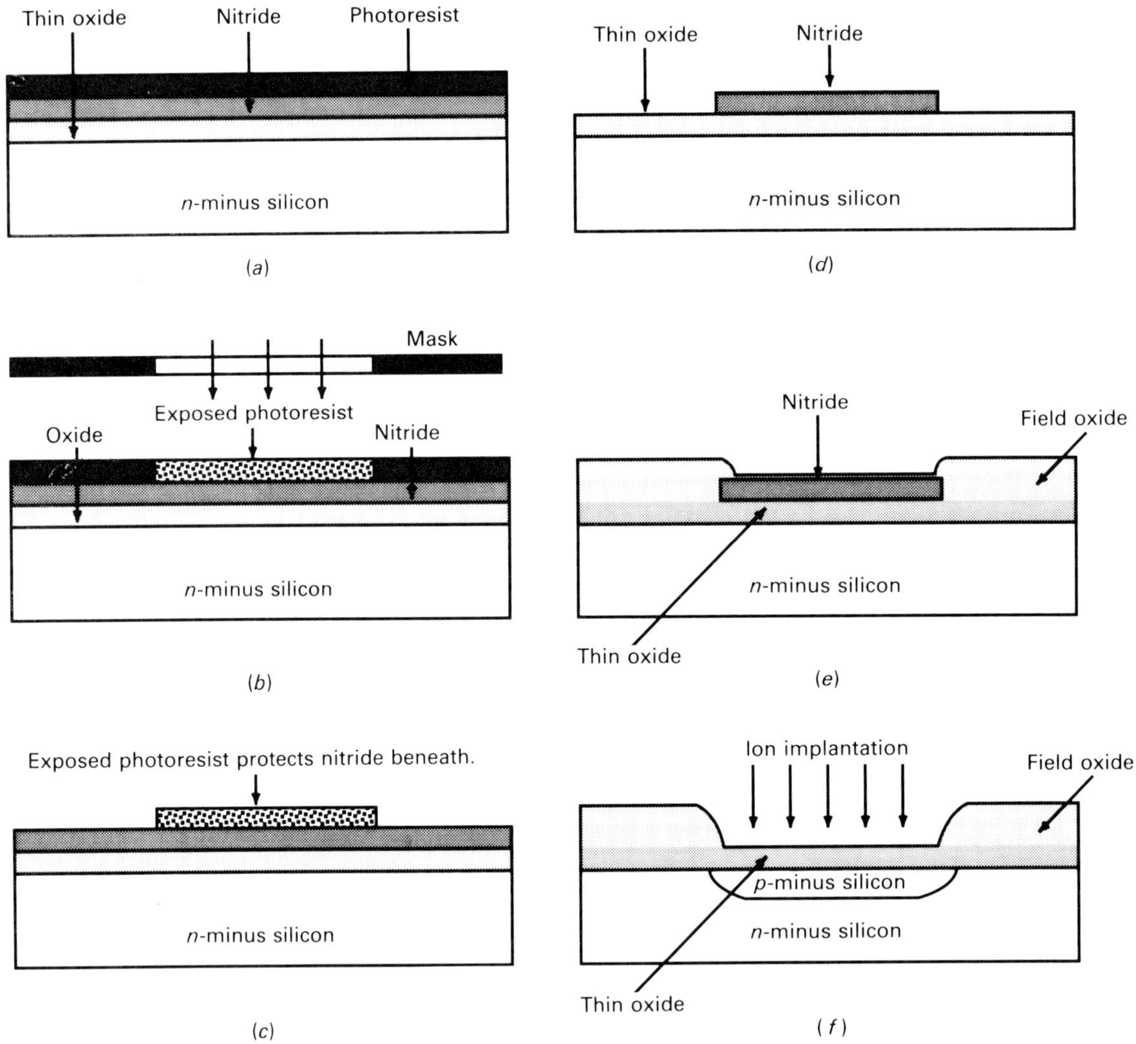

Fig. 2-6 Steps in creating an implant region in silicon.

A *mask*, made of glass, patterned with chromium or iron oxide, is then aligned to the wafer, and the photoresist is exposed with ultraviolet light and developed. In this case, the mask defines the area where a *p*-minus layer will go. If positive photoresist is used, the exposed resist becomes more soluble and more easily removed. If negative resist is used, the exposed region becomes less soluble. Thus the mask can appear either as a positive or a negative of the desired pattern.

The soluble resist is stripped from the wafer, leaving an area of unprotected nitride as shown in Fig. 2-6*c*. The unprotected nitride is etched away while the nitride under the photoresist remains. The hardened photoresist is then removed with a stronger solvent, leaving an island of nitride on a sea of thin oxide as shown in Fig. 2-6*d*. The nitride will prevent the further growth of oxide in the areas it covers. The next step is to grow a layer of thick oxide (called *field oxide*) over the entire area as shown in Fig. 2-6*e*. Finally, the nitride is removed and a region of thin oxide, surrounded by field oxide, is left (Fig 2-6*f*).

The thin oxide allows the *implantation* of acceptor ions into the substrate. The thick field oxide stops these acceptor atoms before they get to the silicon, but the thin oxide allows penetration as shown in Fig. 2-6*f*. A beam of high-energy acceptor ions penetrate the crystal through the thin oxide to a depth which depends upon the energy of the ions. Some lattice damage may result from these collisions, but this damage can be repaired by heating the crystal to a few hundred degrees centigrade. This process is called *annealing*. The result is a *tub* of *p*-type silicon in an *n*-type substrate. An older method of embedding impurities, called *diffusion*, depends upon direct contact of impurity atoms with the substrate and cannot be done through the thin oxide. Diffusion occurs in a high-temperature furnace when the silicon is exposed to a gaseous atmosphere containing high concentrations of impurity atoms. Selective diffusion depends upon openings in the oxide covering. However, even when the older diffusion technique is not actually used, the regions in which impurities are embedded are called *diffusion areas*.

At the border of the two dissimilar substances, a *pn junction* develops. The *p*-minus silicon, doped with acceptor atoms, contacts the *n*-minus silicon which is doped with donor atoms. Fig. 2-7*a* is a simplified representation of the two dissimilar materials just before they meet. We know, from the above discussion, that the joining of the two substances is part of a continuous process and is not like joining two pieces of wood.

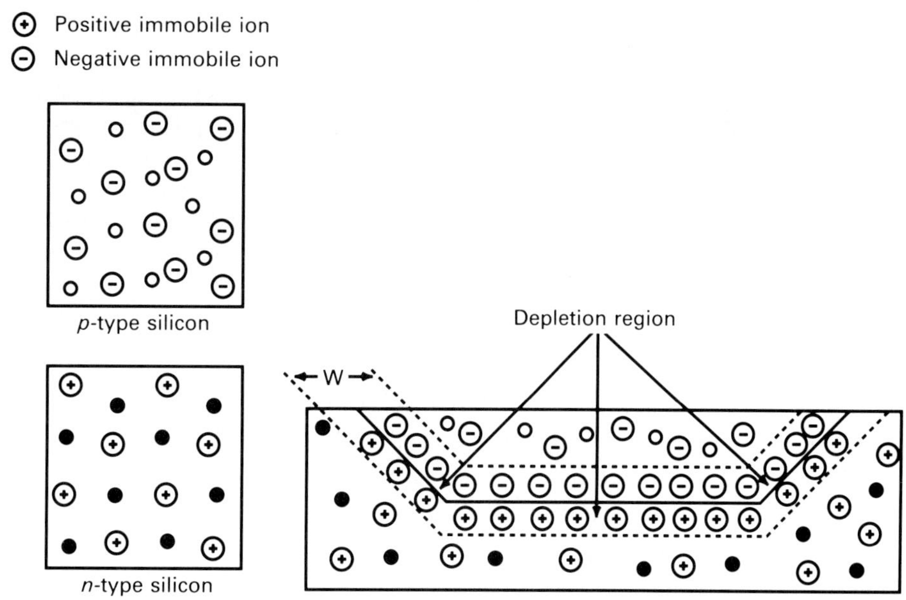

(*a*) (*b*)

Fig. 2-7 The *pn* junction.

But for the sake of discussion, let us pretend they are apart one instant and joined the next. Figure 2-7*a* shows only the impurity atoms with their excess charge carriers. It does not show the silicon atoms or the covalent bonds. Each acceptor ion in the *p*-minus silicon is an *immobile*, negatively charged particle locked in the crystal lattice. It is negative because it is so easy for an electron from a neighboring silicon atom to break away and fill the hole. Each hole, however, is quite mobile and derives its charge by virtue of the fact that wherever it goes, it creates a positively charged silicon ion (unless, of course, it returns to an acceptor atom). We can say that in any microscopic portion of the doped *p*-minus silicon, there are negative, immobile acceptor ions and positive, mobile holes. When viewing Fig. 2-7, it is important to realize that, in practice, there is an extremely small percentage of impurity ions—only about 1 atom of impurity to 10 billion atoms of semiconductor material. While the holes move from atom to atom, the ions never leave their lattice positions. In the absence of an externally applied electric or magnetic field, the movement of the holes is random, so that the holes are distributed uniformly throughout the material, and the bulk material is electrically neutral. The *n*-minus silicon has been doped with pentavalent impurities. Each donor ion is an *immobile*, positively charged ion locked in the lattice while its *mobile*, negatively charged electron wanders randomly in an overall electrically neutral slab.

When the *p*-type material contacts the *n*-type material, mobile holes and mobile electrons begin to move toward one another. As the charged particles meet, they *recombine*, and eliminate each other. In each recombination, one hole and one excess electron are eliminated. One might expect the holes and electrons to recombine until all holes and all excess electrons have been eliminated. But after a relatively few recombinations, the process stops. Why? Because a *depletion region* soon develops in which there are no more holes and no more free electrons, only immobile ions. These immobile ions create opposing electric fields which prevent further recombination. On the *n*-type side, positive donor ions repel any holes which might approach the junction. On the *p*-type side, negative acceptor ions repel any free electrons which might approach the junction. These repelling forces produce what is known as a *barrier potential*. Any electrons or holes which approach this region are swept away by the field. This happens without any external field being applied.

The physical distance from one side of the depletion region to the other is called the *width* (indicated by w in Fig. 2-7*b*.). Physically, this width is almost 0. A more useful measure of the depletion region is its potential in volts. This potential builds up to the point where it opposes any further recombinations. In germanium, this potential is about 0.3 V; in silicon, it is about 0.6 to 0.7 V. The potential difference between one side of the barrier to the other is called the barrier *height*. You could visualize this potential as being caused by an imaginary battery.

Now, what would happen if an external battery were connected to the two materials? If the positive terminal were connected to the *n*-side, and the negative terminal were connected to the *p*-side, as in Fig. 2-8*a*, the

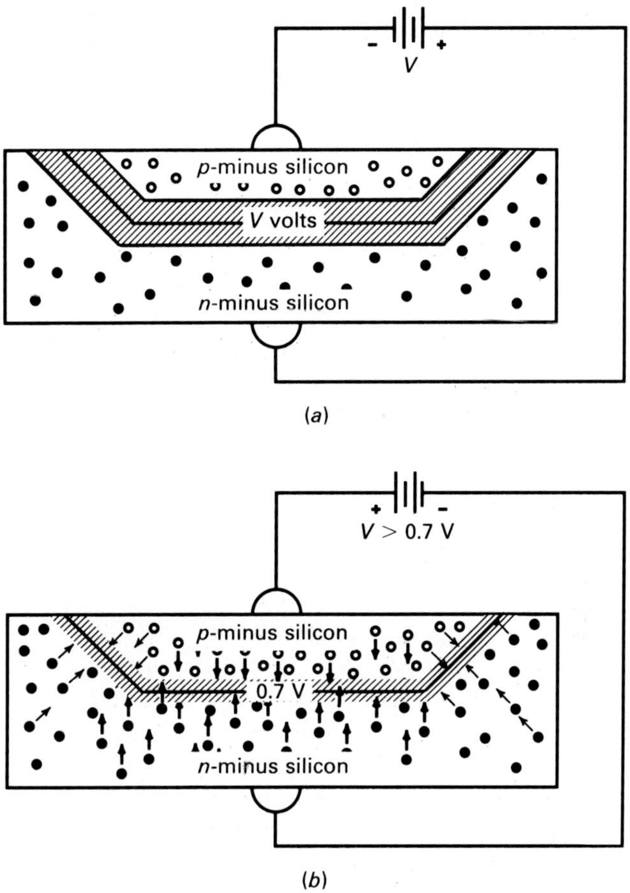

Fig. 2-8 (*a*) Reverse-bias causes holes and electrons to be drawn farther from the junction creating a wider depletion region. (*b*) Conduction requires a forward-bias greater than the barrier height of the junction. In silicon this is approximately 0.7 V.

depletion region would *widen* until the barrier height equaled the potential of the external battery. The reason for this is that the electric field established by the external battery draws holes and electrons farther from the junction. For a short time, charge will flow into the battery terminals, just until the potential across the *pn* junction equals the potential at the battery terminals. Then the *pn* junction will be *reverse-biased*. After the initial instant required to reverse-bias the junction, no more current will flow. It is interesting to note that the depletion region, being completely devoid of mobile charge carriers, is very much like the space between the plates of a parallel plate capacitor. The *pn* junction does, in fact, have a capacitance which is similar to parallel plates:

$$C_j = \epsilon_{si} \frac{A}{w} \tag{2-2}$$

where C_j = capacitance at the *pn* junction
A = cross-sectional area of the junction
w = width of the depletion region
ϵ_{si} = permittivity of the silicon

Note the similarity to the parallel plate capacitor described in Chapter 1.

If the external battery is now reconnected, with the positive end on the *p*-side and the negative end on the *n*-side as in Fig. 2-8*b*, the barrier potential is opposed by the applied electric field, and the depletion region is *narrowed*. The external battery produces an accumulation of electrons on the *n* side and an accumulation of holes on the *p* side which puts pressure on the junction. If the electrical pressure (in volts) of the external battery exceeds the potential of the opposing built-in field of the depletion region, current will flow across the junction. Electrons and holes will be forced across the narrowed depletion region to find each other and recombine. Current on the *n* side is composed of free electrons supplied by the negative terminal of the battery. As the free electrons enter the *p* side, they become valence electrons. But for each electron which is lost to recombination, another enters from the negative terminal, and *current flows*. When current flows in this way, the device is said to be *forward-biased*.

A *pn* junction has a schematic symbol as shown in Fig. 2-9*d* and is commonly called a *diode*. The *p* side of a diode is called the *anode*, and the *n* side is called the *cathode*. This terminology comes from vacuum tube days. If the diode is forward-biased so that the voltage across it exceeds the opposing field at the junction, current will flow. The voltage required to "turn on" the diode is called the *offset voltage*. If the diode is reverse-biased, no current will flow. The schematic symbol suggests this behavior by an arrow shape in the direction of *p* side to *n* side and a blocking line in the opposite direction.

The discussion so far suggests that the current versus voltage characteristic for a diode might look like the graph in Fig. 2-9*b*. There is no conduction until the forward voltage exceeds the barrier height. Fig. 2-9*b* represents the *v-i* characteristic of an *ideal* diode *with an offset voltage*. We mentioned before that this offset voltage is about 0.3 V for germanium and 0.6 to 0.7 V for silicon. There are two important differences, however, between the diode approximation in Figure 2-9*b* and a real diode. Real semiconductor diodes leak when reverse-biased, and real diodes also exhibit an curved slope in the forward characteristic. Moreover, the forward-biased diode conducts a little in the forward region even before the offset voltage is reached. These characteristics are illustrated in the real diode curve of Fig. 2-9*d*.

The curve in Fig. 2-9*d* suggests that the device can be reverse-biased until it breaks. This is true. If the reverse voltage becomes too large, the crystal will break down, and large amounts of current will flow in the reverse direction. There are two types of breakdown, *avalanche* and *zener*. Avalanche breakdown occurs when the electric field is so high that electrons

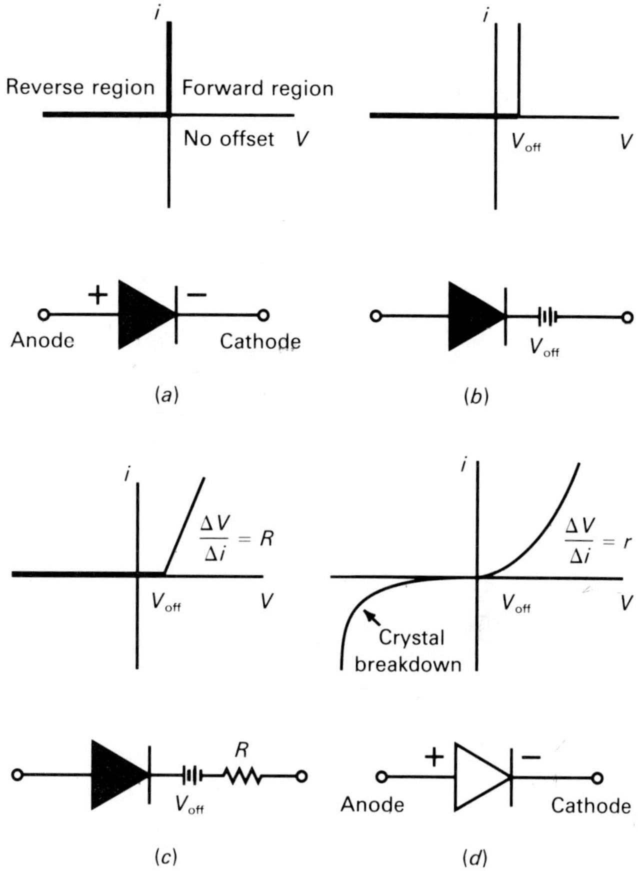

Fig. 2-9 Approximations of a junction diode. (*a*) The ideal diode; (*b*) ideal diode with offset voltage; (*c*) ideal diode with offset voltage and a resistance to account for the slope in the forward characteristic; (*d*) a real diode characteristic.

are able to rip apart covalent bonds when collisions occur in the lattice. Each collision releases another electron so that two electrons are free to accelerate and collide with the lattice. This process continues as a chain reaction until either the current is limited externally or the device is destroyed. Zener breakdown does not involve collisions of electrons with the crystal ions, but occurs when the electric field is strong enough to disrupt the bonds directly. Zener breakdown normally occurs before avalanche breakdown and is not always destructive. Special *zener diodes* are made to safely operate in this region of the curve.

A more interesting portion of the curve is the forward-biased region. In this region the device has forward resistance which decreases with an increase in forward current. This resistance is called *dynamic* resistance because it changes with voltage. Because the resistance of the real diode depends on the voltage across it, the real diode characteristic is not a straight line as shown in Fig. 2-9*c* but a curve as shown in Fig. 2-9*d*. For this reason the diode is called a *nonlinear* device. Real diodes do not obey Ohm's law.

The value of the offset voltage depends upon the substrate material and the impurity concentration. Offset voltages in silicon can vary between less than 0.6 V to greater than 0.85 V, depending on the impurity gradient at the junction. Offset voltages in gallium arsenide can vary from less that 1.0 V to greater than 1.2 V. Some *pn* junctions, such as those used in erasable programmable read-only memories (EPROMS), are made so that their threshold voltage can be altered electrically, by *programming* with surges of current and *erasing* with ultraviolet light. This will be discussed in more detail in Chapter 8.

The MOS Capacitor

A simple device which is important to understanding MOS technology is the MOS capacitor. The important concepts developed here are *electrical isolation, inversion*, and *sheet capacitance*. A top plate of a capacitor can be patterned above the thin oxide in Fig. 2-6*f* using metal or some other conductive material. Although metal is in the name—*metal*-oxide semiconductor—doped *polysilicon*, or simply, *poly*, is more often used. Polysilicon is the polycrystalline form of silicon. Metal has a lower resistivity than poly, but poly has advantages over metal in the fabrication process. The various masking, exposing, and etching steps required to make an MOS capacitor are similar to those needed to fabricate a transistor and are described later. Looking at Fig. 2-10*a* and remembering the discussion

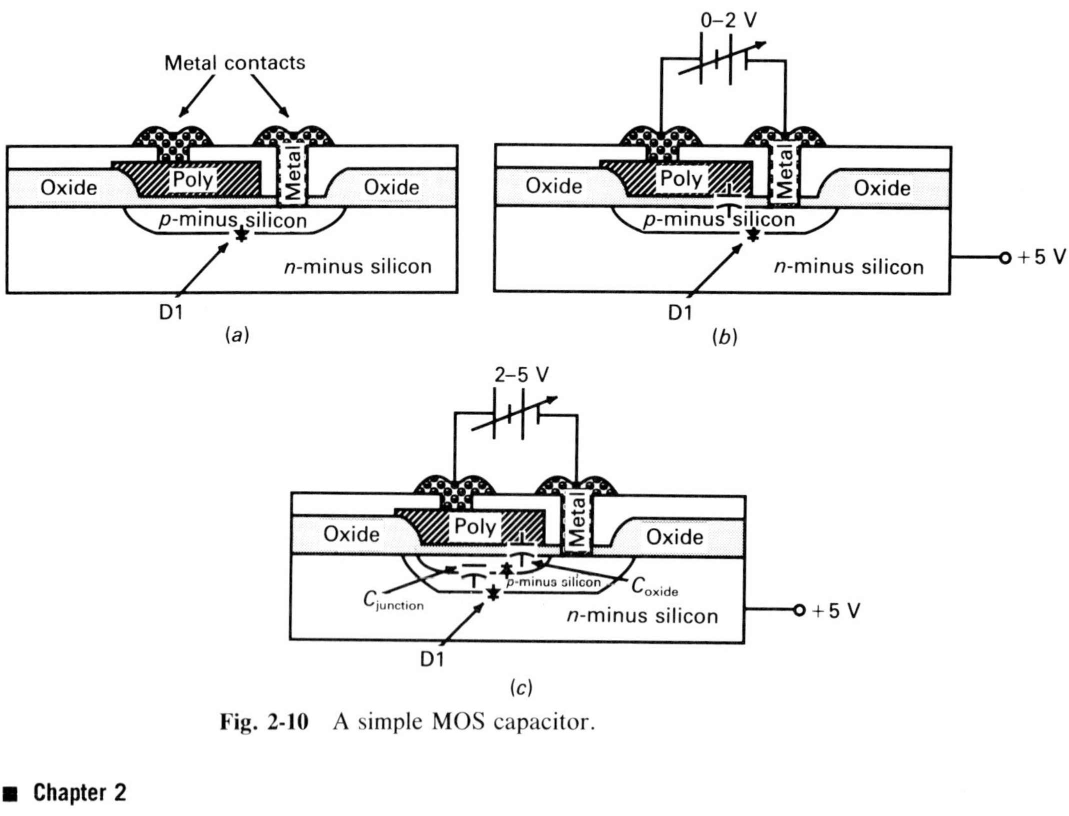

Fig. 2-10 A simple MOS capacitor.

of parallel plate capacitors in Chapter 1, we can view this device as a parallel plate capacitor with the metal or poly as the top plate, the p-type tub as the bottom plate, and the thin oxide as the dielectric material in between (oxide is nonconductive). The bottom plate is isolated from the substrate by virtue of the pn junction. To ensure isolation, we need only tie the substrate to the highest potential in the circuit. The capacitance of the plates on either side of the oxide is:

$$C_{OX} = \epsilon_{OX} \frac{A}{t_{OX}} \tag{2-3}$$

where
C_{OX} = capacitance of the oxide junction
A = area under the top plate
t_{OX} = thickness of the thin oxide
ϵ_{OX} = permittivity of the thin oxide

This analysis is correct as long as we are careful about what voltages are applied across the MOS capacitor. The capacitance of the device is voltage dependent, and to be accurate, its operation must be specified under all conditions. Fig. 2-10b shows a possible circuit configuration with a variable voltage source connected across the device. When the voltage is in the ''flat-band'' region (usually between 0 and 2 V), the formula for a simple parallel plate capacitor does apply. However, as the voltage increases, the mobile holes are repelled, and a depletion region is formed under the top plate. If the voltage source is turned higher, the additional positive bias attracts mobile electrons to the surface, and, if a sufficient number of electrons are attracted to the surface, an *inversion layer* is created. This means that the surface of the p-type silicon has changed to n-type silicon. Fig. 2-10c shows that there are now two capacitors in series. One is an MOS capacitor with poly, oxide, and n-type semiconductor. The other is a pn-junction capacitor. Placing two capacitors in series lowers the overall capacitance.

The inversion process is very important to the understanding of field-effect transistors. The electrical isolation of the bottom capacitor plate, achieved by biasing the p-minus tub, is also important. And the capacitance, associated with laying metal or poly over the substrate, is an undesirable reality which we must be aware of from the start.

Review 2-2

1. **Q.** What is the common name for a pn junction? **A.** diode
2. **Q.** What is the function of silicon nitride in IC fabrication? **A.** to limit the growth of oxide
3. **Q.** Name two methods for implanting impurity atoms. **A.** diffusion, ion implantation

4. **Q.** Explain the following terms: depletion region, barrier height, offset voltage. **A.** region devoid of charge carriers the potential which opposes current through the junction the voltage needed to overcome the barrier height

5. **Q.** Draw a simple circuit with circuit symbols showing forward-bias and reverse-bias.

6. **Q.** Refer to Fig. 2-10. What purpose does the reverse-biased diode, D1, serve? **A.** It isolates the bottom plate from the substrate, otherwise all bottom plates would be connected to the substrate voltage.

7. **Q.** What is an inversion layer? **A.** a layer of opposite polarity created by an electric field

2-3 FIELD-EFFECT TRANSISTORS

The principle of the field-effect transistor dates back to the early 1930s when J. E. Lillienfield in the United States and O. Heil in England proposed that a field-effect device could be used to achieve a solid-state amplifier. This concept was actively pursued at Bell Laboratories, and a group there accidentally discovered the bipolar transistor in the process. Field-effect transistor research suffered a setback because of this discovery, and very large scale integration (VLSI), which is quite difficult with bipolar transistors, had to wait several years for the renewal of interest in MOS technology. Research in the 1960s led to important advances in practical MOS technology.

The processing steps described in Fig. 2-6 are applicable to the fabrication of a *metal-oxide semiconductor field-effect transistor (MOSFET)*. Figure 2-11*a* shows that process taken a step further. A polysilicon layer has been deposited over the entire surface. This layer is heavily doped by the diffusion, or implantation, of donor impurities. The doping reduces the sheet resistance of the poly to an acceptable level, usually 20 to 60 Ω/□ ("ohms per square"). (See Chapter 1 for a discussion of sheet resistance.) A layer of photoresist is applied over the poly, a mask is aligned, and the resist is exposed. (See Fig. 2-11*b*.) The hardened (exposed) resist remains while the soluble resist is removed. The poly beneath the hardened resist is protected from the next step, which is an etching process. The hardened resist is then removed so that a patterned *geometry* of doped poly remains. (See Fig. 2-11*c* and 2-11*d*.)

It is now possible to create diffusions of *n*+ silicon on either side of the poly by means of heavy implantation of donor ions. *N-minus* means lightly doped, and *n*+ means heavily doped. Implantation inverts the *p*-minus silicon into an *n*+ silicon to a depth which is controlled by the energy of the ions. The process is called the n+ *source-drain implant* and leaves two areas of *n*+ diffusion, one on either side of a polysilicon *gate*. The patterned poly is called a *gate* where it separates two areas of diffusion. The resulting device is an *n-channel* transistor. One of the diffusion areas

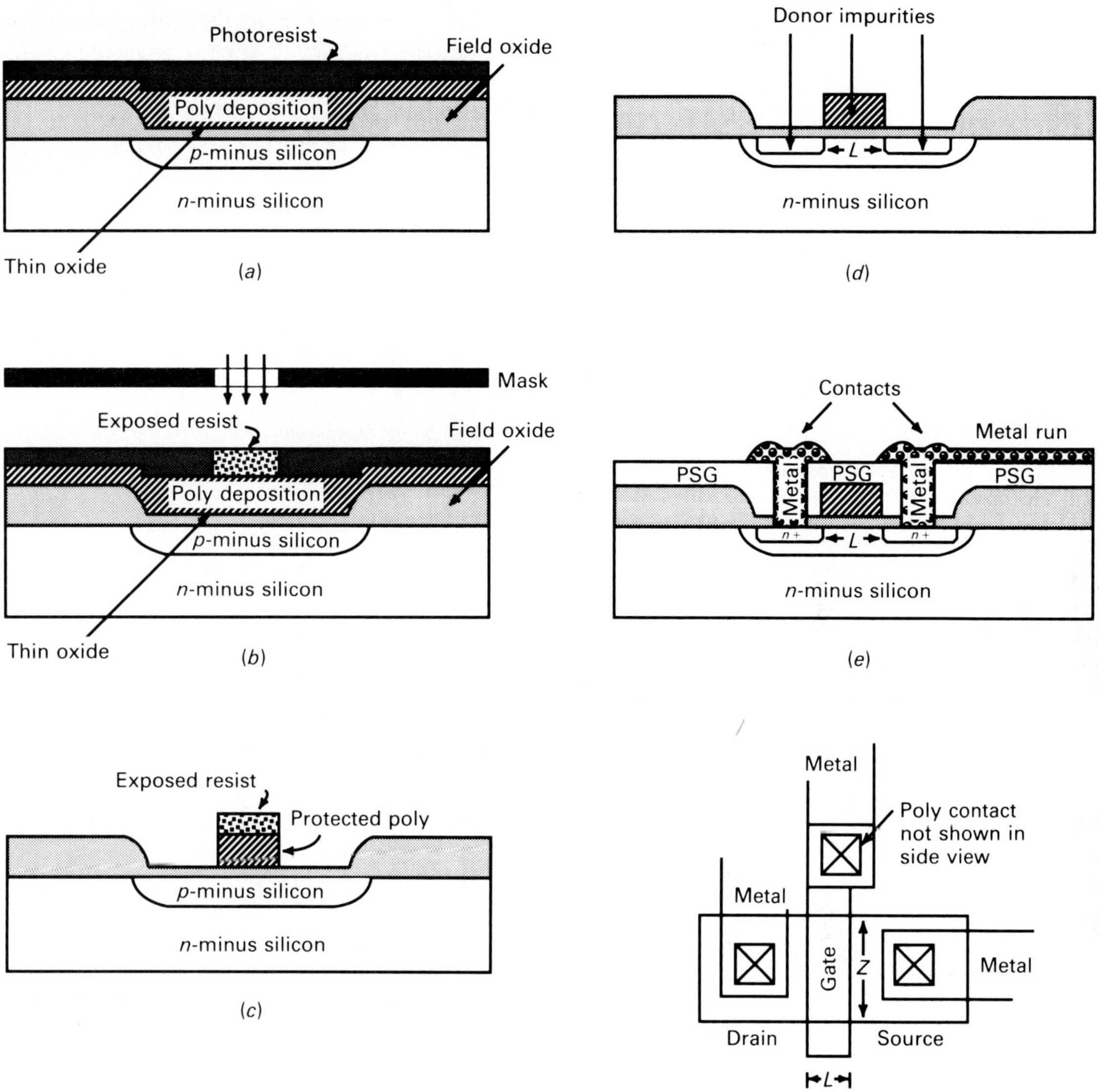

Fig. 2-11 Steps in fabricating a MOSFET, continued from Fig. 2-6*f*. (*f*)

is arbitrarily labeled the source and the other the *drain*. The gate is always insulated from the substrate by means of the thin layer of oxide. The region beneath the gate is called the *channel region*, and the distance from source to drain, indicated by *L* in Fig. 2-11*d*, is the *channel length*. Likewise the distance across the gate, which is equal to the channel length, is called the *gate length*. This step illustrates the reason why poly is favored over metal in the fabrication of transistors. Poly is *self-aligning*, which means that the poly gate will not overlap the diffusion areas since the poly itself prevents implantation directly below it. Metal gates, on the other hand, are patterned as a separate layer *after* the diffusion regions have been created. Thus, registration tolerances almost guarantee some misalignment when metal gates are used, and this misalignment causes *parisitic capacitances* between the gate and the diffusion regions. The result is that metal gate logic has more gate capacitance and is therefore slower. Older metal gate processes used aluminum gates with lengths of 5 μm or larger. Polysilicon is now used for 3- to 5-μm geometries, and refractory metals (such as molybdenum) and polycides (composite metal silicide and poly) are used for smaller gate lengths. These latter materials are self-aligning like poly but are less resistive (around 1 Ω/□). The table in Fig. 2-12 lists typical values for sheet resistance.

Material	Sheet Resistance, Ω/□
Metal	0.03–0.05
Diffusion	30–80
Poly	20–60
Silicide	2–4
Channel	$<$10 k

Fig. 2-12 Typical sheet-resistance values.

Further masking, exposing, and etching steps are needed to make metal contacts to the diffusion regions and to the poly gate. In the end we are left with something resembling Fig. 2-11*e*. *Phosphosilicate glass* (*PSG*) insulates the gate and the field oxide from the metal. The overhead view of this is represented in Fig. 2-11*f*. This overhead view is what the IC designer uses when he or she lays out the geometries needed to produce the desired digital circuit. Notice the distance labeled Z in the overhead view. This is the *gate width*. This terminology is particularly confusing since the "*width*" of a poly gate is almost always longer than the "*length*" labeled L. The term *gate length* derives from *channel length*, which is the distance the inversion layer must bridge to achieve conduction. *Channel width* follows from channel length and also applies to the gate above the channel.

MOSFET Operation

So how does the MOSFET work? First, recall what happens when *n*-type and *p*-type materials are joined. A diode forms at the junction. We can see right away that a diode exists at the boundary of the embedded *p*-minus tub. If the *n*-type substrate is tied to the highest voltage in the circuit, this diode can never be forward-biased, and the *p*-minus tub is electrically isolated from the substrate. This is desirable for reasons which will become clear later. The junctions between the $n+$ diffusions and the *p*-minus tub are also diodes. If thc *p*-minus tub is tied to ground (the lowest potential in the circuit), this region will be isolated from both the source and drain areas. Figure 2-13*a* shows how the diodes are reverse-biased. In this example, we have also tied the source to ground and the drain to a resistor which connects on its other side to supply. Note that extra

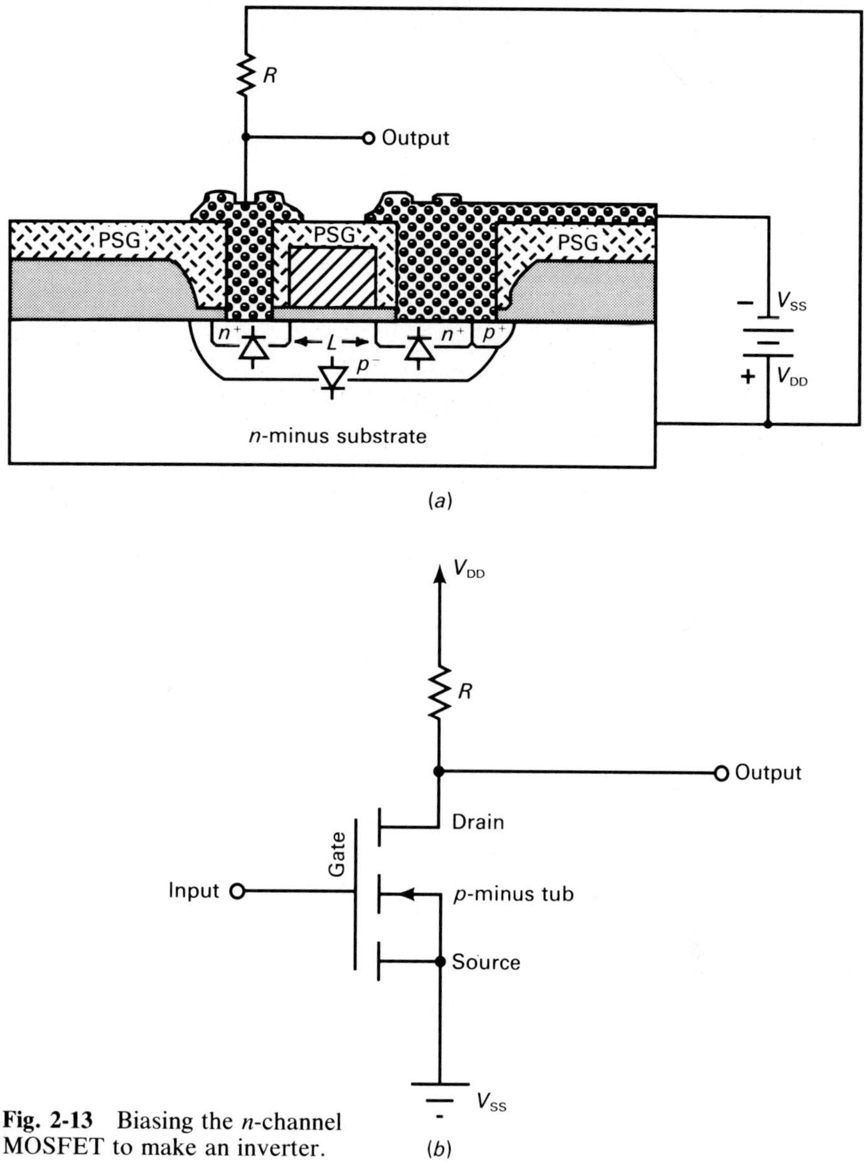

Fig. 2-13 Biasing the *n*-channel MOSFET to make an inverter.

processing steps would be required to create the $p+$ *plug* shown in Fig. 2-13*a*. This plug provides a low-resistance connection between the metal contact and the *p*-minus tub and ensures that the *p*-minus layer will be well grounded. It is not necessary to do this on every transistor since normally many transistors can share a common *p*-minus layer, but it is wise to plug the substrate often (usually every 100 μm is sufficient, 100 μm $\approx 4 \times 10^{-3}$ in). Figure 2-13*b* shows a schematic representation of the circuit.

If the gate were tied *low*, it would have the same potential as the source, and no current would flow between the drain and the source. This is equivalent to saying that $V_{GS} = 0$ (voltage from gate to source is 0). The *p*-minus material below the gate isolates the source from the drain and effectively provides an open circuit. If the gate to source voltage V_{GS} is increased (positively), however, electrons are drawn from the source into the region just below the gate. A very thin wedgelike inversion layer forms on the source side and grows toward the drain side as the voltage V_{GS} is increased. When a *threshold voltage* V_{th} is reached, the inversion layer will have bridged the gap between source and drain to form an *n* channel. After the channel has formed, further increases in V_{GS} will strengthen the channel and cause increases in drain-to-source current. V_{th} is typically less than 1 V but it is a function of impurity concentrations, oxide thickness, and temperature, and it can be much larger. Also, if the source is more positive than the substrate, the depletion region around the source becomes larger and more gate voltage must be applied to turn on the device.

In digital electronics, we like to operate away from V_{th}. Ideally, only one of two possible voltages is applied to the gate, either ground, called V_{SS}, which is the negative terminal of the voltage source or supply, called V_{DD}, which is the positive terminal of the voltage source. This means that we always exceed the threshold voltage so that the device is either fully on or fully off. In this way MOSFETs are used as *switches*.

In the above example, the region directly beneath the gate was doped with a uniform *p*-minus concentration. When a gate bias was applied, electrons were drawn up to form a conductive channel. With no gate bias (0 V), no channel existed and no conduction was possible. This device is an *enhancement-mode*, *n*-channel MOSFET. Had we implanted a thin layer of donor ions just beneath the gate, an *n*-channel would be formed at the time of fabrication, and conduction would occur with $V_{GS} = 0$ V. This latter device is called a *depletion-mode*, *n*-channel MOSFET. When V_{GS} is increased, this device conducts better just like an enhancement-mode transistor, but if V_{GS} is decreased to 0 V, it will not turn off. The gate voltage of a depletion-mode transistor must be considerably lower in potential than the source voltage in order to turn off the device. When sufficient negative gate bias is applied to a depletion-mode MOSFET, free electrons in the channel are repelled, and the region inverts to *p*-minus. The "normally off" behavior of enhancement devices and the "normally on" behavior of depletion devices is illustrated in Fig. 2-14. The word

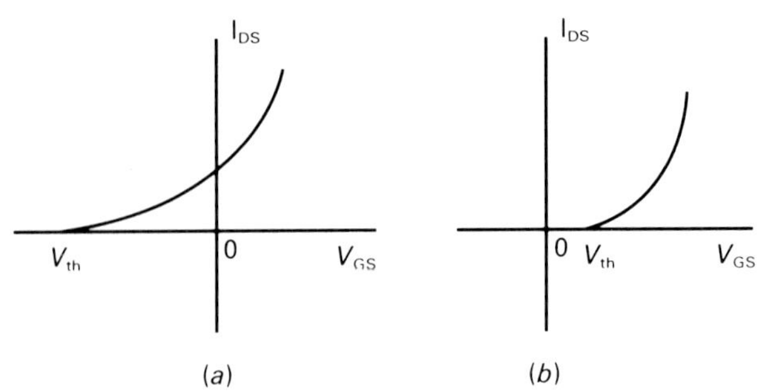

Fig. 2-14 Transconductance curves for *n*-channel MOSFET. (*a*) Depletion mode; (*b*) enhancement mode.

"normal" in both cases signifies V_{GS} = 0 V. The various circuit symbols used for *n*-channel enhancement-mode and depletion-mode MOSFETs are shown in Fig. 2-15.

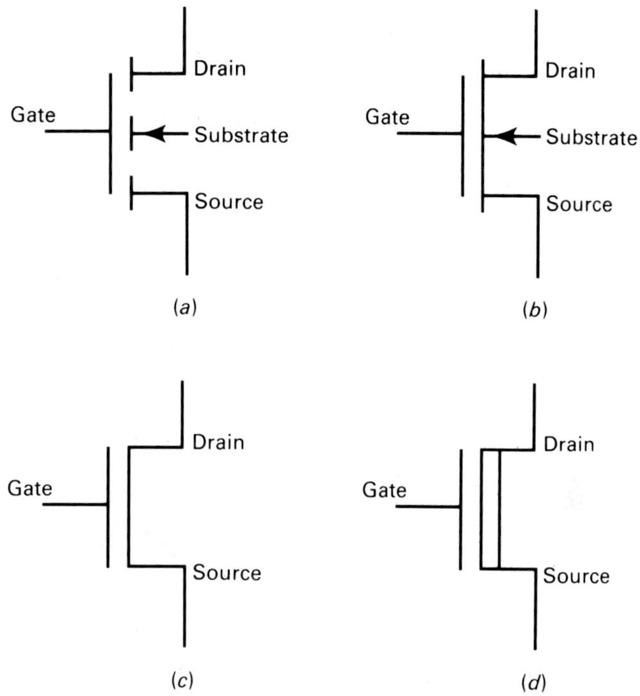

Fig. 2-15 Various circuit symbols for the *n*-channel MOSFET. The symbols in *b* and *c* can be either enhancement mode or depletion mode. The symbol in *a* specifically represents an enhancement-mode device, and *d* specifically represents a depletion-mode device.

JFET Operation

Junction field-effect transistors (JFETs) operate essentially in the same way as depletion-mode MOSFETs. They are fabricated with a built-in channel, and a bias, more negative than the source, is needed to turn them off. The curve in Fig. 2-14*a* could be equally applicable to JFETs. The only difference between a depletion-mode MOSFET and a JFET is that the gate terminal of a JFET is electrically isolated from the channel by a reverse-biased depletion region while a depletion-mode MOSFET achieves isolation with a thin layer of oxide.

JFETs have very little application in digital electronics and are not used at all in VLSI. Their primary application is in radio frequency (RF) and audio amplifiers, mixers, and oscillators. They make vary good voltage-variable resistors for tone controls and automatic gain-control circuits. They can also be found in the front end of some op-amp circuits. They are mentioned here only for completeness.

2-4 INVERTERS

The inverter is the most basic circuit element using digital logic. Its primary function is to change the sense of a logic level. The inverter changes an input logic 0 (low voltage) into an output logic 1 (high voltage), and vice versa. Inverters are also used as *buffers* to extend the drive capability of other logic circuits. Two cascaded inverters may be used to increase current drive without changing the logic sense of a signal. Cascaded inverters of increasing size are used in digital ICs to build the current drive capability of a signal up to levels which are usable off-chip.

The NMOS Inverter

The circuit in Fig. 2-13*b* is functionally equivalent to the circuit shown in Chapter 1, Fig. 1-12*c*. The switch in Fig. 1-12*c* has been replaced by an *n-channel MOSFET (NMOS)*, but in both cases a resistor is used as the upper connection to the power supply. In integrated circuits, resistors are not generally used because the types of resistors which can be easily fabricated on ICs are either too bulky (taking up far more space than transistors) or inexactly specified (their values may vary by 300 percent) or both. Therefore, NMOS inverters typically use depletion-mode transistors instead. The depletion-mode device is connected as shown in Fig. 2-16 so that V_{GS} always equals 0 V and the device will always stay turned on. Functionally, this provides a resistive device which is smaller in physical size and better defined in value than an equivalent resistor. The actual value of the resistance can be controlled by varying the doping concentration beneath the gate and the channel length-to-width ratio.

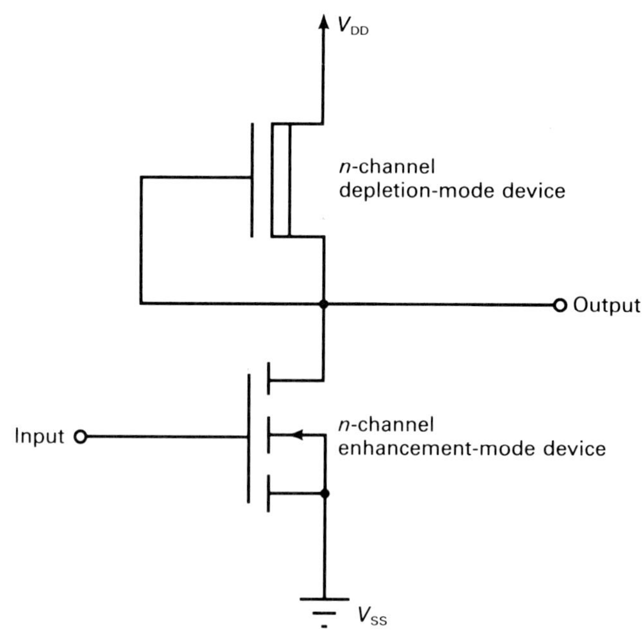

Fig. 2-16 NMOS inverter with depletion-mode connection to V_{DD}.

The inverter just described is called an *NMOS inverter* because only *n*-type transistors are used. The output of the NMOS inverter normally connects to the gate of some other MOS circuit, and the earlier discussion concerning MOS capacitors now becomes relevant. The similarity between poly over doped silicon in Fig. 2-10 (MOS capacitor) and poly over doped silicon in Fig. 2-11 (MOS transistor) should be obvious. In both cases the poly forms the top plate of a capacitor. In the former case it was presumed that the capacitor was fabricated for a reason and its capacitance is useful. In the latter case, however, the capacitance is an undesirable by-product of making transistors. It is undesirable because it slows down transitions between high and low outputs. A review of simple resistance-capacitance (RC) circuits in Chapter 1 might help at this point. Smaller-geometry MOSFETs have smaller gates and therefore less capacitance. Thus, smaller geometries not only save on space but also allow higher speeds.

NMOS inverters suffer from at least two drawbacks. The load device, although fabricated with a transistor, is merely resistive. We may view it in the same way we viewed the resistor in Chapter 1, Fig. 1-12. Like the resistor, it must have considerably more resistance than the *n*-channel switch in order to achieve an output low when the switch is turned on. This means that the *RC* time constant will be considerably longer for a low-to-high transition than for a high-to-low transition. Thus the output waveforms are unsymmetrical, and the worst-case charging time is considerably worse than it would be if an active device were used. Another disadvantage of the NMOS inverter is *quiescent power drain*. Quiescent power drain is the power dissipated when no changes are taking place. When the input is high, the *n*-channel transistor conducts, and there is a path between V_{DD} and V_{SS} through the resistor and the *n*-channel switch. This current contributes nothing to the operation of the circuit but does contribute to power loss and warming of the substrate.

The CMOS Inverter

It would have been simpler, in the preceding sections, to fabricate the *n*-channel MOSFET directly in a *p*-minus substrate rather than sinking it into a *p*-minus tub. The simpler approach *is* normal procedure when fabricating only *n*-channel transistors. However, if both *p*-channel and *n*-channel MOSFETs are to be fabricated on the same wafer, large tubs of opposite polarity are needed. The reason is clear when you look at a larger portion of a substrate and see that an intermediate layer is needed for the *n*-channel transistor if *p*-channel transistors are to be patterned directly into the substrate. The masking, exposing, etching, and implantation steps for the *p*-channel device are part of the same fabrication process but yield a different type of device. Diffusion areas of $p+$ material are embedded directly into the *n*-minus substrate. *N*-minus material lies beneath the poly gate and isolates the source and drain areas. If we had started with a *p*-minus substrate, the *p*-channel transistor would have needed an intermediate layer (an *n*-minus tub) and the *n*-channel device would have been embedded directly into the substrate. Figure 2-17*a* illustrates a *p-tub CMOS*

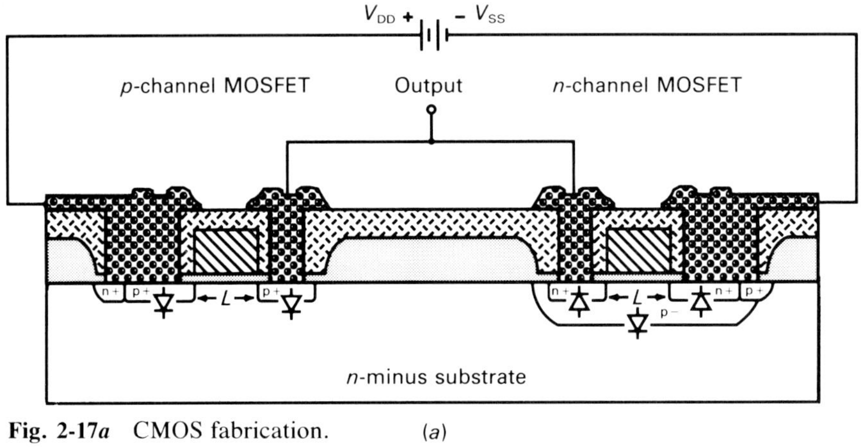

Fig. 2-17*a* CMOS fabrication.

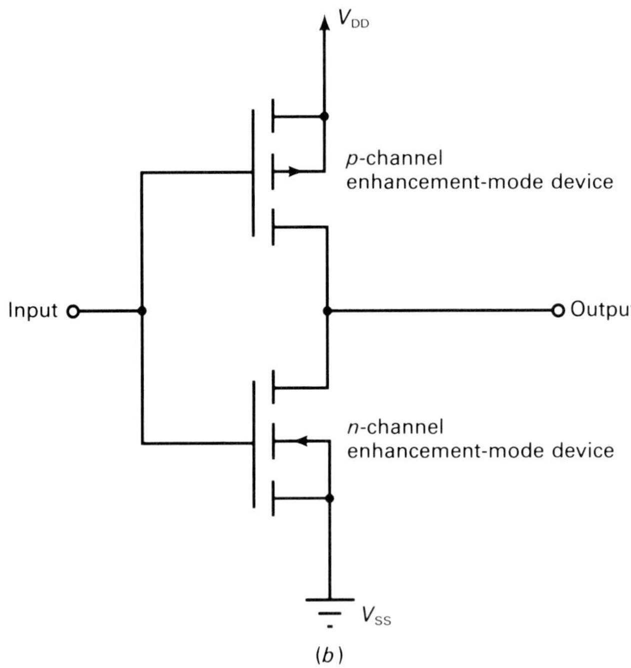

Fig. 2-17*b* CMOS inverter.

process. If a *p*-minus substrate had been used, the fabrication would be an *n-tub process*.

The *p*-channel MOSFET "complements" the *n*-channel device. When the gate to the *p*-channel device is forced high, electrons are drawn upward toward the channel region making the area beneath the gate more heavily *n*-type. This guarantees isolation between source and drain. When the gate is forced low, the electrons are repelled, and an inversion layer forms beneath the gate providing a conductive *p*-type channel connecting the source and drain areas. Thus the *p*-channel transistor operates as a *complementary switch* to the *n*-channel device by closing with a low (V_{SS})

input and opening with a high (V_{DD}) input (just opposite to the *n*-device).

N-channel and *p*-channel enhancement-mode devices can be used together to make the CMOS inverter shown in Fig. 2-17*b*. If the input to a CMOS inverter is high, the *n*-channel transistor turns on and the output is pulled low. If the input to the inverter is forced low, the *p*-channel device turns on and the output is pulled high. The logical operation of the CMOS inverter is identical to that of the NMOS inverter, but physically, the CMOS inverter has advantages. Its output can be pulled all the way to 0 V, and the output transitions can be made symmetrical. Given transistors with equal on-resistance, the CMOS inverter will be faster than the NMOS inverter for reasons discussed in Chapter 1. Furthermore, when the input is high, there will be no quiescent power drain since the path between V_{DD} and V_{SS} is made only during transitions.

Incidentally, if the *p*-channel device were the same size as the *n*-channel device, the output waveforms would not, in fact, be symmetrical. This is because holes have a lower mobility than electrons and therefore do not conduct as well. This imbalance can be corrected only by making the *p*-channel transistor wider. The result will be that the *RC* time constants will be made the same at the expense of silicon area. This is one reason why NMOS designs are denser than equivalent CMOS circuits. The other reason relates to the need for separate tubs in CMOS. The physical size disadvantage of CMOS becomes more obvious in the next section. Fig.

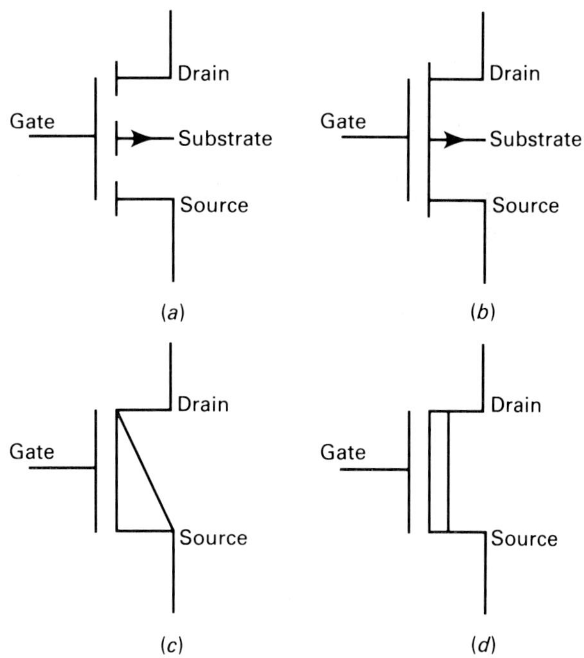

Fig. 2-18 Various circuit symbols for the *p*-channel MOSFET. The symbols in *b* and *c* can be either enhancement mode or depletion mode. The symbol in *a* specifically represents an enhancement-mode device, and *d* specifically represents a depletion-mode device.

2-18 illustates the various circuit symbols for *p*-channel MOSFETs. *P*-channel depletion-mode devices are fabricated with built-in channels, and positive gate bias must be applied to deplete those channels. *P*-channel depletion-mode MOSFETs are rarely used.

Review 2-4

1. **Q.** When is a ribbon of poly called a gate? **A.** when it separates two diffusion regions
2. **Q.** Define channel length and channel width. **A.** Channel length is the distance an inversion layer must bridge to create a conductive channel between the source and drain. Channel width is the width of that inversion layer.
3. **Q.** How do you think channel length and width would relate to the "on resistance" of a MOSFET? **A.** Shorter length means less on-resistance; greater width also means less on-resistance.
4. **Q.** Explain the operation of an enhancement-mode, *n*-channel MOSFET.
5. **Q.** Explain the operation of the depletion-mode, *n*-channel MOSFET. How does it differ, functionally, from an enhancement-mode *p*-channel MOSFET?
6. **Q.** Explain the advantages which CMOS has over NMOS. What advantage does NMOS have over CMOS? **A.** CMOS is faster, uses less power, and permits 0 V outputs. NMOS allows a more compact layout.

2-5 MOSFET SWITCHES

MOSFETs make very good switches, and it is for this reason that they form the cornerstone of VLSI technology. Any digital circuitry, from single-function *small-scale integration* (*SSI*) to complex VLSI, can be implemented using switches. The definition of a good switch, of course, is relative. Simple relays make good switches for many applications, but their size, power requirements, and slowness make them unsuitable for digital circuitry. Bipolar transistors are often used in digital logic, but their size and power requirements, while small compared to relays, are too large for VLSI. Furthermore, bipolar transistors are not true switches. A true switch conducts *bidirectionally*, and this is not possible with bipolar transistors. The fact that MOSFETs can conduct bidirectionally permits simplifications of logic design that translate into denser layout.

There are three important characteristics of switches: off-resistance, on-resistance, and the voltage or current required to turn on and off. MOS transistors have an off-resistance in the range of $10^{12}\ \Omega$. This would allow a leakage of only a few picoamps, which is negligible in a digital circuit. MOSFETs have an on-resistance which is significant. On-resistance is directly related to the width and length of the channel and may be made arbitrarily small. But the minimum gate length is fixed for a given process, and increasing the gate width requires more silicon area. There are practical limits to gate width, and these involve trade-offs between chip size and chip performance. The large signal channel resistance of an *n*-channel MOSFET is given by the following equation:

$$R_{ON} = \frac{1}{(\mu_n C_{OX}' \, W/L)(V_{GS} - V_{th} - V_{DS})} \tag{2-4}$$

where μ_n is the mobility of electrons

C_{OX}' = gate capacitance per square centimeter (which is just C_{ox} from Eq. (2-3) without the area specified)
L and W = length and width of the channel in centimeters
V_{GS} = gate-to-source voltage
V_{th} = threshold voltage
V_{DS} = voltage across the switch (drain-to-source)

This same formula applies to *p*-channel devices except that μ_p, hole mobility, is used in place of μ_n.

If we assume a typical mobility of 750 cm^2/V-s and an oxide thickness of 8.0×10^{-6} cm, the value for $\mu_n C_{OX}'$ will be around 3.3×10^{-5} F/V-s, or 3.3×10^{-5} A/V^2. A good assumption for threshold voltage is between 0.8 and 1.0 V. It is important to realize that these parameters do change from one process to another. There will also be variations within a given process, and these are usually specified in terms of minimum and maximum values.

If the device is being used in an inverter, then V_{DS}, when turned on, should be near 0 V for CMOS (no steady-state current) and no larger than 0.1 V for NMOS. This assumes that the inverter is driving a purely capacitive load.

EXAMPLE 2-1

A CMOS inverter is designed using a 3.0-μm silicon gate process with V_{th} = 0.8 V and mobility and oxide thickness as given above. The channel width is 10.0 μm. If the inverter is driving a purely capacitive load, what is the on-resistance for the *n*-channel MOSFET?

SOLUTION

The width-to-length ratio W/L is 3.3, and the CMOS will provide a steady-state V_{DS} of nearly 0 V. We can assume that the inverter is being driven by another MOS logic gate so that V_{GS} will be 5 V.

$$\begin{aligned} R_{ON} &= \frac{1}{(3.3 \times 10^{-5} \text{A/V}^2)(3.3)(4.2 \text{ V})} \\ &= \frac{1}{4.6 \times 10^{-4} \text{ A/V}} \\ &\approx 2.2 \text{ K}\Omega \end{aligned}$$

The resistance in Example 2-1 is considerably larger than the 25- to 1000-Ω range discussed in Chapter 1 for transistor switches. The reason is that the gate width, 10.0 μm, is very small. Such small gate widths are practical for only those inverters which drive other circuits *on-chip*. It would be impractical for driving off-chip loads. Larger transistors are required at the final outputs of integrated circuits in order to drive the inputs of other integrated circuits. For example, the CMOS inverter could conceivably be required to drive a 74-LS (*low-powered Schottky*) TTL gate. LS TTL logic requires that the driver sink approximately 0.36 mA when the input is pulled down to 0.4 V. Let us assume that the CMOS inverter in the above example can drive this load. If we assume that $V_{DS} = 0.4$ V, how much current is the *n*-channel device capable of sinking? Replacing V_{DS} in Eq. (2-4) with 0.4 V, we get $R_{ON} \approx 2.4$ KΩ. With $I = V/R = 0.4\ \text{V}/2.4\ \text{K}\Omega = 0.17$ mA, we cannot guarantee that this inverter will pull the LS input low. A wider gate would be needed to lower the on-resistance so that:

$$R_{ON} = \frac{V_{IL}}{I_{max}} = \frac{0.4\ \text{V}}{0.36\ \text{mA}} \approx 1.1\ \text{K}\Omega$$

Even if the load were purely capacitive, the transistor in Example 2-1 would be too small for most off-chip applications. Off-chip capacitances are on the order of 10 to 50 pF, and an on-resistance of around 2 KΩ would result in full transition times of between 100 to 500 ns.

The third parameter used to determine the suitability of a switch is the voltage or current needed to turn it on or off. MOS switches need no steady-state current to turn them on or off. The gate of an MOS device is electrically isolated (by thin oxide) from the substrate, and the only current which can flow to and from the input is the *transient* current required to charge or discharge the gate capacitance. The "turn-on" or "pinch-off" voltage of the device is V_{th} which is normally in the range of 1 V or less.

The speed of MOS logic is determined by R_{ON}, gate capacitance, and V_{th}. Figure 2-19 shows an inverter driving a larger inverter. This is not as

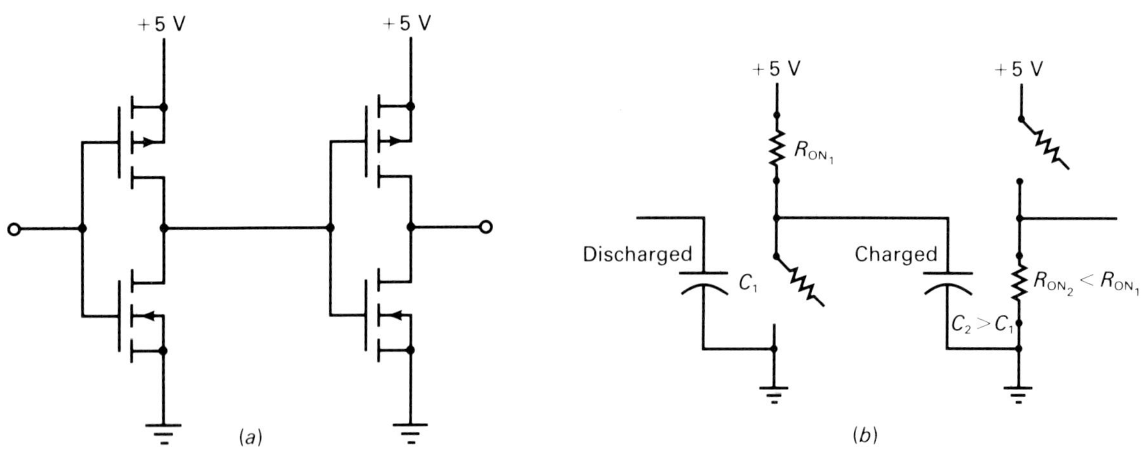

Fig. 2-19 (*a*) Cascaded inverters; (*b*) equivalent circuit.

useless as it might seem. At some point digital signals must leave the IC if they are to perform useful work. The smallest possible geometries are used on-chip in order to minimize silicon area, but off-chip loads require much more drive capability, and at some point there must be some *build-up* to get off-chip. Figure 2-19*b* shows the cascaded inverters in terms of their equivalent resistance and capacitance. The speed of this circuitry depends on the R_{ON} of the first inverter, the capacitance of the second inverter's gate, and the threshold voltage of the second inverter.

EXAMPLE 2-2

Determine how long it will take to just turn on the *p*-device in the second inverter of Fig. 2-19 assuming that its input capacitance of 0.25 pF is fully charged and the *n*-device of the first inverter is fully on at $t = 0$. R_{ON} of the first inverter is 3 KΩ, $V_{DD} = 5.0$ V, and $V_{th} = 0.8$ V.

SOLUTION

The RC time constant is 3 KΩ × 0.25 pF (1 pF = 10^{-12} F). V_0 is V_{DD} = 5.0 V. The voltage required to just turn on the *p*-channel device in the second inverter is V_{th} = 0.8 V *below* V_{DD}.

$$V = V_0 e^{-t/RC}$$

$$4.2 = 5.0\, e^{-t/0.75 \times 10^{-9}}$$

$$0.84 = e^{-t/0.75 \times 10^{-9}}$$

$$\ln 0.84 = \frac{-t}{0.75 \times 10^{-9}}$$

$$t = 0.17 \times 0.75 \times 10^{-9}$$

$$t \approx 0.13 \text{ ns}$$

Smaller geometries mean smaller gate lengths, and this in turn means less on-resistance and less capacitive top plate area over the channel. Thus minimum geometries are always desirable in the design process.

Rise Time and Propagation Delay

Inverters are often used as *drivers* (or buffers) in logic circuits. *Drivers* are inverters which are designed to provide low output resistance so that large line capacitances can be charged and discharged rapidly. This is needed whenever a large number of inputs must be driven, when the receiver inputs are far away, or when a signal must be driven off-chip to face a variety of possible capacitive loads. The output resistance of an MOS inverter is the on-resistance of the *n*- or *p*-channel device which is supplying the charging or discharging current.

There are two important measures of a driver operation, *transition time* (rise or fall time), and *propagation delay*. Normally, transition time is specified as the time required to charge or discharge a given capacitance between 10 and 90 percent of full voltage. Propagation delay is the time interval between input and output crossings of a nominal trip point voltage. The *trip point voltage* is usually defined as the voltage which, when applied to the input, will produce the same voltage on the output. This voltage in CMOS will depend on the relative sizes of the *n*- and *p*-channel devices. If both transistors of a CMOS inverter have the same threshold voltage, and if the *p*-device is sized to have the same on-resistance as the complementary *n*-device, then the trip point will be ½ V_{DD}. This is because with a ½ V_{DD} input, the *p*-device sees the same input voltage *below* V_{DD} that the *n*-device sees *above* V_{SS}. They will be turned on by equal amounts of input voltage and therefore will have the same on-resistance. The output of the inverter, at the trip point, will look like a resistive voltage divider providing ½ V_{DD}. Transition times depend mainly on the size of the output devices and the amount of capacitance which must be charged or discharged. Propagation delays depend on this *and* on internal gate delays. As the complexity of the internal logic increases, propagation delay lengthens and becomes more dependent on the internal gate delays.

The maximum frequency of operation of a driver can be derived from its transition times. The maximum data rate of any logic is:

$$\text{data rate}_{\max} = \frac{1}{(\pi t_r)} = \frac{0.318}{t_r} \qquad (2\text{-}5)$$

Where t_r is the rise time from 10 to 90 percent of full voltage. Rise time is used here because it is normally greater than or equal to the fall time. This means that logic levels cannot change any faster than this. Referring to Fig. 2-20., the Electronic Industries Association (EIA) specification RS-485 specifies that t_r be less than or equal to 0.3 times one bit interval for valid transmission of data. This translates to a maximum data rate of 0.3/t_r which is a close and useful approximation.

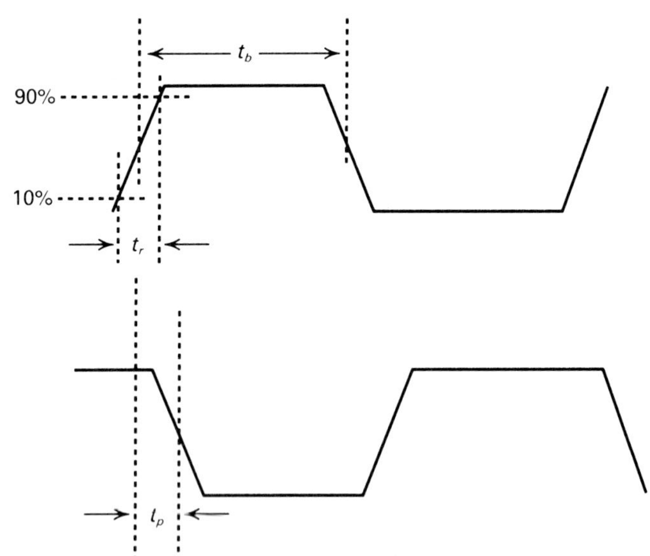

Fig. 2-20 Output waveforms for cascaded inverters showing rise time and propagation delay.

CMOS designs can be quite fast. The smaller-geometry CMOS processes rival their bipolar cousins in speed while cutting power and size requirements substantially. Digital logic circuits are available as packaged parts in CMOS and bipolar technologies. The 74xx series of packaged parts provides CMOS and bipolar families of logic which are pin compatible. Figure 2-21 shows a speed comparison of packaged inverters from major

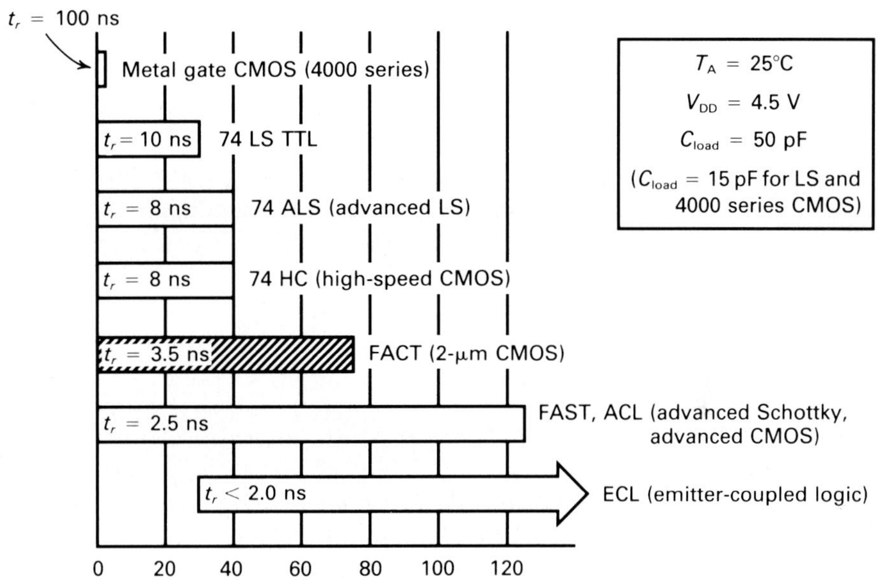

Fig. 2-21 Typical operating frequencies of packaged inverters from major logic families.

logic families. Two- and three-micron CMOS logic is also available in the form of CMOS *gate arrays*. *Gate arrays* are prefabricated arrangements of *p* and *n* devices which may be interconnected to form any type of digital circuitry. All but the last two or three layers are predefined, and the user supplies the information needed to complete the design. Gate arrays permit speed and circuit densities unattainable using packaged parts on printed circuit boards. Gate arrays are part of a new market which has emerged under the name of *Application-specific integrated circuits (ASIC)*. ASIC designs are most often used in high-volume applications where the savings involved in bypassing printed circuit (PC) board assemblies offer a competitive edge.

Review 2-5

1. **Q.** Why do MOSFETs make such good switches? **A.** They conduct bidirectionally and require very little power to turn on and off.

2. **Q.** How does the on-resistance of a MOSFET relate to its threshold voltage? **A.** Larger threshold voltages mean greater on-resistance for the same amount of gate-to-source voltage.

3. **Q.** Define propagation delay, transition time, and trip point. **A.** Propagation delay is the delay between a changing input and the resulting output change. Transition time is the time needed for an output to make some specified change, either from low to high or high to low. The trip point is the voltage level which, when applied to the input of an inverter, will produce the same voltage on the output.

4. **Q.** Which two factors determine output transition time? **A.** the on-resistance of the driver and the amount of capacitance being charged or discharged

5. **Q.** What factors contribute to propagation delay? **A.** internal gate delay, output resistance, and the amount of capacitance being driven

6. **Q.** Describe what happens to the *n*-channel and *p*-channel devices of an inverter during the transition time. **A.** The resistance of one device grows larger while the resistance of the other device grows smaller.

2-6 DESIGN RULES

When an IC designer lays out a circuit, he or she is constrained by a set of design rules which are furnished by the people who run the "fab line." The designer always wants to use the smallest geometries allowable because small geometries mean fast logic and dense layout. But he or she must be told how small the geometries can be and how closely they can be packed. If metal and poly runs are too narrow, it is possible for the runs to become discontinuous in places. Likewise, if they are packed too closely, they might short together in places. The gate length defines the entire process. When we speak of a "3-μm silicon gate process," we mean that the minimum-size gate is 3 μm in length. The design rules will specify the gate length, the minimum metal width, minimum spacing between poly runs and between metal runs, minimum size of contacts, and other limits relating to the registration and etching tolerances of the process.

A color code is normally used to identify each layer. Poly runs and gates, diffusion areas, and metal runs and contacts will each have a color or outline pattern to distinguish them and associate them with a given mask number. Figure 2-22 uses fill patterns to indicate the various colors in a typical layout code. Design usually starts with "stick diagrams" to indicate where everything goes. Then layouts are drawn with colored pencils on gridded paper or on gridded mylar and transferred to a *computer-aided design (CAD)* system by means of a *digitizer*. Alternately, the layout can be created directly on the system using various CAD tools available on workstations and graphics terminals. Figure 2-23 provides an example of how a CMOS inverter might be designed and laid out using the codes

Layer	Mask Number	Drafting	Function
p-minus tub	01	Shaded yellow	Tub for *n*-channel transistors
Nitride	02	Solid blue line	Defines transistor regions
Poly	03	Green diagonal	Defines poly gates and poly runs
Contacts	06	Solid black with cross	Defines metal contact area
Metal	08	Red dashed line	Defines interconnect metal

Fig. 2-22 Color coding for layout rules. If color were available in this illustration, fill patterns would not be necessary.

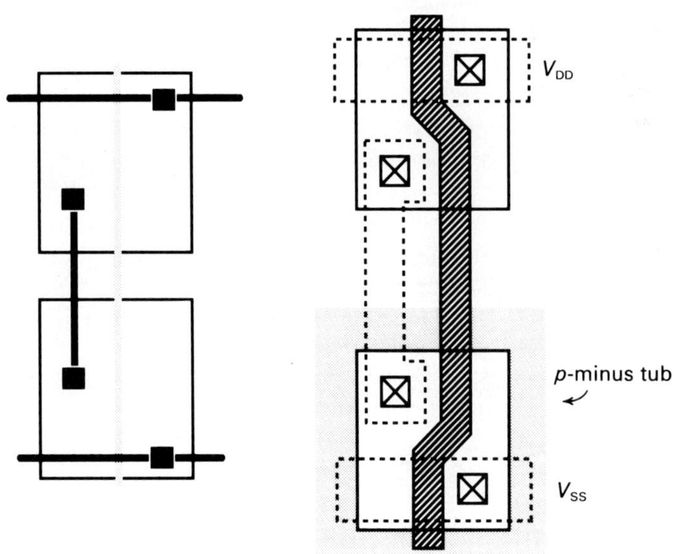

(*a*) (*b*)

Fig. 2-23 (*a*) Stick diagram, and (*b*) layout of a CMOS inverter.

Design Rules:

A.	Minimum poly gate length	4 μm
B.	Minimum poly spacing	5 μm
C.	Minimum width of nitride	6 μm
D.	Minimum spacing of nitride	4 μm
E.	Poly gate overlap of nitride into field oxide	3 μm
F.	Poly gate to nitride edge	5 μm
G.	Minimum metal width	5 μm
H.	Poly contact to edge of poly	2 μm
I.	Diffusion contact to nitride edge	3 μm
J.	Minimum contact size	4 × 4 μm
K.	Poly contact to any nitride	4 μm

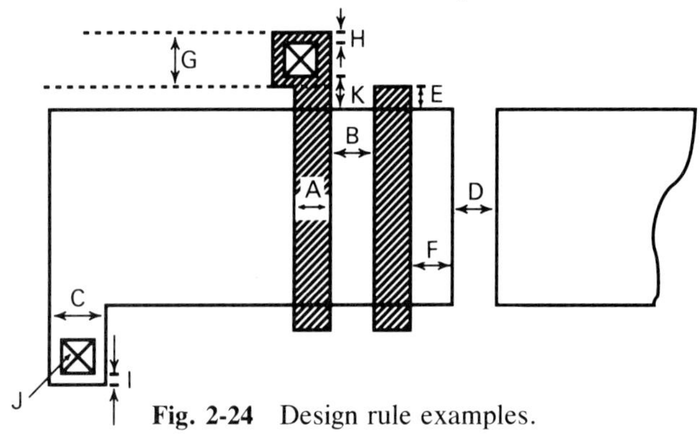

Fig. 2-24 Design rule examples.

given in Fig. 2-22 and the design rules given in Figure 2-24. The inverter layout is called a *cell*. A cell is any functional block that can be used over and over in the circuit design. Note that the inverter in Fig. 2-23 is not identical to the inverter in Fig. 2-13. The difference is that Fig. 2-23 does not have plugs for the substrate. Plugs can be implemented with oversized contacts straddling two opposite polarity diffusions as shown in Fig. 2-13, or they can be made with separate metal contacts elsewhere.

Some processes allow a second layer of metal (or of poly), an obvious advantage since layout will always be easier if conductive runs can cross over one another. The second layer of metal makes *autorouting* possible and thus speeds the layout process. *Auto-routers* are computer programs which automatically determine how metal and poly runs should be laid out to connect cells. Generally the second layer of metal is coarser than the first and is used to distribute power and ground.

REVIEW QUESTIONS

1. Give two reasons why smaller gate geometries mean faster output transitions.

2. Explain why depletion-mode, *n*-channel devices are not complementary to enhancement-mode, *n*-channel transistors and cannot be used like *p*-channel, enhancement-mode transistors to make CMOS inverters.

3. CMOS logic is faster and runs cooler than NMOS. Assuming that the rest of the logic families behave like inverters, explain why this is true.

4. What is the main advantage of poly over metal in the fabrication of gates? What is the disadvantage?

5. How does threshold voltage relate to propagation delay?

REVIEW PROBLEMS

1. The capacitance of a generalized parallel plate capacitor is given as $C = \epsilon(A/d)$, where ϵ is the permittivity constant of the material between the plates, A is the area of the plates, and d is the distance between the plates. Explain how this relates to eqs. (2-2) and (2-3) in Sect. 2.2.

2. Use Eq. (2-3) to find the capacitance of a 1-mil² MOS capacitor with $t_{OX} = 6.0 \times 10^{-6}$ cm and $\epsilon_{OX} = 3.5 \times 10^{-13}$ F/cm (1 mil = 0.001 in.).

3. An n-channel device with a gate length of 5 μm and a gate width of 10 μm is used in a CMOS inverter. The electron mobility is 800 cm²/V-s the threshhold voltage is 0.8 V, and the oxide thickness is 9.0×10^{-6} cm. If the inverter drives a purely capacitive load, what is R_{ON} for the n-device?

4. How would the value of R_{ON} change if the n-channel device in Prob. 3 were scaled down to a 3 μm gate length?

5. An output driver needs to sink 1.6 mA with an output voltage no greater than 0.4 V. Using the parameters given in Prob. 3 and a gate length of 3 μm, how wide must the gate be drawn?

6. What is the 90 to 10 percent fall time of the driver in Prob. 3 assuming a 50-pF load and $V_{DD} = 5.0$ V?

7. Assuming that the p-channel device in Prob. 3 is sized to match the n-channel device, what is the maximum data rate with a 50-pF load?

8. A CMOS inverter is fabricated with equal-sized p- and n-channel transistors in a process where the electron mobility is typically three times greater than the hole mobility. With all other processing parameters being equal, what voltage will appear at the output when $\frac{1}{2}V_{DD}$ is applied to the input of the inverter?

9. What can we say about the rise/fall times of the inverter in Prob. 8?

10. A 2-μm process has the following parameters: $t_{OX} = 6.0 \times 10^{-6}$ cm, $\epsilon_{OX} = 3.5 \times 10^{-13}$ F/cm, $\mu_n = 650$ cm²/V-s, and $\mu_p = 250$ cm²/V-s. If the threshold voltage is typically 0.8 V, what size n-device is needed to sink 24 mA at 0.4 V? What size p-device is needed to source 24mA at 2.4 V.

3 SWITCHING LOGIC

This chapter explains how MOSFETs are combined to implement digital logic functions. Any digital circuit can be built with simple switches. If size, speed, and power were not important, logic circuits could be built with mechanical switches. But the driving concept behind modern technology is economy. Switches must be small and fast, and they must consume very little power. Chapter 2 explained how MOSFETs are used as switches; the objectives of Chapter 3 are to:

1 Introduce switching algebra concepts
2 Explain how basic logic functions are implemented in NMOS and CMOS integrated circuits
3 Outline a method of gate-level logic design using Karnaugh maps
4 Describe how digital logic is packaged for use in board-level circuits

3-1 SWITCHING ALGEBRA

When discussing MOS logic design, it is convenient to define a *connection function* which specifies whether there is, or is not, a switch connection between two points. This function is represented by the letter g in Fig. 3-1. If switch x is open, points A and B are not connected, and $g = 0$. If

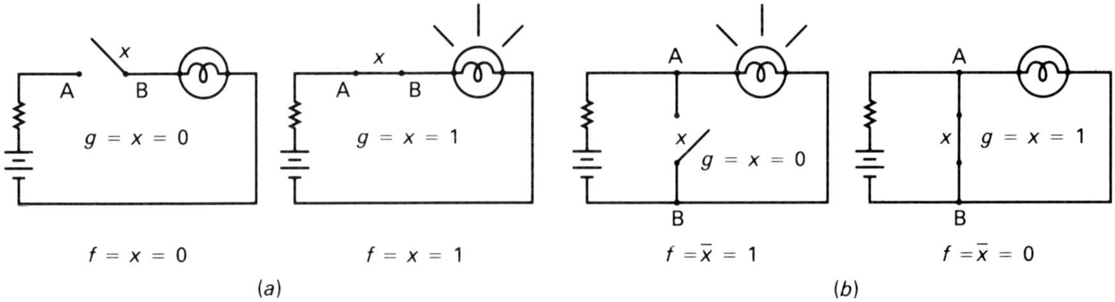

Fig. 3-1 Switch configurations showing the difference between the transmission function g and the logic function f.

switch x is closed, then points A and B *are* connected, and $g = 1$. In this example, g is a function of only one switch x. The function is $g = x$. This means that if x is open ($x = 0$), then $g = 0$. If x is closed ($x = 1$), then $g = 1$. In more complex circuits, g will be a function of two or more switches.

The connection function g tells us only whether points A and B are connected by a switch. It does not tell us the state of the output. Figures 3-1*a* and 3-1*b* illustrate this point. In both cases, when the switch is closed, g equals 1, meaning that points A and B are connected. But in Fig. 3-1*a* closing the switch turns on the lamp, and in Fig. 3-1*b* closing the switch turns off the lamp. We use a different function f to indicate whether the lamp is turned on. The function represented by g is the *connection function*, and it tells us whether connection is made. The function represented by f is the *logic function*, and it tells us the state of the output.

When more than one switch separates two points, some simple mathematics is needed to keep track of the connections. This mathematics is called *switching algebra*, and it was developed by Claude Shannon and published in 1938. Shannon based his work on an earlier algebra developed by George Boole and published in 1854. Boole's algebra was meant to facilitate the study of propositional logic, but Shannon saw that it could be applied to relay switching circuits for use in the telephone industry. Shannon's switching algebra uses three basic operators: *AND, OR,* and *NOT*. Figure 3-2 illustrates the physical meaning of the AND and OR

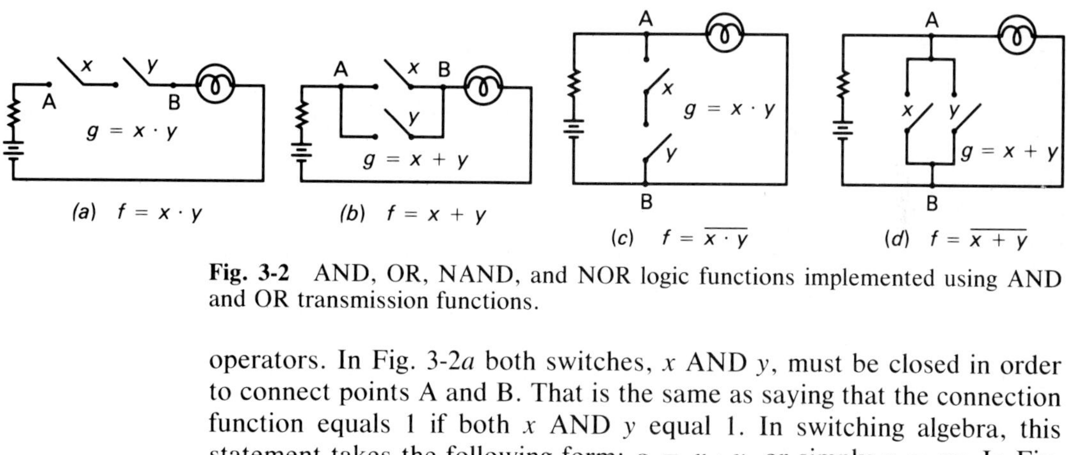

(*a*) $f = x \cdot y$ (*b*) $f = x + y$ (*c*) $f = \overline{x \cdot y}$ (*d*) $f = \overline{x + y}$

Fig. 3-2 AND, OR, NAND, and NOR logic functions implemented using AND and OR transmission functions.

operators. In Fig. 3-2*a* both switches, x AND y, must be closed in order to connect points A and B. That is the same as saying that the connection function equals 1 if both x AND y equal 1. In switching algebra, this statement takes the following form: $g = x \cdot y$, or simply $g = xy$. In Fig. 3-2*b* closing *either* switch x OR switch y will make the connection. This is expressed as $g = x + y$. The $\cdot$ sign is the AND operator, and the $+$ sign is the OR operator. In both of these cases, the connection function equals the logic function. In Fig. 3-2*a*, $f = g = x \cdot y$, and in Fig. 3-2*b*, $f = g = x + y$. In Figs. 3-2*c* and 3-2*d*, however, the connection functions do not equal the logic functions. They are, in fact, opposite. This can be expressed using the NOT symbol, which is a bar over a variable or over an entire expression. When the NOT operator is applied to an OR function, the result is called a NOT OR, or NOR function. When the NOT operator is applied to an AND function, the result is called a NOT AND, or NAND function. In Fig. 3-2*c*, g equals the AND of x and y, but f equals the NAND of x and y. In Fig. 3-2*d*, g equals the OR of x and y, but f equals the NOR of x and y.

Switching variables are letters such as x and y which represent the state of a given switch. A switch is *binary*, which means that a switching variable may assume only one of two discrete values, 0 or 1. Zero means that the switch is open, and one means that the switch is closed. Likewise, functions g and f are binary. When referring to g, a 0 means no connection, and a one means that a connection has been made. When referring to f, a 0 means a low output voltage is produced, and a 1 means that a high output voltage is produced. With these definitions in mind we can construct truth tables for the three operators. A *truth table* is a list of output values for all possible input combinations. Figure 3-3 illustrates the truth tables for the circuits in Fig. 3-2.

x	y	f
0	0	0
0	1	0
1	0	0
1	1	1

(*a*) AND

x	y	f
0	0	0
0	1	1
1	0	1
1	1	1

(*b*) OR

x	y	f
0	0	1
0	1	1
1	0	1
1	1	0

(*c*) NAND

x	y	f
0	0	1
0	1	0
1	0	0
1	1	0

(*d*) NOR

Fig. 3-3 Truth tables for the basic operations AND, OR, NAND, and NOR.

Switching algebra is similar to, but much simpler than, the algebra learned in high school. High school algebra has 4 operators—$+$, $-$, $\times$, $\div$—and 10 elements—0 through 9. Switching algebra has only 2 operators—$\cdot$, $+$—and only 2 elements 0, 1. In both cases there are tables to memorize and rules to learn, but, in switching algebra, these tables and rules are quite simple. The truth table in Fig. 3-3 for the AND function expresses the four possibilities of ANDing 0 and 1.

$$\begin{aligned} 0 \cdot 0 &= 0 \\ 0 \cdot 1 &= 0 \\ 1 \cdot 0 &= 0 \\ 1 \cdot 1 &= 1 \end{aligned} \tag{3-1}$$

This looks very much like ordinary multiplication and is often referred to as "logical multiplication." Likewise, there are only four possible outcomes of the OR function between 0 and 1:

$$\begin{aligned} 0 + 0 &= 0 \\ 0 + 1 &= 1 \\ 1 + 0 &= 1 \\ 1 + 1 &= 1 \end{aligned} \tag{3-2}$$

The first three entries look a lot like ordinary addition, but the last entry $1 + 1 = 1$ is clearly different. It is easy to see, however, why this must be true. In the OR configuration of Fig. 3-2*b*, *if both switches are closed, there will always be a connection,* and f must equal 1.

Connection functions are defined only for AND, OR, or some combination of AND and OR configurations. The AND and OR configurations may have any number of switch variables and may be combined to produce complex functions. *Logic functions* are built with connection functions and may appear as AND, OR, NAND, NOR, NOT, and more complex configurations. Figure 3-4*a*, for example, shows an AND-OR logic function

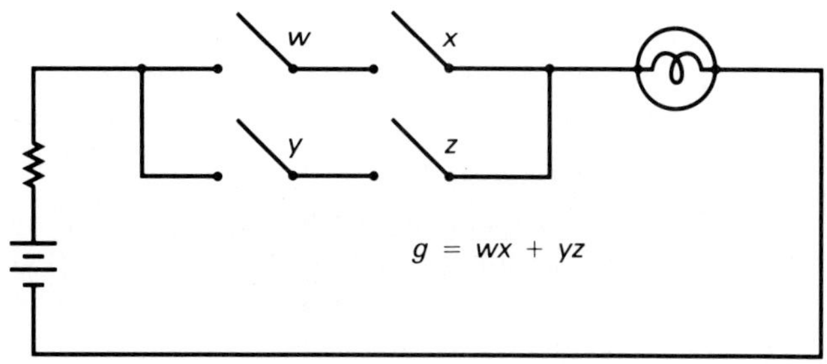

(*a*) AND-OR: $f = wx + yz$

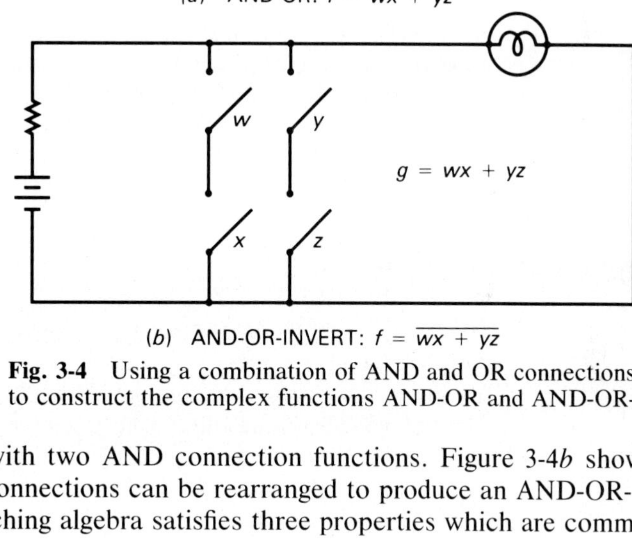

(*b*) AND-OR-INVERT: $f = \overline{wx + yz}$

Fig. 3-4 Using a combination of AND and OR connections to construct the complex functions AND-OR and AND-OR-INVERT.

made with two AND connection functions. Figure 3-4*b* shows how the AND connections can be rearranged to produce an AND-OR-INVERT.

Switching algebra satisfies three properties which are common to other algebras:

$$
\begin{array}{lllr}
\text{The operators are commutative:} & x + y & = y + x & (a) \\
 & x \cdot y & = y \cdot x & (b) \\
\text{The operators are associative:} & (x + y) + z & = x + (y + z) & (a) \\
 & (x \cdot y) \cdot z & = x \cdot (y \cdot z) & (b) \\
\text{The operators are distributive:} & x \cdot (y + z) & = x \cdot y + x \cdot z & (a) \\
 & x + (y \cdot z) & = (x + y) \cdot (x + z) & (b)
\end{array}
\qquad (3\text{-}3)
$$

Note that in part *b* of the distributive property, the OR function distributes over the AND expression in the same way that the AND function distributes over the OR function. This is peculiar to switching algebra and is possible because only two elements, 0 and 1, are used. The best way to convince yourself that this property is valid is to construct a truth table for each side of the equation and compare them.

REVIEW 3-1

1. **Q.** Explain the difference between a connection function and a logic function. **A.** A connection function indicates whether a connection is made between two points; a logic function indicates whether an output voltage is high or low.
2. **Q.** In what way are the numbers 0 and 1 assigned to switch positions? **A.** An open switch is indicated by a 0 while a closed switch is indicated by a 1.
3. **Q.** In what way are the numbers 0 and 1 assigned to connection and logic functions? **A.** Connection between two points is indicated by a 1 while a lack of continuity is indicated by a 0; logic function output voltages are normally indicated as 1 for a high voltage output and 0 for a low voltage output.
4. **Q.** Why is 1 + 1 equal to 1 in switching algebra? **A.** because if both switches are closed in an OR configuration, continuity is made just as if only one switch were closed.
5. **Q.** In addition to the fact that 1 + 1 = 1, what other property is peculiar to boolean algebra? **A.** The OR distributes over AND functions.

3-2 NMOS LOGIC FUNCTIONS

There is a simple procedure for translating logic functions into NMOS circuits. Logic circuits, in general, are constructed with two connection functions, an "upper" connection for pulling the output high and a "lower" connection for pulling the output low. The upper connection switches the output to V_{DD} and the lower connection switches the output to V_{SS} (ground). This is illustrated in Fig. 3-5. The upper connection is labeled h, and the lower connection is labeled g. NMOS logic uses a resistive device for the upper connection so that h is always 1.

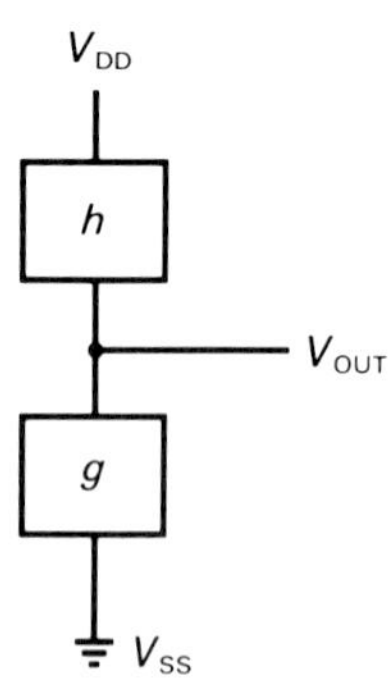

Fig. 3-5 In general, MOS logic consists of two connection functions: an "upper" connection which pulls the output high and a "lower" connection which pulls the output low.

The lower connection of an NMOS circuit is always some combination of n-channel transistors. To determine the appropriate configuration, use the following procedure:

1. Determine the logic function required f.

2. The lower connection g equals $\overline{f}$.

3. The upper connection h is 1, so use a resistor or depletion-mode transistor to make the connection. (3-4)

EXAMPLE 3-1

Implement the logic function $f = \overline{x}$ using an NMOS circuit.

$$f = \overline{x}$$
$$g = \overline{f} = x$$
$$h = 1$$

The resulting circuit is shown in Fig. 3-6. When a variable is inverted, the *opposite value* is produced. In this example, when $x = 0, f = 1$ and when

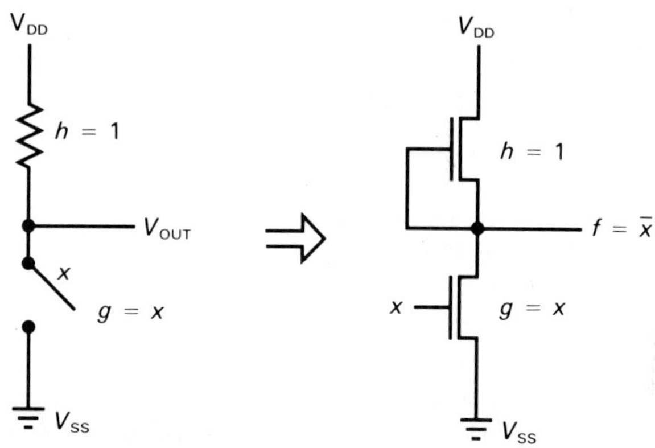

Fig. 3-6 Design of an NMOS inverter.

$x = 1, f = 0$. This inversion is represented by a bar over the variable. The circuit *inverts* the input value x by producing an opposite output value f. The implementation in this example is an NMOS inverter.

EXAMPLE 3-2

Implement the logic function $f = \overline{x + y + z}$ in an NMOS circuit.

$$f = \overline{x + y + z}$$
$$g = \overline{f} = x + y + z$$
$$h = 1$$

The resulting logic circuit, shown in Fig. 3-7, is a three- input NOR.

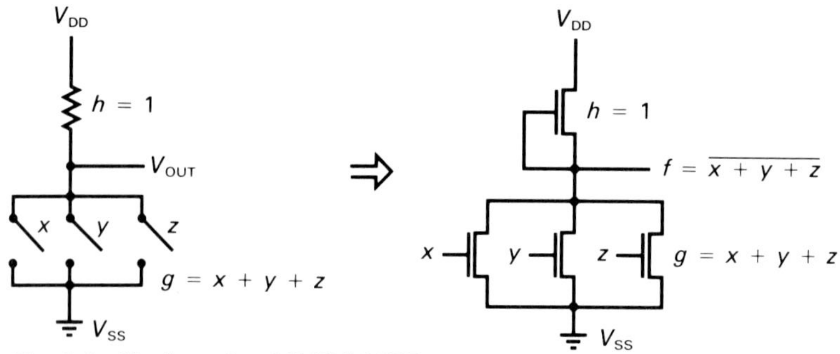

Fig. 3-7 Design of an NMOS NOR gate.

NMOS logic always uses *n*-channel transistors to pull the outputs low and some type of resistive device to pull the outputs high. A high voltage level is required to turn on an *n*-channel switch, and since a turned-on *n*-channel switch pulls the output low, an NMOS easily makes inverting-type logic functions. Noninverting logic functions are not so easily made. Consider the case of a simple AND function.

$$f = x \cdot y$$
$$g = \overline{x \cdot y}$$

Connection functions are defined only for OR configurations, AND configurations, or some combination of the two. In order to implement g, we need to change its form so that it looks like a simple AND or OR. This is done with *DeMorgan's theorem*. DeMorgan's theorem can be summarized with the following rules:

$$\textbf{1.}\quad \overline{x + y} = \overline{x} \cdot \overline{y} \qquad (3\text{-}5)$$
$$\textbf{2.}\quad \overline{x \cdot y} = \overline{x} + \overline{y}$$

Rule 2 can be used to simplify the above example,

$$f = x \cdot y$$
$$g = \overline{x \cdot y} = \overline{x} + \overline{y}$$

The resulting NMOS circuit, shown in Fig. 3-8, is actually a NOR function with inverted inputs. In many cases the inverted forms of x and y are available, and extra inverters are not needed. However, if the inverted forms of x and y are not available, it is more economical to use a NAND function followed by an inverter. In any case, noninverting functions require at least one extra stage and will therefore either be slower than or larger than comparable inverting functions. This is also true of bipolar logic where *npn* transistors are used in a fashion similar to the use of MOS *n* transistors. The extra stage required to make noninverting logic is especially noticeable in bipolar line drivers where propagation delays in

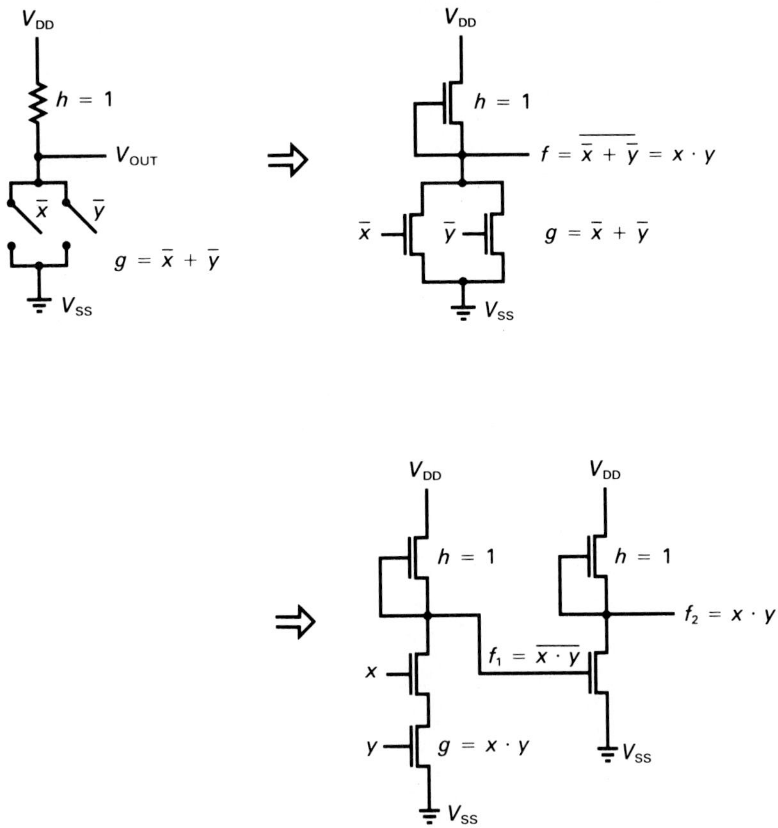

Fig. 3-8 Two ways to make an AND function.

inverting and noninverting drivers are equal. Line drivers are designed to drive large capacitive loads, and in order to make the propagation delays equal for both inverting and noninverting drivers, the noninverting drivers are made larger. This compensates for the added delay but contributes noticeably to the quiescent power drain.

The question may arise: Why not use an *n*-channel device between V_{DD} and the output? Then an AND function could be made as easily as a NAND. You may recall that in Chapter 2 a reference was made to the effect of source-to-substrate voltage on threshold. If the source is more positive than the substrate, the depletion region around the source becomes larger, and more gate voltage must be applied to turn on the device. Using typical processing parameters, it can be shown that, with a 5-V V_{DD}, the threshold voltage would increase by a factor of two or greater. This alone would be discouraging, but a more serious problem is that a positive gate-to-source voltage greater than V_{DD} would be needed to turn on the device. To achieve a V_{GS} of 5 V, we would need to apply 10 V to the gate while maintaining a 5 V V_{DD} and a 5 V output high. This would require voltage doubling and other circuit complexities which would more than offset the benefit of simple AND implementation.

REVIEW 3-2

1. **Q.** What, generally speaking, is NMOS logic? **A.** It is an implementation of logic functions using a combination of *n*-channel switches to pull an output low and a resistive device to pull the output high.

2. **Q.** Why are there no easy implementations of noninverting functions? **A.** because *n*-type devices are used to pull outputs low and they turn on with high inputs

3. **Q.** How would you expect noninverting logic to differ from inverting logic? **A.** Noninverting logic will have at least one extra stage and will therefore be either slower or larger. If larger, then it will consume more power.

4. **Q.** Why are *n*-channel devices used as lower connection switches but not as upper connection switches? **A.** *n*-channel devices require a positive gate-to-source voltage to turn on, and this would be difficult if the device were connected between V_{DD} and the output. Second, source-to-substrate voltage would be large, and this would raise the threshold of the device, causing even larger voltage requirements at the gate.

3-3 CMOS LOGIC FUNCTIONS

CMOS logic design differs from NMOS logic design *only* in the implementation of the upper connection function. The CMOS uses *p*-channel transistors to pull the output high. The advantages of CMOS logic are faster response, less quiescent power drain, and symmetrical output waveforms. The disadvantages of logic CMOS relate to circuit density. One device is sufficient to serve as the upper connection element in an NMOS logic function, but a CMOS requires as many switches in the upper connection as there are in the lower connection. Moreover, the additional switches are *p* devices which cannot share a tub with the *n*-channel devices. The *n* and *p* transistors must be separated by some distance, which adds significantly to the size of the chip. These disadvantages are diminished as circuit geometries become smaller.

The procedure for CMOS design is as follows:

1. Determine the logic function f.
2. The lower connection g equals $\bar{f}$.
3. The upper connection h is the *dual* of g. (3-6)

The *dual* of an algebraic expression is achieved by interchanging AND and OR operators. This is best illustrated by example:

$$\begin{aligned} &\text{If } g = x + y, && \text{then } g^d = x \cdot y \\ &\text{If } g = x \cdot y, && \text{then } g^d = x + y \\ &\text{If } g = x \cdot y + wz, && \text{then } g^d = (x + y)(w + z) \end{aligned} \tag{3-7}$$

If a connection function is valid for n-channel devices, then its dual is valid for p-channel devices within the same CMOS circuit. Notice that the · was dropped in the last example. The AND operator is seldom used explicitly; instead, variables are combined as in ordinary multiplication.

EXAMPLE 3-3

Implement the logic function $f = \bar{x}$ using a CMOS circuit.

$$f = \bar{x}$$
$$g = \bar{f} = x$$
$$h = g^d = x$$

The resulting CMOS inverter is shown in Fig. 3-9. In this simple case, the dual of the function equals the function itself and the switch connections

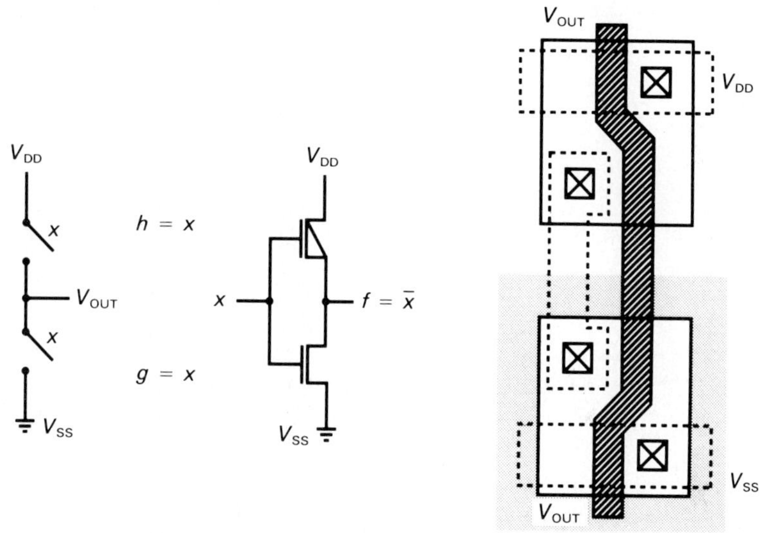

(a) (b)

Fig. 3-9(*a*) CMOS inverter design. (*b*) Layout.

are identical. Notice that the cell outputs are not connected to anything. This is because the cell may be used in different ways in different areas of the integrated circuit. Connection is made at the time of placement by extending the diffusion area above or below the power rails and making a contact with either metal or poly.

EXAMPLE 3-4

Implement the logic function $f = \overline{x + y + z}$ with a CMOS circuit.

$$f = \overline{x + y + z}$$
$$g = \bar{f} = x + y + z$$
$$h = g^d = xyz$$

The resulting three-input NOR is shown in Fig. 3-10. When verifying the operation of this circuit, remember that p devices turn on with low input voltages.

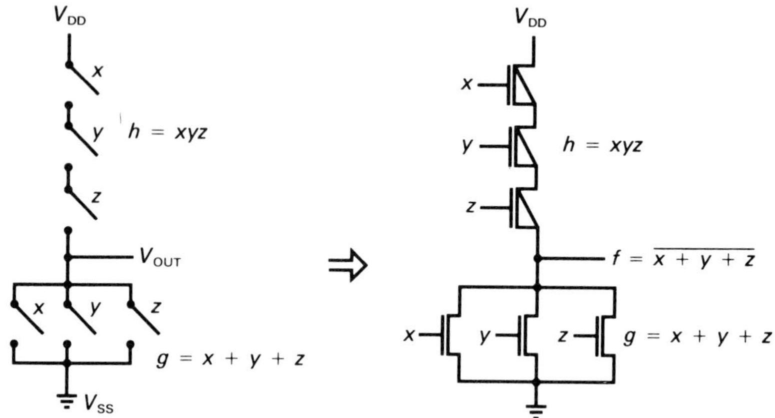

Fig. 3-10(*a*) Three-input CMOS NOR gate design.

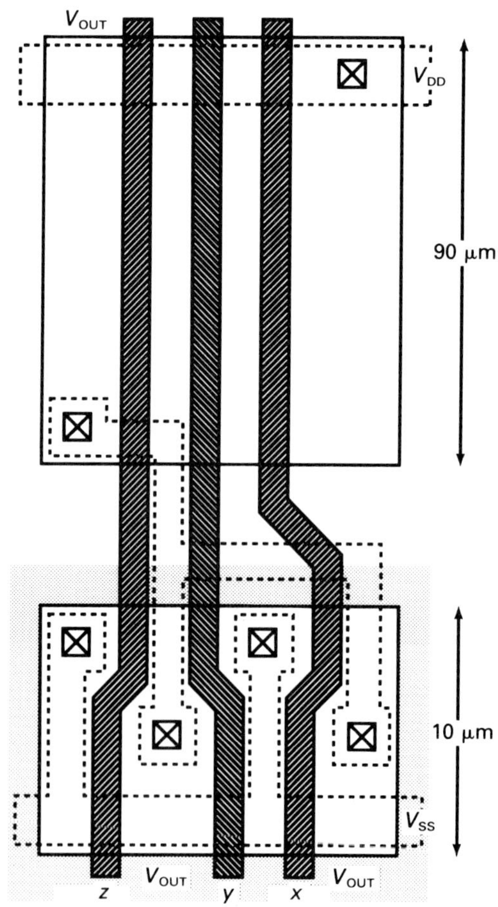

Fig. 3-10(*b*) Layout of three-input NOR gate.

Sizing CMOS Devices

The switching characteristics of CMOS logic are determined by the relative sizes of the n and p devices. The devices must be large enough to charge and discharge the capacitive load within some required time t_r. The on-resistance is the critical parameter and will need to be $\approx t_r/3C_{\text{load}}$ since three time constants are needed to reach 95 percent of final value. The worst-case parameters—threshold, mobility, and C_{OX}—are provided by the process characterization.* The minimum gate length can be found in the design rules.

Rearranging the equation for on-resistance given in Chapter 2 results in an equation for the transistor width W:

$$W = \frac{1}{(\mu\, C'_{\text{OX}} R_{\text{ON}}/L)\,(V_{\text{GS}} - V_{\text{th}} - V_{\text{DS}})} \tag{3-8}$$

V_{DS} will be small and may be approximated as zero if a purely capacitive load is being driven. The thresholds, gate oxide thickness, and minimum geometries are normally equal for n and p devices in a given process. Only the mobility differs. This means that we can calculate the n-channel width using eq. (3-8) and find the p-channel width by multiplying the result by the ratio μ_n/μ_p. The ratio μ_n/μ_p is normally in the range of 2 or 3.

As an example, suppose that a 10-μm n-channel gate width is sufficient to discharge some load capacitance within the required amount of time. If the electron mobility is three times greater than the hole mobility, then the p-channel width must be 30 μm in order to achieve equivalent drive. With this sizing of n- and p-channel devices, we can expect the rise and fall times of an inverter to be equal and meet the specification. Suppose, however, that the circuit must perform a NOR function. Referring to Fig. 3-10, we see that the V_{DD} supply must charge a load capacitance through three p-devices in a zero-to-one transition but may discharge the same load capacitance through only one n-device during a one-to-zero transition. If a 30 μm width p-channel has the equivalent resistance of a 10-μm n-channel, then a series of three 90-μm p-channels will be needed to achieve the same on-resistance as one 10 micron n-channel. This makes the NOR function inefficient in terms of silicon area. In order to construct a three-input NOR with the drive capability of a 10-μm n-channel, we must use 90-μm p-channels.

Consider, on the other hand, a three-input NAND function. Figure 3-11 shows that a NAND circuit uses n-channel devices in series. All assumptions being the same, we now have a requirement for 30-μm n-channel widths and 30-μm p-channel widths. This represents a sizable savings in silicon area and explains why NAND functions are preferred over NOR functions in MOS IC design. It is possible to design any digital logic network with NAND functions only, although this is not always the most

*In order to maintain quality control, space is allowed on each wafer for process control circuitry. This is dummy circuitry used to characterize the process.

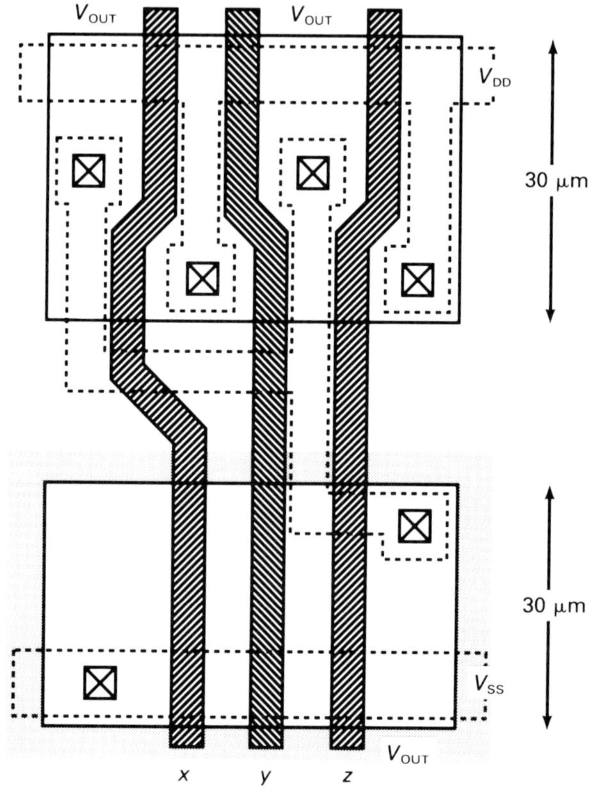

Fig. 3-11 Layout of three-input NAND gate.

economical approach. In many cases load capacitances are so small that equal-sized p and n devices will do the job in the NOR configuration. This means that the smallest-sized p devices often have adequate drive, and the n devices have excess drive capability. In such cases the output waveform will not be symmetrical, but symmetry is not always necessary. When equal sized p and n devices are used, NAND and NOR logic cells are identical except for the positions of V_{DD} and V_{SS}.

REVIEW 3-3

1. **Q.** What is the fundamental difference between NMOS and CMOS logic? **A.** NMOS uses a single resistive pull-up device for the upper connection while CMOS uses a combination of MOSFET switches.

2. **Q.** How are the upper and lower transmissions of CMOS logic related? **A.** They are duals.

3. **Q.** How is the dual of a function formed? **A.** by interchanging the · and + operators

4. **Q.** What type of output waveforms might we expect from a CMOS inverter if n- and p-channel devices are of equal length? **A.** Rising edges will be slower than falling edges.

5. **Q.** Why are NAND functions preferred to NOR functions in CMOS logic design? **A.** NAND functions use less silicon area for equivalent drive requirements.

6. **Q.** How can a CMOS NAND cell be changed into a CMOS NOR by simple relabeling? **A.** Swap V_{DD} and V_{SS}.

3-4 LOGIC SYMBOLS

The symbols currently used to represent basic logic functions are specified in ANSI Standard Y32.14 and Military Standard 806B. The circuits represented by these schematic symbols are commonly referred to as *logic gates*. There have been repeated attempts by various standards committees to replace these symbols but, to this day, their acceptance has not been seriously challenged. Their popularity is no doubt due to their simplicity and distinctiveness. They can be easily drawn by hand or with a template, and they are not easily confused with one another. These symbols are shown in Fig. 3-12. Also shown are the latest attempt at new symbols. The new symbols are part of ANSI/IEEE Standard 91-1984.

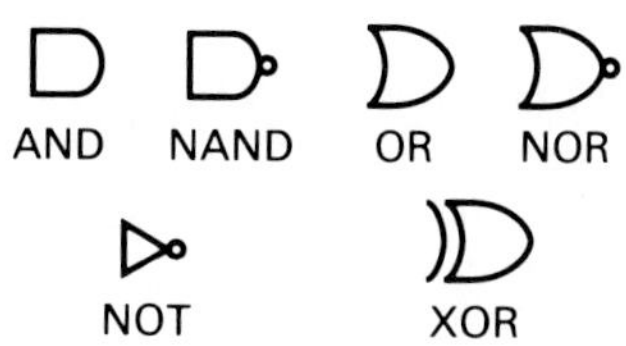

(*a*) Traditional logic symbols

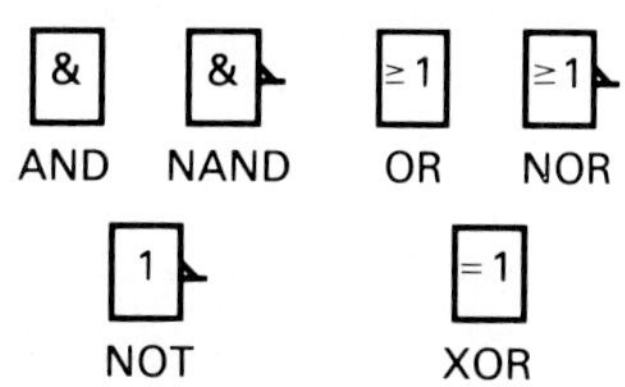

(*b*) IEEE standard 91-1984 symbols

Fig. 3-12 Graphic symbols for logic functions.

There are pockets of enthusiasm in industry for the new standard, and the new symbols have appeared in textbooks and data books. It is more than just a set of symbols, however, it is a complex and extensive international symbolic language which describes the functional behavior of logic circuits with minimal reference to data books. This is meant to be a solution to the problem of representing complex circuits symbolically. Currently, complex functions are represented as boxes labeled with part numbers, and data books must be consulted to determine their exact functions. The full text of ANSI/IEEE Standard 91-1984, entitled *IEEE Standard Graphic Symbols for Logic Functions,* may be obtained from the IEEE at 345 East 47th Street, New York, NY 10017.

Logic Conversion Using Symbols

Some algebraic manipulation can be performed directly on the traditional logic symbols. The application of DeMorgan's theorem is illustrated graphically in Fig. 3-13. The *bubbles*, small circles, indicate inversions. If

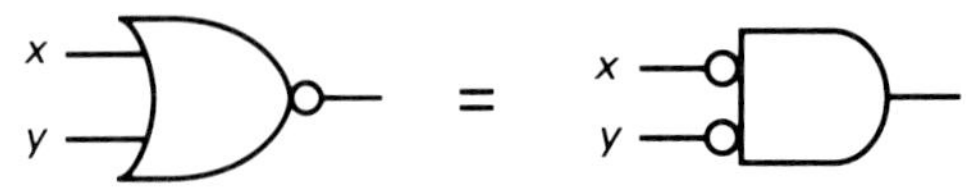

$$\overline{x + y} = \bar{x} \cdot \bar{y}$$

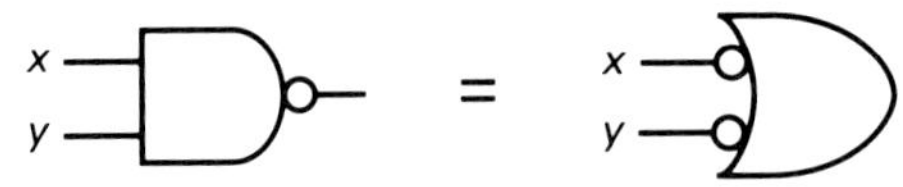

$$\overline{x \cdot y} = \bar{x} + \bar{y}$$

Fig. 3-13 DeMorgan's theorem illustrated graphically.

the variable x enters a bubble, it means that the variable x is being inverted before being applied to the logic function. The addition of a second bubble in sequence nullifies the effect of the first bubble. Thus we can add two bubbles anywhere *in series* without changing the operation of the network. This fact, along with the DeMorgan equivalencies shown in Fig. 3-13, can be used to simplify networks without resorting to algebraic equations.

EXAMPLE 3-5

Use the graphic application of DeMorgan's theorem to convert the circuit in Fig. 3-14 into a functionally equivalent network using only NAND gates.

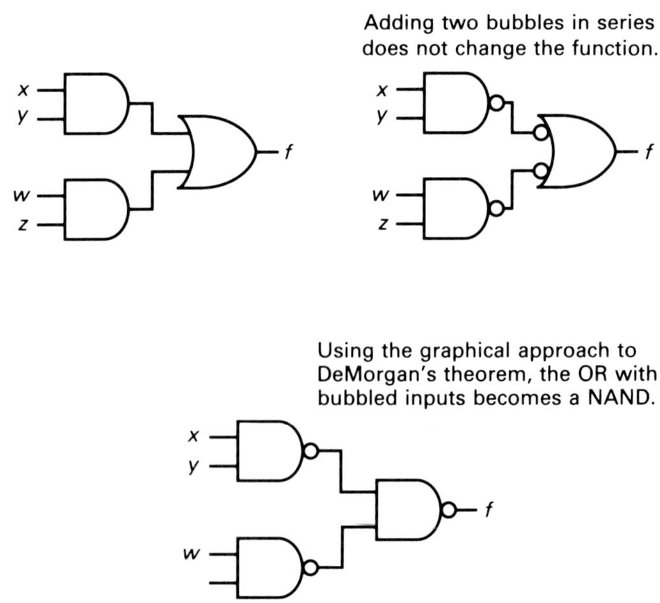

Fig. 3-14 Graphical approach to logic conversion.

SOLUTION

This is solved graphically in Fig. 3-14.

3-5 THE DESIGN OF COMBINATIONAL CIRCUITS

Combinational logic can be defined as logic functions without memory. *The outputs of a combinational network depend only on the present state of the inputs and not on any previous states.* Combinational logic can be designed in two ways, by *method* and by *intuition*. Successful designers use both. There are various techniques available for the methodical approach, and they fall into two categories: algebraic and graphical. Algebraic techniques are best suited to computer-aided design since they can be time-consuming and prone to error. Graphical techniques can be easily applied without the aid of a computer but are useful only for problems involving five or fewer input variables. It can be argued that any problem involving more that four or five input variables is too complex and should be broken down into simpler parts. This is good design practice anyway and often leads to a simpler overall solution.

Design methodology usually involves the following steps:

1 Statement of the problem in terms of a truth table
2 Minimization using algebraic or graphical techniques
3 Implementation of the simplified function in logic gates

Minimization means reducing a logic function to its simplest form. Algebraic minimization requires the knowledge of the theorems and properties already presented. In addition, the following theorems are important:

$$
\begin{array}{llll}
\textit{Idempotency:} & x + x = x & (a) & \\
 & x \cdot x = x & (b) & \\
\textit{Absorption:} & x + xy = x(1 + y) = x & (a) & \\
 & x \cdot (x + y) = xx + xy = x + xy = x & (b) & (3\text{-}9) \\
\textit{Simplification:} & x(\bar{x} + y) = x\bar{x} + xy = xy & (a) & \\
 & x + \bar{x}y = (x + \bar{x}) \cdot (x + y) & (b) & \\
 & \quad = 1 \cdot (x + y) = x + y & &
\end{array}
$$

Since the variable x can assume the values of only 0 or 1, *idempotency* simply restates the relationships $1 + 1 = 1$, $0 + 0 = 0$, and $1 \cdot 1 = 1$, $0 \cdot 0 = 0$. *Absorption* states, in *a*, that if $x = 1$, the functions given must equal 1 since only one closed switch will make an OR connection, and if $x = 0$, the value of y does not matter since $0 + 0 \cdot y$ will always equal 0. If the variable x, in absorption part *b*, is multiplied through, the result is identical to the first. These follow from the definitions of the operators $\cdot$ and $+$ and the properties described earlier. Part *a* of the *simplification* theorem can be proven by simply multiplying through by x and noting that any variable ANDed with its own inverse must always equal 0. Part *b*, however, is tricky, and you must refer back to the distributive property in Eq. (3-3) to convince yourself of its validity.

These theorems and properties of boolean algebra can be used directly to minimize algebraic expressions, and they form the basis of graphical minimization techniques. The following example uses a truth table and algebraic manipulation to solve a simple logic problem.

EXAMPLE 3-6

Design and minimize a circuit which will choose between two input variables on the basis of a *select* input. Variables x and y are inputs, s is the select input, and f is the output value. When $s = 0$, output $f = x$, and when $s = 1$, $f = y$.

SOLUTION

First construct a truth table:

s	x	y	f
0	0	0	0
0	0	1	0
0	1	0	1
0	1	1	1
1	0	0	0
1	0	1	1
1	1	0	0
1	1	1	1

From the truth table we can extract the following:

$$\begin{aligned} f &= \bar{s}x\bar{y} + \bar{s}xy + s\bar{x}y + syx && (a) \\ &= \bar{s}x(\bar{y} + y) + sy(\bar{x} + x) && (b) \\ &= \bar{s}x + sy && (c) \end{aligned}$$

The combinations of s, x, and y and their inverses in a are called *minterms, standard products*, or *product terms*. The sum-of-products expression (a) comes directly from the truth table and can be implemented using a combination of AND gates, OR gates, and inverters. Usually the sum-of-products expression can be simplified. In this example the distributive property is used to simplify the original function. The terms in parentheses will equal 1 because any variable ORed with its inverse must equal 1. In more complex problems any of the theorems and properties discussed earlier might be helpful in simplifying the original expression. Computer programs are also available to minimize logic expressions using similar algebraic techniques.

Karnaugh Maps

A graphical approach, *Karnaugh mapping*, may be used to achieve the same results. The basis for the Karnaugh map technique is the boolean identity used in the example above:

$$\begin{aligned} f &= xyz + xy\bar{z} \\ &= xy(z + \bar{z}) \\ &= xy \end{aligned} \qquad (3\text{-}10)$$

The Karnaugh map is actually just a retabulation of the information in the truth table, but its format aids the minimization process. Figure 3-15 illustrates three truth tables and their corresponding Karnaugh maps.

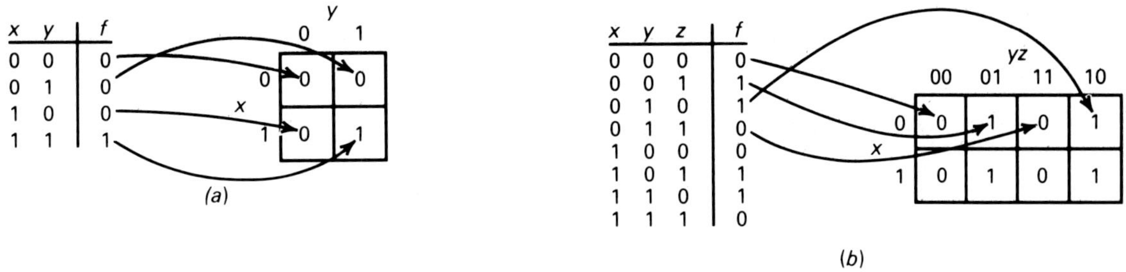

w	x	y	z	f
0	0	0	0	1
0	0	0	1	1
0	0	1	0	0
0	0	1	1	0
0	1	0	0	1
0	1	0	1	1
0	1	1	0	0
0	1	1	1	0
1	0	0	0	0
1	0	0	1	0
1	0	1	0	1
1	0	1	1	1
1	1	0	0	0
1	1	0	1	0
1	1	1	0	1
1	1	1	1	1

wx \ *yz*	00	01	11	10
00	1	1	0	0
01	1	1	0	0
11	0	0	1	1
10	0	0	1	1

(*c*)

Fig. 3-15 Karnaugh map examples.

Notice that the digits at the end of each row or column represent states of the input variables and the digits inside the squares represent the corresponding output values. The map is constructed in the following way:

1 Sketch a box with 2^n squares, where n is the number of input variables in the truth table. The number of squares equals the number of output values in the truth table.

2 Label the rows and columns so that adjacent rows and columns differ by only one bit position.

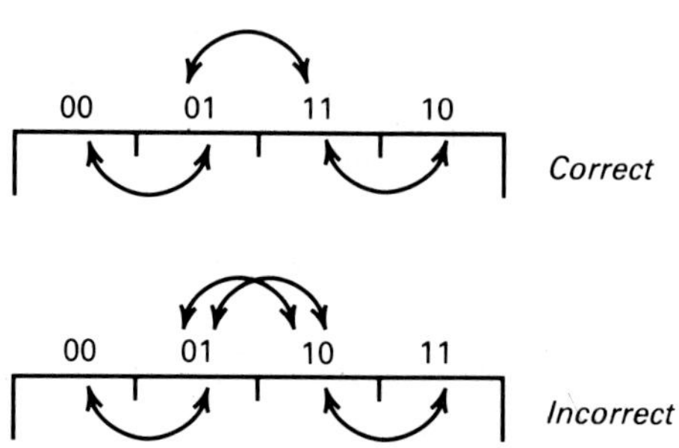

3 Assign the input variables to the digits at the end of rows and columns so that they correspond to the truth table when read from left to right. You may actually make these assignments any way you like but transferring the output values from truth table to Karnaugh map is facilitated by keeping the same order, left to right.
4 Fill in the table by copying the output values into the corresponding squares.
5 Group two- four- and eight-square *implicants* by circling.
6 Extract the simplified sum-of-products expression from the map.

The first four steps can be best understood by comparing the truth tables in Fig. 3-15 with their Karnaugh maps. The last two steps require some explanation. *Adjacent* pairs of squares containing 1s are called *two-square implicants. Adjacent* groups of four squares containing 1s are *four-square implicants*, and so on. The word *adjacent* is italicized because it has a broader meaning than might be expected. Any two squares which differ by only one input variable bit position are considered adjacent. This includes squares at opposite ends of a row or column. In some cases you may want to visualize the map wrapped around a cylinder so that the edges meet. This mental trick does not help, however, when there is a four-square implicant at the corners. Figure 3-16 provides some examples of two-, four-, and eight-square implicants.

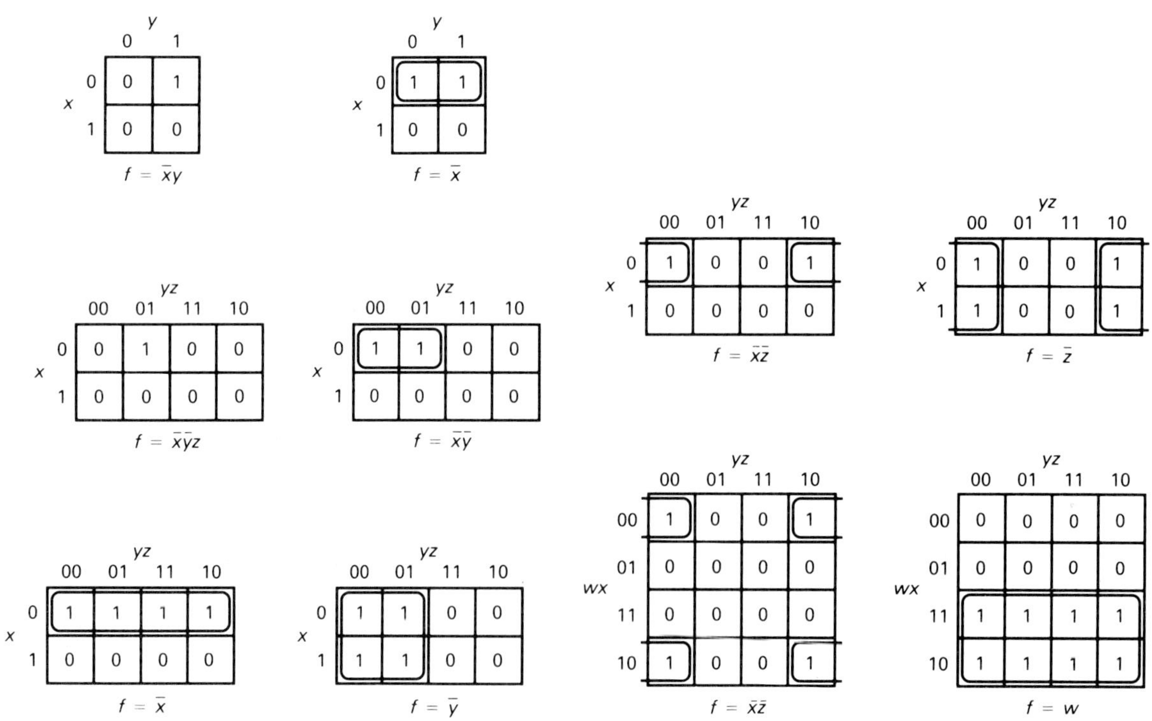

Fig. 3-16 Examples of one-, two-, and four-square implicants.

When an implicant includes both states of an input variable, that input variable is redundant and may be removed from the corresponding product term as in Eq. (3-10) above. To illustrate this, we will repeat Example 3-6 using a Karnaugh map. The truth table outputs are copied into the Karnaugh map in Fig. 3-17. The circuit may be implemented using AND and OR gates as shown or converted to NANDs with graphical DeMorgan techniques. This circuit is called a *multiplexer*, or *mux*. The term to *multiplex* means to share one output between two or more inputs.

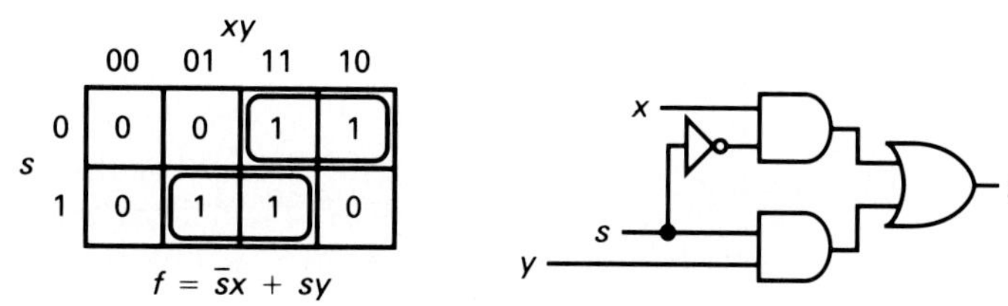

Fig. 3-17 Karnaugh map and logic implementation of Example 3-6.

Essential Implicants

Consider the Karnaugh map for the multiplexer in Fig. 3-18. There are actually three possible implicants in this map yielding the expression:

$$f = \bar{x}z + xy + yz$$

One of the implicants, yz, however, covers 1s in the map which are already covered by other implicants. This implicant is redundant and may be removed from the expression. The corresponding AND gate is marked

x	y	z	f
0	0	0	0
0	0	1	1
0	1	0	0
0	1	1	1
1	0	0	0
1	0	1	0
1	1	0	1
1	1	1	1

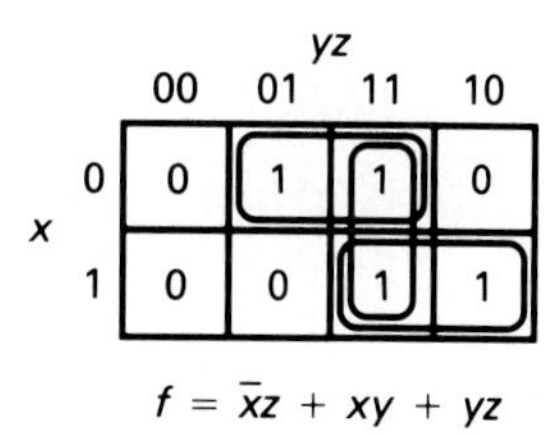

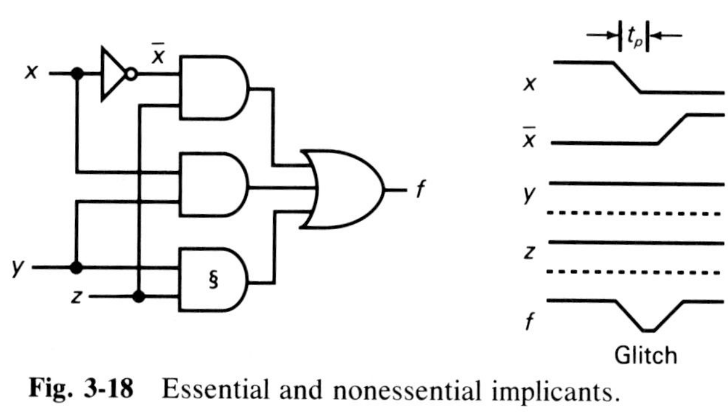

Fig. 3-18 Essential and nonessential implicants.

with a § in the logic network and may also be removed to achieve a simpler circuit. Only *essential* implicants are needed to guarantee the correct output. *If an implicant covers at least one 1 in the map which is not covered by any other implicant, then it is essential.*

We will find, however, that it is sometimes necessary to include a nonessential implicant in a logic network. The inclusion of a nonessential implicant may prevent a glitch from occurring. A *glitch* is a false output of short duration. In the above example a glitch will occur when the inputs are $x = 1, y = 1, z = 1$ and change to $x = 0, y = 1, z = 1$. The output of the circuit would be expected to remain high since both the old state and the new state both produce outputs of 1, but the propagation delay through the inverter causes an intermediate state where the lower essential AND gate is turned off while the upper essential AND gate has not yet turned on. For the length of time equal to the propagation delay through the inverter, the output will be 0. Thus the output will be high, go low for a short time, and then return to a high value. This is shown in Fig. 3-18. If the nonessential implicant is used, it will hold up the output during the transition. Malfunctions of this type, due to delays in the paths of input variables, are called *hazards*. In the next chapter we will discuss hazards and ways of dealing with them in logic circuits.

Don't-Care Conditions

Sometimes there are combinations of input variables which can never occur. In such cases the square corresponding to the output is marked with an X to indicate that its value does not matter. An X means that we can assign either 0 or 1 to the square. The choice depends on which value produces the simplest implementation. The value we choose will not affect the operation of the circuit because the corresponding inputs will never occur.

EXAMPLE 3-7

A *thumbwheel switch* is a rotary switch with thumb indentations. Its purpose is to convert decimal numbers into binary code for digital control. A single-digit thumbwheel switch produces all combinations between 0000 and 1001 but not 1010, 1011, 1100, 1101, 1110, or 1111. Design a circuit which will produce an output high whenever the codes 1000 or 1001 are selected.

SOLUTION

Six input combinations have don't-care outputs as shown in Fig. 3-19. These don't-care output values are chosen so as to make the largest possible implicant. It turns out that x_1 can be used directly to detect the codes 1000 and 1001.

Finding the Inverse of a Function

Sometimes it is easier to group 0s into implicants instead of grouping 1s. The resulting sum-of-products expression will represent the inverse of the desired function. The resulting implementation may be followed by an

x_1	x_2	x_3	x_4	f
0	0	0	0	0
0	0	0	1	0
0	0	1	0	0
0	0	1	1	0
0	1	0	0	0
0	1	0	1	0
0	1	1	0	0
0	1	1	1	0
1	0	0	0	1
1	0	0	1	1
1	0	1	0	X
1	0	1	1	X
1	1	0	0	X
1	1	0	1	X
1	1	1	0	X
1	1	1	1	X

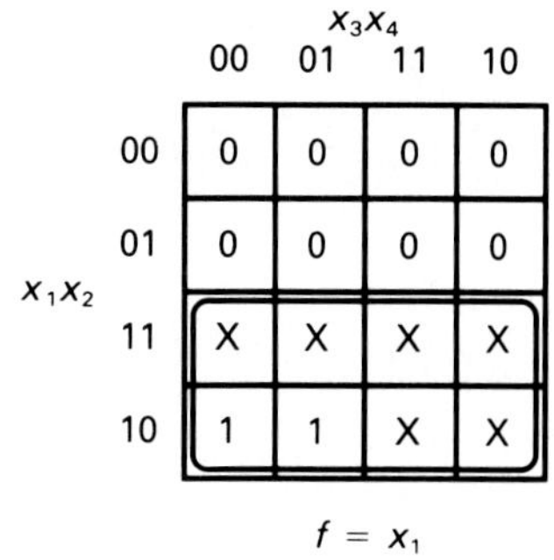

Fig. 3-19 Thumbwheel switch decoder.

inverter to achieve the correct output. Often no inverter is necessary. Complex functions, like simple functions, are easier to implement in their inverting form. Consider, for example, the Karnaugh map and implementation in Fig. 3-20. The AND-OR-INVERT function is actually a single

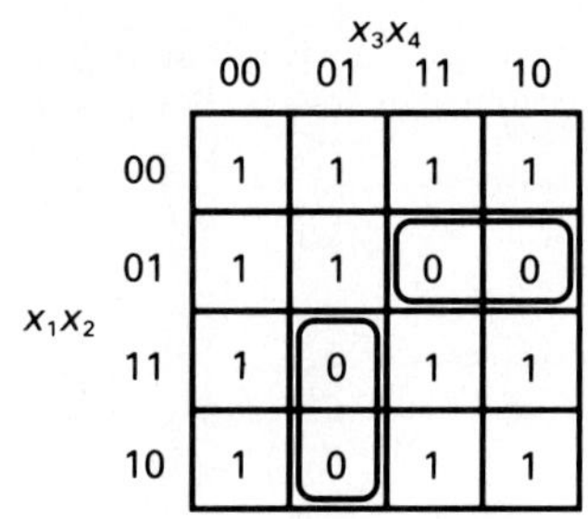

$$\bar{f} = x_1\bar{x}_3x_4 + \bar{x}_1x_2x_3$$

$$f = \overline{x_1\bar{x}_3x_4 + \bar{x}_1x_2x_3}$$

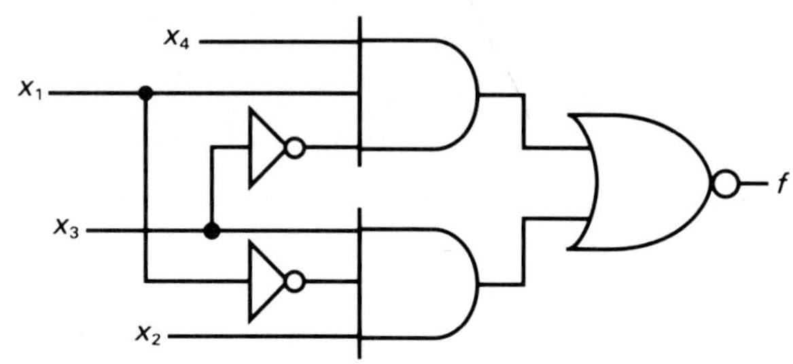

Fig. 3-20 Using 0s to find the inverse of a function.

complex gate which lays out easily as one logic cell. It is important to realize that we could have derived the same expression and logical implementation by grouping 1s instead of 0s. The only difference is that more algebraic manipulation would have been necessary if we had used the 1s.

Complex Gates

Complex gates can be constructed using the rules for CMOS logic design given in Eq. (3-6). For instance, the AND-OR-INVERT part of the network shown in Fig. 3-20. can be designed and laid out in the following manner:

$$f = \overline{x_1\bar{x}_3x_4 + \bar{x}_1x_2x_3}$$
$$g = \bar{f} = x_1\bar{x}_3x_4 + \bar{x}_1x_2x_3$$
$$h = g^d = (x_1 + \bar{x}_3 + x_4)(\bar{x}_1 + x_2 + x_3)$$

The cell layout for this function is shown in Fig. 3-21. Complex gates are

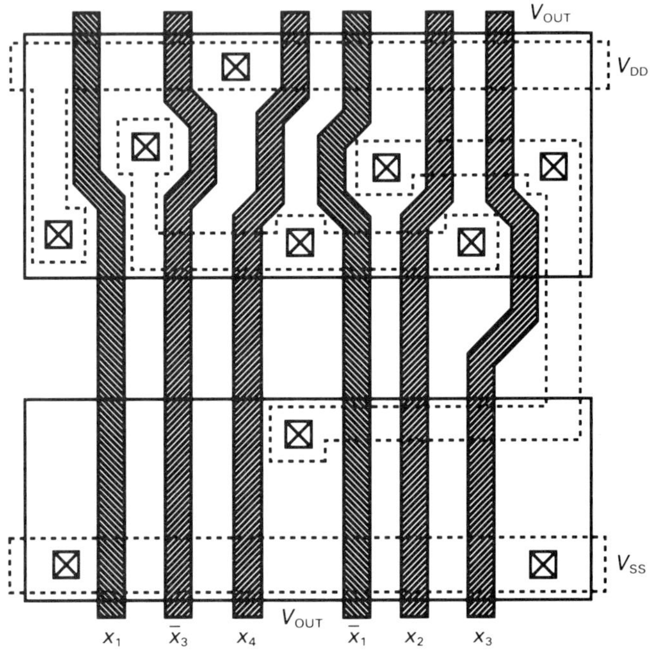

Fig. 3-21 AND-OR-INVERT cell layout.

useful because they reduce the number of transistors required to implement a function. If we had implemented the AND-OR-INVERT with individual logic gates, 16 transistors would have been needed. The complex gate implementation requires only 12.

Sometimes functions will have no two- or four-square implicants in the Karnaugh map but can be simplified anyway using a complex gate called an EXCLUSIVE OR. Consider the truth table and Karnaugh map illustrated in Fig. 3-22. No real grouping can be made, and the implementation seems to require an AND-OR-INVERT gate followed by an inverter. The real problem with this is that both true and inverted forms of the input variables are needed. However, the EXCLUSIVE OR implementation has only two inputs, x and y. The graphical symbol for this function appears in Fig. 3-22*b*. with the abbreviation XOR. The sum of products expression for an

x	y	f
0	0	0
0	1	1
1	0	1
1	1	0

x \ y	0	1
0	0	1
1	1	0

$f = \bar{x}y + x\bar{y} = x \oplus y$

(*a*) XOR function

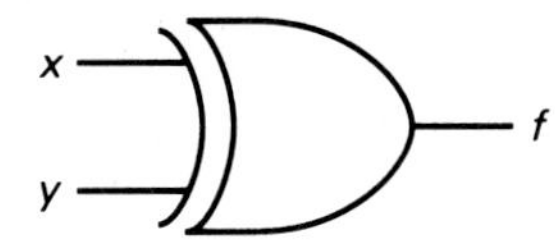

(*b*) XOR symbol

Fig. 3-22 The EXCLUSIVE OR function.

XOR is rarely used; instead, a special symbol is placed between the variables:

$$f = \bar{x}y + x\bar{y} = x \oplus y \tag{3-11}$$

In general, whenever a checkerboard pattern appears in a Karnaugh map, the XOR function may be used to simplify the design. Fig. 3-23 illustrates three XOR functions and their maps. Notice that in every case there are an equal number of 1s and 0s.

x	y	z	f
0	0	0	0
0	0	1	1
0	1	0	1
0	1	1	0
1	0	0	1
1	0	1	0
1	1	0	0
1	1	1	1

x \ yz	00	01	11	10
0	0	1	0	1
1	1	0	1	0

(*a*) $f = x \oplus y \oplus z$

x	y	z	f
0	0	0	0
0	0	1	1
0	1	0	1
0	1	1	0
1	0	0	0
1	0	1	1
1	1	0	1
1	1	1	0

x \ yz	00	01	11	10
0	0	1	0	1
1	0	1	0	1

(*b*) $f = y \oplus z$

x	y	z	f
0	0	0	0
0	0	1	0
0	1	0	1
0	1	1	1
1	0	0	1
1	0	1	1
1	1	0	0
1	1	1	0

x \ yz	00	01	11	10
0	0	0	1	1
1	1	1	0	0

(*c*) $f = x \oplus y$

Fig. 3-23 Some EXCLUSIVE OR patterns.

EXCLUSIVE ORs are, by definition, two input functions. Three or more inputs require multiple XOR gates. Moreover, the EXCLUSIVE OR function, like the AND and OR functions, is commutative and associative. Thus,

$$x \oplus y = y \oplus x$$

and

$$(x \oplus y) \oplus z = x \oplus (y \oplus z) \tag{3-12}$$

The design of the EXCLUSIVE OR logic gate requires a combination of intuition and method. The function could be implemented using an AND-OR-INVERT followed by an inverter. The cost would be 10 transistors. However, the fact that both true and inverted forms of the input variables are required brings the total transistor count to 14. If, instead of starting with the EXCLUSIVE OR function, we started with the inverse of the function, then:

$$\overline{f} = xy + \overline{x}\,\overline{y}$$

so

$$f = \overline{xy + \overline{x}\,\overline{y}}$$

and, using DeMorgan's theorem:

$$f = \overline{xy + (x + y)}$$

In this way we can eliminate the need for inverters at the inputs and at the output. The inverse of the function was derived directly from the truth table in Fig. 3-22 by swapping 1s with 0s. The resulting logic gate implementation is shown in Fig. 3-24*a*. Notice that this is a special case of an AND-OR-INVERT with a NOR feeding one of the inputs. Using the design rules for the CMOS, we can implement the two parts separately with MOSFETs and then combine them to get the MOSFET design shown in Fig. 3-24*b*. Note that the transistor count has been reduced from 14 to 10. Figure 3-24*c* illustrates one possible cell layout for this design.

REVIEW 3-5

1. **Q.** When may two 1s in a Karnaugh map be grouped to form a two-square implicant? **A.** when their inputs differ by only one bit position

2. **Q.** What is an essential implicant? **A.** any implicant which covers at least one 1 in the map not covered by another implicant

3. **Q.** What is a don't-care condition? **A.** an output corresponding to an input combination which can never occur

4. **Q.** Why are don't-care conditions important? **A.** they can be helpful in simplifying logic expressions

5. **Q.** What is a complex gate? **A.** a logic function composed of two or more logic gates which can be implemented as one gate

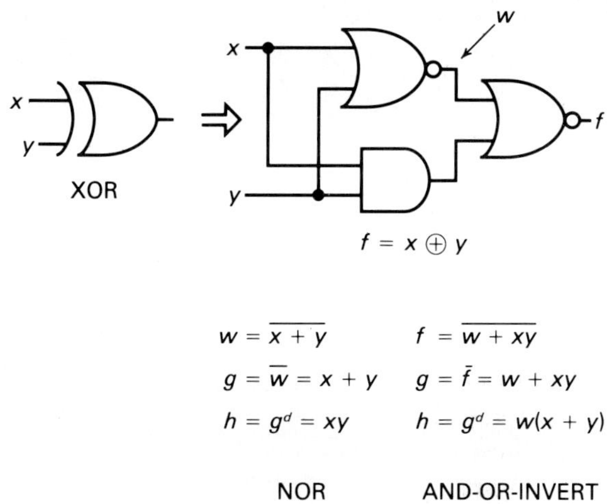

Fig. 3-24(*a*) Implementation of an EXCLUSIVE OR function.

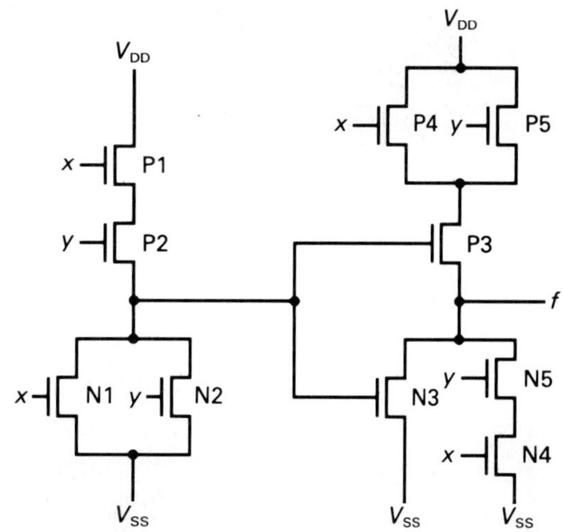

Fig. 3-24(*b*) MOSFET implementation of the XOR function.

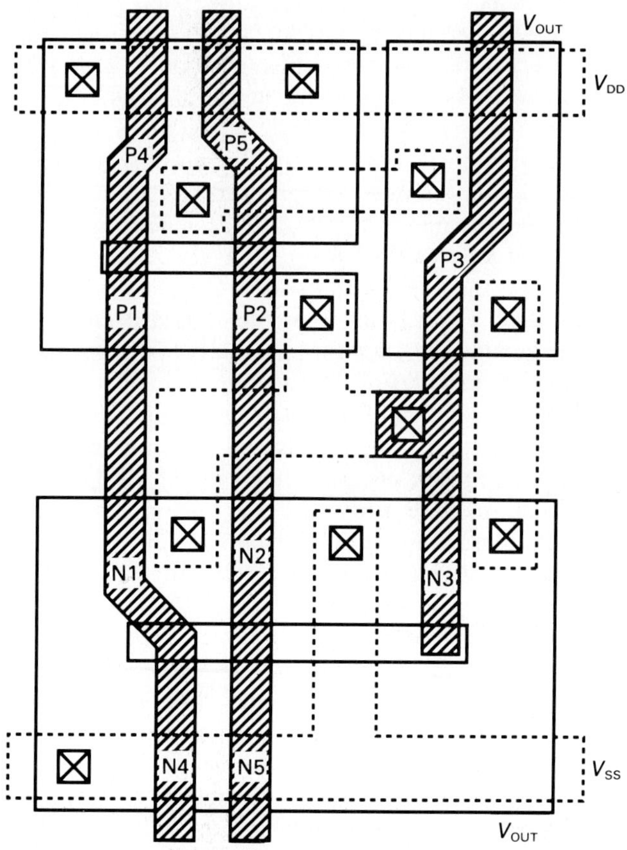

Fig. 3-24(*c*) Cell layout for EXCLUSIVE OR function.

6. **Q.** How can we implement a three-input EXCLUSIVE OR function? **A.** one EXCLUSIVE OR gate with two inputs followed by a second EXCLUSIVE OR gate with the third input and the output of the first gate

7. **Q.** Why might we want to start a design using the inverse of a function? **A.** because inverting logic is easier to design

3-6 PACKAGING

The logical question at this point is: How are the logic gates described in this chapter actually used? Obviously, we do not lay out and fabricate an EXCLUSIVE OR gate every time we need one. The answer to this question has at least four levels. At the most basic level, *standard cell libraries* are used which contain all the simple and complex gates needed for digital circuit design. These cell libraries are sold by semiconductor manufacturers to customers with high-volume sales. When large quantities of a product are produced, it becomes economical to design semicustom ICs which contain thousands of logic cells like the one in Fig. 3-24*c*. This approach reduces power consumption and space requirements and improves reliability. It also has the advantage of protecting proprietary designs because it is very difficult to copy a semicustom IC. This approach is not economical, however, when sales figures are low or when designs change often.

The next higher level of design uses gate arrays. *Gate arrays* are prefabricated ICs which contain *p* and *n* devices in rows and columns. Software is supplied by gate array manufacturers to aid the layout of interconnections. Circuits are designed and laid out by choosing logic function macrocells from a library. A *macrocell* is a metal pattern which connects the *p* and *n* transistors to produce a cataloged logic element. The macrocell is different than the standard cell described above. Standard cells include information for all layers, diffusion, nitride, poly, and metal. Standard cell design starts with a blank silicon wafer which must be produced individually for each customer. Gate arrays start with predesigned masks. All but the metal layers are in place. This generic approach has cost advantages because much of the design is standardized. However, because the gate array masks are generic, circuit densities will always be lower than standard cell designs. Even though gate array manufacturers produce a variety of arrays for various needs, there is always some waste involved in this approach.

The third level of logic design involves the use of *programmable-logic devices (PLDs)* and *erasable programmable logic devices (EPLDs)*. These devices are generic like the gate arrays above. However, gate arrays are unconnected *p* and *n* devices which must be connected with metal before fabrication and packaging. PLDs and EPLDs are already packaged, and the transistors are already connected to form logic elements. The user does not have the freedom to connect the transistors to form any logic element but instead connects the predefined logic elements to form digital circuits. All the inputs in PLDs are preconnected with fuses, and the user must selectively remove them by blowing the fuses. EPLDs are similar but use programmable diodes instead of fuses. These may be erased with ultraviolet light and reprogrammed. Figure 3-25 shows the logic diagram for a PLD device called a PAL. PAL stands for programmable-array-logic.

16L8

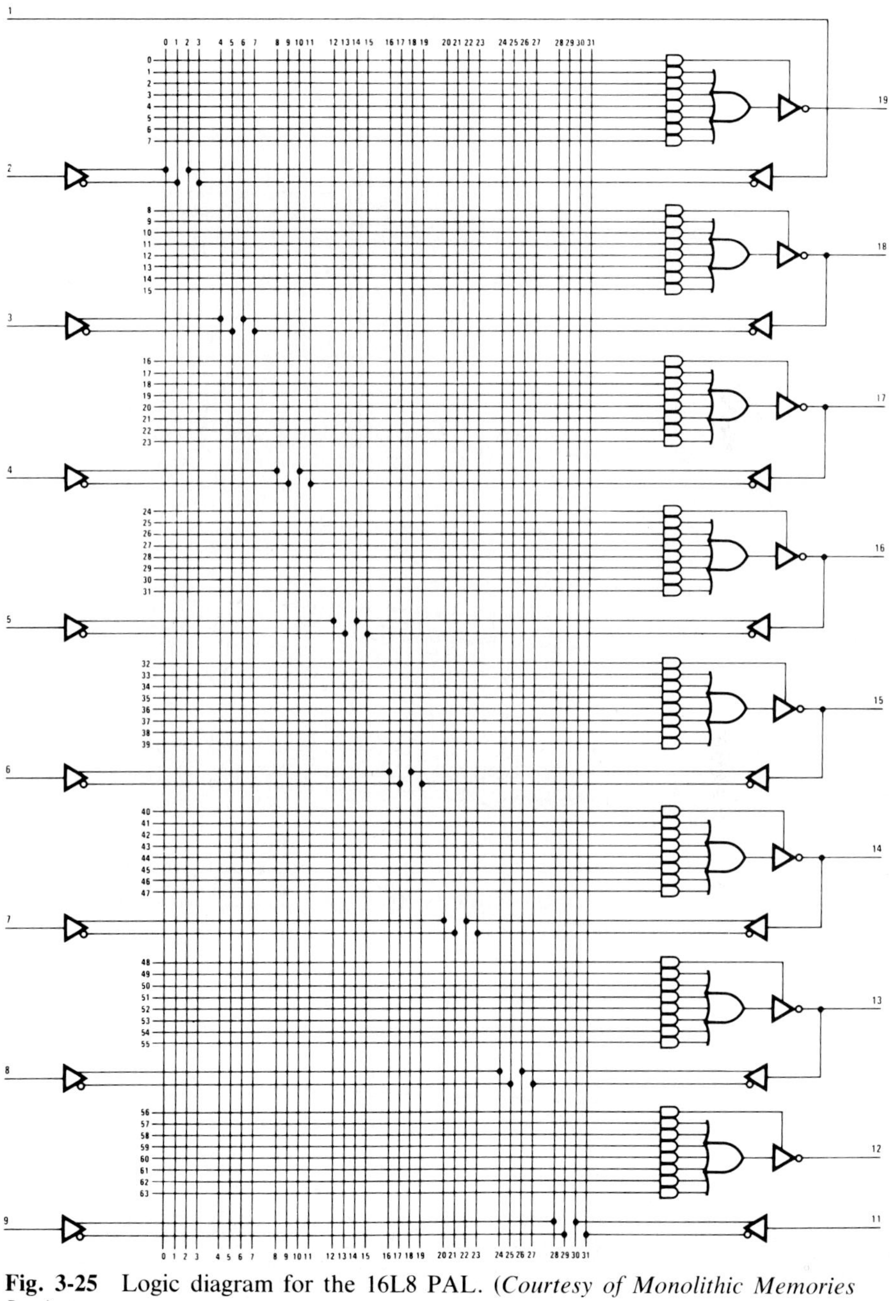

Fig. 3-25 Logic diagram for the 16L8 PAL. (*Courtesy of Monolithic Memories Inc.*)

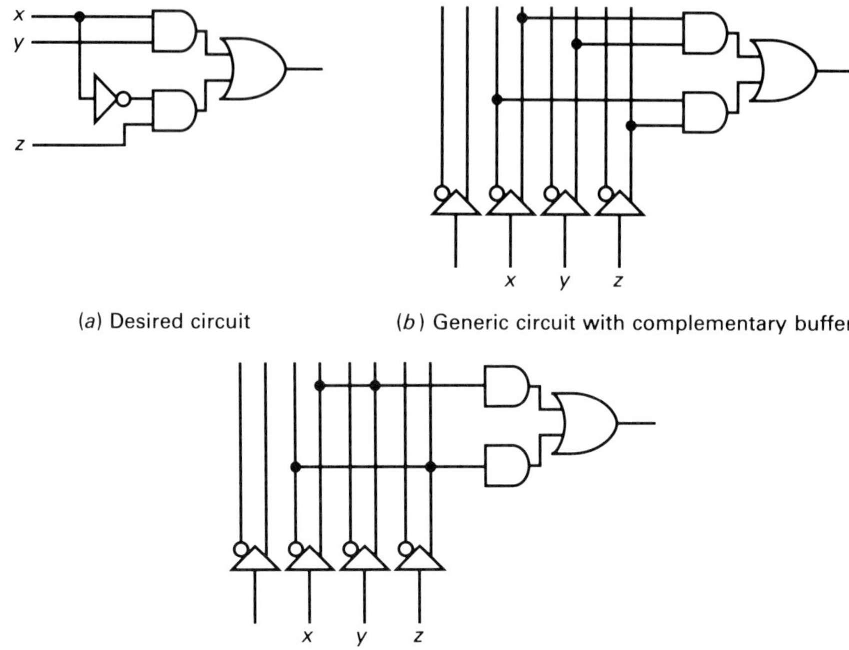

(*a*) Desired circuit (*b*) Generic circuit with complementary buffers

(*c*) AND-array notation

Fig. 3-26 AND-array notation used in PLD diagrams. AND gate inputs are represented as a single line with multiple connections made to it.

PALs use a compact form of input representation called *AND-array notation*. Figure 3-26 shows how AND-array notation relates to actual circuit diagrams. Gate arrays, PLDs, and EPLDs are discussed in greater detail in Chapter 9.

A fourth level of logic design uses *standard packaged parts*. The most commonly used standard parts are found in the 54/74 logic series. Part numbers with the 54 prefix are manufactured to military specifications while the 74 prefix indicates a commercial-grade product. The one to three characters following the prefix indicate the technology, or logic family, the part belongs to. Examples are LS (low-powered Schottky) and AC (advanced CMOS). The final two or three numbers indicate the function of the part. For example, the 74LS00 is a package containing four low-power Schottky NAND gates manufactured to commercial specifications. The 74F00 (FAST) is the logic equivalent of the 74LS00 and is pin-for-pin compatible but operates at much higher speeds. The same pinout, package, and logic function is also available in two CMOS families with part numbers 74HC00 (high-speed CMOS) and 74AC00 (Advanced CMOS).

The major advantage of standard parts is cost. Standard packages are manufactured in large quantities and are available at low cost. The least expensive of the standard packages are the *small scale integrated circuits* (*SSI*) containing 12 or fewer logic gate equivalents. One logic gate equivalent is any circuitry with the same number of transistors as a two-input NAND gate. Figure 3-27 shows the contents of a typical SSI package. *Medium-*

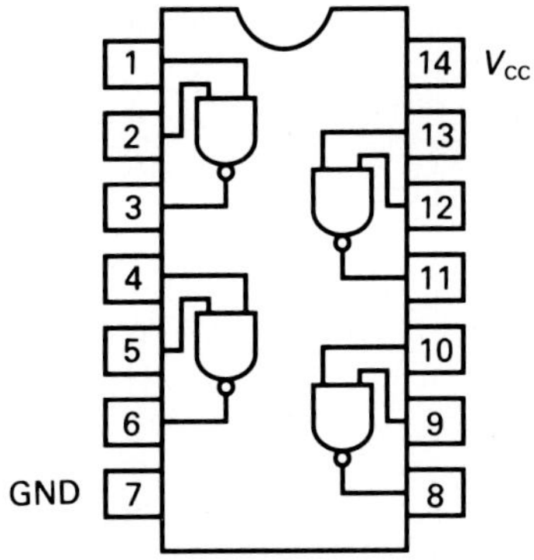

Fig. 3-27 Pin configuration for a standard SSI package. This example is a 74xx00 quad two-input NAND gate.

scale integrated circuit (*MSI*) packages contain gate equivalencies between 12 and 100. MSI packages are more expensive than SSI packages but cost *less per gate*. *Large-scale integrated circuit* (*LSI*) packages contain more than 100 equivalent gates at an even lower cost per gate. *Very large scale integrated circuits* (*VLSI*) contain thousands of equivalent gates and provide the most economical scale of standard part integration. Standard cell and gate array designs are classified as *semicustom* LSIs or VLSIs.

The 54/74 series of standard parts are available in *dual in-line packages* (*DIP*), and *surface-mounted-device packages* (*SMD*). DIPs and SMDs provide a means of interfacing the very small input and output pads of the silicon die to the much larger world of printed circuit boards. Figure 3-28

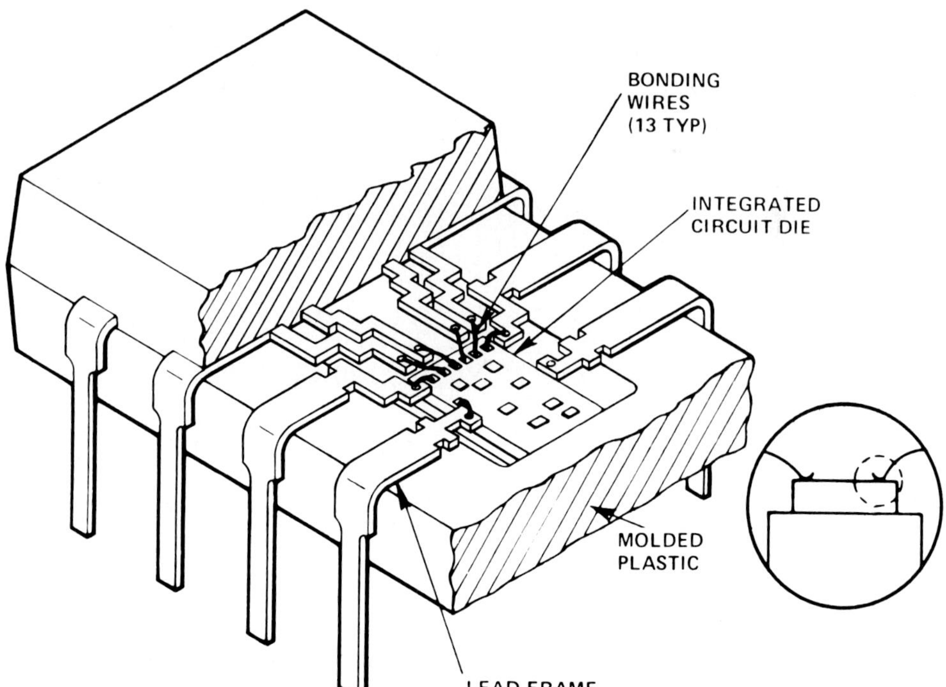

Fig. 3.28(*a*) Dual In-line Package (DIP). (*Courtesy of Fairchild Semiconductor Corporation*)

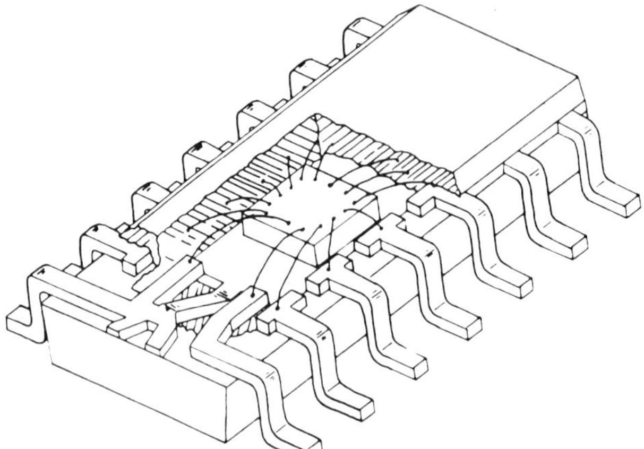

Number	Leads Pitch	Tip to Tip	Length	Body Width	Height
14	.050	.236	.340	.154	.061
16	.050	.236	.390	.154	.061

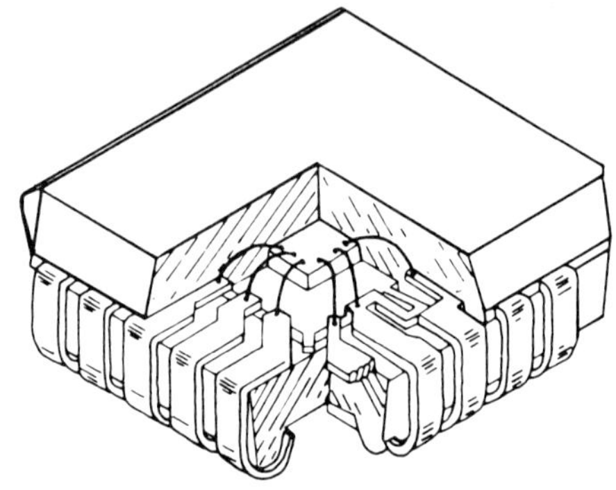

Surface Attached Plastic Chip Carrier (PCC)

Number of Leads	Lead Pitch	Dimensions Side x Thickness
20	.050	.385 Sq. x .172
28	.050	.485 Sq. x .172
44	.050	.685 Sq. x .172

Fig. 3.28(*b*) Small Outline Integrated Circuit (SOIC). (*c*) Leadless Chip Carrier (LCC) or Plastic Chip Carrier (PCC). (Courtesy of Fairchild Semiconductor Corporation)

shows a comparison between the DIP and two types of SMD packages. There is very little difference between the DIP and the *small outline integrated circuit* (*SOIC*). Both DIPs and SOICs use the same materials and assembly technology. The main differences are that the SOIC is smaller and has a lead bend which allows it to be mounted on the surface of a circuit board instead of being inserted through drilled holes. The SOIC is roughly one-fourth as large as the DIP, and this permits greater circuit density. The other SMD package shown in Fig. 3-28, the *leadless chip carrier* (*LCC*), is square with contacts on all four sides. Special sockets are required to mount these packages. All the parts described come in both plastic and ceramic packages, but when plastic is used for the LCC, it is called a *plastic chip carrier* (*PCC*).

REVIEW QUESTIONS

1. What does it mean to say that a switch is *binary*?
2. What is another name for V_{ss}?
3. Explain DeMorgan's theorem using algebraic examples.
4. What is the *dual* of a function?
5. What is the function of a logic "bubble"?
6. Explain the algebraic basis of the Karnaugh map technique of logic minimization.

7. What is a *nonessential implicant*?

8. How can an EXCLUSIVE OR type function be spotted in a Karnaugh map?

9. What is a PLD? How is it different from an EPLD?

10. How do gate arrays differ from semicustom ICs designed with standard cells?

11. Explain the numbering system for standard parts in the 54/74 series.

12. Define the following abbreviations: DIP, SMD, SOIC, and LCC.

REVIEW PROBLEMS

1. A two-way switch commonly used in house wiring is illustrated in Fig. 3-29. The switch variables are defined as 0 when the switch in is the

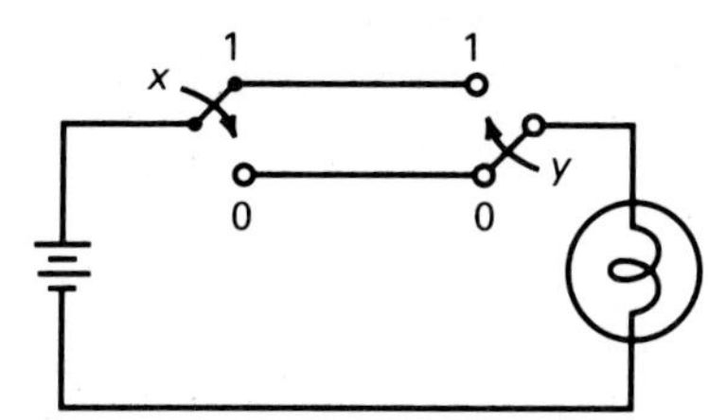

Fig. 3-29 Two-way switch for Prob. 1.

down position and 1 when the switch is in the up position. Construct a truth table for logic function f such that $f = 1$ when the light is on and $f = 0$ when the light is off.

2. Use simple switching circuits to show why:

$$\begin{aligned} &\text{(a) } x + x = x \\ &\text{(b) } x \cdot x = x \\ &\text{(c) } x + xy = x \end{aligned}$$

3. Look at the two Karnaugh maps shown in Fig. 3-30. Which map is labeled *incorrectly*? Why does it matter?

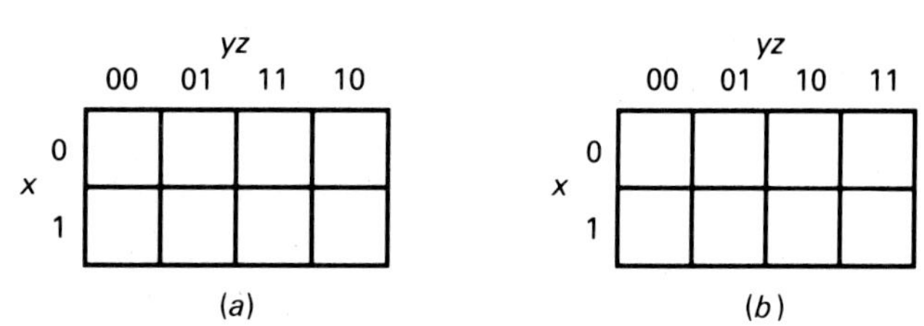

Fig. 3-30 Karnaugh maps for Prob. 3.

x	y	z	f
0	0	0	1
0	0	1	0
0	1	0	1
0	1	1	0
1	0	0	1
1	0	1	0
1	1	0	0
1	1	1	0

Fig. 3-31 Truth table for Prob. 4.

4. Simplify the truth table in Fig. 3-31 with a Karnaugh map. Draw the logic circuit using AND, OR, and inverter gates.

5. Simplify the truth table in Fig. 3-31 using the inverse function. The simplified sum-of-products expression is found by grouping 0s instead of 1s and inverting the result. Implement this logic circuit with only two gates. Is the resulting circuit equivalent to the circuit in Prob. 4?

6. Use the graphical DeMorgan's technique to reduce the number of inverters needed in the circuit implementation developed in Prob. 4.

7. Suppose we are discussing whether two people, Mr. A and Mr. B, are in a room. Use a truth table to demonstrate why the following statements are equivalent: (*a*) "Either A is *not* here *or* B is *not* here," and (*b*) "A *and* B are *not* both here." What boolean property does this verify?

8. Implement the multiplexer circuit in Fig. 3-17 entirely with NAND gates. What 74 series standard package could be used for this?

wx \ yz	00	01	11	10
00	0	0	0	0
01	1	1	0	1
11	0	1	1	0
10	0	1	1	0

Fig. 3-32 Karnaugh map.

9. Given the Karnaugh map in Fig. 3-32, find the minimum sum-of-products expression and implement this expression with logic gates. Eliminate nonessential implicants.

10. Repeat Prob. 8 for the Karnaugh map in Fig. 3-33.

11. Design a *majority voting gate* which looks at three inputs and produces a 1 whenever the majority of the inputs equal 1. Implement with logic gates.

12. Design and implement a *parity checker* which looks at four inputs and produces two outputs, *ODD* and *EVEN*. $ODD = 1$ when there is an odd number of 1s in the input and $EVEN = 1$ when there is an even number of 1s in the input. How are the two outputs related to each other?

wx \ yz	00	01	11	10
00	1	1	0	0
01	1	1	1	1
11	0	1	1	0
10	0	1	1	0

Fig. 3-33 Karnaugh map.

13. Design and implement a *2-bit magnitude comparator*. This circuit will have four inputs, A_1, A_0, B_1, and B_0 and three outputs. The circuit will compare the 2-bit binary numbers and determine which is larger on the basis of the following inequality: $11 > 10 > 01 > 00$. The three outputs will correspond to the following:

$$A_1A_0 = B_1B_0$$
$$A_1A_0 > B_1B_0$$
$$A_1A_0 < B_1B_0$$

14. Draw a stick diagram layout for the EXCLUSIVE OR gate in Fig. 3-24*b*.

15. Design an EXCLUSIVE NOR gate and draw a stick layout. How can the cell layout for an EXCLUSIVE OR be easily changed into an EXCLUSIVE NOR?

16. Find three EXCLUSIVE OR Karnaugh map patterns not shown in Fig. 3-23.

4 SEQUENTIAL LOGIC

The next level of circuit complexity, beyond transistor switches and combinational logic, is sequential logic. *Combinational logic circuits*, introduced in Chapter 3, accept changes in external inputs and produce new outputs based on those changes. *Sequential logic circuits* also accept new external inputs but produce new outputs based on both the external inputs *and* the present state of the circuitry. Sequential networks are *state machines* which use the present state of the circuit to determine the next state. Sequential machines use *delays* to preserve the present state of the machine long enough to be used as input for the next state.

There are two categories of sequential logic, asynchronous and synchronous. *Asynchronous* sequential logic uses propagation delay to hold the outputs long enough to be used as inputs. This can be tricky. Asychronous logic design must be used with care. A safer approach is to use synchronous sequential logic. *Synchronous logic* uses a clock to measure and equalize the delays. A *clock* is a waveform which makes high and low excursions at a fixed periodic rate. Clocked delay devices can be used to hold the present state of an output until the next transition of the clock. In this way the delays are fixed and of equal length. The objectives of Chapter 4 are to:

1. Clarify the concept of *delay* in sequential logic.
2. Illustrate asynchronous sequential design with a simple example.
3. Present a methodology which will make it easy to design synchronous sequential networks with assurance that they will work reliably.

4-1 ASYNCHRONOUS SEQUENTIAL LOGIC

Asynchronous logic design techniques, in general, are beyond the scope of this text. Asynchronous design is practiced mainly by two groups: those who know a great deal about logic design and those who know very little. Great emphasis must be placed on eliminating hazards and timing abiguities in asynchronous logic, and the resulting circuits are usually more complex than equivalent synchronous networks. In asynchronous design, cost and logic minimization are secondary considerations. At the printed circuit board level, asynchronous design is rarely practiced because very little control is possible over the delays in packaged parts and interconnections. At the chip level, asychronous design is sometimes practiced but only when available clocks are not fast enough for synchronous activity and then only when inputs can be carefully controlled. In this section, we will present a very simple example of asynchronous design, the *RS latch*.

Design techniques for all sequential networks were developed by D. A. Hoffman in 1954 and E. J. Moore in 1956. Since then a standard model has emerged for building *any* sequential state machine. A *sequential state machine* may be synchronous or asynchronous, depending upon the type of delays used. The key concept in sequential logic is *delay*. Figure 4-1

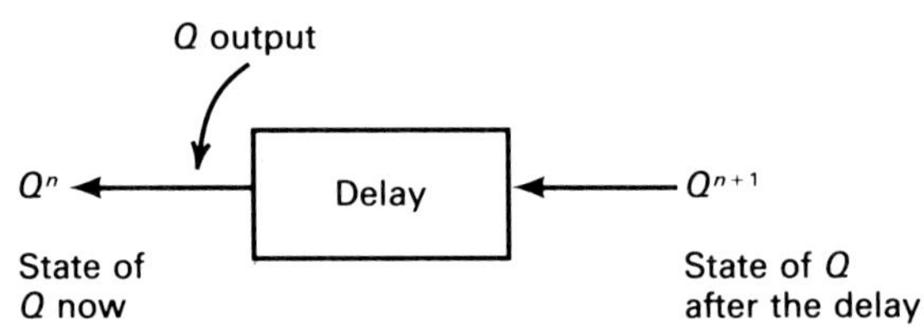

Fig. 4-1 The delay element causes any changes at its input to be delayed by a fixed amount of time.

illustrates this concept with the terminology commonly used for the Hoffman-Moore model. A logic signal changes at the input of a digital delay device at time $t = n$ but the change will not affect the output until time $t = n + 1$. The letter n represents the time now, and $n + 1$ represents the time now *plus one delay*. The time now is called the *present state*, and the time after one delay is called the *next state*. The output of the device is called Q, and it is labeled Q^n because it is the output now. The input to the device is labeled Q^{n+1} because it is the value the output will assume after one delay. The device that causes this delay could be a simple logic gate. We know that all logic gates exhibit propagation delay. However, the propagation delay through a gate is not a well-controlled quantity. It varies widely between parts of the same type. The uncertainty associated with delays through logic gates makes the design of complex asynchonous logic very difficult.

Figure 4-2 shows the Hoffman-Moore model for a machine with one delay device, two external inputs, and one output. In general, there may

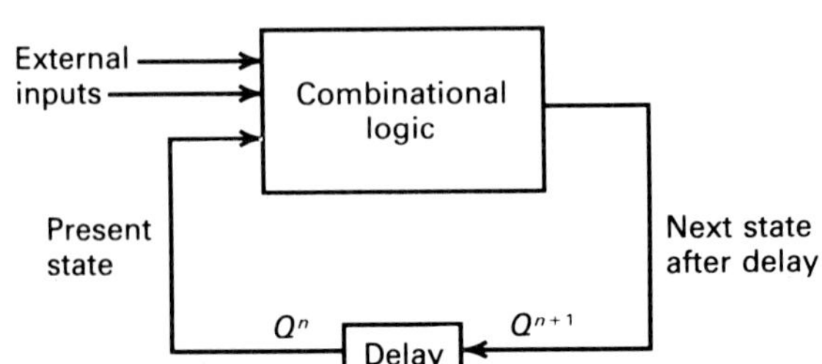

Fig. 4-2 The Hoffman-Moore model of a sequential state machine. This example has only one state variable and two external inputs.

be any number of inputs and outputs, and the number of delay devices depends only on the number of states a machine must sequence through. The *state* of a sequential machine is the set of values held by the delay devices. The present state represents the outputs of the delay elements now, and the next state represents the output of the delay devices after one delay. The *state variables* of a machine are the outputs of the delay devices. The state machine in Fig. 4-2 has one state variable, Q. The present state of Q is labeled Q^n, and the next state of Q is labeled Q^{n+1}.

Once the number of state variables is determined, a machine may be designed using truth tables and Karnaugh maps. The truth table lists all possible combinations of external inputs and present state values under the heading of "inputs." The combinations are listed this way because they are inputs to the combinational logic which drives the delay elements. Under "outputs" are listed the *next* state values of the machine and any special *decoded outputs* which may be needed. Karnaugh maps are used to derive a simplified sum-of-products expression for each delay element.

EXAMPLE 4-1

Design a state machine which has two external inputs, R (*RESET*) and S (*SET*), and two states, 0 and 1. A momentary low on the R input should produce a 0 state, and a momentary low on the S input should produce a 1 state. Simultaneous lows on both R and S are not allowed. Implement this state machine using NAND gates.

SOLUTION

Only two states are required, 0 and 1, so only one state variable is needed. The truth table in Fig. 4-3 lists all possible combinations of external inputs

External Inputs		Present State	Next State
S	R	Q^n	Q^{n+1}
0	0	0	x
0	0	1	x
0	1	0	1
0	1	1	1
1	0	0	0
1	0	1	0
1	1	0	0
1	1	1	1

S \ RQ^n	00	01	11	10
0	x	x	1	1
1	0	0	1	0

$$Q^{n+1} = \overline{S} + RQ^n$$

Fig. 4-3 Design of a simple asynchronous circuit.

and present-state outputs. The desired combinational logic will produce the next-state outputs. The truth table is copied over into a Karnaugh map for minimization and reduced to a simplified sum-of-products expression. The resulting logic can be implemented in NAND gates. Figure 4-4 shows the direct logic implementation together with the steps necessary to convert to NAND implementation. The resulting asynchronous state machine is an active-low RESET/SET (RS) *latch*. It is "active-low" because only lows at the input will cause state changes. High inputs cause no change.

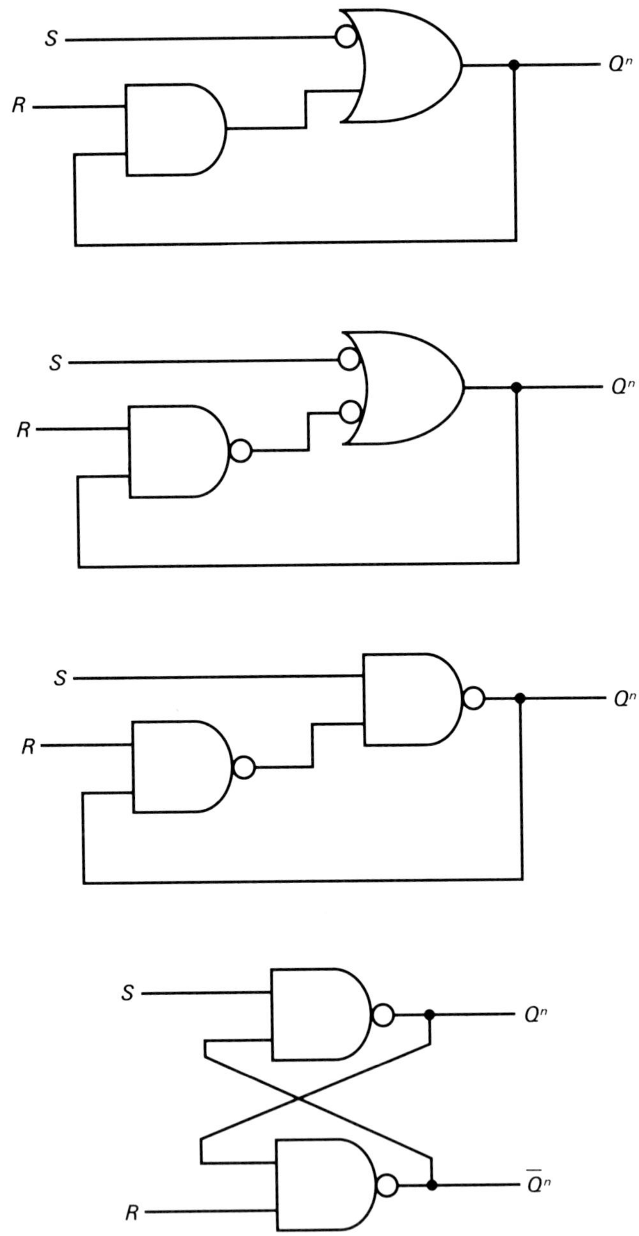

Fig. 4-4 Implementation of an RS flip-flop using NAND gates.

The RS latch is a memory storage device in the sense that an input can return to its high state and the output will remember that the momentary low had occurred. The output will remain in the desired state even after the active-low input has been removed.

Despite the simplicity of the RS latch, it already exhibits one of the drawbacks of asynchronous logic. If two inputs change at the same time, the result is not predictable. If the exact same momentary low were applied to both inputs, the output would be *indeterminate*. This means that after the momentary low inputs were removed, we would have no way of knowing which state the circuit would settle to. For as long as both R and S are low, the outputs of both NAND gates will be high; but after the low input is removed, there will be highs on all NAND inputs, and a race will ensue between the gates. The gate with the shortest propagation delay will force the other high. Both NANDs cannot ultimately achieve the same state. Oscillations may occur, with outputs changing back and forth, but in the end, one must settle low and the other high. The problem is that we have no way of knowing how it will turn out. This circuit can be used only when we are sure that only one input can change at a time.

When the two inputs of the RS latch change at the same time, the network suffers a race condition. A *race condition* is a particular type of *hazard*. In Chapter 3 we described one type of hazard. It was caused by a single changing input encountering unequal delays. In general, all hazards are real, or potential, malfunctions caused by unequalized delays in a network. Whenever two or more changing inputs race toward a common point, there is a potential for temporary false outputs. Even if the paths contain an equal number of logic gates, unequal switching times of the gates themselves may cause false outputs of short duration. Since the present state of the machine will determine the next state, false outputs of even short duration can cause the machine to sequence to the wrong state. In asynchronous logic, temporary outputs can determine the final state of the machine. Thus asynchronous logic must be carefully designed so that temporarily false outputs can never cause the machine to sequence to an incorrect final state.

REVIEW 4-1

1. **Q.** What is the present state of a delay device? **A.** The present state is the value of the output at any instant.

2. **Q.** What is the next state of a delay device? **A.** The next state is the value of the output after some delay.

3. **Q.** What is the state of a sequential machine? **A.** The state of a machine is the combined outputs of all the delay devices in the machine.

4. **Q.** What is a hazard? **A.** A hazard is a real or potential malfunction caused by unequalized delays in a network.

5. **Q.** What is the basic problem with asynchronous sequential logic? **A.** Any temporarily false outputs may cause the machine to malfunction.

4-2 SYNCHRONOUS LOGIC

One way of solving the problems of predictability in the RS latch is to make sure that the two inputs never change at the same time. This may not always be possible. A better solution is to eliminate predictability problems by requiring that inputs and outputs change only at specified times. A signal called a *clock* is used to define exact moments when the machine may change state. False outputs of short duration will still occur, but only during the moments immediately after the change, and they will settle out before the next change is allowed. Thus the false outputs will not influence the next change of state. The only requirement for reliable design is that all intermediate states have time to settle between clock transitions. If this condition is met, hazards and race conditions will not cause malfunction.

Figure 4-5 shows a clock waveform. Usually the clock has a 50 percent duty cycle, which means that half the time the signal is low and the other

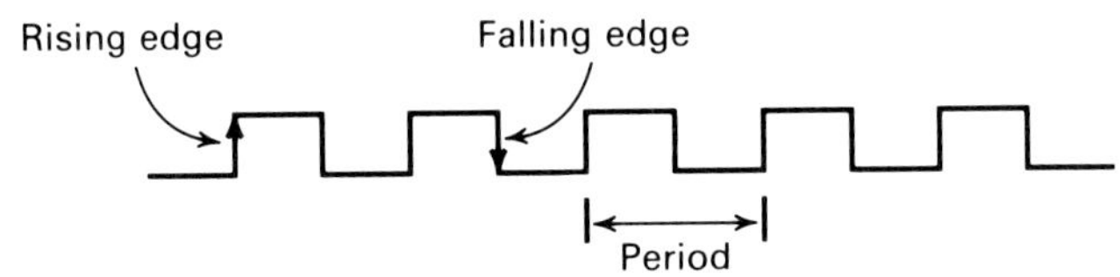

Fig. 4-5 Clock waveform.

half it is high. A 50 percent duty cycle is seldom necessary, however, because in fully synchronous logic design, changes occur only on clock edges. A system may be designed so that all changes occur *either* on the *rising edge* or on the *falling edge* of the clock. Delay devices in such a system are called *edge-triggered flip-flops*. They are "edge-triggered" because they change only on either the rising or falling edge of the clock, and they are called "flip-flops" because their outputs flip and flop between two states. In this text we will assume that changes occur only on the rising edge of the clock. When the term *clock edge* is used, it will refer to the rising edge. The *period* of the clock refers to the time between rising clock edges. In a fully synchronous system, *one delay* is equal to the period of the clock.

The most basic clocked delay element used in synchronous sequential networks is the *delay flip-flop* (also called a "*D flip-flop*" or *D flop*). Flip-flops are also referred to as "storage elements" because they *store* logic values for specified amounts of time. For this reason, the D flip-flop is sometimes called a "data flip-flop" because it stores data. The D flop is constructed with two clocked D latches. A clocked *D latch* is just a special case of the clocked RS latch shown in Fig. 4-6. Notice that the use of a clock as an input to the RS latch limits the time that the R and S inputs can influence the output. In this case the outputs change on momentary high inputs while the clock is high. When the S input is inverted and applied to the R input, we have a D latch. During the time the clock is high, the output of the D latch will follow the input. When the clock falls, whatever input was last applied will determine the final state of the latch until the next rising edge occurs. In order to guarantee that the output holds the correct D input, the desired value at the input must be present

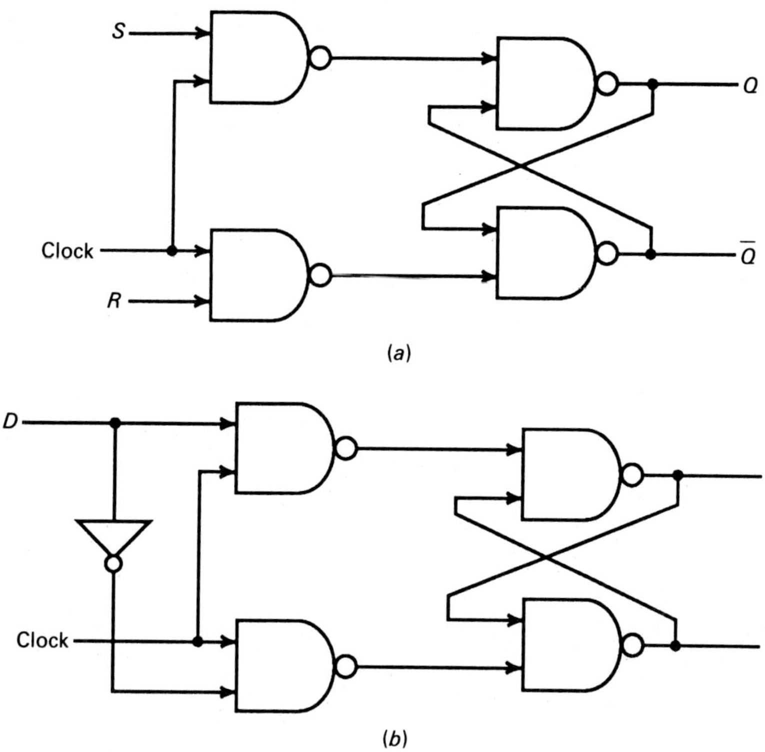

Fig. 4-6(*a*) Clocked RS latch. Output may change only when the clock is high. (*b*) Clocked D latch. The output follows the input while the clock is high. When the clock falls, the last value set up on the input is held until the next rising edge.

long enough before the fall of the clock to set *up* the output. The amount of time that the *D* input must be stable before the clock transition is called the *set-up time*. This time is a function of the gate delays in the network.

The D latch is clocked, but it is not edge-triggered. Its output changes soon after the clock rises and may continue to change for as long as the clock is high. The D latch is sometimes referred to as a *transparent latch* because while the clock is high, the circuitry between input and output is transparent. If the clock in Fig. 4-6 were inverted, the latch would be transparent while the clock was low. Transparent latches have applications in the storage of data, but they are not suitable for use as delay devices in synchronous networks. In synchronous networks outputs must change only on clock edges. This can be achieved by placing two clocked latches in series and using both phases of the clock as shown in Fig. 4-7. When the first latch is loading a new input, the second is holding the old output. Referring to Fig. 4-7, if the *clock* and the *inverted clock* switch at the same time, the effect is that the new value is passed to the output on the rising clock edge and is held there until the next rising edge. Thus it appears to the outside world that data is being entered only on rising edges. The input latch is often called the *master*, and the output latch, the *slave*. This circuit is sometimes referred to as a *master-slave flip-flop*.

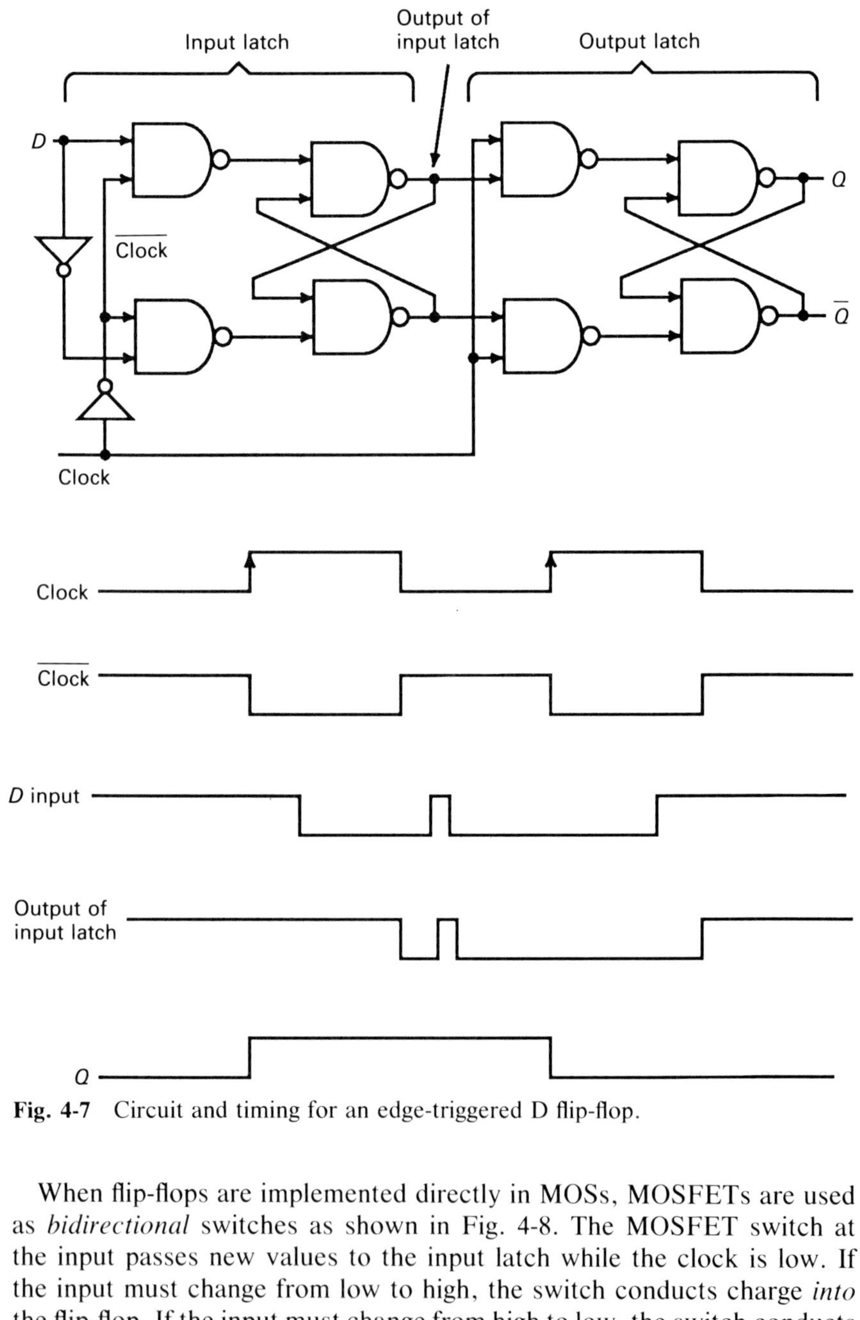

Fig. 4-7 Circuit and timing for an edge-triggered D flip-flop.

When flip-flops are implemented directly in MOSs, MOSFETs are used as *bidirectional* switches as shown in Fig. 4-8. The MOSFET switch at the input passes new values to the input latch while the clock is low. If the input must change from low to high, the switch conducts charge *into* the flip-flop. If the input must change from high to low, the switch conducts charge *out of* the flip-flop. This approach to building flip-flops greatly simplifies the design and reduces space and power requirements.

The D flip-flop is the basic building block for all sequential state machines. Any state machine which sequences through 2^m states or fewer can be designed with *m D flip-flops* and some combinational logic. The next-state values serve as inputs to the D flip-flops, and the D flip-flop outputs serve as inputs to the combinational logic. One flip-flop can assume one of two

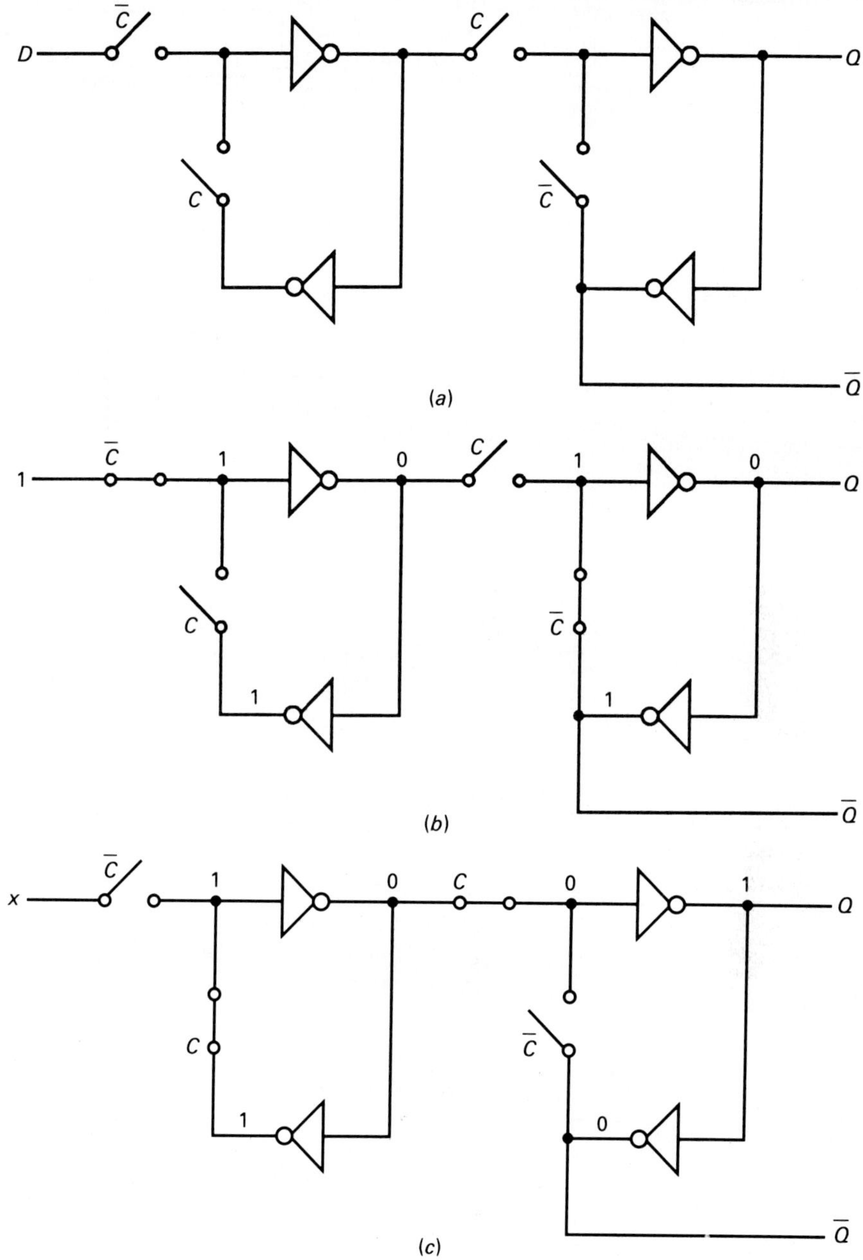

Fig. 4-8(*a*) MOS edge-triggered D flip-flop constructed with inverters and switches. (*b*) While the clock is low, the input latch accepts new data, *but the old output is unaffected.* (*c*) On the rising edge of the clock, the new data is passed on to the output, *while changes at the input are ignored.*

possible states, 0 or 1. Two flip-flops can assume four possible states, 00, 01, 10, and 11. In general, m delay devices can assume 2^m possible states, and m delay devices are required for any number of states less than or equal to 2^m. This will become clear as you work through the examples. This is also true for other, more complex types of flip-flops such as JK

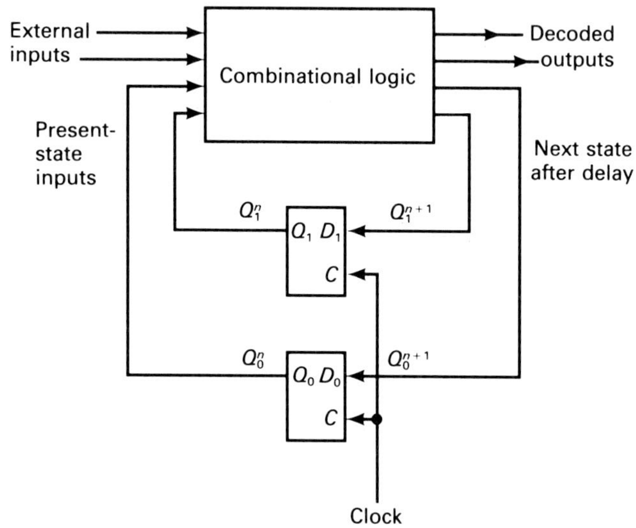

Fig. 4-9 Hoffman-Moore model using D flip-flops.

flip-flops and T flip-flops. Figure 4-9 shows the Hoffman-Moore model as it applies to the D flip-flop . The design procedure, using this model, is as follows:

1 Define the problem in words, pictures, or timing diagrams.
2 Draw a state diagram.
3 Make a binary state assignment.
4 Construct a truth table.
5 Use a minimization technique, such as Karnaugh maps, to derive state equations (simplified sum-of-products expressions for the next state).
6 Translate the state equations into a circuit diagram.

This methodology can be illustrated with a simplified traffic light example. Once the process is understood, a more realistic traffic light problem can be approached as an exercise. In the simplest case, three lights turn on in sequence: green, then yellow, then red, then green again. This statement satisfies step 1 in the design process. The second step involves drawing a sketch showing how the lights change. This is called a *state diagram* because each output—green, yellow, and red—represents one state of the machine. Drawing a state diagram can be very helpful, especially when complex problems must be reduced to simple digital logic. It forces us to state the problem logically, in terms of machine states. We must then make a *binary state assignment* to associate each state, or color in this case, with a pattern of 1s and 0s. Figure 4-10 shows the state diagram and one possible state assignment. Since there are three possible states, two delay devices are needed, each represented by a binary digit. Each binary digit in the state assignment corresponds to the output of a D flip-flop.

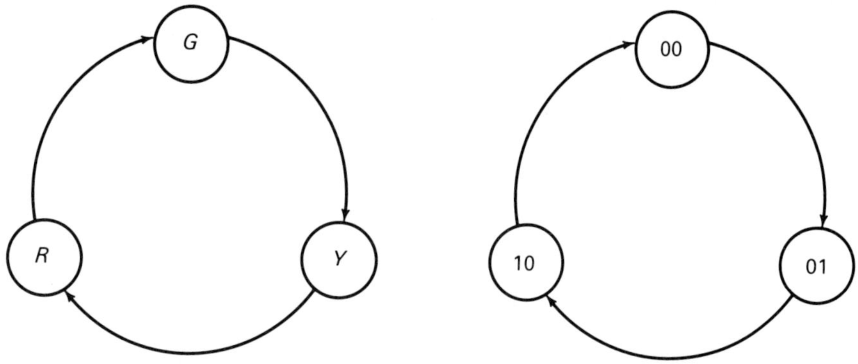

(*a*) State Diagram

(*b*) State Assignment:

Green = 00
Yellow = 01
Red = 10

Fig. 4-10 State diagram and binary state assignment.

Figure 4-11 shows the state table and Karnaugh maps for this example. A state table is just a truth table used to derive sum-of-products expressions for state machines. Notice that one possible state is not needed, and this one is marked as a don't-care condition in the next-state column. The outputs of the combinational logic are listed under "next state" and

External Inputs (None)	Present State Q_1^n	Q_0^n	Next State Q_1^{n+1}	Q_0^{n+1}	Decoded Outputs G	Y	R
X	0	0	0	1	0	1	1
	0	1	1	0	1	0	1
	1	0	0	0	1	1	0
	1	1	x	x	1	1	1

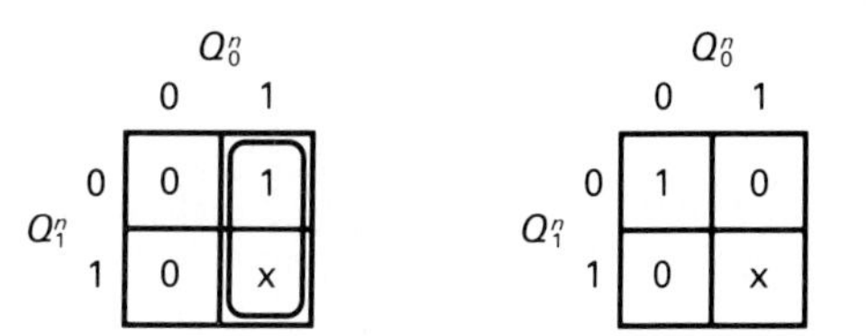

$Q_1^{n+1} = Q_0^n$

$Q_0^{n+1} = \overline{Q_1^n} \cdot \overline{Q_0^n}$

$= \overline{Q_1^n + Q_0^n}$

Fig. 4-11 Truth table and Karnaugh maps for simplified traffic light sequencer.

"decoded outputs." The next-state outputs are used as inputs to the flip-flops while the decoded outputs are used to convert the binary present-state values into colored lights. If we used light-emitting diodes (*LEDs*) as indicators, we could turn on a colored light with the circuit shown in Fig. 4-12. The use of LEDs is discussed in greater detail in the appendix. When

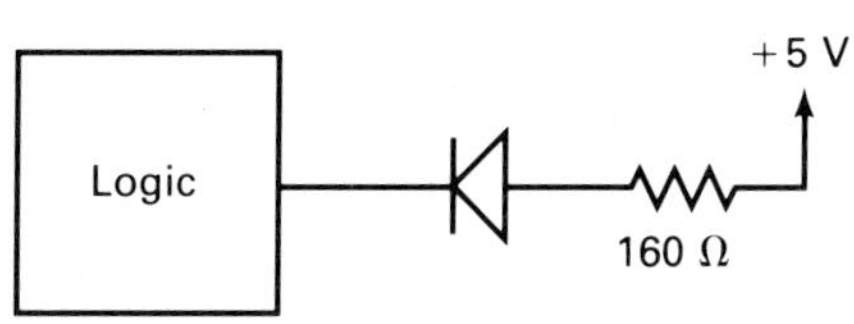

Fig. 4-12 Circuit for turning on light-emitting diodes. When logic output is low, the LED turns on. When logic output is high, the LED turns off.

the present state is the green state, the logic pulls the green decoded output low and turns on the green LED. There is no need, in this case, to construct Karnaugh maps for the decoded output logic. The complete implementation of the simplified traffic light problem is shown in Fig. 4-13.

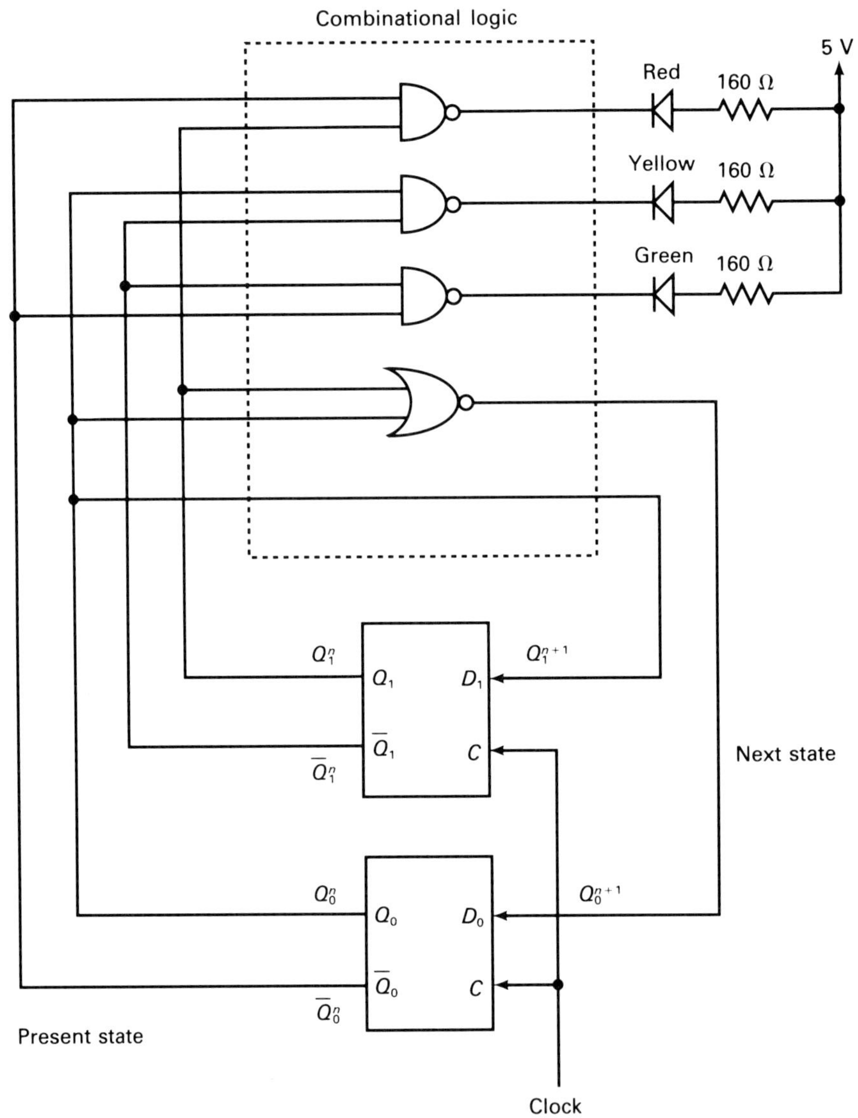

Fig. 4-13 Implementation of the simplified traffic light sequencer using the Hoffman-Moore model.

There is a part of the design strategy which is implicit in step 5 but not actually spelled out. What do we do with the don't-care conditions? We mentioned in Chapter 3 that these could be assigned to achieve the simplest implementation. The justification for this is that, since the don't-care combination of inputs will never occur, we don't really care what outputs they produce. This is true of purely combinational logic, but in sequential circuitry there is a possibility that illegal states *could* occur at power-up or during power line glitches. There is no way to know how the flip-flops will "come up" when power is turned on or how they will be affected by power line perturbations. Since we cannot control the flip-flops during power transitions, the circuit must be designed so that it cannot stay in an illegal state for very long after power has been restored. In general, it is permissible to start out in an illegal state as long as the proper sequence is entered within a reasonable number of clock periods. We must allow for these possibilities in the state diagram. In the traffic light, for example, if both don't-care conditions had been replaced with 1s, there would have been a *hang-up state*. The state table shows that the 11 state would have sequenced back into itself without ever entering the proper sequence. Figure 4-14 is a better state diagram than Fig. 4-10 because it accounts for

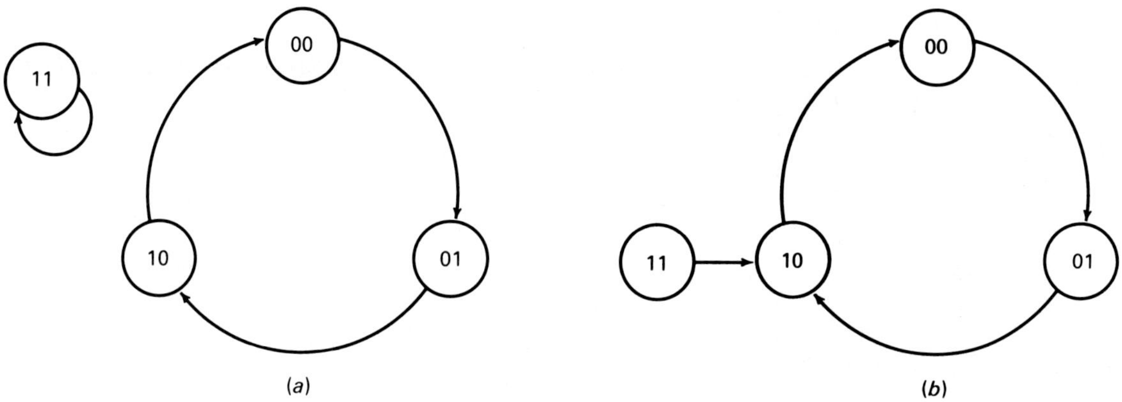

Fig. 4-14 Hang-up states. (*a*) Poor design contains a hang-up state. (*b*) Proper design prevents hang-up.

all possibilities. Figure 4-14*b* shows that the implementation actually chosen was a good one since state 11 will return to the proper sequence on the next clock edge. It is permissible to sequence through one or more illegal states as long as they lead to the proper sequence within a reasonable amount of time.

Many state machines have external inputs which help to determine the next state. These can be accounted for in the state diagram with the notation shown in Fig. 4-15. The digit (or digits) before the slash indicates

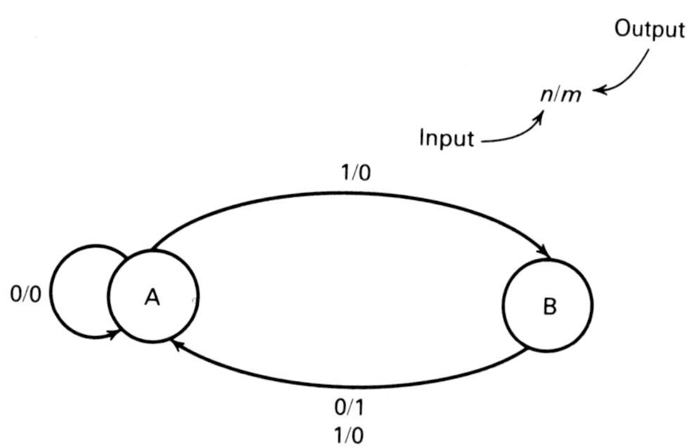

Fig. 4-15 State diagram notation.

the input conditions. The digit (or digits) after the slash indicates the output conditions. For example, the directed line segment from state A to state B is labeled 1/0. This indicates that if the circuit is in the present state A, with an input of 1, then its output is 0, and on the next clock edge it will advance to state B. Likewise, the input may be 0 with an output of 0, and on the next clock edge the circuit will stay in state A. Once in state B, either input will cause a change of state on the next edge, but the present output will depend on the present input.

EXAMPLE 4-2

Design a state machine which has the state diagram shown in Fig. 4-16.

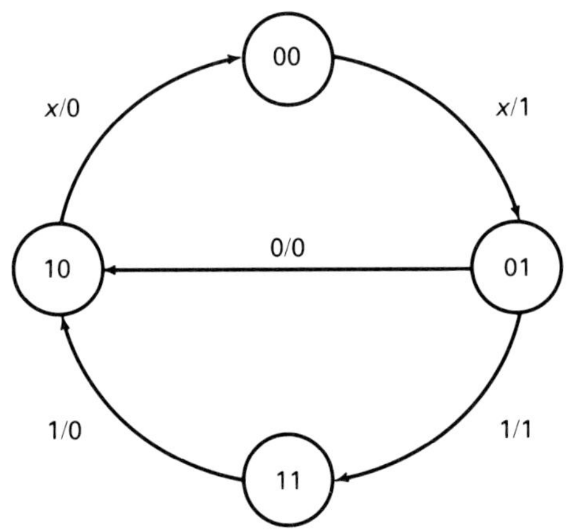

Fig. 4-16 State diagram for Example 4-2.

SOLUTION

The machine has one input, one output, and four possible states. This tells us that two flip-flops are needed. All we need to do is construct a truth table which satisfies the conditions specified in the state diagram. The truth table in Fig. 4-17 meets this requirement. (Figure 4-18 gives the

External Inputs X	Present State Q_1^n	Q_0^n	Next State Q_1^{n+1}	Q_0^{n+1}	Decoded Outputs Z
0	0	0	0	1	1
0	0	1	1	0	0
0	1	0	0	0	0
0	1	1	x	x	x
1	0	0	0	1	1
1	0	1	1	1	1
1	1	0	0	0	0
1	1	1	1	0	0

Fig. 4-17 Truth table for Example 4-2.

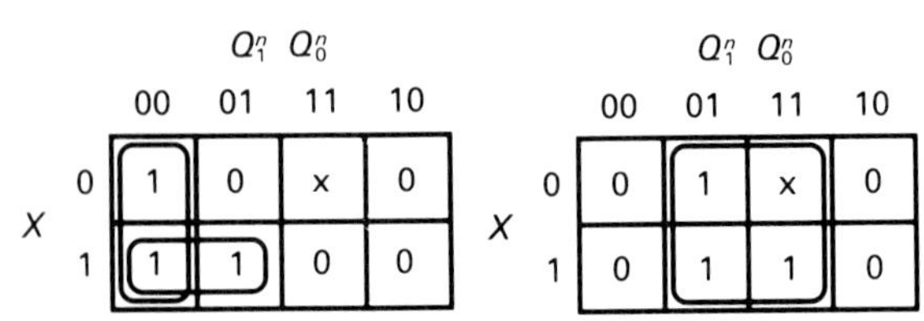

$Q_0^{n+1} = \overline{Q_1^n} \cdot \overline{Q_0^n} + X \cdot \overline{Q_1^n}$ $\qquad$ $Q_1^{n+1} = Q_0^n$

Fig. 4-18 Karnaugh maps for Example 4-2.

Karnaugh maps for this example.) The x's in the state diagram mean that either a 0 or a 1 will cause a state change. There are no hang-up states since every possible state leads to a legal sequence. The complete circuit is shown in Fig. 4-19.

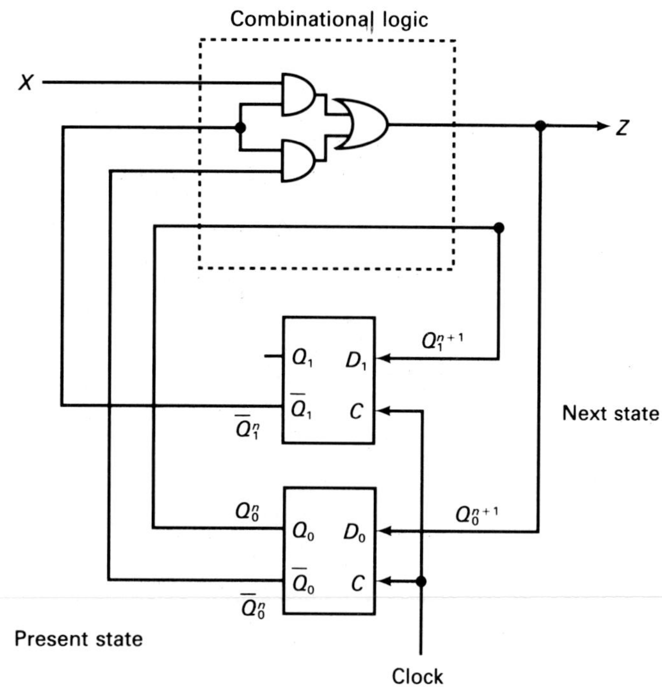

Fig. 4-19 Implementation of Example 4-2.

REVIEW 4-2

1. **Q.** How does sequential design solve the problem of temporarily false outputs? **A.** Outputs may change only on a clock edge and cannot influence the state of the machine until the next clock edge. As long as the outputs are stable at least one setup time before the next edge, no malfunction will occur.

2. **Q.** What signal measures and equalizes delays in a synchronous sequential circuit? **A.** the clock signal

3. **Q.** What device is used to hold the present state of the circuit between clock edges? **A.** an edge-triggered flip-flop

4. **Q.** What is the difference between a D flip-flop and a D latch? **A.** A D flip-flop output may change only on a clock edge; a D-latch output follows the input during one phase of the clock.

5. **Q.** Why is a D latch sometimes called transparent? **A.** because while the input is enabled, the output follows the input and the latch appears transparent

6. **Q.** How is a D flip-flop constructed? **A.** The D flip-flop is constructed with two D latches, each enabled on a different phase of the clock.

7. **Q.** Why does the use of MOSFETs simplify D flip-flop design? **A.** because MOS transistors conduct bidirectionally

8. **Q.** Why are state diagrams useful? **A.** They force us to define our problem in terms of machine states.

9. **Q.** Why must a binary state assignment be made? **A.** because machine states must be reduced to 1s and 0s before we can design the boolean logic needed to produce them

10. **Q.** If five states are needed, how many flip-flops are required? **A.** three

11. **Q.** What are hang-up states? **A.** illegal states which return only to themselves or to other illegal states without reentering the desired sequence

12. **Q.** How do hang-up states occur? **A.** by careless use of don't-care conditions

13. **Q.** How can hang-up states be avoided? **A.** by checking don't-care assignments after the Karnaugh map assignments have been made to see that every state leads to the desired sequence

14. **Q.** What is a state table? **A.** A state table is a truth table used to find the next state of each flip-flop as a function of the present state of the machine. The next state is applied to the inputs of the flip-flops.

15. **Q.** What are state equations? **A.** the simplified sum-of-products expressions which define the next state of each flip-flop

4-3 COMPLEX FLIP-FLOPS

Just as complex logic gates can sometimes be used to simplify combinational problems, complex flip-flops can sometimes be used to simplify sequential logic design. The generic complex flip-flop is the *enabled D flip-flop*. Other complex flops, such as *toggle* (or *T*) *flip-flops* and *JK flip-flops*, are just special cases of the enabled D flop. Figure 4-20 shows the state tables for the D flip-flop and the enabled D flip-flop. The D flip-flop accepts new values on every clock edge. The enabled D flip-flop accepts new values only when the *enable input* is equal to 1 (although it could be designed to enable on a 0). The state equation of the enabled D flop is an AND-OR function. Either the next state is a new value, or the next state is the present value. This allows us to hold data until an enable signal indicates that new data should be accepted. Enabled D flip-flops can be useful in digital systems where not every flip-flop changes states on every clock edge. Computers use this flip-flop structure extensively.

External Input x	Present State Q^n	Next State Q^{n+1}
0	0	0
0	1	0
1	0	1
1	1	1

$Q^{n+1} = x$

External Inputs E	x	Present State Q^n	Next State Q^{n+1}
0	0	0	0
0	0	1	1
0	1	0	0
0	1	1	1
1	0	0	0
1	0	1	0
1	1	0	1
1	1	1	1

$Q^{n+1} = \bar{E}Q^n + Ex$

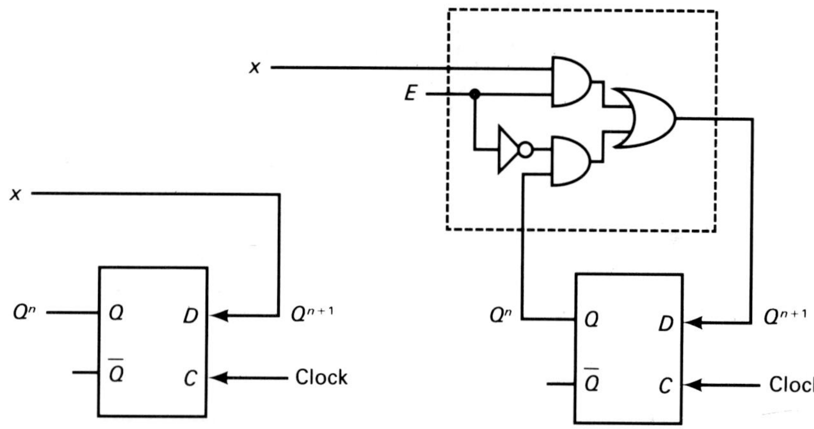

Fig. 4-20 Comparison of a D flip-flop and an enabled D flip-flop.

JK Flip-Flops

A special case of the enabled D flip-flop is the *JK flip-flop*. The state table and implementation of this circuit are shown in Fig. 4-21. A high at the J input while the K input is low will always cause an output high on the next clock edge. A high at the K input while the J input is low will always cause an output low on the next clock edge. If both inputs are high, the output will toggle on the next clock edge. *To toggle* means to change to

	External Inputs J	K	Present State Q^n	Next State Q^{n+1}
No change	0	0	0	0
	0	0	1	1
Output 0	0	1	0	0
	0	1	1	0
Output 1	1	0	0	1
	1	0	1	1
Toggle	1	1	0	1
	1	1	1	0

$Q^{n+1} = J\bar{Q}^n + \bar{K}Q^n$

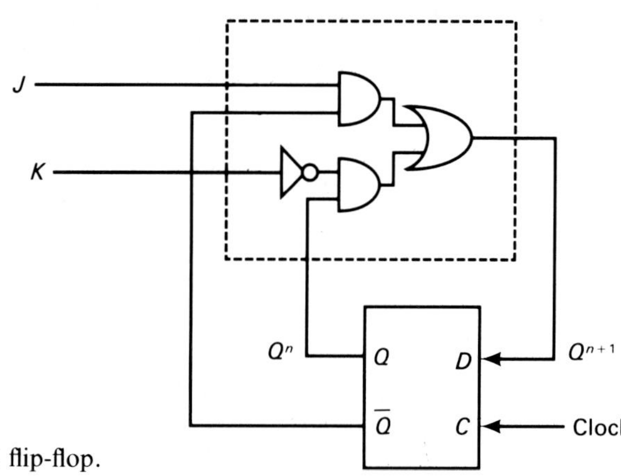

Fig. 4-21 Design of a JK flip-flop.

the opposite state. The Karnaugh map and gate implementation of this device show that it has the same AND-OR structure as the enabled D. The J input enables changes from 0 to 1, and the K input enables changes from 1 to 0. When both inputs are high a change will always occur. When both inputs are low, a change will never occur.

How does the JK flip-flop simplify logic design? Let us design the traffic light sequencer again but this time with JK flops. The problem and state diagram need not be repeated, but the state table will change somewhat. In Fig. 4-22, notice that the J and K inputs for each flip-flop must be

External Inputs (None)	Present State Q_1^n	Q_0^n	Next State Q_1^{n+1}	Q_0^{n+1}	JK Inputs J_1	K_1	J_0	K_0	Outputs G	Y	R
X	0	0	0	1	0	x	1	x	0	1	1
	0	1	1	0	1	x	x	1	1	0	1
	1	0	0	0	x	1	0	x	1	1	0
	1	1	x → 1	x → 0	x	0	x	1	1	1	1

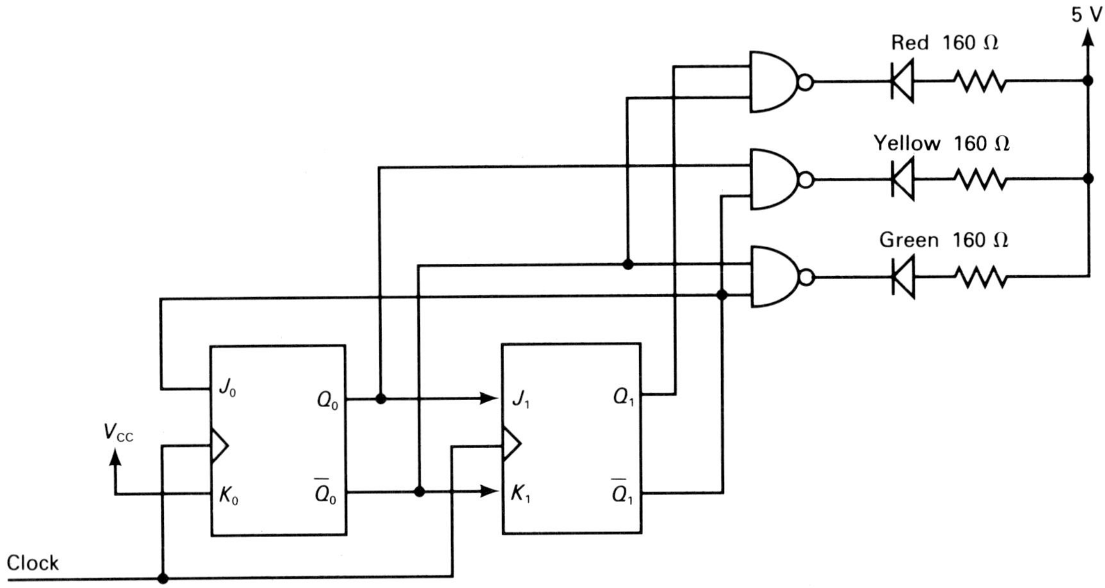

Fig. 4-22 Traffic light sequencer implemented with JK flip-flops.

treated as decoded outputs. Since J and K state equations must be obtained for each flip-flop, it appears that twice as many Karnaugh maps might be needed. But the large number of don't-care conditions often make it possible to avoid drawing a Karnaugh map altogether. For instance, the K_0 input could be tied high if 1s were substituted for x's in that column. The other inputs can be made to look like present-state outputs or their complements. Don't-care substitutions in the J and K columns will not lead to hang-up states, and so they may be used in any way that simplifies the design.

How did we arrive at the don't-care conditions? Well, with a little thought you can convince yourself that the following is true:

a If a JK flip-flop must change from a 0 to a 1, then the J input must be high but the K input may be either high or low.
b If a JK flip-flop must change from a 1 to a 0, then the K input must be high but the J input may be either high or low.
c If an output is 0 and must stay 0, then the J input must be low but the K input may be either high or low.
d If an output is 1 and must stay 1, then the K input must be low but the J input may be either high or low.

After the 1s, 0s, and x's are tabulated for each J and K input, try to make each J and K column look like one of the present-state columns (or its complement) so that existing flip-flop outputs can be used with no additional logic gates. When this is done for the traffic light problem, the NOR gate is no longer needed for sequencing the machine. Now, you may argue that we just replaced an external NOR gate with two complex internal AND-OR functions and actually complicated the design. This is true, and for this reason IC designers seldom use JK flip-flops. But at the packaged-parts level of logic design, it does not matter whether D flops or JK flops are used because the packages are the same size and dissipate the same amount of power. The internal complexity is transparent.

Toggle Flip-Flops

Another special case of the enabled D flip-flop is the *toggle* (or *T*) *flip-flop*. Figure 4-23 shows the state table for the T flip-flop. A high at the T

External Input T	Present State Q^n	Next State Q^{n+1}
0	0	0
0	1	1
1	0	1
1	1	0

$$Q^{n+1} = T\bar{Q}^n + \bar{T}Q^n$$
$$= T \oplus Q^n$$

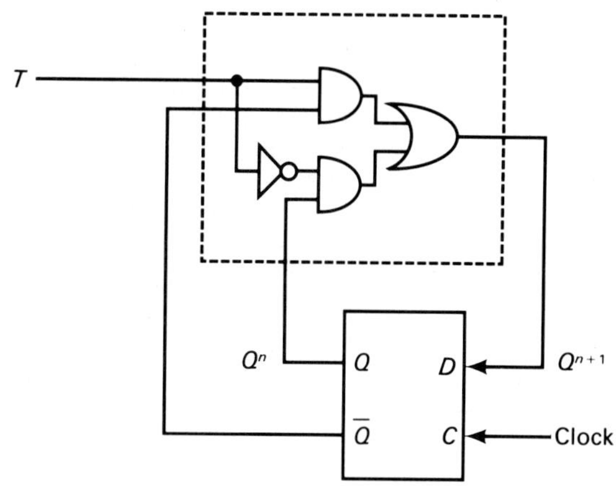

Fig. 4-23 Design of a toggle flip-flop.

input will cause the output to toggle on the next clock edge. The structure of the combinational logic needed to implement this flip-flop is identical to that of the enabled D and JK flip-flops. Designing with the T flip-flop is similar to designing with JK flops. The state table is drawn with T inputs for each flip-flop decoded from the present state. Whenever a flip-flop output must change, a 1 is entered in the T column. When no change is desired, a 0 is entered.

Toggle flip-flops are very useful in the design of binary counters. A *binary counter* is a sequential machine whose outputs follow the binary number sequence. For a two-stage counter this is 00, 01, 10, 11. A three-stage binary counter would count 000, 001, 010, 011, 100, 101, 110, 111. These are the binary number equivalents of the decimal integers 0, 1, 2, 3, etc. The simplified traffic controller machine is a counter but not quite a full binary counter. It counts the binary equivalent of 0, 1, 2, 0, 1, 2, etc., leaving out the 3 state. If we were to build the traffic controller with toggle flip-flops, we would find that the T flip-flops only complicate the design. However, T flip-flops greatly simplify the design of a full binary counter. Figure 4-24 illustrates the design of a two-stage binary counter built with T flip-flops. This is a fully synchronous binary counter. All changes occur on the rising clock edge.

External Inputs (None)	Present State Q_1^n	Present State Q_0^n	Next State Q_1^{n+1}	Next State Q_0^{n+1}	T Inputs T_1	T Inputs T_0
	0	0	0	1	0	1
	0	1	1	0	1	1
	1	0	1	1	0	1
	1	1	0	0	1	1

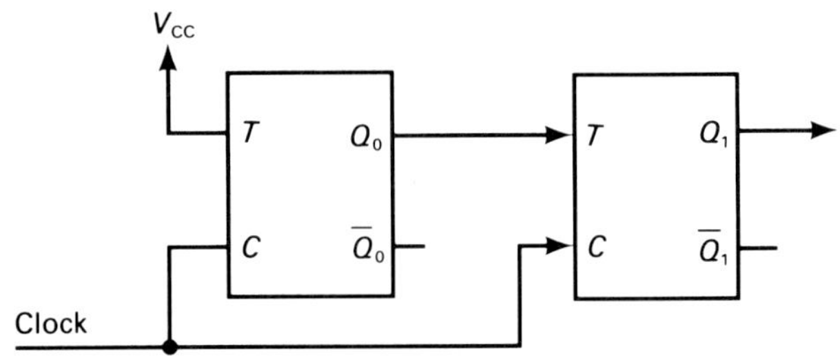

Fig. 4-24 Fully synchronous, two-stage binary counter.

Another type of binary counter, which is not synchronous, is called a *ripple counter*. A ripple counter uses T flip-flops with the T inputs always enabled. Figure 4-25 shows that this is equivalent to a D flip-flop with the

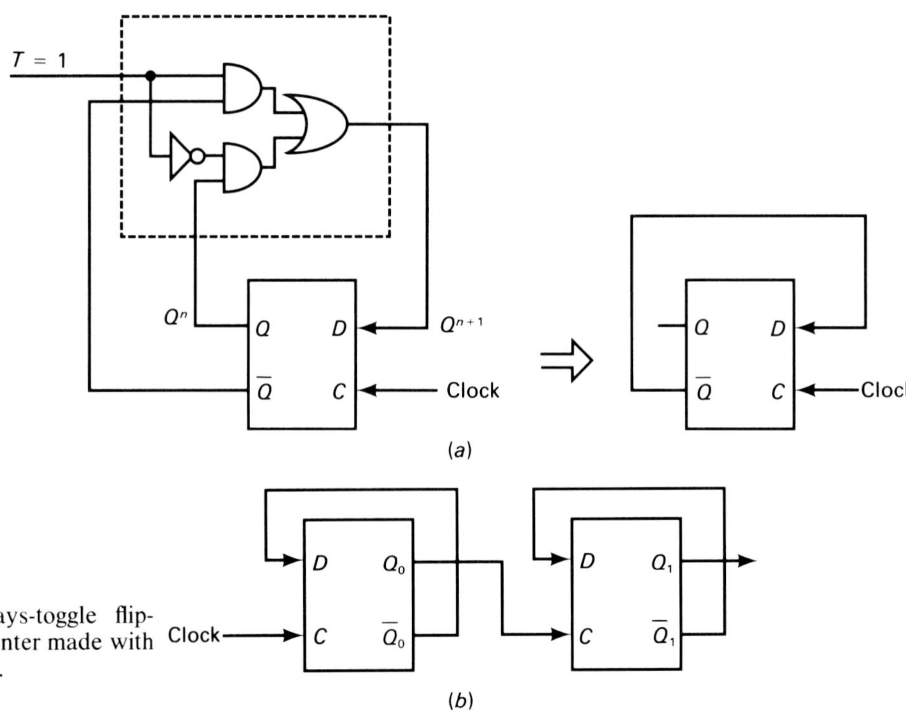

Fig. 4-25(*a*) Always-toggle flip-flop. (*b*) Ripple counter made with always-toggle flops.

Q NOT output connected to the *D* input. The ripple counter uses the output of each preceding stage as the clock for the next stage. The result is an asynchronous binary sequence. In a multistage ripple counter, each flip-flop changes one propagation delay after the previous stage. As the signal propagates, or ripples, through all the flip-flops, the delays accumulate so that it is possible, in large counters, for the total delay to exceed the clock period. When all the flip-flops must change from 1s to 0s, the total delay is equal to the number of flip-flops times the propagation delay. This interval is called the *turnover time*. Whenever some outputs are changing, the counter will indicate false values. If the outputs must be decoded, then it is important that the turnover time is sufficiently short so that the new state is always stable long enough to be useful. In a synchronous counter the turnover time is always equal to one propagation delay no matter how many stages are used. Counters are discussed in greater detail in Chapter 6.

Programmable Logic Cells

Programmable combinational logic was discussed in Chapter 3 under "Packaging." Programmable logic devices (PLDs) are also available with registered outputs. *Registered outputs* are flip-flop outputs built into general-purpose logic cells called *macrocells*. Each macro-cell is a flip-flop preceded by some generic AND-OR type of logic. The pioneer in the development of these devices was Monolithic Memories Inc. with their registered *programmable array logic* (*PAL*). National Semiconductors and Signetics followed with equivalent devices. PALs are programmable only once. Altera developed the first reusable PLDs which they called "*Erasable Programmable Logic Devices*," or EPLDs. These are CMOS devices which can be erased with ultraviolet light and reprogrammed. A further step was taken by Xilinx Corp. when they brought out a line of CMOS *logic cell arrays*, or LCAs, which can be reprogrammed while still in the circuit.

Figure 4-26 shows a typical macrocell commonly used in programmable logic. The special AND-OR array notation introduced in Chapter 3 is used

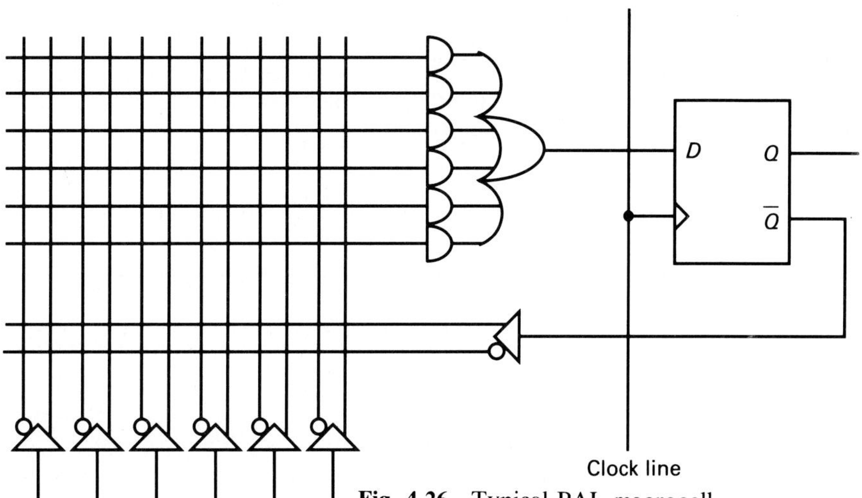

Fig. 4-26 Typical PAL macrocell.

in the data books to illustrate the macrocells. Complementary buffers provide both senses of each input signal, and the devices are programmed by making or breaking AND input connections. All the flip-flops previously mentioned can be implemented with these cells. Computer software is available which allows the designer to enter state tables, Karnaugh maps, state equations, and, in some cases, even state diagrams. The software then provides the outputs needed to program the devices. These products will be discussed in more detail in Chapter 9.

REVIEW 4-3

1. **Q.** What is the generic structure of a complex flip-flop? **A.** a D flip-flop with AND-OR logic

2. **Q.** Which complex flip-flop is used to hold data? **A.** the enabled D flip-flop

3. **Q.** Why do JK flip-flops often simplify design? **A.** because they create an abundance of don't-care conditions in the state table

4. **Q.** How can we avoid the use of Karnaugh maps when using JK flip-flops? **A.** by assigning don't-care conditions in such a way that the J and K inputs can be formed with existing outputs

5. **Q.** When are toggle flip-flops used? **A.** when designing binary counters

6. **Q.** What does the T input do? **A.** It enables the Q NOT input when high, and it enables the Q input when low.

7. **Q.** What is a ripple counter? **A.** It is a binary counter in which the T inputs are always high, and the output of each stage serves as the clock input to the next stage.

8. **Q.** What is turnover time? **A.** It is the time required for a counter to change from the all-1s state to the all-0s state or vice versa.

9. **Q.** What is the turnover time for a ripple counter? **A.** It is the propagation delay times the number of stages.

10. **Q.** What is a fully synchronous binary counter? **A.** It is a binary counter in which each stage is enabled by the output of the previous stage and every stage is clocked with the same edge.

11. **Q.** What is the main problem with ripple counters? **A.** The turnover time may approach or exceed the clock period, making the present-state outputs unusable.

REVIEW PROBLEMS

1. Design an asynchronous RS latch which triggers on a momentary high input. Show truth table, Karnaugh map, and logic gate implementation.

2. Examine the transparent D latch shown in Fig. 4-27*a*. Assuming that each gate has an equal propagation delay of 10 ns, complete the timing diagrams in Fig. 4-27*b* showing what happens when the enable input *E* pulses high for 50 ns while the data input *D* remains high. What hazards do you see in this circuit? How does it differ from the circuit in Figure 4-6*b*?

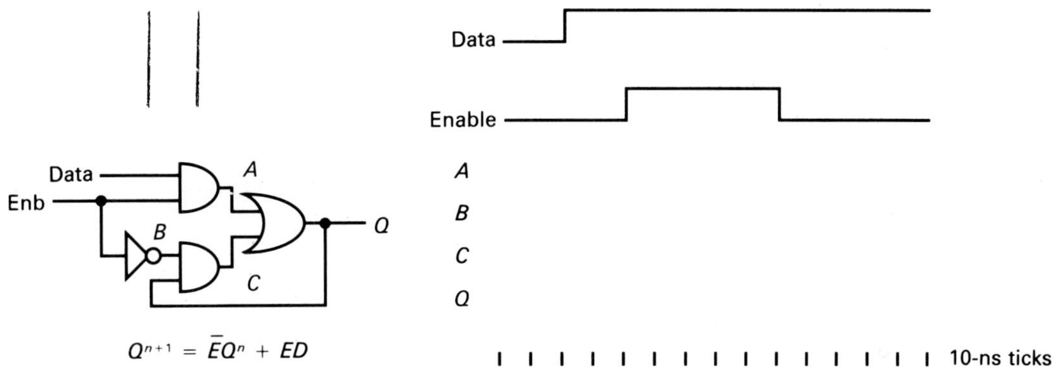

(*a*) (*b*)

Fig. 4-27 Transparent D latch.

3. Design a two-state machine with the state diagram shown in Fig. 4-28. Use a D flip-flop.

4. Design the two-state machine of Prob. 3 using a JK flip-flop.

5. Design the four-state machine of Fig. 4-29 using D flip-flops.

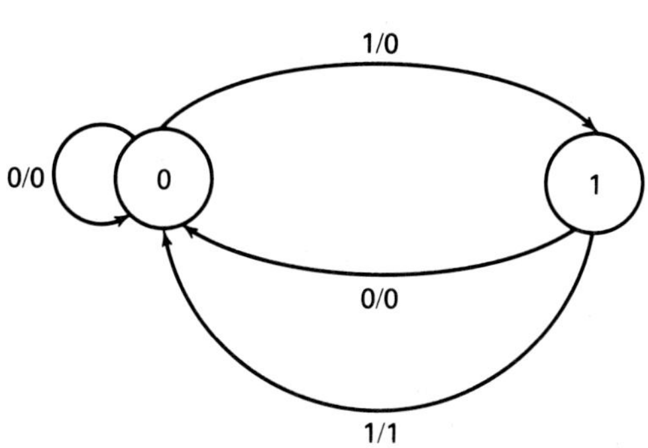

Fig. 4-28 State diagram for Prob. 3.

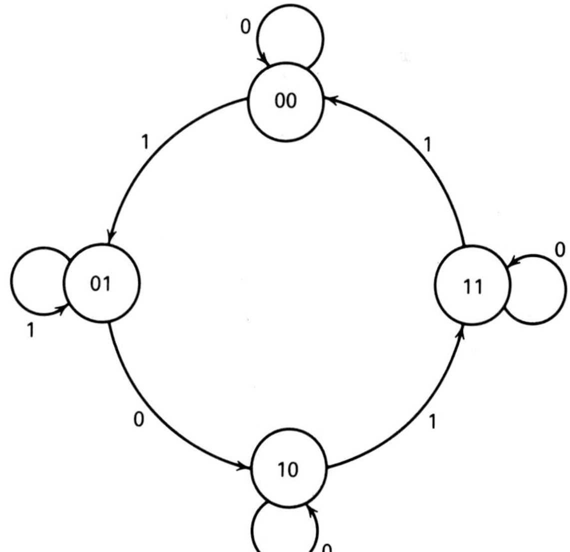

Fig. 4-29 State diagram for Prob. 5.

6. Repeat Prob. 5 using JK flip-flops.

7. Design a simple traffic light using LEDs. The red and yellow LEDs should each stay lit for one clock period, but the green should stay on for two clock periods.

8. Design a synchronous state machine, using D flip-flops, which looks for an input pattern of *at least* two 1s followed by a 0. When the pattern is detected, the output will go high; otherwise, the output will remain low.

9. Design a four-state machine with two inputs, x and y, which will sequence in the following manner:

If $x = 0$ and $y = 1$, then:

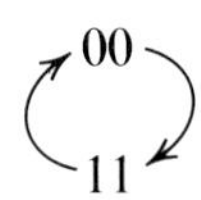

If $x = 1$ and $y = 1$, then:

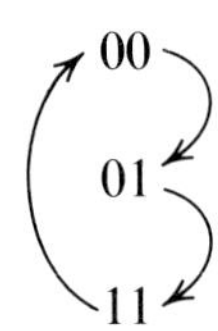

If $x = 1$ and $y = 0$, then:

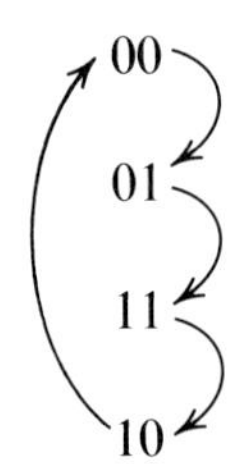

If both x and y are low, the machine stays in the 11 state, with 00→01→11 or 10→01→11 and 11→11.

10. Design the machine shown in Fig. 4-30 using JK flip-flops.

11. Design a three-stage binary counter.

12. Design a more complex traffic light controller which includes a sensor input S to indicate the presence of oncoming vehicles. The state diagram for this controller is shown in Fig. 4-31.

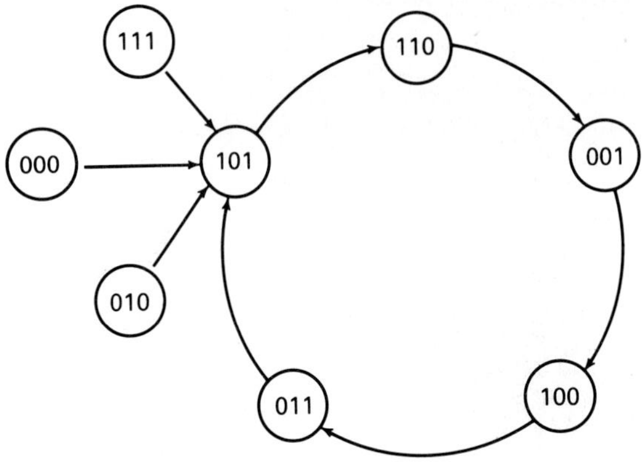

Fig. 4-30 State diagram for Prob. 10.

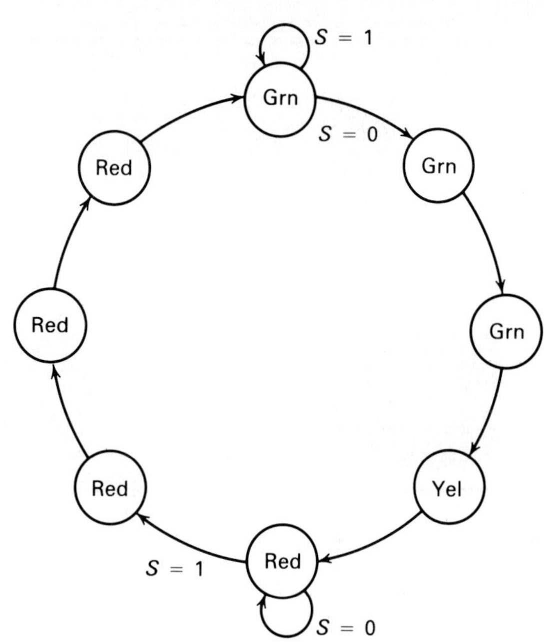

Fig. 4-31 State diagram for Prob. 12.

13. The network in Fig. 4-32 has no inputs other than the clock. (a) Sketch its output waveforms. (b) Derive the state diagram from the waveforms.

14. Design a traffic light controller with two sensor inputs S_1 and S_2 for two directions of traffic as shown in Fig. 4-33. Show state diagram, state table, and Karnaugh maps.

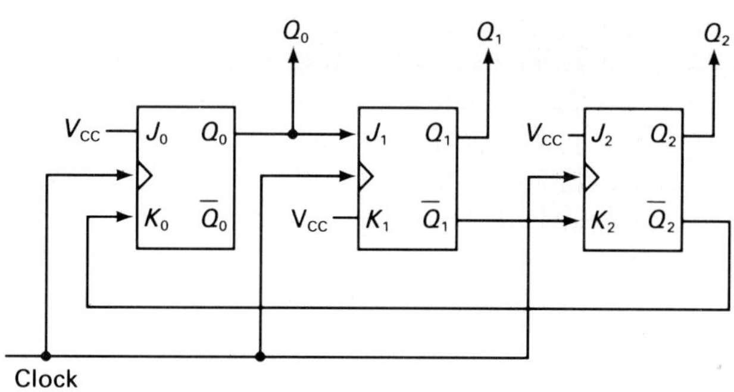

Fig. 4-32 Network for Prob. 13.

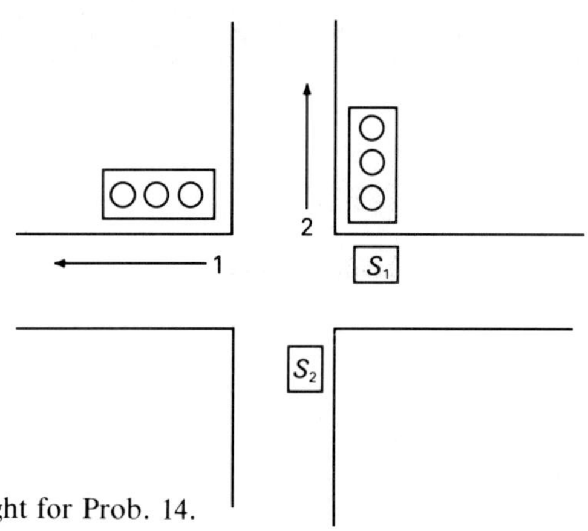

Fig. 4-33 Two-way traffic light for Prob. 14.

5 STORING AND SHIFTING DATA

A computer is, among other things, a machine which *stores* and *shifts* data. This is also a fair description of the high-speed synchronous circuitry which handles telephone calls. Soon this description will apply to television sets as well. Storing and shifting data are, in fact, basic and important functions in many electronic systems. But before examining these operations in detail, we should clarify our notions of what *data* actually is. Chapter 5 objectives are to introduce digital forms of information and show how they are stored and transferred. Specifically, we would like to know:

1. What is digital data and how is it organized?
2. How are numbers and letters represented digitally?
3. How is working data stored for *serial* and *parallel* transfer?
4. How can we implement storing and shifting operations in programmable logic devices and in packaged parts?

5-1 DIGITAL FORMS OF INFORMATION

Most digital networks, from simple electronic combination locks to high-speed digital signal processors, are designed to receive and process information. The meaning of the information itself depends on what functions are to be performed. But in every case the data is represented by voltages, high and low, which correspond to 1s and 0s in a boolean system. These voltages, which must be stable for at least one setup time before they are used, are called bits of information. A *bit* is the smallest unit of digital data.

Bits are grouped into bytes or words, which are, in turn, grouped into messages or instructions. A *byte* is specifically a group of 8 bits, and a *nibble* is half a byte, or 4 bits. A *word* usually refers to 2 or more bytes but may mean any number of bits. These groupings of bits are used to represent numbers, written language, voice levels, and even picture intensities. Their positions in time, relative to some synchronizing signal, determine their exact meaning. Circuitry is needed to collect the bits and hold them until the right moment so they can either be used or transferred to another system. The collection and transfer circuits are called *registers*. Other circuitry counts the bits, maintains word boundaries, and provides timing signals to indicate when the next transfer should occur. These circuits are called *counters*.

The digital information itself may be classified as either binary numbers or binary codes. *Binary numbers* are very much like decimal numbers except that only two symbols, 0 and 1, are used. Binary numbers are *real* numbers. *Binary codes* are only *representations* of numbers. Binary codes may also represent written language or special symbols.

Binary Numbers

Binary numbers are *positional*, in the same way that decimal numbers are positional. Each position is *weighted* by a power of the *base*, or *radix*. In the decimal system the base is 10; in the binary system the base is 2. If we use the letter r to represent the radix, then the following will define the weight of each position for any number of radix r:

$$r^n \ldots r^3 r^2 r^1 r^0$$

In decimal notation, this means that the right-most position is weighted by $r^0 = 1$ and is called the "1's position"; the next position to the left is weighted by $r^1 = 10$, and is called the "10's position," and so on. The right-most position of every positional number system is the 1's position; but the positions to the left depend on the radix. As we move to the left, the weight is always increased by a power of the radix.

In order to find the decimal equivalent of a number represented in positional notation, we multiply each digit by its positional weight and sum it with the others. The digits themselves are called *coefficients*. If we use the letter d to denote the coefficients, then:

$$\text{Decimal value} = \Sigma\, d_n r^n + \cdots d_3 r^3 + d_2 r^2 + d_1 r^1 + d_0 r^0 \qquad (5\text{-}1)$$

For example, the number 111 in base 10 would represent:

$$1 \times 10^2 + 1 \times 10^1 + 1 \times 10^0 = 100 + 10 + 1 = 111_{10}$$

In binary, this same sequence of digits would mean:

$$1 \times 2^2 + 1 \times 2^1 + 1 \times 2^0 = 4 + 2 + 1 = 7_{10}$$

The subscript 10 indicates base 10.

Conversion from binary to decimal is performed by adding the powers of 2 which have nonzero coefficients. Conversion from decimal to binary can be achieved by repeated division of the decimal number by 2. After each division, the remainder becomes the corresponding binary coefficient, 1 or 0. The repeated division is finished when the integer quotient has been reduced to 0. For example, conversion of the decimal number 37 to binary is performed as follows:

$$
\begin{array}{rcccccc}
 & & \textbf{Quotient} & & \textbf{Remainder} & & \textbf{Coefficient} \\
\frac{37}{2} & = & 18 & + & \frac{\mathbf{1}}{2} & & d_0 = 1 \\
\frac{18}{2} & = & 9 & + & \frac{\mathbf{0}}{2} & & d_1 = 0 \\
\frac{9}{2} & = & 4 & + & \frac{\mathbf{1}}{2} & & d_2 = 1 \\
\frac{4}{2} & = & 2 & + & \frac{\mathbf{0}}{2} & & d_3 = 0 \\
\frac{2}{2} & = & 1 & + & \frac{\mathbf{0}}{2} & & d_4 = 0 \\
\frac{1}{2} & = & 0 & + & \frac{\mathbf{1}}{2} & & d_5 = 1
\end{array}
\qquad (5\text{-}2)
$$

The resulting binary equivalent of 37 is 100101. In practice it is not necessary to tabulate the process so formally. Simply divide the number by 2, record the remainder, divide the quotient by 2, record the remainder, and repeat until the integer quotient is 0.

Integer	**Remainder**
37	1
18	0
9	1
4	0
2	0
1	1
0	

Binary numbers may be grouped, for convenience, into octal or hexadecimal equivalents. An *octal digit* represents a cluster of three binary digits, and a *hexadecimal digit* corresponds to a cluster of four binary digits. There are 8 octal symbols, 0, 1, 2, . . ., 7, corresponding to 000, 001, 010, . . ., 111 in binary. There are 16 hexadecimal symbols, and the alphabet must be pressed into service to supplement the 10 symbols which are normally used as numbers. Hexadecimal symbols are called *alphanumeric characters* because they include letters from the alphabet. Figure 5-1 illustrates the correspondence between binary, octal, and hex. The binary equivalent to the decimal number 37 is 100101, but it is written as 45_8 in octal and 25_{16} in hex. When there is no doubt about which base the number is represented in, the subscripts may be dropped.

Decimal	Binary	Octal	Hex
0	0000	0	0
1	0001	1	1
2	0010	2	2
3	0011	3	3
4	0100	4	4
5	0101	5	5
6	0110	6	6
7	0111	7	7
8	1000	10	8
9	1001	11	9
10	1010	12	A
11	1011	13	B
12	1100	14	C
13	1101	15	D
14	1110	16	E
15	1111	17	F
16	1 0000	20	10

Fig. 5-1 Decimal, binary, hex, and octal equivalents.

Octal and hex are easier to use than binary because they condense long strings of 1s and 0s into more manageable units. Even though octal, hex, and binary represent separate and distinct number systems, it is easy to change from one to another because they all have bases which are powers of 2. Converting from binary to octal is simply a matter of combining the binary digits into groups of three, starting from the least significant bit. For instance, consider the 16-bit binary word 0010110010100111. It can be converted into octal in the following manner:

```
        0 2 6 2 4 7    Octal
Binary  0010110010100111
```

Likewise, conversion from binary to hexadecimal is made with groups of four:

```
         2   C   A   7     Hex
Binary  0010 1100 1010 0111
```

Conversion from octal or hex to binary is achieved by disassembling each digit into its binary equivalent.

Binary Codes

While binary numbers are actual real numbers which can be added, subtracted, multiplied, and divided, *binary codes* are only *representations* of numbers or letters or other characters. Binary codes are used because digital systems can operate with only 0s and 1s but need to represent other symbols. Decimal digits, for instance, are often represented by a 4-bit binary code called *binary-coded decimal*, or *BCD*. BCD is used for nontechnical human interface, when numbers must be entered and displayed in decimal notation. Binary numbers could be displayed using single LEDs for each bit. But this is clumsy and uneconomical. Hexadecimal keypads

Decimal	Binary	BCD	Seven -Segment a	b	c	d	e	f	g
0	0000	0000	1	1	1	1	1	1	0
1	0001	0001	0	1	1	0	0	0	0
2	0010	0010	1	1	0	1	1	0	1
3	0011	0011	1	1	1	1	0	0	1
4	0100	0100	0	1	1	0	0	1	1
5	0101	0101	1	0	1	1	0	1	1
6	0110	0110	1	0	1	1	1	1	1
7	0111	0111	1	1	1	0	0	0	0
8	1000	1000	1	1	1	1	1	1	1
9	1001	1001	1	1	1	1	0	1	1
10	1010	1 0000							
11	1011	1 0001							
12	1100	1 0010							
13	1101	1 0011							
14	1110	1 0100							
15	1111	1 0101							
16	1 0000	1 0110							

Fig. 5-2 Decimal, binary, and BCD equivalents with seven-segment code. A 1 in the segment columns means that a segment is turned on.

and displays are more economical but are foreign to the average person. The BCD code, as shown in Fig. 5-2, is a straightforward assignment of the binary equivalent for each decimal digit. This simplifies the display of decimal numbers. The BCD code may be converted into the *seven-segment code* which is used to drive the individual LED segments as shown in Fig. 5-3. ICs are available which contain the decoder and drivers for BCD displays, but higher-priced displays are also available which require no external decoder or drivers at all. The JEDEC type 4N57 BCD display

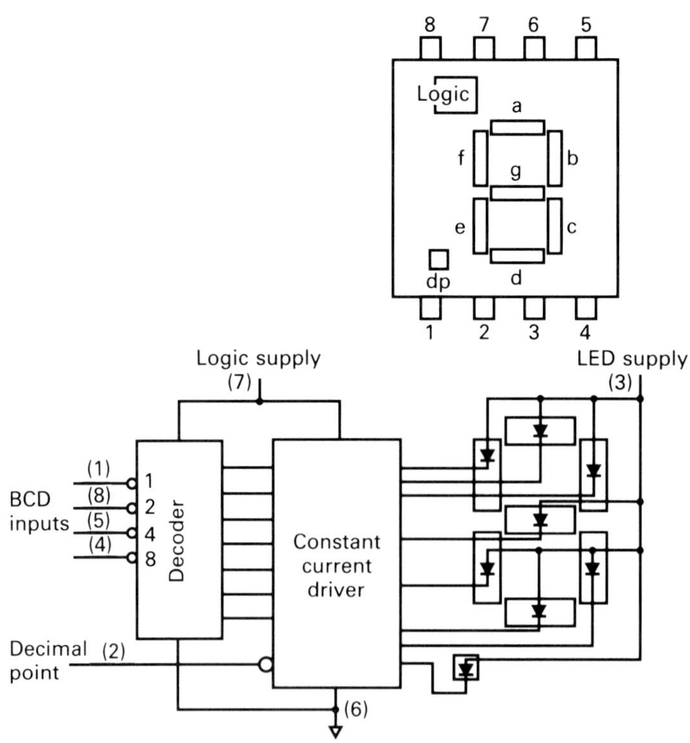

Fig. 5-3 JEDEC type 4N57 BCD display with on-board logic.

shown in Fig. 5-3 combines display, decoders, and drivers in a single package and interfaces directly to biplolar and CMOS logic families. The TIL308 and TIL309 BCD displays also provide on-board latches to hold BCD data when it is present for only a short time. As we will see in Chapter 7, the BCD digits may be added, subtracted, and displayed without conversion to binary.

Another common binary code represents English letters and characters in data communications equipment. It is called the *American Standard Code for Information Interchange,* or *ASCII* for short. The ASCII (pronounced as-ski) is a 7-bit code which has received almost universal acceptance. Figure 5-4*a* illustrates the correspondence between the various

ASCII Character Set (7-bit code)								
M.S. Char. / L.S. Char.	0 000	1 001	2 010	3 011	4 100	5 101	6 110	7 111
0 0000	NUL	DLE	SP	0	@	P	\	p
1 0001	SOH	DC1	!	1	A	Q	a	q
2 0010	STX	DC2	"	2	B	R	b	r
3 0011	ETX	DC3	#	3	C	S	c	s
4 0100	EOT	DC4	$	4	D	T	d	t
5 0101	ENQ	NAK	%	5	E	U	e	u
6 0110	ACK	SYN	&	6	F	V	f	v
7 0111	BEL	ETB	'	7	G	W	g	w
8 1000	BS	CAN	(	8	H	X	h	x
9 1001	HT	EM	)	9	I	Y	i	y
A 1010	LF	SUB	*	:	J	Z	j	z
B 1011	VT	ESC	+	;	K	[	k	{
C 1100	FF	FS	,	<	L	\	l	¦
D 1101	CR	GS	–	=	M	]	m	}
E 1110	SO	RS	•	>	N	↑	n	~
F 1111	SI	US	/	?	O	←	o	DEL

(*a*)

NUL	Null	DLE	Data link escape
SOH	Start of heading	DC1	Device control 1
STX	Start of text	DC2	Device control 2
ETX	End of text	DC3	Device control 3
EOT	End of transmission	DC4	Device control 4
ENQ	Enquiry	NAK	Negative acknowledge
ACK	Acknowledge	SYN	Synchronous idle
BEL	Bell (audible signal)	ETB	End of transmission block
BS	Backspace	CAN	Cancel
HT	Horizontal tab	EM	End of medium
LF	Line feed	SUB	Substitute
VT	Vertical tab	ESC	Escape
FF	Form feed	FS	File separator
CR	Carriage return	GS	Group separator
SO	Shift out	RS	Record separator
SI	Shift in	US	Unit separator
SP	Space	DEL	Delete

(*b*)

Fig. 5-4(*a*) The ASCII character set. (*b*) The key to the abbreviations for the ASCII chart.

symbols and their ASCII equivalents. The ASCII is used in the popular RS-232 interface between computers, data terminals, and data communications equipment. But while the ASCII is a standard for passing information between computers, many computers maintain their own code assignments internally. For example, IBM uses an 8-bit internal code called the *Extended BCD Interchange Code*, or *EBCDIC* (pronounced eb-sa-dic).

ASCII characters require only 7 bits, but the complete code is actually 8 bits, or 1 byte. The most significant bit is often used as a *parity bit*, or *check bit*. This bit is added to the 7-bit code to provide an indication of data integrity. There are two forms of 1-bit parity, *even* and *odd*. If *even parity* is used, the check bit ensures that the entire byte contains only an even number of 1s. If *odd parity* is used, the check bit ensures that the entire byte contains only an odd number of 1s. A parity-generating circuit is used to supply the proper bit, and a parity-checking circuit is used to verify the count after the byte has been received by another system. A system is configured to accept either even parity, odd parity, or no parity. When parity is used and the parity checker finds the wrong number of 1s in a data byte, it knows that the byte has been altered during transfer. The parity bit itself becomes part of the code. Data that is particularly important or especially prone to damage may contain several parity bits for each word. *Error correction codes* use multiple parity bits to detect, and even correct, errors in systems where data integrity is particularly important.

Some binary codes are used for counting. *Cyclic*, or *Gray*, codes are binary codes which are used when state machine outputs must never produce glitches or temporarily false values. Adjacent code assignments in a cyclic sequence will never change by more than 1 bit. Since no more than 1 bit may change on each clock edge, no false decoded outputs can occur. This is important when the outputs of state machines are used to clock other circuits. Figure 5-5 illustrates a method for finding Gray codes using Karnaugh maps. Adjacent boxes in a Karnaugh map represent binary codes differing by only 1 bit. This makes it easy to find code sequences which differ by only 1 bit. If vertical and horizontal lines are used to trace a path covering all squares in a Karnaugh map, the resulting sequence will be a Gray code. Figure 5-5 shows two possible paths and their corresponding code sequences.

Another type of binary code is used to represent voice levels for digital telephone transmission and switching. Voice waveforms are quantitized into levels which can be represented with 8-bit binary numbers. This is called *pulse-coded modulation*, or *PCM*. Quantitizing techniques will be discussed in Chapter 10. High-speed synchronous circuitry *multiplexes* 24 telephone conversations, each represented by 1s and 0s, onto a single line. Within the switching network, the 24-channel lines are further multiplexed onto lines containing 96 and 768 channels. A *multiplexer* is a digital circuit which switches information units from a larger number of lines onto a smaller number of lines. This requires changing clock speeds so that more information can be transferred over fewer lines in the same amount of

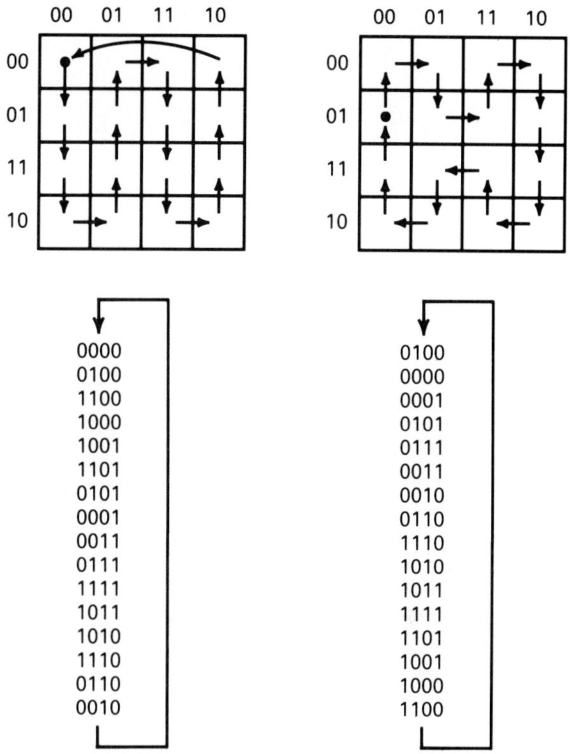

Fig. 5-5 Using Karnaugh maps to find Gray codes.

time. At the receiving end, the information is *demultiplexed* and distributed again over the original number of lines with a slower clock. The multiplexer *selects* data from *n* separate lines and connects those lines, in sequence, to a single common transmission line. The demultiplexer *distributes* data from the common transmission line to one of *n* separate data lines at the other end. The data on the common line must move *n* times faster than the data on either end. The concentration of information saves transmission lines and is not noticeable at either end. These operations are performed with exactly the types of storage circuits decribed in this chapter.

The above examples are meant to illustrate the variety of binary-coded information. Just as the written word may represent many different languages using the same set of letters, binary information may have many different meanings while using the same 1s and 0s. We seldom even need to know the meaning of the information being processed as long as we understand its format and timing.

REVIEW 5-1

1. **Q.** What is the difference between a byte and a word? **A.** A byte consists of exactly 8 bits; a word may be more than 8 bits.

2. **Q.** What is the difference between a binary number and a binary code? **A.** A binary number is a real base 2 number; a binary code is a *representation* of a number or some other character.

3. **Q.** What is a radix? **A.** It is the base of a number; decimal numbers are base 10, binary numbers are base 2, hexadecimal numbers are base 16.

4. **Q.** What do we mean when we say that numbers are "positional"? **A.** We mean that each position is weighted by a power of the radix, and a given position always has the same weight.

5. **Q.** Can you think of a number system which is not positional? **A.** Roman numerals are not positional.

6. **Q.** Why is it so easy to convert between binary, octal, and hexadecimal notations? **A.** because they all have bases which are powers of 2

7. **Q.** When is BCD preferred over straight binary? **A.** when data entries and displays must be made with decimal numbers

8. **Q.** What is ASCII? **A.** It is a standard code for representing English letters, numbers, and special characters.

9. **Q.** What is parity? **A.** Parity is one or more bits added to a data word for the detection of errors.

10. **Q.** When are Gray codes useful? **A.** They are useful whenever state machine outputs are used to clock other circuits. Gray code counting sequences never change by more than 1 bit on each clock edge. For this reason, false outputs due to more than one stage changing at a time cannot occur.

5-2 STORING AND SHIFTING DATA

Electronic circuits often need to provide temporary storage for working data. Very often, this data is stored in 8-, 9-, 10-, or 16-bit registers. A *register* is a group of flip-flops which perform the same task and use the same clock. The simplest type of register is the *parallel-in–parallel-out* register consisting of D flip-flops with a common clock. A more useful parallel-in–parallel-out register would include a *load* or *enable* input which allows data entry at specified times rather than on every clock edge. The *enabled D flip-flop* introduced in Chapter 4 and shown in Fig. 5-6*a* may be used as the basic cell for this type of register. The flip-flop triggers on each rising edge, but the enable signal determines whether it should accept a new value or retain its present state. Figure 5-6*b* illustrates another approach. Here the flip-flop triggers only when the enable signal passes a rising edge. This technique, called *clock gating*, has the advantage of using less logic since only one pair of gates is needed for an entire register. However, clock gating does have drawbacks. Gate delays are introduced in the clock signal reaching the flip-flops. These delays skew the gated clock relative to the master clock. *Skew* is the phase difference between

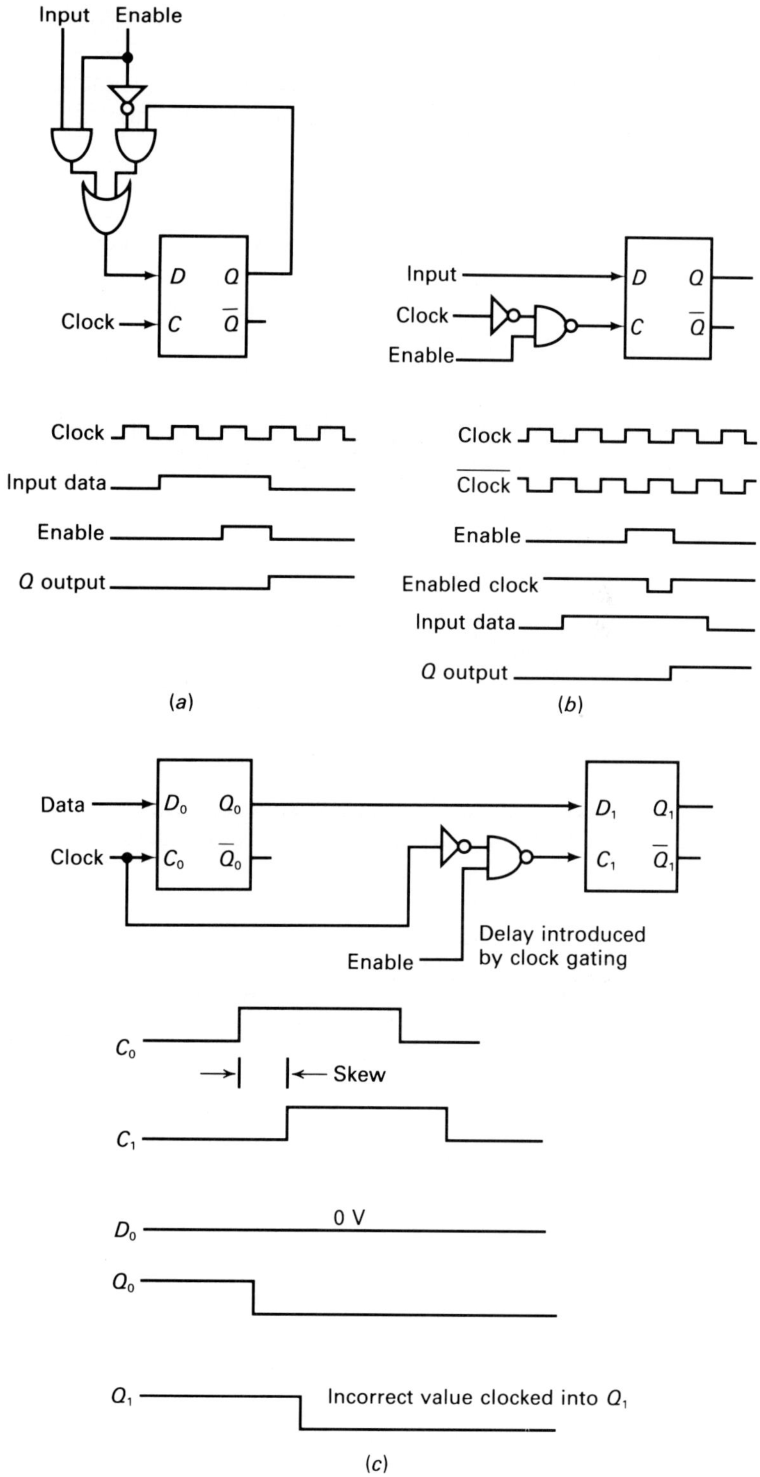

Fig. 5-6(*a*) and (*b*) Enabling input data to register cells. (*c*) The effect of clock skew on data transfer.

two signals which should change at the same moment. Two clock signals are skewed if one rises or falls before the other. Figure 5-6*c* shows how clock skew can cause circuit malfunction. If the first flip-flop receives its rising edge sooner, its output may change before the second flip-flop has time to latch the correct value. This problem will not occur if the first flip-flop receives its rising edge later or at the same time as the second flip-flop. Another point which must be watched when clock gating is used is the phase relationship between the master clock and the enable signal. The enable signal must always change *just after* the clock edge. If the enable signal is not synchronous with the gated clock, glitches may occur which could cause false clocking.

Data Buses

Serial data storage and data transfer involves *shifting* bits, one at a time, onto a single line. The line is called a *serial bus* or *serial link*. If 8 bits are to be transferred, then 8 clock edges are used to shift the bits, one at a time, onto the bus. At the receiving end, the logic level on the bus is sampled on 8 clock edges to shift the data from the bus into another register. *Parallel* data transfer of 8 bits is accomplished with only 1 clock edge using 8 separate lines. The separate lines are grouped to form a *parallel bus*. Eight bits are shifted onto the parallel bus with 1 clock edge and clocked into a register at the other end using the next clock edge. *Serial data transfer* saves lines and connectors at the expense of time; *parallel transfer* saves time at the expense of lines and connectors. Within a small system, such as a printed circuit board or IC, parallel transfer is preferred. But when signals must be sent over long distances, the line drivers, receivers, connectors, and channel material (wire or optical fiber) become so expensive that serial transfer is nearly always used. If the data transfer is to be synchronous, as described, a common clock must be sent along with the data.

A data bus is physically just a connection or group of connections shared by two or more logic subsystems. But there are special problems associated with data buses. Shared buses generally present more capacitive loading to the driving logic. They are often longer than other connections, and the propagation delays from one end to the other may result in troublesome reflections on the line. Grounding problems and electrical noise are also associated with data buses. In addition, when more that one output can drive a connection, there must be some way to disable other outputs in order to prevent contention. *Contention* means that one output is trying to pull a line high while another is trying to pull the same line low. All of these problems, associated with grouping and sharing connections, are explained and analyzed in Chapter 11.

Shifting Data

There are four basic combinations of serial and parallel data storage used in bus applications. They are illustrated in Fig. 5-7 along with their generic designations. The clock inputs are not actually shown in Fig. 5-7 but are implied. At least three of these operations involve *shifting* data from one flip-flop to another. The basic shift operation can be illustrated with only two flip-flops. The two D flip-flops in Fig. 5-8*a*. constitute a simple *shift register*. On each clock edge the data present at the D_0 input is transferred

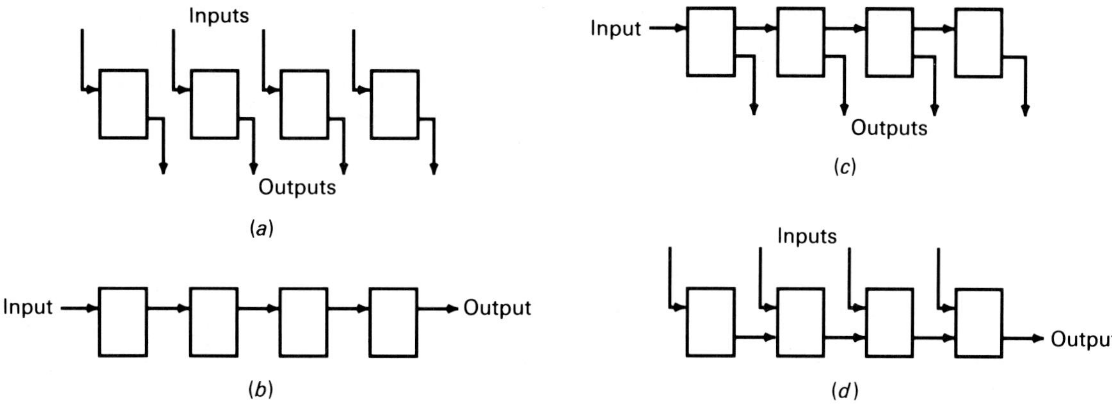

Fig. 5-7 Four types of storage registers. (*a*) Parallel in, parallel out; (*b*) serial in, serial out; (*c*) serial in, parallel out; (*d*) parallel in, serial out.

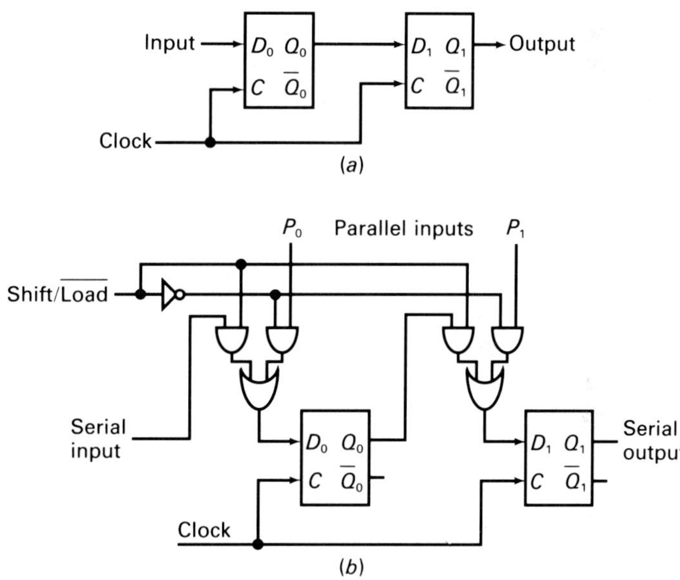

Shift/$\overline{\text{Load}}$	Serial Input	Parallel Inputs	Q^{n+1}
0	0	0	0
0	0	1	1
0	1	0	0
0	1	1	1
1	0	0	0
1	0	1	0
1	1	0	1
1	1	1	1

(*c*)

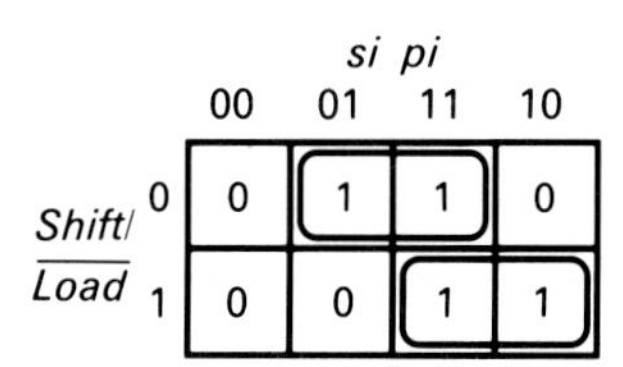

$$Q^{n+1} = \overline{S/\overline{L}} \cdot pi + S/\overline{L} \cdot si$$

(*d*)

Fig. 5-8 Two-stage shift registers. (*a*) Simple serial in, serial out; (*b*) parallel in, serial out; (*c*) truth table; and (*d*) Karnaugh map for parallel-in–serial-out shift register.

to the Q_0. output. At the same time, the present value of Q_0 is transferred to Q_1. This is possible because the data on the D input of each flip-flop is present long enough before the clock edge to *set up* the respective input latch. When the clock edge occurs, the input latch disallows any further changes to its own input state and transfers the new information to its output latch. New data on the input of each flip-flop must remain stable only for as long as it takes to disable the input switch. This time is called "data hold time." *Data hold time* is the amount of time *just after* the clock edge for which the input must remain stable. Data hold time is 0 ns (nanoseconds) for most CMOS parts but may be 1 to 5 ns or more for bipolar parts. If the flip-flops supplying the new data are designed with propagation delays exceeding the data hold time and if there is no *clock skew* between stages, then reliable shifting is guaranteed. This will always be the case when shift registers within the same logic family are used and share the same clock signal. However, if shift registers from a faster logic family must transfer data to shift registers from a slower technology, then the *guaranteed minimum hold time* of the slower register must be compared with the *worst-case minimum propagation delay* of the faster register. If the propagation delay of the sending register can possibly be less than the data hold time requirement of a receiving register, or if *clock skew* can cause a late clock edge, then reliable data transfer cannot be guaranteed.

Setup time is the time *just before* a clock edge during which the input must be stable. In a synchronous system, the setup time *we get* will be equal to the clock period minus the propagation delay of the preceding logic. This must be greater than or equal to the setup time *we need*. *Preceding logic* refers to the flip-flop supplying new data plus any gating between the two stages. If this amount of time does not meet the requirements of the succeeding stage, then the clock rate is too fast for the logic. This situation can be remedied by either using faster logic or a slower clock. The *maximum clock frequency* specifications in manufacturers' data sheets do not take gating delays into account, and since most systems require some gating in the signal path, these specifications have little value. Instead we must look at *worst-case propagation delays, minimum set-up time*, and *minimum data hold time*. Any timing violations involving setup or hold times can cause partial switching. This means that the output of the succeeding stage may hang up, for some time, in a transition region called the *metastable state*. While the output will eventually settle out to a high or low value, it may not be the correct value. Moreover, the time spent in the metastable state may be long enough to cause metastability in the following stage.

EXAMPLE 5-1

Suppose we wish to transfer information between two systems using a serial bus operating at a 12-MHz clock rate. Figure 5-9 shows the general configuration and the worst-case ac parameters for each register. Using the information given, evaluate the reliability of the data transfer on the basis of (a) setup time and (b) hold time.

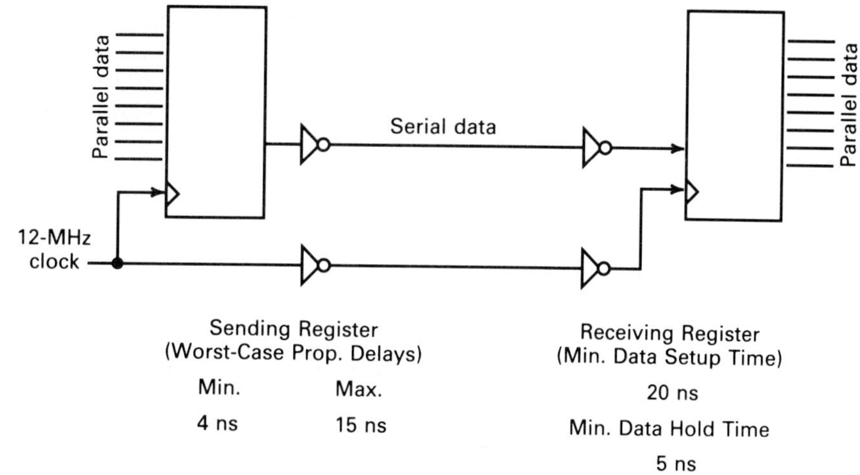

Fig. 5-9 Example 5-1.

SOLUTION

The clock period is 83 ns. In that amount of time we must accommodate the propagation delay of the sending flip-flop, the delay of the links connecting the two systems, and the setup time of the receiving flip-flop. Assuming that the clock and the data are equally delayed by drivers, receivers, and connecting lines, we may concentrate only on the difference in the delay between their paths. Assuming that the clock and data links provide equal delays, the setup time requirement is easily met. The worst-case *maximum* propagation delay of the sending flip-flip is only 15 ns, and since there is no difference in line delays, the data arrives no later than 15 ns after the received clock edge, leaving at least 68 ns of setup time before the next clock edge. However, the worst-case *minimum* propagation delay for the sending flip-flop is only 4 ns, which means that the data may change too soon after the received clock edge to satisfy the minimum data hold time of the receiving flip-flop. Reliable data transfer cannot be guaranteed because of a hold time violation.

Storage Register Design

There are essentially four operations that may be expected of a storage register, *parallel input* (*pi*), *shift right* (*SR*), *shift left* (*SL*), and *HOLD*. One or more of these functions may be needed in a single register. Parallel output (po) is often required, but it is not actually an operation and requires no additional combinational logic. Serial input (si) *is* an operation, but it is included in the shift-right and shift-left functions. Fortunately, the circuitry needed to implement all four operations is simple and has been discussed already. The basic AND-OR logic introduced in Chapter 4 may be applied and expanded to accommodate any or all of these functions. Moreover, when designing a storage register, only one stage needs to be considered since the logic is essentially the same for all.

Figure 5-8*b* illustrates the design of a parallel-in–serial-out (piso) register. The AND-OR logic multiplexes one of two values to the input of each flip-flop, depending on the level of the control signal. If the control signal is high, the new value comes from the preceding stage; if the control signal is low, then parallel data is loaded in. Note that serial-in–serial-out (siso) operation is also possible since, when shifting, the first stage will receive its input from outside. The word *load* normally refers to parallel data entry. Since this particular circuit accepts parallel data when the load input is *low,* it is an "active-low" enabled register.

The design in Fig. 5-8*b* could be converted to a bidirectional shift register by rewiring the Q_1 output to the P_0 input and labeling P_1 as "shift-left serial input." Q_0 would then be the *shift-left serial output*, and the control signal would be labeled "shift-right/shift-left NOT." Control signal notation using a slash usually means that the function preceding the slash will occur when the control signal is high, and the function after the slash will occur when the signal is low.

More functions may be added to this basic design with additional gates and control signals. It is sometimes desirable to HOLD the present value of a register through several clock periods. This could be accomplished by gating the clock with a HOLD signal as shown in Fig. 5-10. This circuit

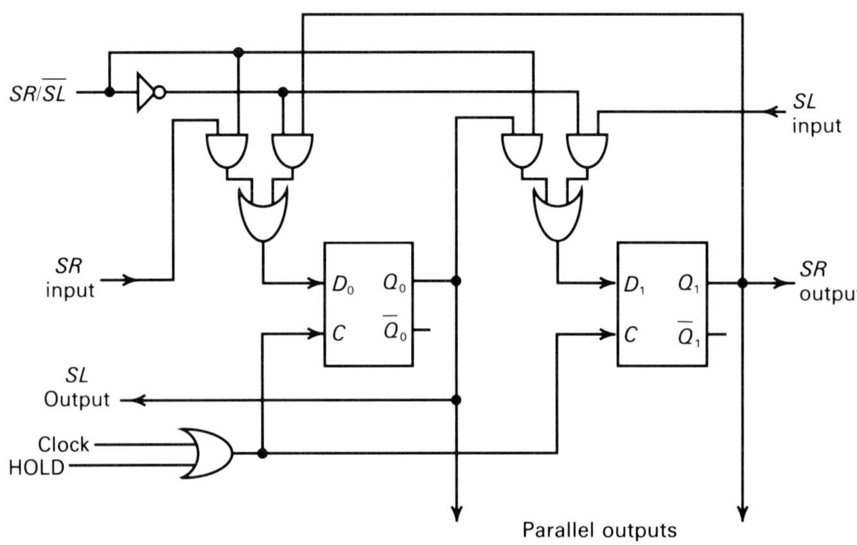

Fig. 5-10 Two-stage bidirectional register with HOLD.

is a serial-in–parallel-out storage register with shift left, shift right, and HOLD. No special logic is needed to provide parallel outputs. The only requirement is that pins and buffers correspond to each Q output. No special logic is needed to provide a serial output either, since the parallel output of the final stage provides this feature. While the HOLD function was implemented with clock gating in this example, it could have also been included in the truth table as a control input. In the latter case, the state table would be complicated by two additional input variables, HOLD and the present-state Q_n. This would also require *three* three-input ANDs in the logic of each flip-flop. Both approaches have their place. The IC designer may wish to economize by using clock gating, but the board-level designer, having the AND-OR functions available in PLD or PAL macrocells, might choose input gating. Clock gating in a PAL could introduce considerable clock skew since the derived clock would need to be driven off the chip and back onto the chip.

At the system level, storage register stages can be combined to form any size register. Parallel-in–parallel-out packaged parts are available in 4, 8, 9, 10, and 16 stages. Packaged shift registers are available in 4, 8, and 16 stages. These parts may be cascaded to form longer registers. An example of cascading four 8-stage shift registers is given in Fig. 5-11. This application, which is fairly common in the telephone industry, involves multiplexing *n* low-speed serial data buses onto a single high-speed serial bus using an *n times* data clock. The low-speed buses are fully synchronous

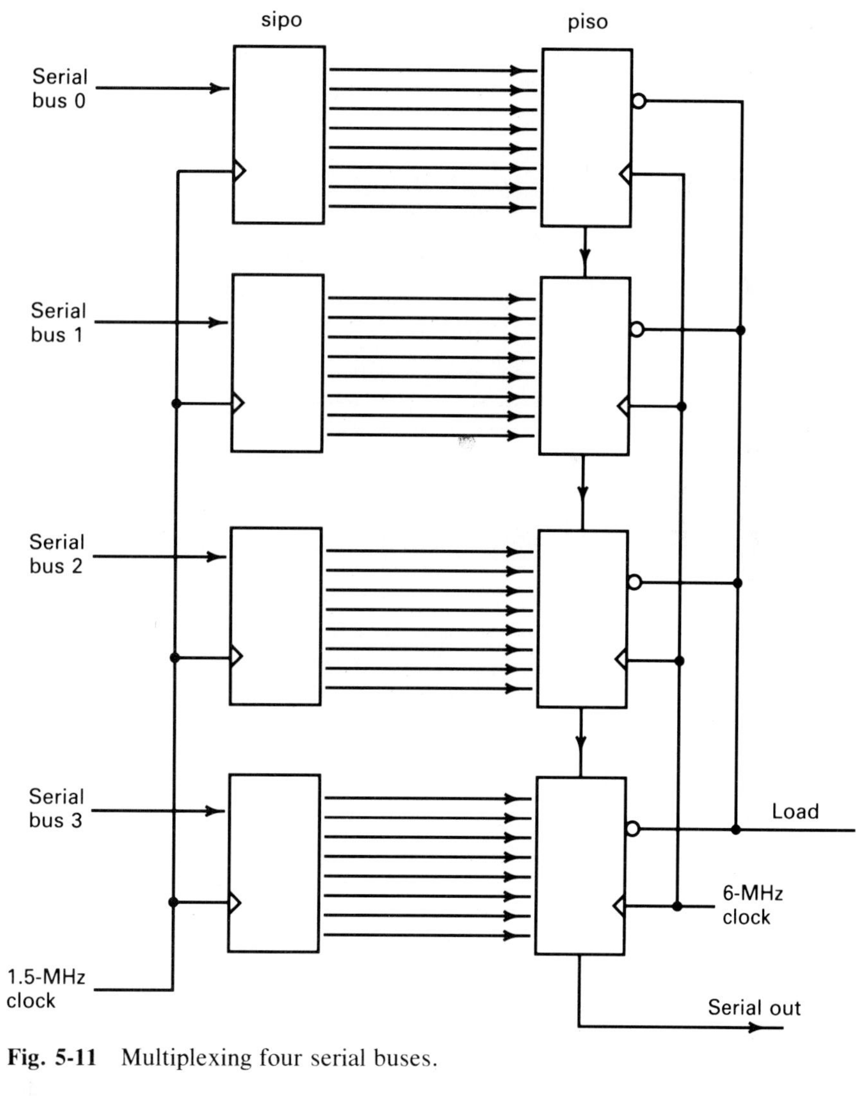

Fig. 5-11 Multiplexing four serial buses.

with the high-speed bus but operate at a scaled-down clock rate. In Fig. 5-11, for example, it takes 5.33 μs to clock in each low-level byte. When 8 bits have been shifted into each of the low-level serial-in–parallel-out registers, a load signal causes the new bytes to be transferred to the parallel-in–serial-out registers on the next clock edge. While 8 more bits are shifting into the sipo's, the piso's are shifting out the 32 bits which were received.

The shift registers described above provide one shift for each clock period. This is desirable in most applications, but sometimes multiple shifts are desired in one clock period to speed the shifting operation. This can be accomplished with a *barrel shifter*. Barrel shifting logic utilizes the same basic AND-OR structure common to the other circuits. However, the choice of shifting up to 2^n-1 places necessitates n control lines, more AND functions, and more inputs for each AND. Figure 5-12 illustrates a barrel shifter which uses two control lines to shift data 0, 1, 2, or 3 places to the right. Barrel shifters are often used to increase computational speed in computers.

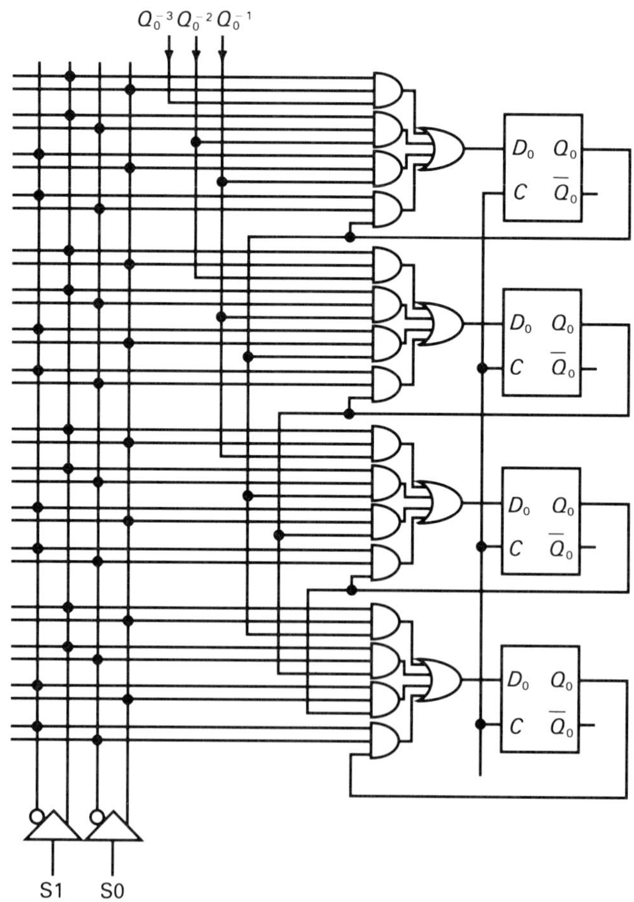

S1	S0	Shift
0	0	0
0	1	1
1	0	2
1	1	3

Fig. 5-12 Four-bit barrel shifter.

5-3 STANDARD REGISTER PARTS

Packaged parts are available which perform the four storage operations illustrated in Fig. 5-7. These represent a fairly efficient implementation at the PC board level of design. Sometimes specialized clearing, loading, or initializing requirements will make PALs and PLDs attractive for storing and shifting, but in most cases standard packages are more economical in terms of cost and power dissipation. Using Fig. 5-7 as a guide, we will discuss examples of the four basic storage configurations labeled types *a*, *b*, *c*, and *d*. We are aided in this task by the 54/74 numbering convention. Part numbers using this convention are preceded by either a 54 or a 74 indicating whether they are manufactured according to military or commercial specifications. The next one to three characters indicate the technology, or logic family, the part belongs to. And finally, two or three numbers at the end indicate the function of the part.

Type A Registers

The type *a* parallel-in–parallel-out register is represented by several packaged parts. They are listed in data books as multiple flip-flops, registers, and latches. While the word *register* applies in all cases, the words *flip-flop* and *latch* are more specific. For example, the 74xx373 and 74xx374 in Fig. 5-13 are both 8-bit, parallel-in–parallel-out registers. The '373, listed as a latch, is transparent, meaning that its present-state outputs may change during the time that the *latch enable* (*LE*) input is high. The '374, on the other hand, classified as multiple flip-flops, is edge-triggered. Output values of the '374 will change only on a rising clock edge. Both have tristate outputs. The tristate drivers on the outputs allow the devices to be disconnected from a parallel bus when it is shared with other logic. When the *output enable* (*OE*) NOT pin is pulled high, the output drivers go into a *high-impedance*, or high resistance, state. In this state they cannot source or sink any appreciable amount of current and therefore cannot interfere with other activity on the shared bus. These functions are also available with broadside pinouts. *Broadside* packages provide all inputs on one side of the package and all outputs on the opposite sides of the package. This allows easy interface with shared buses. Broadside versions of the '373 and '374 are numbered '573 and '574.

Type B Registers

Type *b* registers are somewhat specialized since their function, in most cases, can be provided by type *c* registers. There are relatively few applications where shifting is required, but examination of the intermediate results is not required. One such application arises when multiple copies of the same message are sent over a serial link and recovered by *majority voting* on each bit. If, for example, five repeats of a 96-bit message are transmitted, then four 96-bit serial-in–serial-out registers might be used as receivers. Figure 5-14 shows how the five versions of each message bit are examined with majority voting logic. The final value is determined on the basis of a 3-out-of-5 majority. This technique is used when data is particularly prone to corruption, such as in radiotelephone signaling data transmission. Since we need to see only the final-stage output, and since

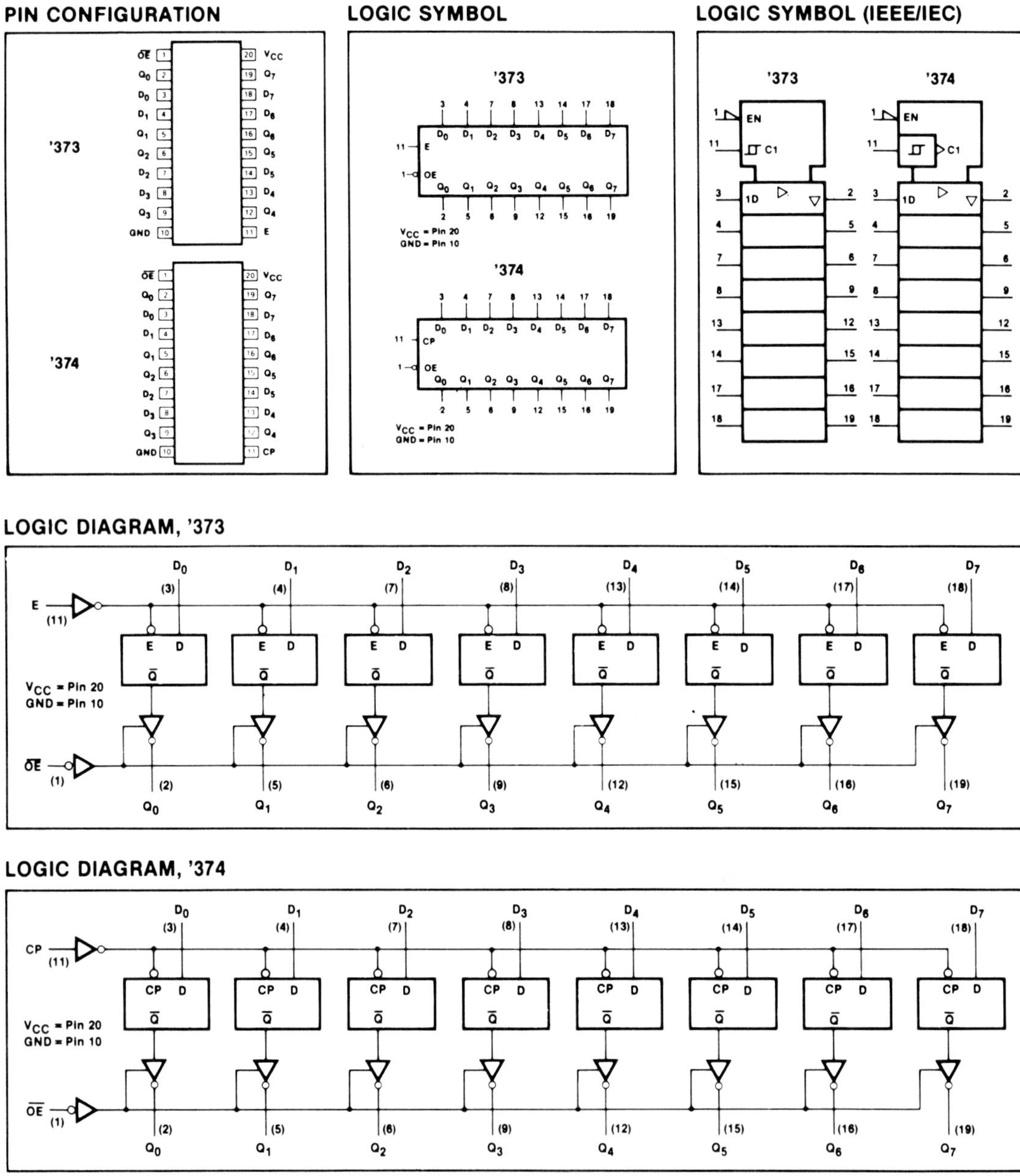

Fig. 5-13 Logic diagrams and pin configurations for the '373 and '374. (*Courtesy Signetics Corp.*)

providing 96 parallel outputs on a single IC would be impractical anyway, the type *b* register is clearly indicated. There are presently no long shift registers of this type in the 54/74xx series logic. However, the 4000 series metal gate logic provides a 128-bit shift register with output taps at Q_{16}, Q_{32}, Q_{48}, Q_{64}, Q_{80}, Q_{96}, Q_{112}, and Q_{128}. This part, the '4562, is available in a 14-pin DIP.

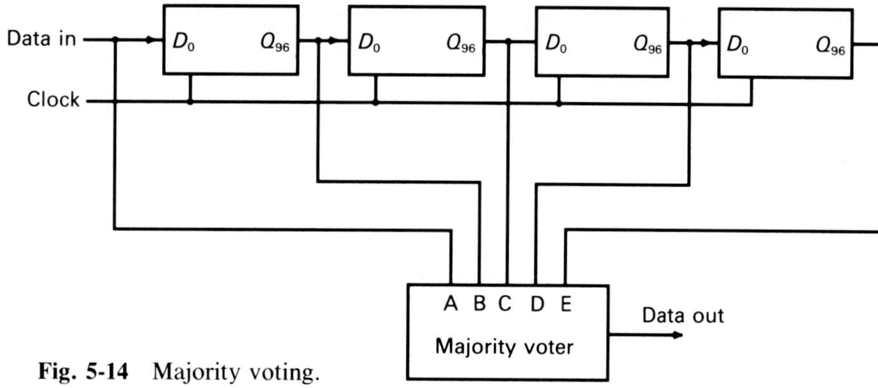

Fig. 5-14 Majority voting.

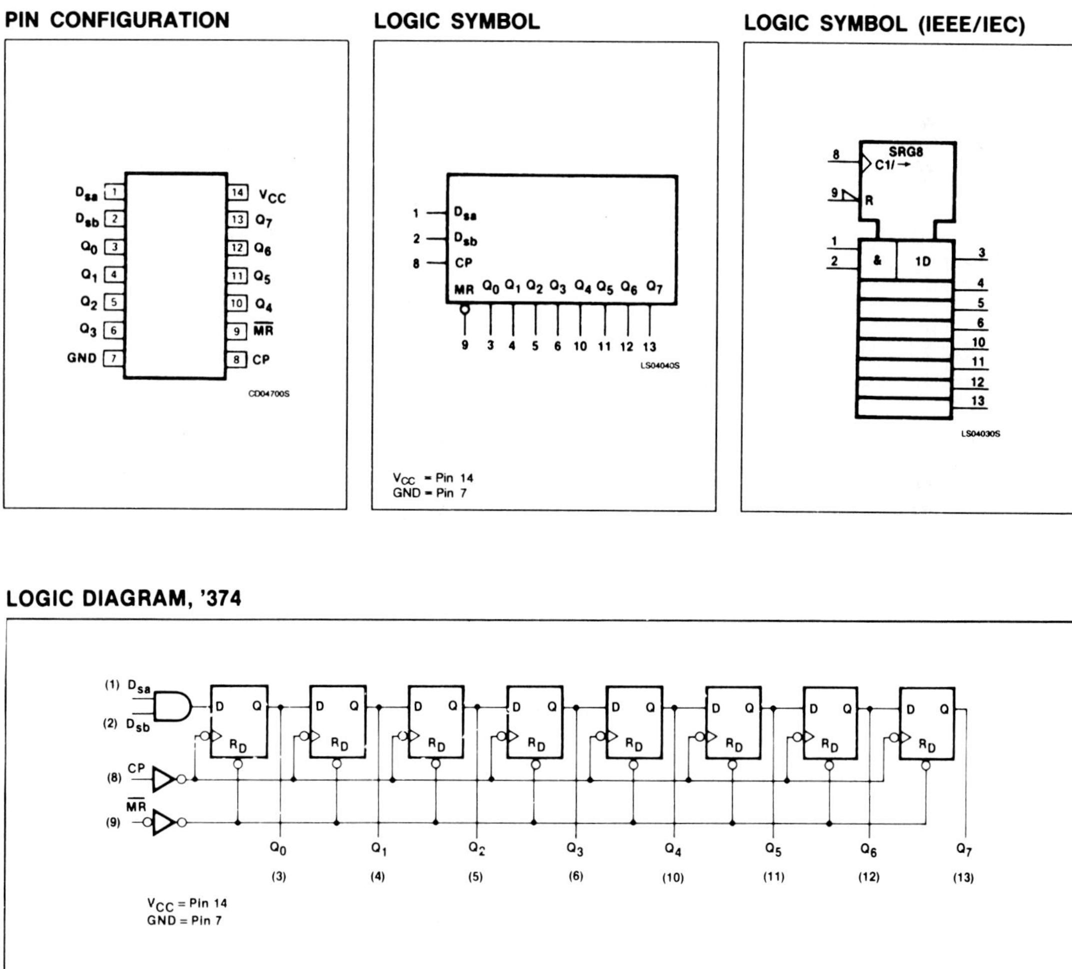

Fig. 5-15 Logic diagram and pin configuration for the '164. (*Courtesy Signetics Corp.*)

Type C and D Registers

Examples of type *c* and type *d* storage registers can be found in the 74xx164 and 74xx166. The '164 is an 8-bit serial-in–parallel-out shift register with two serial inputs. Data is entered through one of the two inputs, and either input may be used as an active-high enable for data entry through the other. This does not mean that present-state data will be held while one input is low. The logic diagram in Fig. 5-15 shows that if one input is low, 0s will be shifted into the register. There is no true enable-HOLD input to the '164. All stages may be reset asynchronously with an active-low input to pin 9.

The complement to the '164 is the 74xx166. This is an 8-bit serial or parallel-in–serial-out shift register with clock enable. When the PE NOT input is low for one setup time before the rising clock edge, parallel data is entered into the register. When the PE NOT input is high, data is shifted right on each rising clock edge. Serial data may be entered via pin 1. The clock input is a gated OR structure which allows the CE NOT input to be used as an active-low clock enable. Figure 5-16 shows the pin configuration and logic diagram for this part.

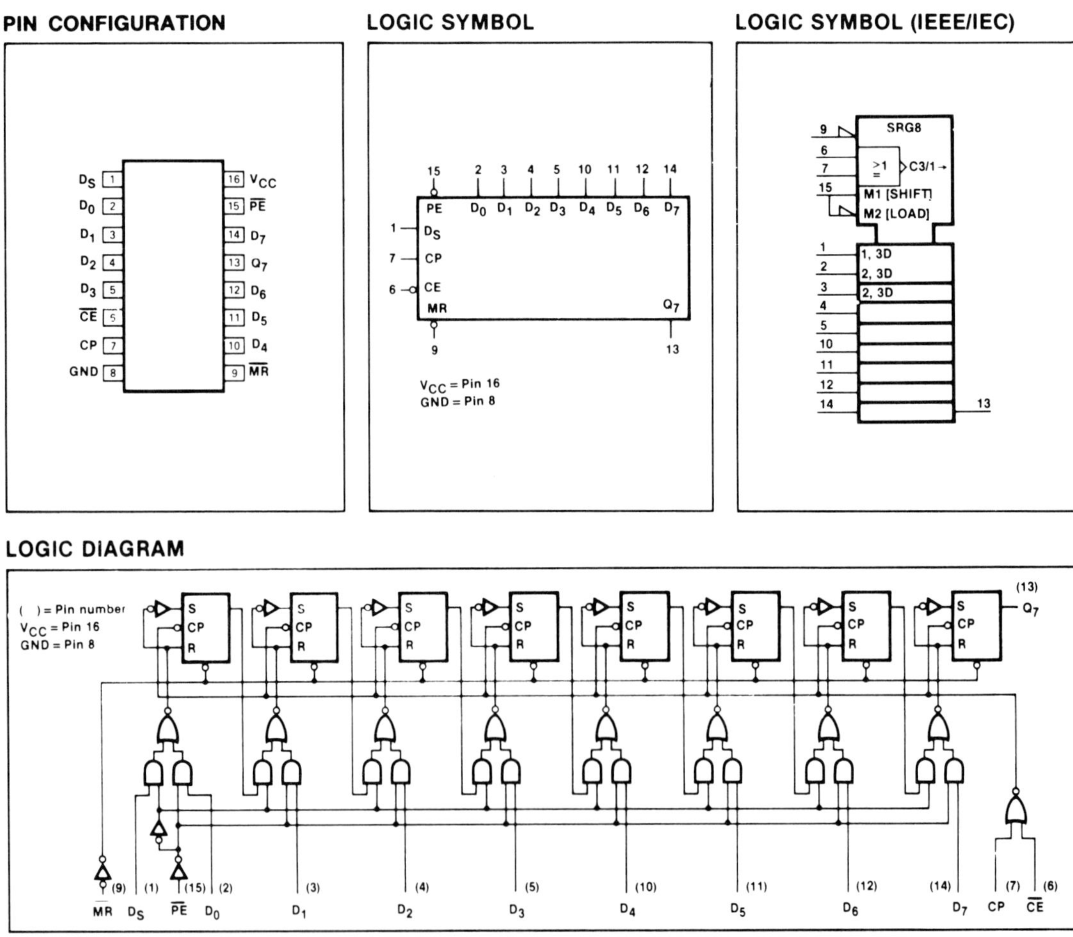

Fig. 5-16 Logic diagram and pin configuration for the '166. (*Courtesy Signetics Corp.*)

The 74xx299 is an 8-bit universal shift storage register. Figure 5-17 shows that this part performs all the previously mentioned functions—parallel load, shift left, shift right, and HOLD. The '299 also has tristate outputs for bus-oriented applications. The parallel inputs are multiplexed with the outputs to reduce pin count. A HIGH on both S_0 and S_1 disables the tristate buffers and allows parallel data to be loaded into the register. A

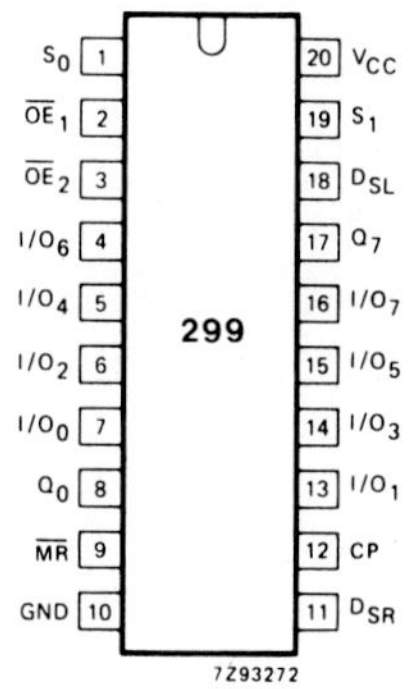

Fig. 1 Pin configuration.

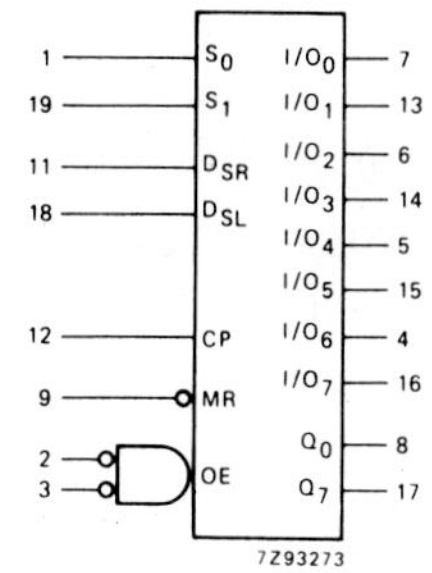

Fig. 2 Logic symbol.

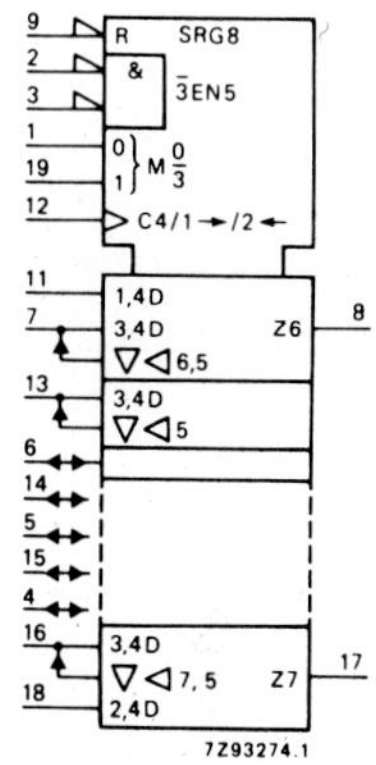

Fig. 3 IEC logic symbol.

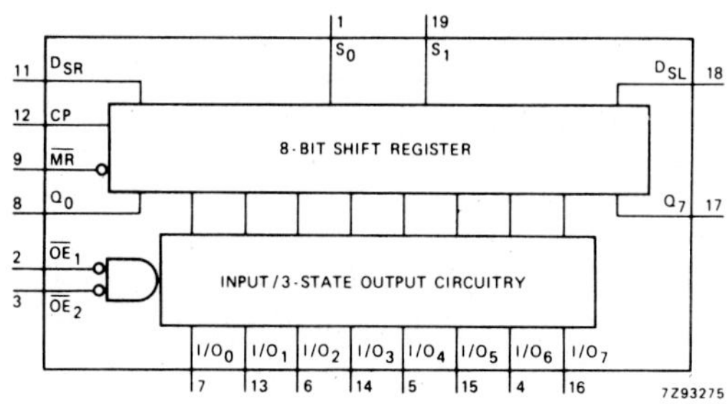

Fig. 4 Functional diagram.

MODE SELECT TABLE

INPUTS				RESPONSE
$\overline{MR}$	S_1	S_0	CP	
L	X	X	X	asynchronous reset; Q_0-Q_7 = LOW
H	H	H	↑	parallel load; $I/O_n \rightarrow Q_n$
H	L	H	↑	shift right; $D_{SR} \rightarrow Q_0$, $Q_0 \rightarrow Q_1$ etc.
H	H	L	↑	shift left; $D_{SL} \rightarrow Q_7$, $Q_7 \rightarrow Q_6$ etc.
H	L	L	X	hold

H = HIGH voltage level
L = LOW voltage level
X = don't care
↑ = LOW-to-HIGH CP transition

Fig. 5-17 Logic diagram and pin configuration for the '299. (*Courtesy Signetics Corp.*)

HIGH signal on either OE_1 NOT or OE_2 NOT will also disable the output buffers and put the I/O pins in the high-impedance state. In this condition the shift, load, HOLD, and RESET operations can still occur.

REVIEW 5-3

1. **Q.** Name the four basic types of storage register. **A.** parallel-in–parallel out, serial-in–serial-out, serial-in–parallel out, and parallel-in–serial-out

2. **Q.** What is an enabled D flip-flop? **A.** It is a flip-flop which either stores a new input or retains its present value. The choice is made on the basis of a control signal called *load* or *enable*.

3. **Q.** Describe two ways in which enabled D flip-flops can be implemented. **A.** The new input may be multiplexed with the present value, or the clock can be gated.

4. **Q.** What is clock skew? **A.** Clock skew is the difference in arrival time of two clock edges which are meant to be coincident. It occurs when the master clock is gated for some, but not all, of the logic, or whenever two clock signal paths encounter differing delays.

5. **Q.** What are the trade-off's between serial and parallel data transfer? **A.** Serial data transfer saves lines and connectors while parallel data transfer saves time.

6. **Q.** When is serial data transfer most often used? **A.** whenever signals are to be sent over long distances

7. **Q.** Define setup time and hold time. **A.** Setup time is the time that input data must remain stable just before a clock edge in order to ensure proper latching. Hold time is the time just after the clock edge that the data must remain stable.

8. **Q.** What parameters must be considered to ensure proper data setup between register stages? **A.** maximum propagation delay, clock speed, and minimum setup time

9. **Q.** What parameters must be considered to ensure proper data hold time at the receiving register? **A.** minimum propagation delay of the sending register and minimum data hold time for the receiving register

10. **Q.** What can happen if set-up or hold times are violated? **A.** The flip-flop can become metastable.

11. **Q.** Name four register operations which can be implemented using flip-flops together with combinational logic. **A.** parallel input, shift right, shift left, and HOLD

12. **Q.** Which type of storage registers are used to multiplex slower serial buses onto a faster bus? **A.** Serial-in–parallel-out registers transfer data to faster parallel-in–serial-out registers.

13. **Q.** What is a barrel shifter? **A.** It is a shift register which allows multiple shifts on a single clock edge.

14. **Q.** Where are barrel shifters used? **A.** in computational circuitry

5-4 DATA SELECTION AND DISTRIBUTION

There are a number of applications in communications and computing systems for circuits which select and distribute data. The simplest data selector is the *2-1 multiplexer* designed in Chapter 3 and shown in Fig. 3-17. With one *select* input, that circuit chooses between two data sources for an output. In general, a data selector chooses between 2^N data sources using N select lines. Each of the 2^N source inputs may be applied, in turn, to the output by sequencing the select lines through all possible combinations. The sharing of one output line between several source inputs is a form of *multiplexing*. Data which has been multiplexed may be sent over a single transmission line and *de* multiplexed at the other end. Demultiplexing data from one line to 2^N lines requires the same N control lines which were used to multiplex the data in the first place. Proper distribution of data requires knowledge of when each source is present on the transmission line. Figure 5-18 shows the selection and distribution of data from four sources. The select lines are labeled A_1 and A_0. Select control lines are commonly labeled with either an A (for address) or an S (for select), followed by a subscript.

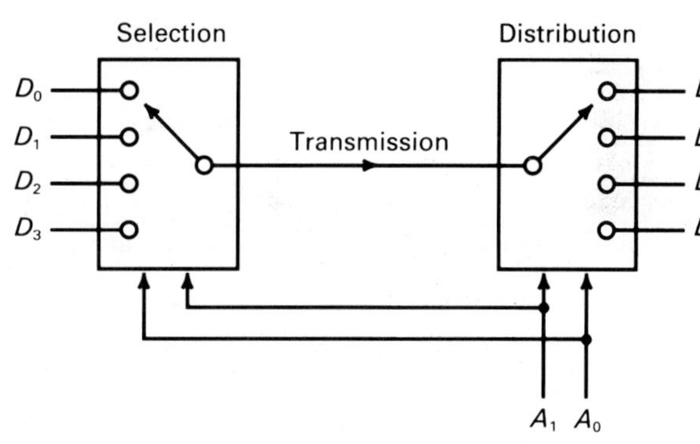

Fig. 5-18 Selection and distribution of data.

Standard parts which correspond to the circuits shown in Fig. 5-18 are the 54/74xx151 and the 54/74xx138. The '151 is an 8-input data selector-multiplexer. It is the digital equivalent of an 8-position switch controlled by the state of the three select inputs. Figure 5-19 shows the HC version of this part. Figure 5-20 shows the HC version of the '138, a 1-of-8 decoder-demultiplexer. The '138 uses three select inputs to produce an active LOW on one of its 8 outputs. The '138 acts as a demultiplexer when one of its active-LOW chip selects (also called "enables") is used as a data input. When the data on the chip select is low, the chip is enabled and the select (address) lines pull the appropriate output LOW. When the data is high,

MOTOROLA

MC54/74HC151

Advance Information

HIGH-PERFORMANCE
CMOS
LOW-POWER COMPLEMENTARY MOS
SILICON-GATE

8-INPUT DATA SELECTOR/ MULTIPLEXER

8-INPUT DATA SELECTOR/MULTIPLEXER

The MC54/74HC151 is identical in pinout to the LS151. The device inputs are compatible with standard CMOS outputs; with pullup resistors, they are compatible with LSTTL outputs.

This device selects one of the eight binary Data Inputs, as determined by the Address Inputs. The Output Enable pin must be at a low logic level for the selected data to appear at the outputs. If Output Enable is high, the Y output is forced to a low logic level and the $\overline{Y}$ output is forced to a high logic level.

The HC151 is similar in function to the HC251 which has 3-state outputs.

- Low Power Consumption Characteristic of CMOS Devices
- Output Drive Capability: 10 LSTTL Loads Minimum
- Operating Speeds Similar to LSTTL
- Wide Operating Voltage Range: 2 to 6 Volts
- Low Input Current: 1 μA Maximum
- Low Quiescent Current: 80 μA Maximum (74HC Series)
- High Noise Immunity Characteristic of CMOS Devices
- Diode Protection on All Inputs
- Selects One of Eight Binary Data Sources
- May Be Used In Parallel-to-Serial Conversion

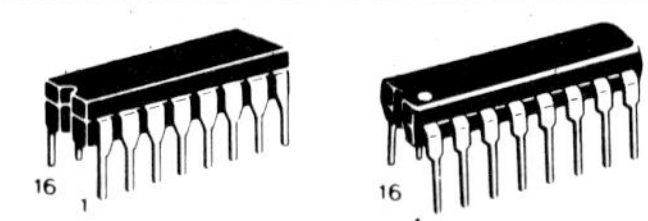

J SUFFIX
CERAMIC PACKAGE
CASE 620

N SUFFIX
PLASTIC PACKAGE
CASE 648

ORDERING INFORMATION

54 Series: −55°C to +125°C
MC54HCXXXJ (Ceramic Package Only)

74 Series: −40°C to +85°C
MC74HCXXXN (Plastic Package)
MC74HCXXXJ (Ceramic Package)

BLOCK DIAGRAM

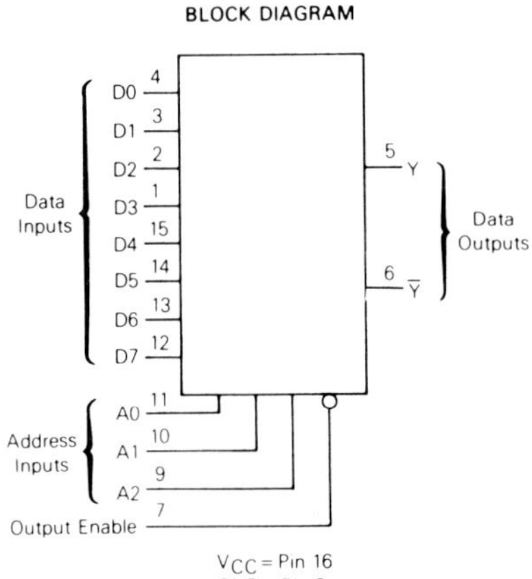

V_{CC} = Pin 16
GND = Pin 8

PIN ASSIGNMENT

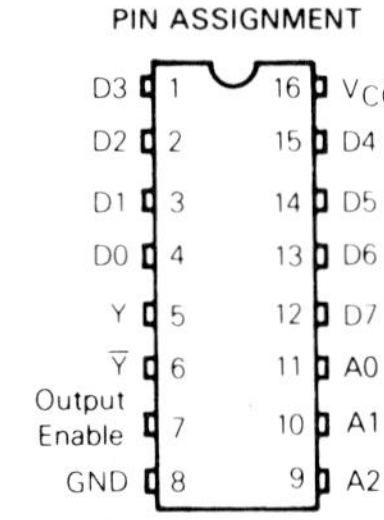

FUNCTION TABLE

Inputs				Outputs	
A2	A1	A0	Output Enable	Y	$\overline{Y}$
X	X	X	H	L	H
L	L	L	L	D0	$\overline{D0}$
L	L	H	L	D1	$\overline{D1}$
L	H	L	L	D2	$\overline{D2}$
L	H	H	L	D3	$\overline{D3}$
H	L	L	L	D4	$\overline{D4}$
H	L	H	L	D5	$\overline{D5}$
H	H	L	L	D6	$\overline{D6}$
H	H	H	L	D7	$\overline{D7}$

D0, D1 ... D7 = the level of the respective D input

Fig. 5-19 The 54/74HC 151 data selector-multiplexor. (*Courtesy Motorola Semiconductor Products, Inc.*)

MOTOROLA

MC54/74HC138

Advance Information

HIGH-PERFORMANCE
CMOS
LOW-POWER COMPLEMENTARY MOS
SILICON-GATE

1-OF-8 DECODER/ DEMULTIPLEXER

1-OF-8 DECODER/DEMULTIPLEXER

The MC54/74HC138 is identical in pinout to the LS138. The device inputs are compatible with standard CMOS outputs; with pullup resistors, they are compatible with LSTTL outputs.

The HC138 decodes a three-bit Address to one-of-eight active-low outputs. This device features three chip select inputs, two active low and one active high to facilitate the demultiplexing, cascading, and chip-selecting functions. The demultiplexing function is accomplished by using the Address inputs to select the desired device output; one of the Chip Selects is used as a data input while the other Chip Selects are held in their active states.

- Low Power Consumption Characteristic of CMOS Devices
- Output Drive Capability: 10 LSTTL Loads Minimum
- Operating Speeds Similar to LSTTL
- Wide Operating Voltage Range: 2 to 6 Volts
- Low Input Current: 1 μA Maximum
- Low Quiescent Current: 80 μA Maximum (74HC series)
- High Noise Immunity Characteristic of CMOS Devices
- Diode Protection on All Inputs

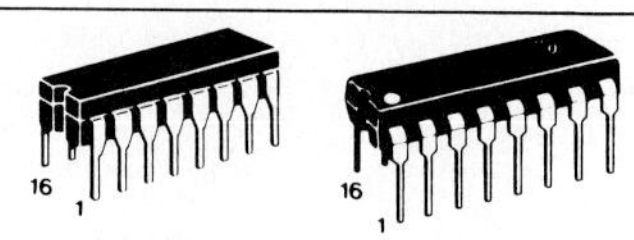

J SUFFIX
CERAMIC PACKAGE
CASE 620

N SUFFIX
PLASTIC PACKAGE
CASE 648

ORDERING INFORMATION

54 Series: −55°C to +125°C
MC54HCXXJ (Ceramic Package Only)

74 Series: −40°C to +85°C
MC74HCXXN (Plastic Package)
MC74HCXXJ (Ceramic Package)

PIN ASSIGNMENT

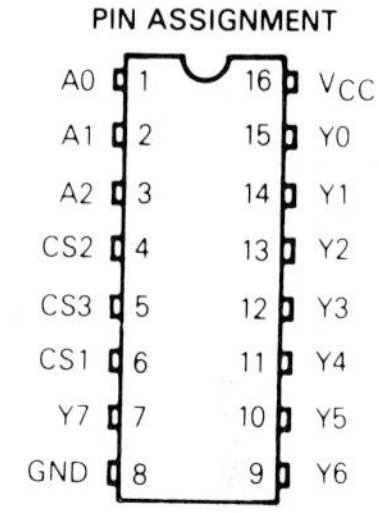

BLOCK DIAGRAM

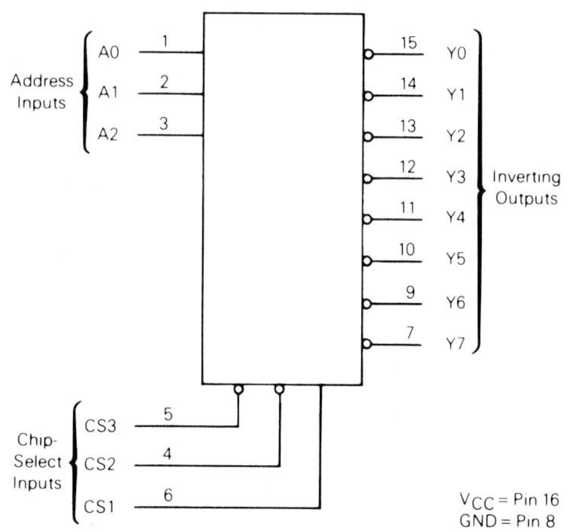

TRUTH TABLE

Inputs						Outputs							
CS1	CS2	CS3	A2	A1	A0	Y0	Y1	Y2	Y3	Y4	Y5	Y6	Y7
X	X	H	X	X	X	H	H	H	H	H	H	H	H
X	H	X	X	X	X	H	H	H	H	H	H	H	H
L	X	X	X	X	X	H	H	H	H	H	H	H	H
H	L	L	L	L	L	L	H	H	H	H	H	H	H
H	L	L	L	L	H	H	L	H	H	H	H	H	H
H	L	L	L	H	L	H	H	L	H	H	H	H	H
H	L	L	L	H	H	H	H	H	L	H	H	H	H
H	L	L	H	L	L	H	H	H	H	L	H	H	H
H	L	L	H	L	H	H	H	H	H	H	L	H	H
H	L	L	H	H	L	H	H	H	H	H	H	L	H
H	L	L	H	H	H	H	H	H	H	H	H	H	L

H = high level (steady state)
L = low level (steady state)
X = don't care

This document contains information on a new product. Specifications and information herein are subject to change without notice.

Fig. 5-20 The 54/74HC 138 decoder/demultiplexor. (*Courtesy Motorola Semiconductor Products, Inc.*)

the chip remains disabled with all outputs HIGH. The unused chip select inputs must be permanently tied to the appropriate supply. Figure 5-21 shows how the two standard parts may be used together for data selection and distribution.

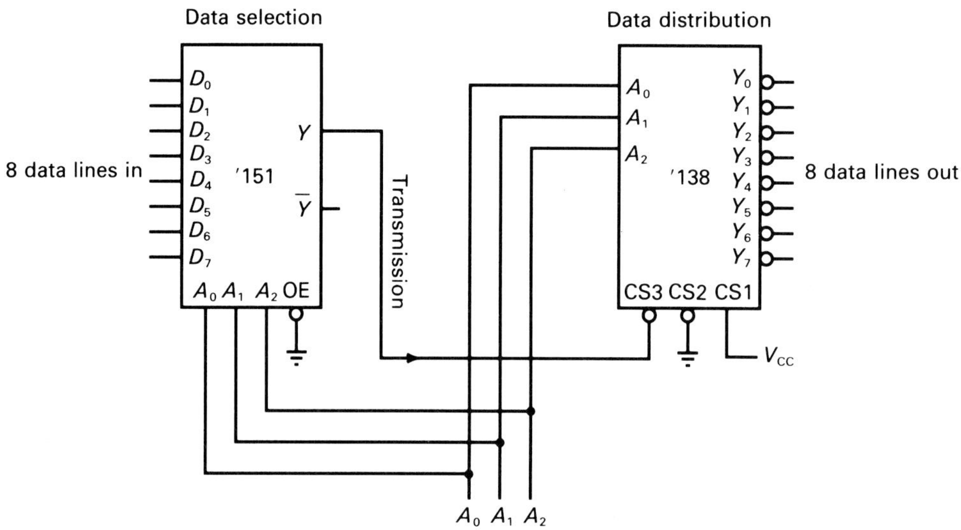

Fig. 5-21 Using standard parts for the selection and distribution of data. The '151 acts as a multiplexer, and the '138 is used as a demultiplexer.

These two parts have other uses besides multiplexing and demultiplexing serial data. The '151 may be used as a *universal function generator.* Any truth table with three or four variables may be implemented with the '151 without Karnaugh mapping or algebraic minimization. In the case of a three-variable truth table, logic 1s (V_{cc}) and 0s (GND) are placed at the eight data inputs to correspond with the truth table output column. These inputs are permanently connected for any given truth table. Figure 5-22*a* shows a three-variable map with input variables x, y, and z. Data input D_0 corresponds to the first entry in the table and is therefore tied low to represent $Y = 0$. Data input D_1 corresponds to the second entry and is tied high to represent $Y = 1$, and so on. When $x = 0$, $y = 0$, and $z = 0$, the data selector will connect the D_0 input to the Y output and produce an output LOW. This is exactly what is called for in the truth table. Since each of the Y outputs in the truth table is simply transferred over to the D inputs of the data selector, any combination of input variables x, y, and z will produce the output value called for in the truth table. The '151 is universal, or functionally complete, because any function in three variables may be implemented.

The '151 may also generate functions in four variables using a technique called *folding*. The four-variable map in Fig. 5-22*b* may be conceptually divided into two halves. The upper-half inputs are identical to the lower-half inputs *except* for the variable w and the outputs. If we cut the table in two, lay one half over the other, and compare the outputs, we will see that in some cases the outputs are the same regardless of whether $w = 0$ or $w = 1$. In these cases we can apply the table output to the appropriate

x	y	z	Y
0	0	0	0
0	0	1	1
0	1	0	0
0	1	1	1
1	0	0	0
1	0	1	0
1	1	0	1
1	1	1	1

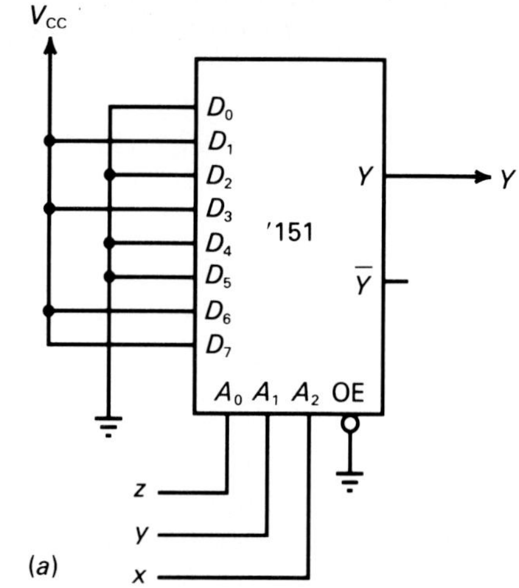

(a)

w	x	y	z	Y
0	0	0	0	0
0	0	0	1	0
0	0	1	0	0
0	0	1	1	1
0	1	0	0	0
0	1	0	1	1
0	1	1	0	1
0	1	1	1	0
1	0	0	0	0
1	0	0	1	1
1	0	1	0	0
1	0	1	1	1
1	1	0	0	1
1	1	0	1	0
1	1	1	0	1
1	1	1	1	0

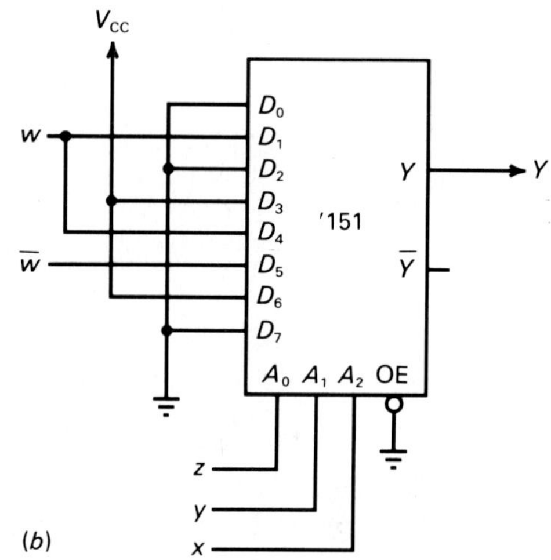

(b)

Fig. 5-22 Using the '151 as a universal function generator.

chip input as if the table contained only the input variables x, y, and z. In some cases, however, we need an output which is a 1 when $w = 1$ and a 0 when $w = 0$, so we connect w to this input. In other cases, we need an output which is a 0 when $w = 1$, and a 1 when $w = 0$, and we connect w NOT to this input. The '151 is thus functionally complete in four variables as well.

The '138 may also be used as a function generator but only when the function represents a single product term. For instance, the following product terms may be decoded by applying x, y, and z to the address inputs and w to CS1:

$$\overline{Y}_0 = w \cdot \overline{x} \cdot \overline{y} \cdot \overline{z}$$
$$\overline{Y}_1 = w \cdot \overline{x} \cdot \overline{y} \cdot z$$
$$\overline{Y}_7 = w \cdot x \cdot y \cdot z$$

The '138 provides eight functional outputs, each corresponding to one product term. This feature makes it useful for decoding address lines into 1-of-8 chip selects for memory and other bus-compatible logic. This application will be discussed in Chapter 8. Sum-of-product expressions cannot be generated with the '138, and it is not functionally complete.

REVIEW 5-4

1. **Q.** What is a data selector? **A.** A data selector is the digital equivalent of an 2^N position switch with N control lines used to select one input to be applied to the output.

2. **Q.** What is a multiplexer? **A.** *Multiplexing* means sharing one output line with several source inputs. This term is applied to data selectors because they perform this function.

3. **Q.** Why is the '151 called a *universal function generator?* **A.** because it can be used to implement any three- or four-variable truth table

4. **Q.** Can the '138 generate sum-of-product expressions? **A.** No.

5. **Q.** What are the two primary uses for the '138? **A.** The '138 can be used to demultiplex serial data and to decode three-variable addresses into 1-of-8 active-low outputs.

REVIEW PROBLEMS

1. Convert the following decimal numbers to binary: 41, 64, 127.

2. Convert the following binary numbers to hex: 00111100, 01101011, 11111001.

3. Convert the hex numbers in Prob. 2 to decimal.

4. Write the first 12 decimal digits in base 3.

5. Represent the following decimal numbers in BCD: 9, 10, 99, 100.

6. Find the ASCII equivalent to the following characters: A, 1, +, a.

7. Design a simple circuit that will check the parity of serial data. The state diagram for this circuit is shown in Fig. 5-23. Use a state table to find the D flip-flop implementation, and then duplicate the design using a JK flip-flop.

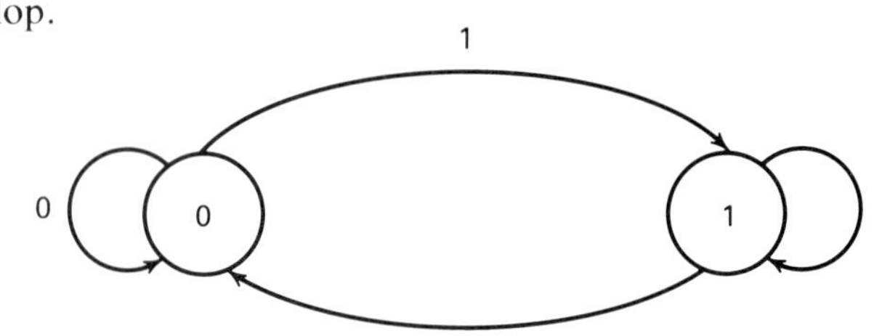

Fig. 5-23 State diagram for parity checker. 1

8. Find a 4-bit cyclic code not shown in Fig. 5-5.

9. Use a state table and Karnaugh map to arrive at the circuit in Fig. 5-6*a*.

10. How can the load signal in Fig. 5-8*b* be changed to active high?

11. Examine the specifications in Fig. 5-24. If an ac shift register is to transfer data to an LS shift register, what concerns might we have?

$V_{CC} = 5$ V, $T_A = 25°C$

Parameter		74LS164 $C_L = 15$ pF, $R_L = 2$ KΩ Min.	74LS164 $C_L = 15$ pF, $R_L = 2$ KΩ Max.	Unit
f_{max},	Max. Shift Freq.	25		MHz
t_{PLH}	Prop. delay Clock to output (Low to high		27	ns
t_{PHL}	and high to low)		32	ns
t_s	Setup time Data to clock	15		ns
t_h	HOLD time Data to clock	5.0		ns

$V_{CC} = 5$ V, $T_A = 25°C$

Parameter		74AC164 $C_L = 50$ pF, $R_L = 500$ Ω Min.	74AC164 $C_L = 50$ pF, $R_L = 500$ Ω Max.	Unit
f_{max},	Max. Shift Freq.	80		MHz
t_{PLH} t_{PHL}	Prop. delay Clock to output (Low to high and high to low)	1.0	8	ns
t_s	Setup time Data to clock	7		ns
t_h	HOLD time Data to clock	0		ns

(*a*)

Fig. 5-24(*a*) AC specifications for logic used in problems.

$V_{CC} = 5$ V, $T_A = 25°C$

Parameter		74AC74 $C_L = 50$ pF, $R_L = 500$ Ω Min.	74AC74 $C_L = 50$ pF, $R_L = 500$ Ω Max.	Unit
f_{max},	Max. Shift Freq.	100		MHz
t_{PLH} t_{PHL}	Prop. delay Clock to output (Low to high and high to low)	1.0	8	ns
t_s	Setup time Data to clock	4.0		ns
t_h	HOLD time Data to clock	0		ns

$V_{CC} = 5$ V, $T_A = 25°C$

Parameter		74AC157 $C_L = 50$ pF, $R_L = 500$ Ω Typical	74AC157 $C_L = 50$ pF, $R_L = 500$ Ω Max.	Unit
t_{PLH} t_{PHL}	Prop. delay Clock to output (Low to high and high to low)	10	15	ns

(*b*)

Fig. 5-24(*b*) AC specifications continued.

12. What is the maximum clock rate that can be used to shift data in Fig. 5-8*b* if an AC74 is used for the D flip-flop and an AC157 is used for the AND/OR logic? Use the specifications given Fig. 5-24.

13. What is the maximum clock speed possible for the circuit in Fig. 5-11 if AC164s and AC166s are used? Assume that the specifications are the same for the '164 and the '166.

14. How is the parallel-out feature implemented on a shift register?

15. Use a state table and Karnaugh map to design a universal shift register with PL, SR, SL, and HOLD.

16. Redesign the universal shift register of Prob. 15 using clock gating.

17. Sketch the true relationship between the data coming out of a shift register onto a serial link and the data clocked into another shift register on the receiver side of the same serial link. Show how each clock edge delivers new data onto the link to be accepted on the following clock edge by the receiver.

18. Show how '164 and '166 shift registers can be used to multiplex four 1.5-MHz serial data buses onto a single 6-MHz serial bus as shown in Fig. 5-11. Show timing diagrams and actual circuit schematic with pin numbers.

19. Show how the 11-bit *sync word*, 11100010010, can be detected in a serial data stream using two '164s, six inverters, and a '133 13-input NAND gate.

20. Implement the truth table in Fig. 5-25 using a Karnaugh map and logic gates. Implement the same truth table using a '151 data selector.

w	*x*	*y*	*z*	*Y*
0	0	0	0	1
0	0	0	1	0
0	0	1	0	0
0	0	1	1	0
0	1	0	0	0
0	1	0	1	1
0	1	1	0	0
0	1	1	1	0
1	0	0	0	0
1	0	0	1	0
1	0	1	0	1
1	0	1	1	0
1	1	0	0	0
1	1	0	1	0
1	1	1	0	0
1	1	1	1	1

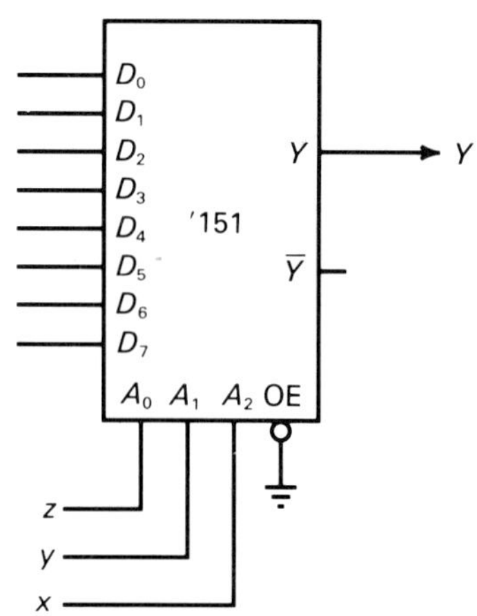

Fig. 5-25 Truth table and data selector for Prob. 20.

6

COUNTERS

Counters are the most commonly used of the sequential state machines. Their present-state outputs are used to measure time, frequency, speed, events, and even distance. They can be used to keep track of serial or parallel data transfers by counting bits and bytes. They may be used to *scale* down the frequency of a master clock for use in slower parts of a circuit. They are used in computers to keep track of instructions and data in memory. They are used in control sequencers to generate timed control signals. *Frequency counters* measure unknown frequencies by counting the time between edges. Digital timepieces use counters to measure elapsed time and the time of day. Radar and sonar systems use counters to measure speed and distance. Counters are, in fact, so commonly used that any list of applications would probably be incomplete.

There are several different types of counters. They include *binary counters,* which produce a binary-coded sequence. 0, 1, 2, 3, . . ., $2^n - 1$, 0, 1, . . .; *decade,* or *BCD, counters,* which follow a binary-coded decimal sequence, 0, 1, 2, 3, . . ., $10^n - 1$, 0, 1, . . .; *scalers*, which produce arbitrary sequences with specific periods; and *Gray-code counters* which generate cyclic codes. The objective of this chapter is to understand the operation of various types of counters so that we can build and troubleshoot these circuits at the board level and at the programmable-logic device (PLD) level of design. After completing this chapter, the reader should be able to:

1. Design reliable counters using the techniques introduced here and in Chapter 4
2. Understand the pitfalls and limitations associated with counter design
3. Determine what type of counter is applicable to a given situation

6-1 BINARY COUNTERS

Binary counters may be easily designed with any type of flip-flop using techniques already described in Chapter 4. Experience will show, however, that the T flip-flop is best suited to binary count sequences. In fact, when designing straight binary counters with D or JK flip-flops, you will find that you are actually building T flip-flops into the circuit. T flip-flops may be used for either synchronous or asynchronous binary counters. As mentioned earlier in Chapter 4, a fully sychronous circuit uses the same clock for every flip-flop. Since all flip-flops change state on the same edge, a synchronous counter will yield stable outputs after only one propagation delay. Asynchronous counters use the output of one flip-flop to trigger the next. Each flip-flop sees the delays of the counter stages preceding it.

Normally counters are made to sequence from an initial value to a final value and then *repeat*. Sometimes, however, they may sequence from some initial value to some final value and then *stop*. In both cases, the number of states they sequence through is called the *modulus* of the counter. A *modulo-16 counter*, for instance, sequences through 16 states before either repeating or stopping. The design of a repeating, fully synchronous, modulo-16 binary counter is illustrated in Fig. 6-1. Each stage uses the decoded outputs of the previous stages to enable a toggle. As long as there is sufficient time between clock edges to fully decode the present state of the previous stages, with time left for setup, the counter

Present State				Next State				*T* Inputs			
Q_3^n	Q_2^n	Q_1^n	Q_0^n	Q_3^{n+1}	Q_2^{n+1}	Q_1^{n+1}	Q_0^{n+1}	T_3	T_2	T_1	T_0
0	0	0	0	0	0	0	1	0	0	0	1
0	0	0	1	0	0	1	0	0	0	1	1
0	0	1	0	0	0	1	1	0	0	0	1
0	0	1	1	0	1	0	0	0	1	1	1
0	1	0	0	0	1	0	1	0	0	0	1
0	1	0	1	0	1	1	0	0	0	1	1
0	1	1	0	0	1	1	1	0	0	0	1
0	1	1	1	1	0	0	0	1	1	1	1
1	0	0	0	1	0	0	1	0	0	0	1
1	0	0	1	1	0	1	0	0	0	1	1
1	0	1	0	1	0	1	1	0	0	0	1
1	0	1	1	1	1	0	0	0	1	1	1
1	1	0	0	1	1	0	1	0	0	0	1
1	1	0	1	1	1	1	0	0	0	1	1
1	1	1	0	1	1	1	1	0	0	0	1
1	1	1	1	0	0	0	0	1	1	1	1

(*a*)

Fig. 6-1(*a*) Design of a 4-bit synchronous binary counter using T flip-flops. Karnaugh maps are not required to see that:

$$T_0 = 1 \quad \text{and} \quad T_1 = Q_0^n$$

Further examination of the truth table or Karnaugh mapping would reveal the following:

$$T_2 = Q_1^n Q_0^n \quad \text{and} \quad T_3 = Q_2^n Q_1^n Q_0^n$$

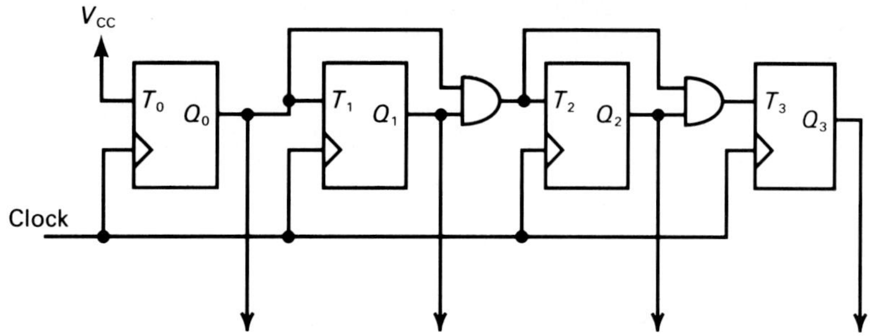

(*b*)

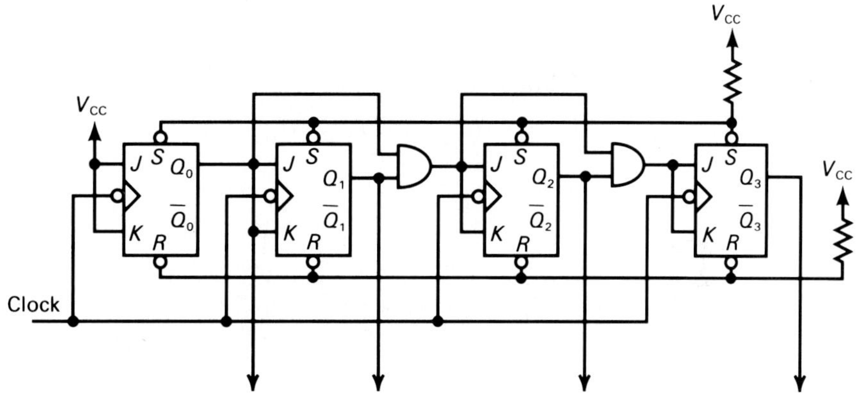

(*c*)

Fig. 6-1(*b*) T flip-flop implementation; (*c*) JK flip-flop implementation.

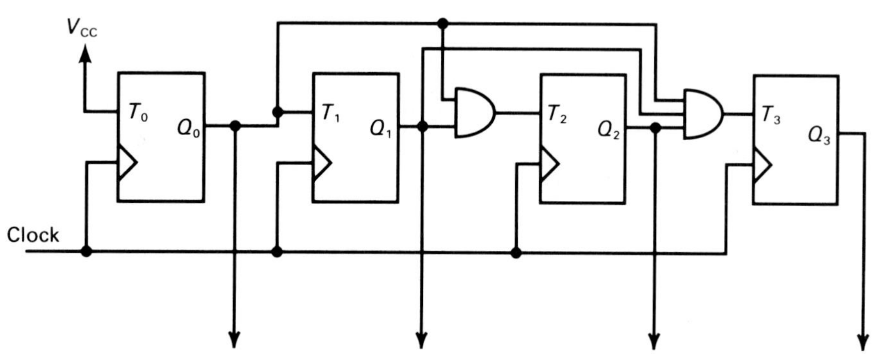

(*d*)

Fig. 6-1(*d*) The logic propagation delay can be reduced by using only one level of gating logic between the present-state outputs and the last flip-flop input. This increases the maximum speed of the counter at the expense of a three-input gate.

will work reliably. The maximum frequency of operation is:

$$f_{\max} = \frac{1}{t_{p(ff)} + t_{p(g)} + t_s} \tag{6-1}$$

where $t_{p(ff)}$ is the worst-case maximum propagation delay through one flip-flop, $t_{p(g)}$ is the worst-case maximum propagation delay through the gating logic, and t_s is the worst-case minimum setup time required by the flip-flops. The example illustrated in Fig. 6-1*b* has two levels of gating logic between the present-state outputs and the last input. This is because *two* AND gates were used to decode $T_3 = Q_2 \cdot Q_1 \cdot Q_0$. If necessary, the logic propagation delay could be reduced by using a single three-input AND gate as shown in Fig. 6-1*d*.

EXAMPLE 6-1

Assuming that the synchronous binary counter in Fig. 6-1 is constructed with 74LS-type JK flip-flops and two-input AND gates, use the data sheet information provided in Fig. 6-2 to determine the maximum frequency of operation for this circuit. If the 74LS packages were replaced with 74HC parts, how fast could this circuit count?

$V_{CC} = 5$ V, $T_A = 25°$C

Parameter		74LS112 $C_L = 15$ pF, $R_L = 2$ KΩ Min.	74LS112 $C_L = 15$ pF, $R_L = 2$ KΩ Max.	Unit
f_{max},	Max. Clock Freq.	30		MHz
t_{PLH} / t_{PHL}	Prop. delay clock to output (Low to high and high to low)		20 / 30	ns
t_s	Setup time J or K to clock	20		ns

$V_{CC} = 5$ V, $T_A = 25°$C

Parameter		74LS08 $C_L = 15$ pF, $R_L = 2$ KΩ Typical	74LS08 $C_L = 15$ pF, $R_L = 2$ KΩ Max.	Unit
t_{PLH} / t_{PHL}	Prop. delay clock to output (Low to high and high to low)	7 / 12	15 / 20	ns

(*a*)

$V_{CC} = 5$ V, $T_A = 25°$C

Parameter		74HC112 $C_L = 15$ pF, $R_L = 2$ KΩ Min.	74HC112 $C_L = 15$ pF, $R_L = 2$ KΩ Max.	Unit
f_{max},	Max. Clock Freq.	30		MHz
t_{PLH} / t_{PHL}	Prop. delay clock to output (Low to high and high to low)		21	ns
t_s	Setup time J or K to clock	20		ns

$V_{CC} = 5$ V, $T_A = 25°$C

Parameter		74HC08 $C_L = 15$ pF, $R_L = 2$ KΩ Typical	74HC08 $C_L = 15$ pF, $R_L = 2$ KΩ Max.	Unit
t_{PLH} / t_{PHL}	Prop. delay clock to output (Low to high and high to low)	7 / 12	15 / 20	ns

(*b*)

Fig. 6-2(*a*) Switching characteristics for LS. (*b*) Switching characteristics for HC.

SOLUTION

Figure 6-1*c* shows the 74xx112 implementation of this circuit. The data sheet tells us that the maximum propagation delay of the LS flip-flop is 30 ns and the maximum delay through *one* LS08 AND gate is 20 ns. Two LS08 AND gates are required so the total gate delay is 40 ns. Minimum setup time for the '112 is 20 ns. Using the formula given above in Eq. (6-1), the maximum frequency of operation is $1/[(30 + 40 + 20) \times 10^{-9}\ s] = 11.1$ MHz.

The HC version of the '112 shows a 21-ns maximum delay and a minimum setup time of 20 ns. The HC08 has a maximum delay of 20 ns. Using HC logic, the same circuit would operate at $1/[(21 + 40 + 20) \times 10^{-9}\ s] = 12.3$ MHz.

Note that the data sheets state that the maximum frequency of operation of these flip-flops is at least 30 MHz. However, those figures apply only to a single flip-flop operating with no external logic. Furthermore, it is important to realize that the values given in Fig. 6-2 are valid only for room temperature (25°C), with $V_{CC} = 5$ V, and a relatively light load ($C_L = 15$ pF, $R_L = 2$ KΩ). Bipolar logic tends to get slower at lower temperatures while a CMOS gets faster. Larger loads and lower supply voltages slow down both types of logic. Derating curves are supplied in the front part of data books to illustrate the effects of variation in temperature, loading, and supply voltage. Note also that when the propagation delay depends on whether the output is making a high or low transition, the worst-case value must be used.

The '112 JK flip-flop also has asynchronous *set* (*S*) (make logic 1) and *reset* (*R*) (make logic 0) inputs. As with any asynchronous inputs, these must be used with care. When they are driven by logic gate outputs, any small glitch, coming at any time, could cause the flip-flop to set or reset unexpectedly. In the examples above they are unused and tied high through resistors to minimize the possibility of power supply glitches causing malfunction. If they are to be driven by gate outputs, the driving logic must be glitch free. Examples of safe reset logic are illustrated in Fig. 6-3. They include *power-on set* or *reset* logic which can change only when power is first applied, and manual reset logic for restarting a machine with a push button. The power-on circuit takes advantage of an *RC* time constant to hold the set or reset inputs low for a short time after power has been applied and a 74xx14, *Schmitt trigger,* to provide a sharp rising edge to the receiving logic. Schmitt triggers are useful because they have two separate thresholds, a higher threshold for rising inputs and a lower threshold for falling inputs. This causes the output to *snap* when the capacitor voltage reaches either the upper or lower threshold voltage instead of following the relatively slow middle part of the capacitor curve.

The manual reset circuit uses an asynchronous latch to *debounce* the switch inputs. This is necessary because even the best-quality switches exhibit some amount of mechanical contact bounce when they open or close. The voltage delivered by the switch will bounce up and down for some short amount of time, but the latch will set or reset on the first low input and ignore all subsequent bounces. The type of switch normally used

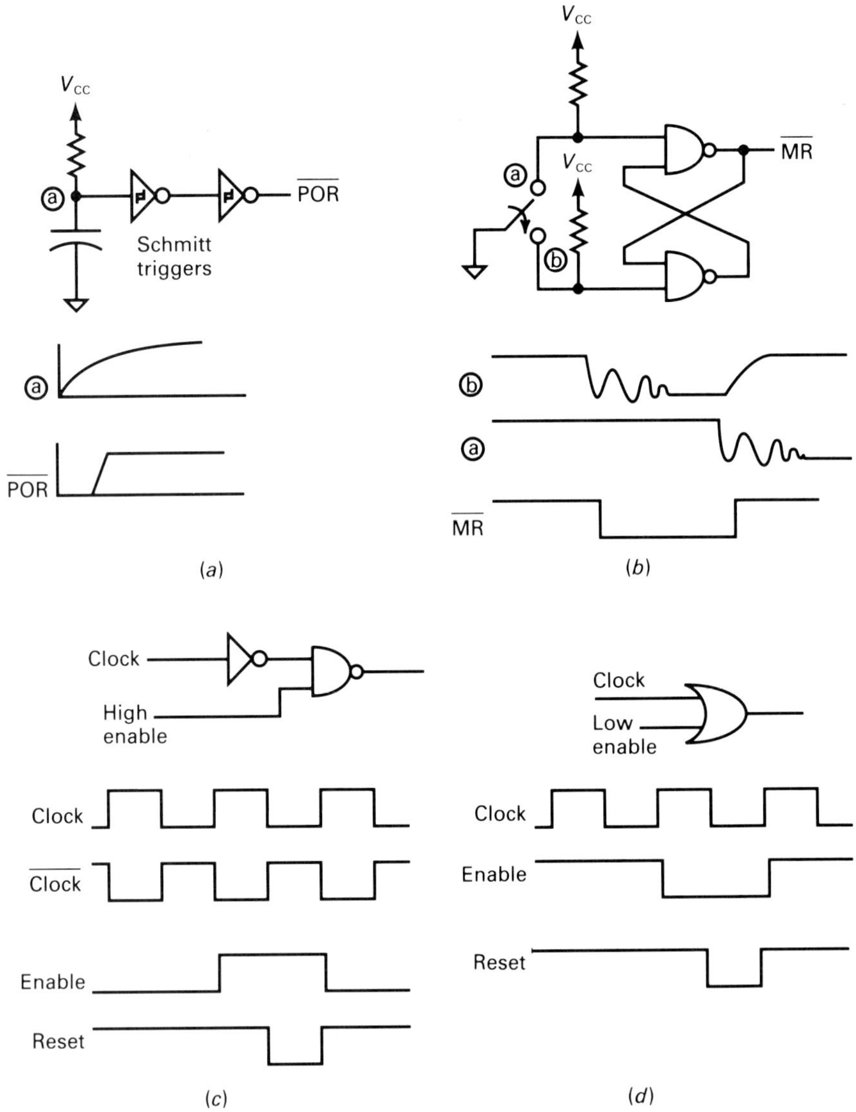

Fig. 6-3 Safe reset circuits. (*a*) Power-on reset (POR); (*b*) manual reset (MR); (*c*) and (*d*), internal signals gated with clock.

here is momentary and returns to its orginal position after the finger is lifted. Clock-gating circuits are also shown, in Fig. 6-3*c* and *d*. These were discussed in Chapter 5.

Ripple Counters

The amount of gating logic between counter stages may be reduced with an asynchronous, ripple counter design as shown in Fig. 6-4. The inputs to each T flip-flop become logic 1s, which simplifies the design of the T flip-flops themselves. A logic 1 input to a T flip-flop causes a toggle on every clock edge. This may be implemented using a D flip-flop with the Q

Not only does it eliminate the need for external gating in the design, but any state may be decoded from its outputs with a single two-input gate. Figure 6-13*a* shows the twisted-tail ring code with the decoding function for each state.

REVIEW 6-2

1. **Q.** What is a scaler? **A.** A scaler in a broad sense is any counter which produces an arbitrary state sequence.
2. **Q.** What is a shift counter? **A.** A shift counter is a type of scaler which uses a shift register with feedback logic to produce a ring-code sequence.
3. **Q.** What advantages do scalers have over straight binary counters? **A.** They are simpler and faster than binary counters.
4. **Q.** What are the disadvantages of scalers? **A.** They produce an irregular output sequence and often do not lend themselves to applications where a square wave is required.
5. **Q.** What is a Gray-code sequence? **A.** It is a sequence of counter outputs in which only 1 bit changes at a time.
6. **Q.** When is a Gray-code counter used? **A.** It is used whenever the outputs must be decoded and used directly to clock or reset other logic. Its outputs are inherently glitch free.
7. **Q.** What are twisted-tail ring counters? **A.** They are shift registers with the Q NOT output of the last stage fed back to the input of the first stage.

6-3 VARIABLE MODULUS COUNTERS

There are many cases in counter design where a variable modulus is needed. One example is when a clock must be scaled down by varying amounts. A variable divide-by-n counter allows other digital circuitry to dynamically change the output frequency. The control circuitry, in some cases, may be looking at a special data format coming in from another system and deciding whether the local clock is too slow or too fast to recover the data. A variable divide-by-n counter is used in these cases to momentarily adjust the beginning of a long counter string so that the final derived clock will be *in phase* with the incoming data. This is the basis of clock and data recovery circuits used in some computer networks and radiotelephone systems. The clock, data, and reset information in these systems are mixed together into a single serial bit stream and separated again at the far end with recovery circuits. Digital implementations of these circuits require variable divide-by-n counters.

The design of a simple divide by-2, -3, or -4 counter is illustrated in Fig. 6-14. Note that the counting sequence is not familiar. Frequency dividers do not need to follow binary-coded sequences. A ring code was selected

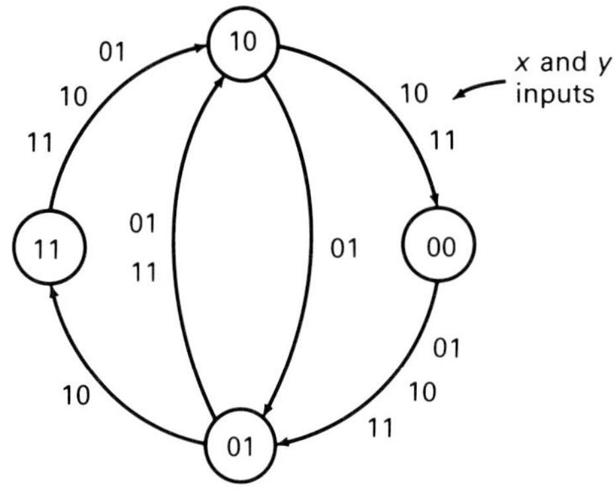

Fig. 6-14(*a*) State diagram for variable modulus divider.

	Inputs x	Inputs y	Present State Q_1^n	Present State Q_0^n	Next State Q_1^{n+1}	Next State Q_0^{n+1}
No count	0	0	0	0	X	X
	0	0	0	1	X	X
	0	0	1	0	X	X
	0	0	1	1	X	X
÷ 2	0	1	0	0	0	1
	0	1	0	1	1	0
	0	1	1	0	0	1
	0	1	1	1	1	0
÷ 4	1	0	0	0	0	1
	1	0	0	1	1	1
	1	0	1	0	0	0
	1	0	1	1	1	0
÷ 3	1	1	0	0	0	1
	1	1	0	1	1	0
	1	1	1	0	0	0
	1	1	1	1	1	0

By inspection:
$Q_1^{n+1} = Q_0^n$

$x\,y$ \ $Q_1^n Q_0^n$	00	01	11	10
00	X	X	X	X
01	1	0	0	1
11	1	0	0	0
10	1	1	0	0

$$\overline{Q_0^{n+1}} = yQ_0^n + xQ_1^n$$

$$Q_0^{n+1} = \overline{yQ_0^n + xQ_1^n}$$

Fig. 6-14(*b*) State table, Karnaugh map, and state equations for variable modulus divider.

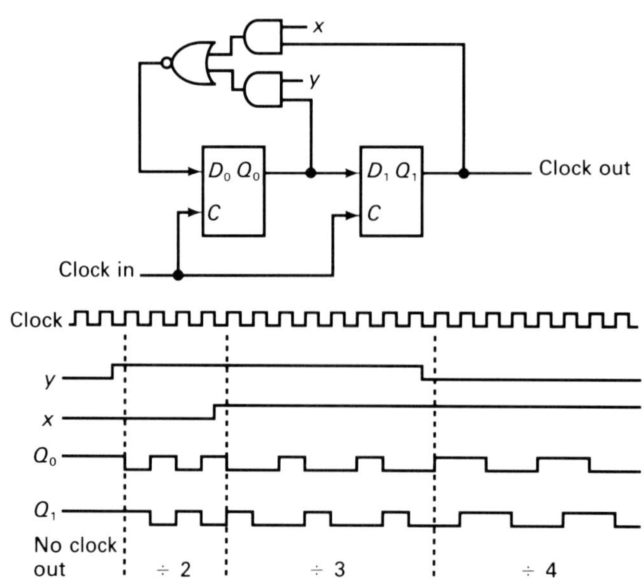

Fig. 6-14(*c*) Circuit and timing diagrams for variable modulus divider.

in this case to simplify the design. Ring codes are often used in variable modulus scalers. The don't-care conditions for the next state of Q_1 were selected so that they would look just like the present state of Q_0. This eliminates gating between the flip-flops. The don't-care conditions for the next state of Q_0 were found by examining the Karnaugh map. In this particular case it was easiest to assign 1s to the don't-care conditions and use the 0s to derive the function. You may recall from Chapter 3 that this produces the inverse of the desired function. The true function is produced by simply inverting the results.

The variable modulus divider in Fig. 6-14 was designed by the rules, using a state diagram, a state table, and a Karnaugh map. This is always a safe approach. However, if variable divide sequences greater than 4 are desired, five- or six-variable maps would be needed. Karnaugh mapping beyond four variables becomes unwieldy, and it is often easier to break the problem into smaller, more manageable parts. We can determine, using four-variable maps, what feedback is required for each separate divide ratio. Using this information, we can then use the familiar AND-OR feedback structure to combine them. For instance, Fig. 6-15a shows divide-

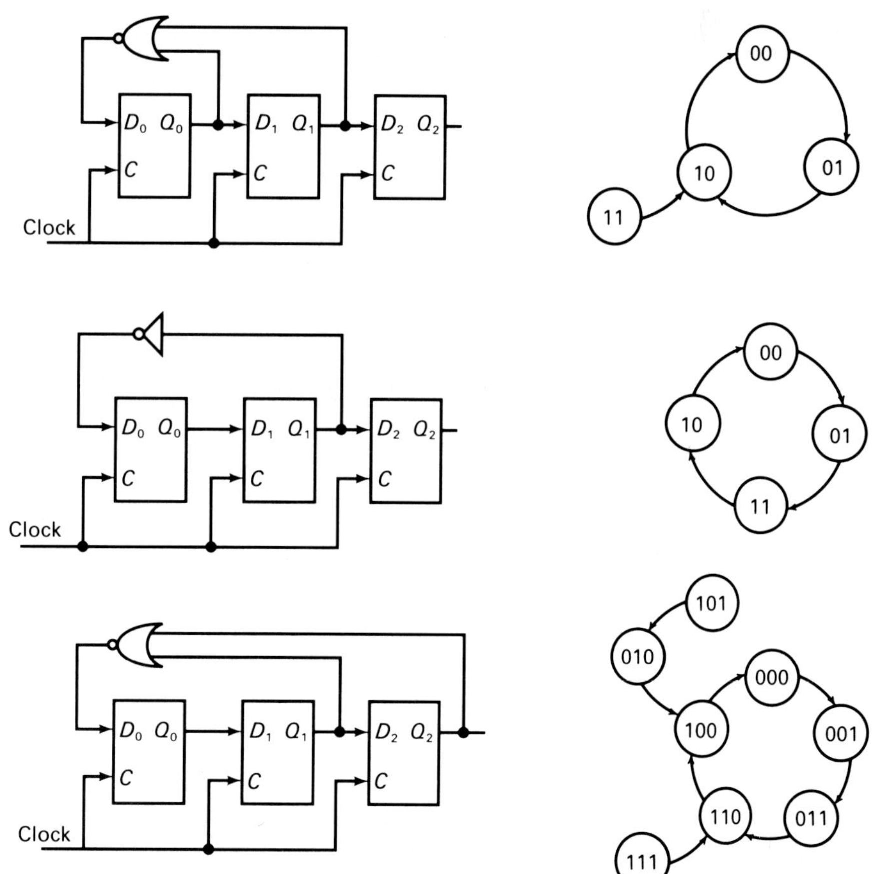

Fig. 6-15(a) Modulo-3, modulo-4, and modulo-5 dividers.

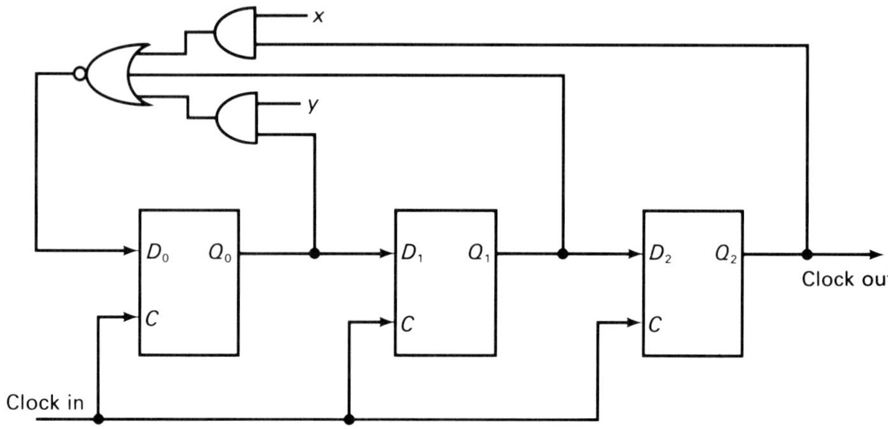

x	y	Function
0	0	÷ 4
0	1	÷ 3
1	0	÷ 5
1	1	÷ 5

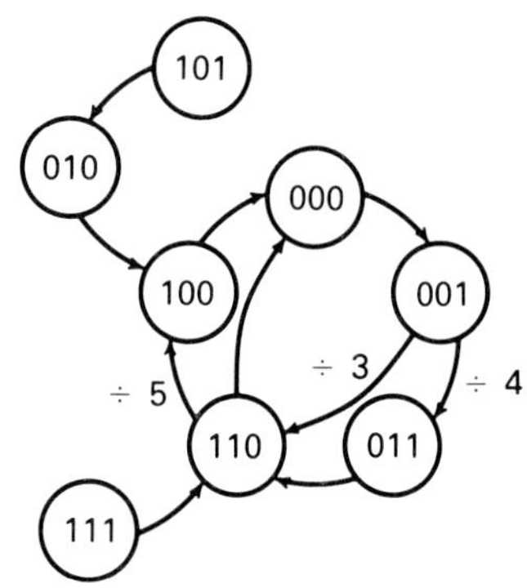

Fig. 6-15(*b*) Modulo-3, 4, 5 divider.

by-3, divide-by-4, and divide-by-5 circuits separately. These can be derived using two- and three-variable maps. Figure 6-15*b* shows how these circuits were combined to produce a variable modulo-3, 4, 5 divider. State or timing diagrams may be used to verify the operation of the divider, and modifications can be made if hang-up states exist. The state diagram in Fig. 6-15*b* shows that the example operates correctly.

6-4 *N*-And-Only-*N* COUNTERS

Another type of counter is called *n-and-only-n* because it counts through *n* states and stops. The final state is decoded and used to synchronously disable any further changes. An external reset is required to restart the counting. One application for this type of counter is in asynchronous data transfer between remote systems. A commonly used circuit, called a *universal asynchronous receiver-transmitter (UART),* looks for the beginning of an 8-bit coded character and clocks in only 8 bits. A *beginning-of-character circuit* resets an *n-and-only-n* counter which produces 8 and only 8 clock edges. After 8 clock edges, it stops and waits for a new character to arrive. Since no clock is sent along with the data, the operation depends on the shortness of the message to ensure reliable reception. If the 2 local clocks are fairly close in frequency (to within one-tenth of a bit time), the difference will not cause malfunction in only 8 bits. Each character is preceded by a *space*, or low level, and followed by at least one *mark*, or high level. Synchronization is achieved at the beginning of each character by the beginning-of-character circuit and loosely maintained through 8 bits by the closeness of the 2 clocks. This type of data transfer is only used for relatively slow data rates (up to 9600 bits per second) and only for noncritical applications.

EXAMPLE 6-5

Design a circuit which sequences through eight states and stops.

SOLUTION

There are two ways to approach this problem. One is to use enabled flip-flops for each stage. This involves decoding the final state and disabling the flip-flops when the final state is reached. Figure 6-16*b* shows how the

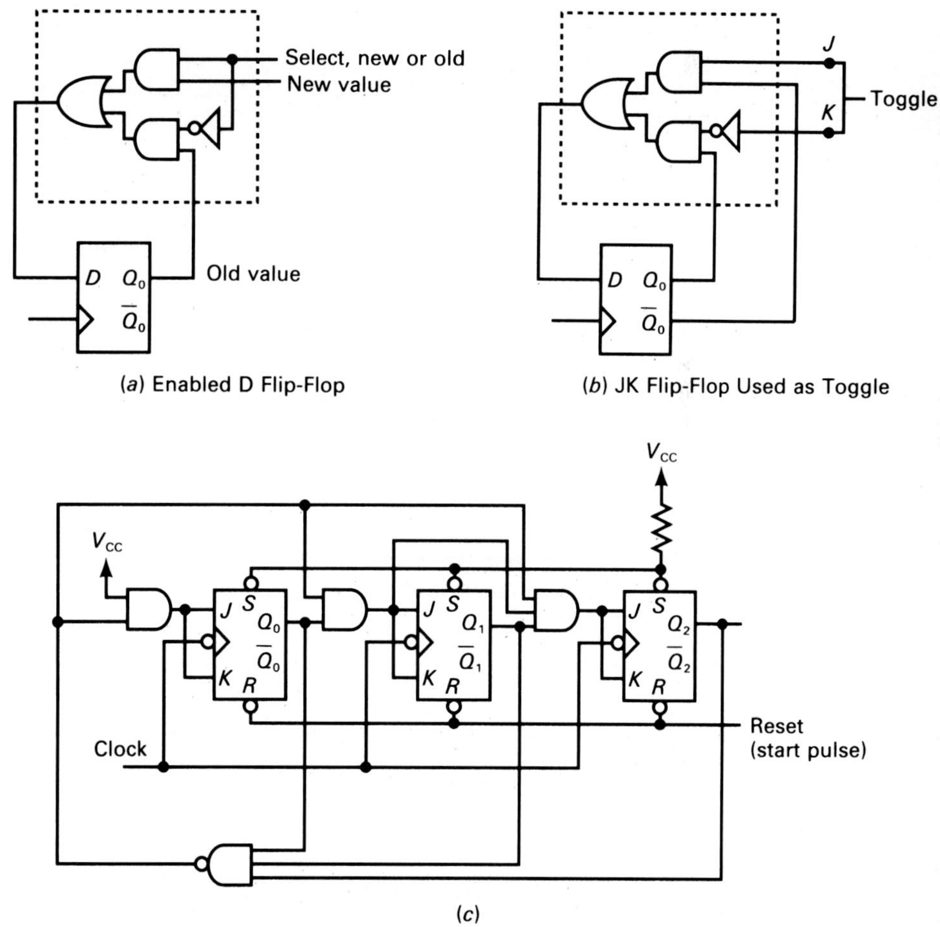

Fig. 6-16(*a–c*) 8-and-only-8 counter using enabled inputs.

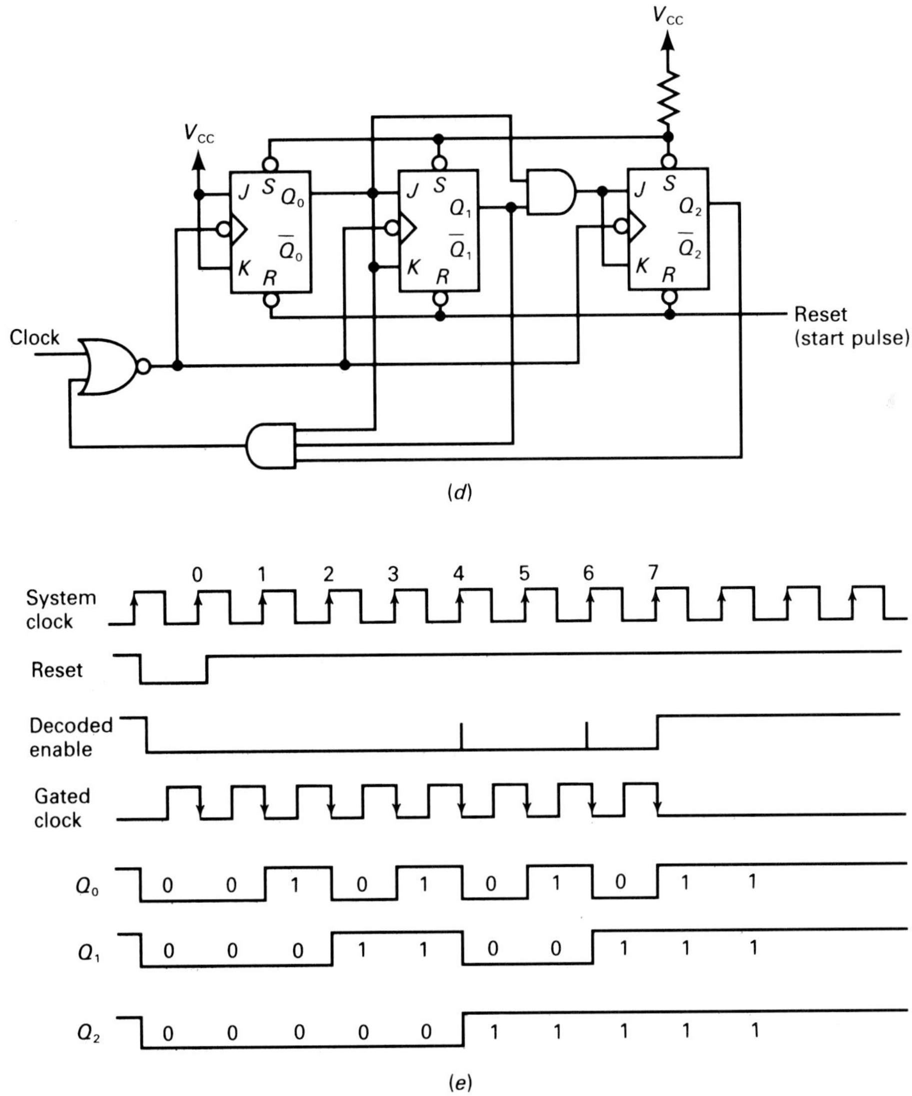

Fig. 6-16(*d*) 8-and-only-8 counter using clock-gating. (*e*) Timing waveforms showing decoding glitches.

JK flip-flop is actually an enabled D flip-flop with the enable control inputs separated. We recombine the enable control inputs when we use the JK flip-flop as a T flop. The choice of counter design is easy since the required modulus is a power of 2. A straight binary-coded sequence is chosen, and a three-input NAND gate is used to decode the final state.

The other approach, previously discussed in Chapters 4 and 5, is *clock gating*. This involves decoding the final state and disabling the clock. There is always a danger involved with enabling and disabling the clock. Whenever the counter outputs change, there is a possibility that they will not all change at exactly the same time. This may be due to differences in external loading, to differences in internal loading, or even to differences in the intrinsic propagation delays of the flip-flops. Consequently the decoding logic may produce momentarily false outputs. And these false outputs may be present long enough to cause glitches on the gated clock. However, these decoding glitches can occur only *after* a clock edge, since it is the clock edge which produces them. The enabled flip-flop method is immune to this problem because by the time the glitches occur, the clocking operation is over. However, when clock gating is used, the decoding glitches may produce an extra clock edge. This can be prevented by gating the clock in such a way that the effect of the glitches is masked by the clock itself. Figure 6-16*e* shows where the decoding glitches might occur and how the counter operation is unaffected because of proper gating. The glitches can occur only after the clock has already changed high, and the NOR gate will not respond to false highs or lows during this time.

It is important to note that in both design examples it was assumed that care was exercised in producing the start pulse. The timing waveforms in Fig. 6-16*e* show the counter restarting just after a clock edge. The logic in the system responsible for resetting the 8-and-only-8 counter must be synchronous with the master clock.

REVIEW 6-4

1. **Q.** What are n-and-only-n counters? **A.** They are counters which stop counting after sequencing through n states.

2. **Q.** How is a counter stopped after sequencing through n states? **A.** The nth state is decoded and used to either hold the present state of each flip-flop with input gating or disable the clock with clock gating.

3. **Q.** What is the danger of clock gating in this application? **A.** Improper clock gating could result in glitches reaching the clock inputs.

4. **Q.** When would this type of counter be used? **A.** whenever we need to clock in a fixed number of bits or bytes

5. **Q.** How is this type of counter started? **A.** with a synchronous reset pulse

6-5 PARALLEL LOAD COUNTERS

Another type of counter features synchronous load inputs similar to those used in parallel-in–serial-out registers. A *load* input permits each flip-flop to accept a parallel input regardless of its present state. This provides variable divide-by-n operation by virtue of the fact that the counter may be forced to any starting value. Naturally more logic is required than was needed for the customized variable divide-by-n circuits, but the result is a more universal design. The additional logic is the familiar AND-OR structure which permits each flip-flop to change either to its next count sequence or to some external preset value. The same type of logic was used in Chapter 5 to achieve parallel-in–serial-out operation with a load input.

Figure 6-17*a* shows one stage of a toggle flip-flop with parallel load. This counter stage may be represented as a box with inputs and outputs shown in Fig. 6-17*b*. The synchronous binary up-counter of the type designed in

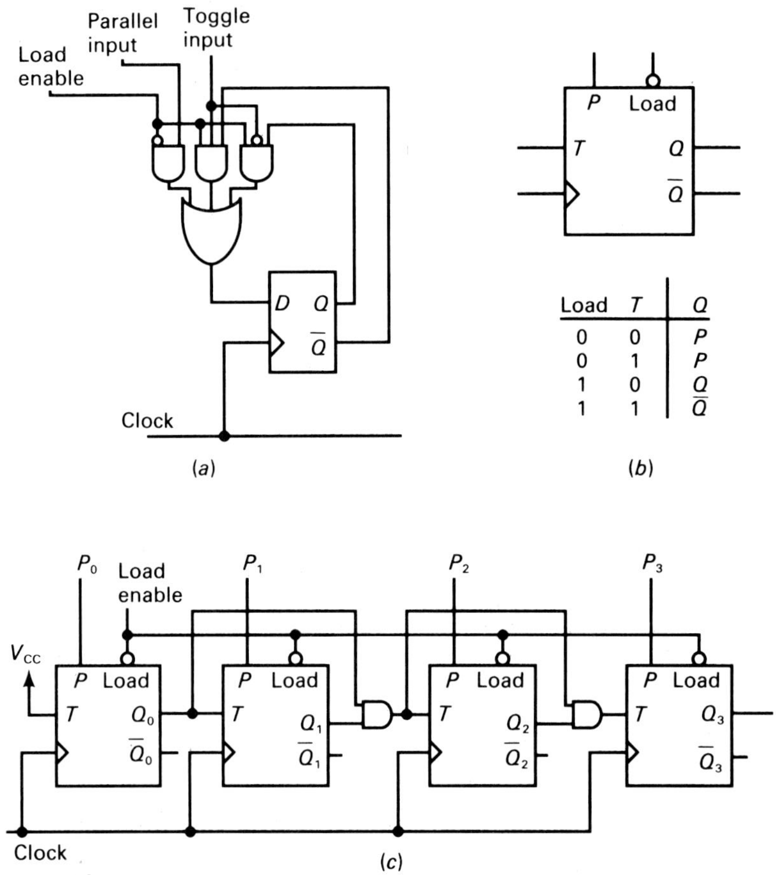

Fig. 6-17(*a–c*) Four-stage binary counter with parallel inputs.

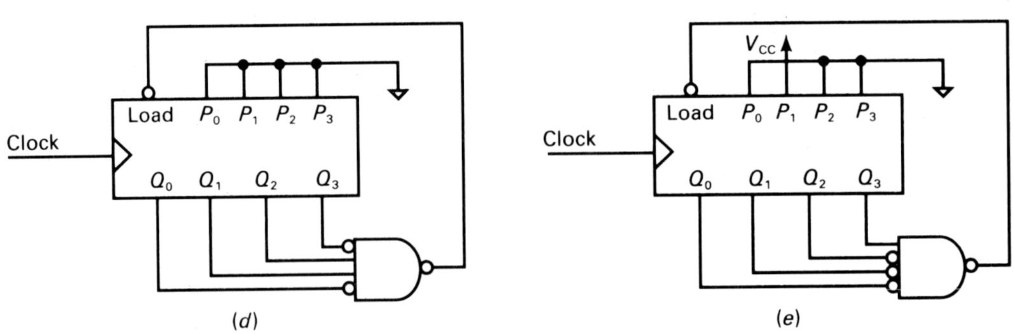

Fig. 6-17(*d, e*) Two modulo-7 counters using parallel load.

Sec. 6-1 may be constructed using four of these boxes. A modulo-n, or divide-by-n, counter may be implemented with this design by decoding a final state and using it to preset each flip-flop. For example, a modulo-7 counter is easily constructed using a NAND gate to decode the binary number 6. The decoded output is then used to load the flip-flops on the next clock edge as shown in Figure 6-17*d*. The counting sequence is . . . 0, 1, 2, 3, 4, 5, 6, 0, ... Other counting sequences can be produced with the same logic. Figure 6-17*e* shows how a modulo-7 counter could be configured to count . . . 2, 3, 4, 5, 6, 7, 8, 2, ... In this case the 1000 state is decoded and used to parallel load the value 0010. The load function is synchronous with the clock. Asserting the load control input does not immediately cause the counter outputs to change. The load control input is asserted just after 1 clock edge and causes the desired change just after the next clock edge. It is also possible to feed the decoded load enable back to asynchronous sets and resets. In this way the flip-flops are *jammed* to predetermined values. But this causes two states to occur in one clock period, and the decoded state may not remain stable long enough to change all the flip-flop stages correctly. Moreover, the decoding logic may produce glitches which could cause intermittent and unpredictable operation. This type of design should be avoided.

REVIEW 6-5

1. **Q.** What actually happens on a clock edge when the parallel load enable is asserted? **A.** The parallel input is clocked into the flip-flop stage.
2. **Q.** How can parallel inputs be used to implement modulo-n operation? **A.** The counter state can be decoded and used to load parallel inputs on the next clock edge.
3. **Q.** Why does the decoded load enable not cause the counter to immediately change state? **A.** The counter does not immediately change states because the load enable is used to gate the input to each flip-flop. Since the load enable occurs only after a clock edge and new input data is accepted only on a clock edge, the parallel inputs will not be seen until the next clock edge arrives.
4. **Q.** Can parallel inputs be loaded asynchronously? **A.** Yes, but this is not good practice since the state which produces the load signal may not remain stable long enough to ensure reliable operation. Furthermore, the decoding logic may produce glitches which could cause unexpected results.

6-6 STANDARD PARTS

Many of the counting functions described above are available in a single IC provided by a family of packaged parts numbered 54/74xx160 to 54/74xx163. Figure 6-18 shows the HC versions of these parts. The '160 series of parts are the workhorses of board-level counter design. The popular

MOTOROLA
SEMICONDUCTOR
TECHNICAL DATA

Presettable Counters

High-Performance Silicon-Gate CMOS

MC54/74HC160
MC54/74HC161
MC54/74HC162
MC54/74HC163

The MC54/74HC160 through HC163 are identical in pinout to the LS160 through LS163, respectively. The device inputs are compatible with standard CMOS outputs; with pullup resistors, they are compatible with LSTTL outputs.

The HC160 and HC162 are programmable BCD counters with asynchronous and synchronous Reset inputs, respectively. The HC161 and HC163 are programmable 4 bit binary counters with asynchronous and synchronous reset, respectively.

- Output Drive Capability: 10 LSTTL Loads
- Outputs Directly Interface to CMOS, NMOS, and TTL
- Operating Voltage Range: 2 to 6 V
- Low Input Current: 1 μA
- High Noise Immunity Characteristic of CMOS Devices
- In Compliance with the Requirements Defined by JEDEC Standard No. 7A
- Chip Complexity: 234 FETs or 58.5 Equivalent Gates

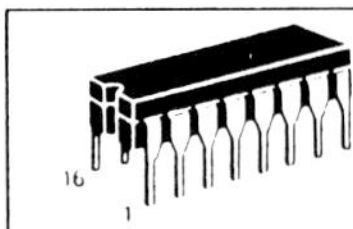

J SUFFIX
CERAMIC
CASE 620

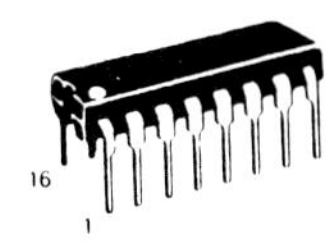

N SUFFIX
PLASTIC
CASE 648

D SUFFIX
SOIC
CASE 751B-01

ORDERING INFORMATION

MC74HCXXXN	Plastic
MC54HCXXXJ	Ceramic
MC74HCXXXD	SOIC

T_A = −55° to 125°C for all packages.
Dimensions in Chapter 7.

LOGIC DIAGRAM

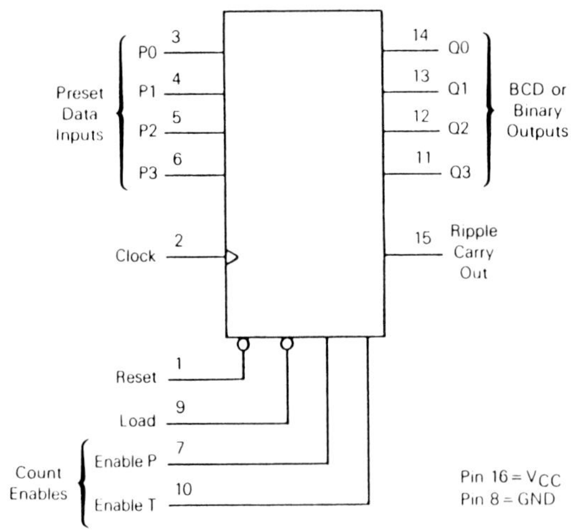

PIN ASSIGNMENT

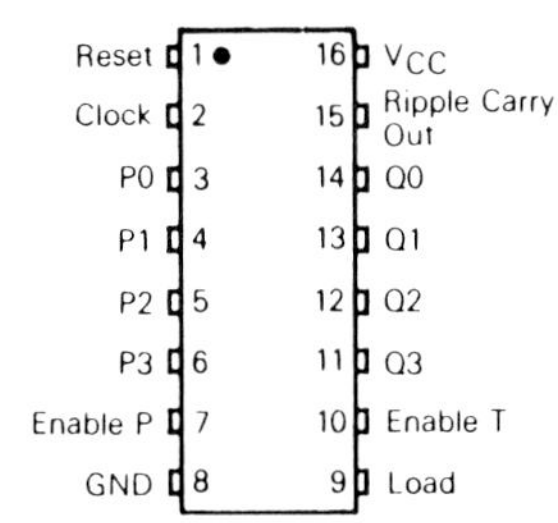

Device	Count Mode	Reset Mode
HC160	BCD	Asynchronous
HC161	Binary	Asynchronous
HC162	BCD	Synchronous
HC163	Binary	Synchronous

FUNCTION TABLE

Inputs					Output
Clock	Reset*	Load	Enable P	Enable T	Q
↑	L	X	X	X	Reset
↑	H	L	X	X	Load Preset Data
↑	H	H	H	H	Count
↑	H	H	L	X	No Count
↑	H	H	X	L	No Count

*HC162 and HC163 only. HC160 and HC161 are Asynchronous-Reset Devices

H = high level
L = low level
X = don't care

Fig. 6-18 Data sheet for the HC 160 series programmable counters. (*Courtesy Motorola Semiconductor Products, Inc.*)

'163, for instance, is a modulo-16 binary counter with parallel load, synchronous reset, count enable, and decoded full count. If the clear and parallel inputs are unused (tied high), and the count is enabled, the '163 will count from 0 to 15 and back to 0. The parallel load permits shorter counts and count sequences starting at numbers other than 0. The synchronous clear allows a parallel load of all 0s independent of the parallel inputs. In addition, there are two enable inputs which allow the counter to change states only when both are asserted.

Any modulo-n sequence for $2 \le n \le 16$ can be produced with the '163 by decoding the desired state and resetting or presetting the counter outputs using the clear or parallel load controls. The modulus may also be changed, dynamically, with control logic applied to the parallel inputs. N-and-only-n operation can be achieved by decoding the final state and applying the decoded signal to one of the counter enables. Finally, '163 packages may be cascaded to form longer synchronous or quasi-synchronous counters.

The '162 is a modulo-10 version of the '163. If the clear and parallel inputs are unused (tied HIGH) and the counts enabled, it will count from 0 to 9 and back. The '162 outputs may be decoded using BCD-to-7 segment decoder-drivers to produce a 0 through 9 decimal display with LED segments. The '160 is identical to the '162, and the '161 is identical to the '163 except for the fact that the lower-numbered parts have *asynchronous* rather than synchronous reset inputs. The '162 and '163 are the most versatile members of the '160 series and will be used in the examples below.

There is some discrepancy between data books in the naming of the load, reset, and enable inputs. Some data books refer to pin 7, one of the count enable inputs, as "PE" while others refer to pin 9, the load input, as "PE." The reset input is variously referred to as clear (spelled out), *master reset* (*MR*), and *synchronous reset* (*SR*). The decoded full count is known as *carry out* (spelled out), *terminal count* (*TC*), and *ripple carry* (*RC*). Since there is no agreement on what these inputs and outputs should be called, we will use terminology consistent with that already introduced. These names are listed with the pin numbers to avoid confusion. Pin 1 is the *reset* (*R*) input and is active low. Pin 9 is the parallel *load* (*LD*) input and is also active low. Pins 7 and 10 are count enable inputs and must both be high in order for any state changes to occur. The count enables are ANDed in the '163, and if either input is low, the counter will not change state. Pin 15 is the *terminal count* (*TC*) output and decodes to an output HIGH when the counter reaches the 1111 state *if* pin 10 is also HIGH. Since the count enable at pin 10 must be HIGH for the terminal count, it is commonly referred to as *count enable terminal* (*CET*). The other count enable, at pin 7, is commonly called the *count enable program* (*CEP*). The HC data sheets shorten these signal names to *enable T (ET)* and *enable P (EP)*. The following examples are meant to illustrate the capabilities of the '163.

EXAMPLE 6-6

Design a modulo-13 counter using the '163.

SOLUTION

This problem involves decoding and parallel loading in the same manner as illustrated in Fig. 6-17. If a . . . 0, 1, 2, . . ., 9, A, B, C, 0, 1, . . . sequence is desired, then the hex number *C*, binary 1100, is decoded with two inverters and a NAND gate and fed back to the reset input. This is equivalent to parallel loading all 0s. Another, simpler, approach can be used if the 0-to-12 sequence is not needed. The TC output can be inverted and applied to the load input to parallel load the number 3. The count would then be . . . 3, 4, 5, 6, 7, 8, 9, A, B, C, D, E, F, 3, ... In general, if a binary counter normally counts to $2^n - 1$ and we want modulo-m operation, we can load the number $2^n - m$ by inverting the terminal count output and using it as the load enable. If the counter starts in an invalid state, it is self-correcting. There is no danger of hang-up states when using the '163 since it is full-count modulo-16 counter. Figure 6-19 shows the design along with timing diagrams.

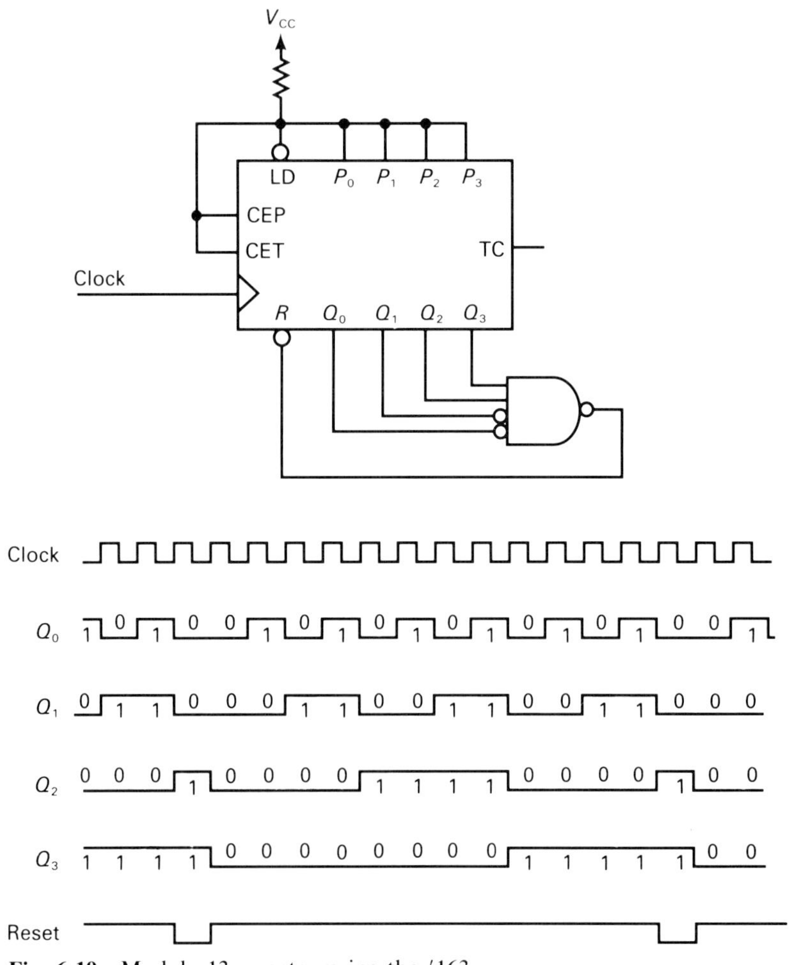

Fig. 6-19 Modulo-13 counter using the '163.

EXAMPLE 6-7

Design a 12-and-only-12 counter using the '163.

SOLUTION

This design is similar to the 8-and-only-8 counter in Fig. 6-16. The '163 makes it easy since it provides a synchronous means of disabling the count. The '163 may be restarted using the synchronous reset input. Figure 6-20 shows how this circuit is implemented using the '163.

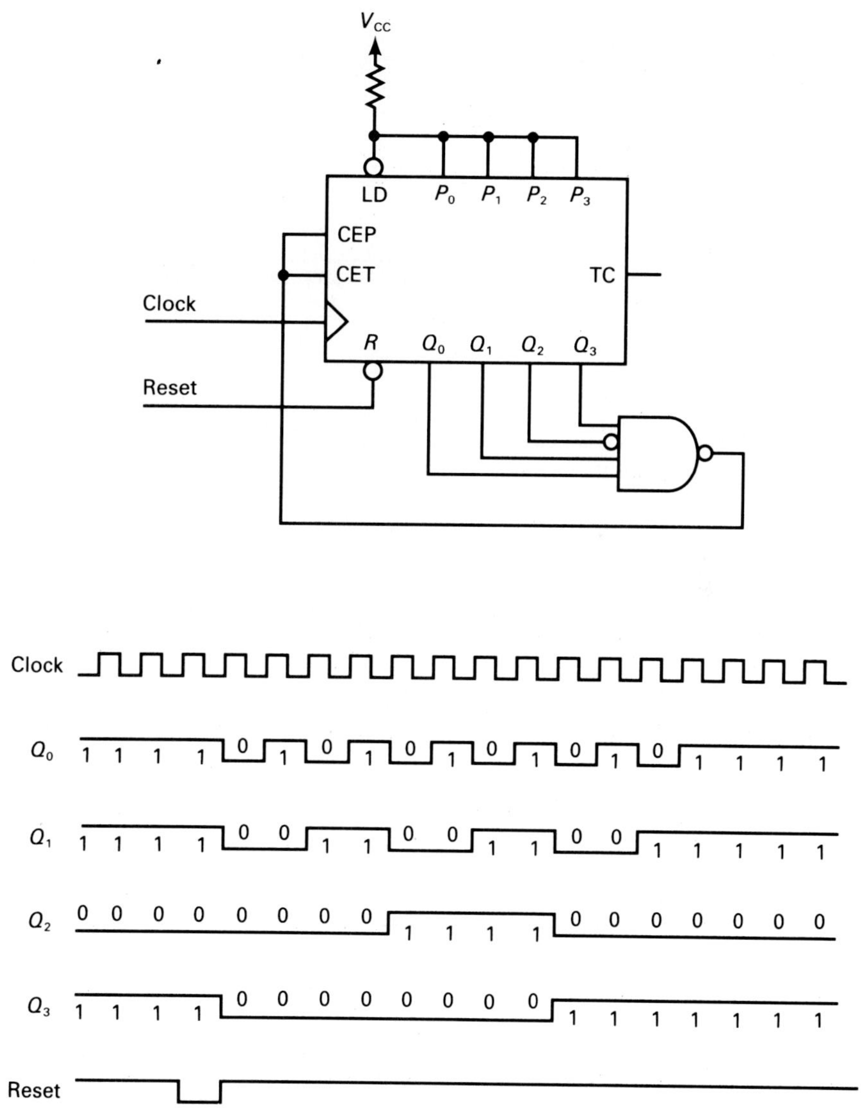

Fig. 6-20 12-and-only-12 counter using the '163.

EXAMPLE 6-8

Design a variable divide-by-3, 4, 5 scaler using the '163.

SOLUTION

Here we take advantage of the *terminal count* output. When the counter reaches the 1111 (hex F) state, the TC output goes HIGH and is inverted to produce a LOW which parallel loads the appropriate inputs. Since the TC is simply an AND of the four Q outputs with CET, glitches may occur in the places shown in Fig. 6-21. The TC output therefore cannot be used to directly clock another circuit but may be used to enable the clock to another circuit.

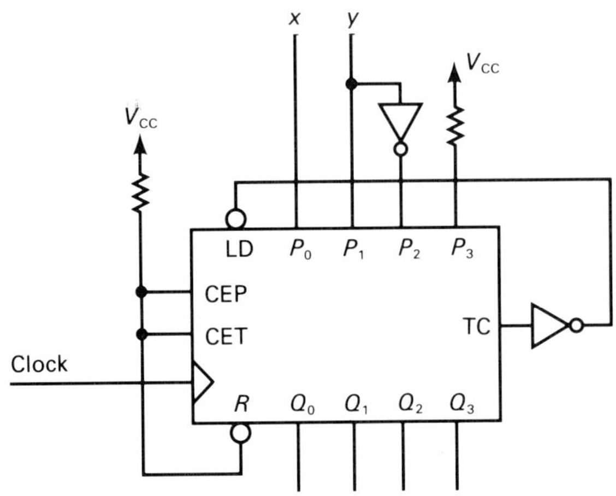

x	y	Ratio
0	0	÷ 4
0	1	÷ 6
1	0	÷ 3
1	1	÷ 5

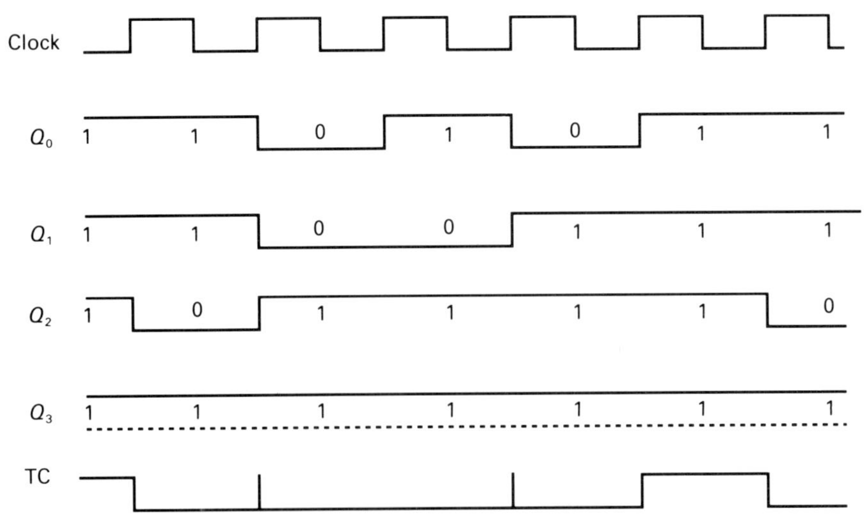

Fig. 6-21 Programmable divide-by-3, 4, 5, 6 scaler with timing for the ÷ 5 mode.

EXAMPLE 6-9

Design a modulo-1000 counter using cascaded '162 counters.

SOLUTION

The individual '162 stages can be cascaded using the TC outputs and the CE inputs as shown in Fig. 6-22. The TC output cannot be used to directly

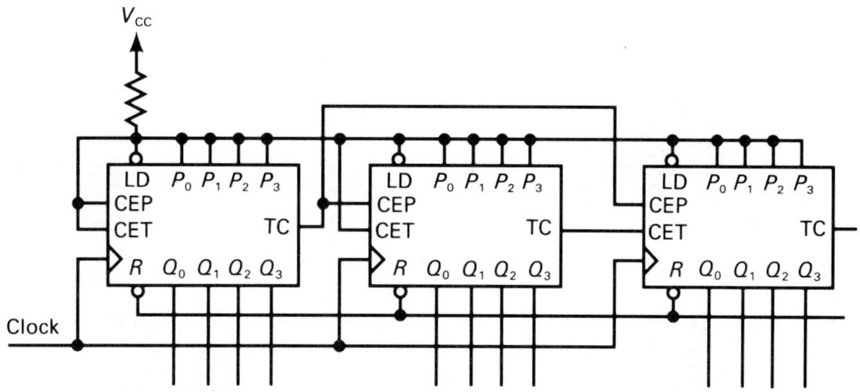

Fig. 6-22 Modulo-1000 counter using '162s.

clock the following stage since it may produce glitches. It is used instead to enable the following stage. After the first stage, each TC output drives the CET input of the succeeding stage, and the first counter TC drives the CEP inputs to all succeeding counters in the chain. Since the first counter is the last counter to come to a full count, all other stages will have been at maximum count for at least nine clock periods before the first stage TC goes high.

EXAMPLE 6-10

A word of warning: The '160 and '161 counters may not be used interchangeably with the '162 and '163. The asynchronous reset can cause unpredictable operation in circuits where the reset input is used. As an example, consider the modulo-13 counter in Fig. 6-19. The timing diagrams in Fig. 6-23 show two possible malfunctions which may result when using the '161. The first involves a decoding glitch which might have reset the counter early. The second malfunction could occur if one of the outputs were slower than the others. In this case a faster output changed soon enough to remove the reset before the slower output had time to stabilize. This put the counter into the wrong state. In addition, since the counter resets immediately after the final count has been reached, there will always be two present states in one clock period. One clock edge will produce 1100 and 0000, and the next edge will produce 0001.

Fractional Dividers

Sometimes fractional divide ratios are useful. It may be necessary, for instance, to provide a divide-by-1.5 or divide-by-2.5 scaler when a single

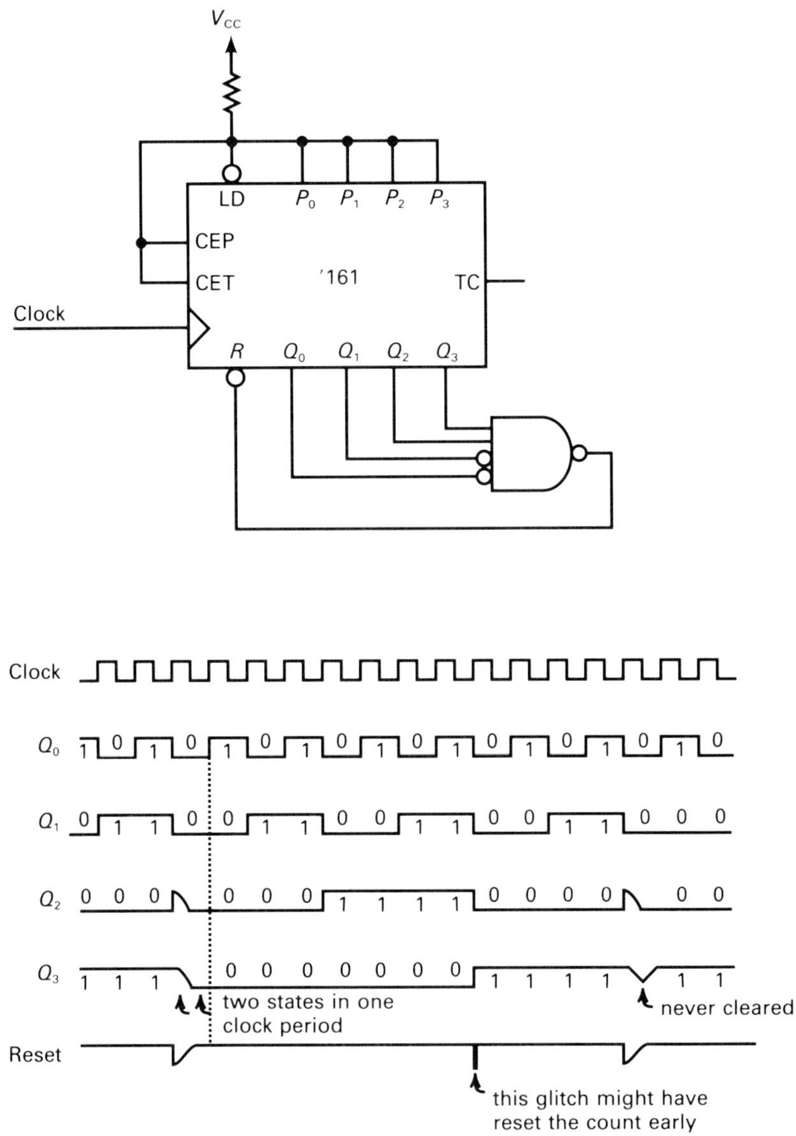

Fig. 6-23 Possible malfunctions caused by incorrect usage of the asynchronous reset on the '161.

master clock is used for several applications. It is often possible to use shift counters to produce $(n + 1/2)$ divide ratios by doubling the frequency of odd modulus counters. The divide-by-3 scaler in Fig. 6-15, for instance, can be modified to produce a 1.5 divide ratio by doubling its output frequency with an extra flip-flop and a NOR gate. Figure 6-24 shows the implementation of this counter along with a double-rate version of the divide-by-5 scaler. These counters are self-starting and glitch free.

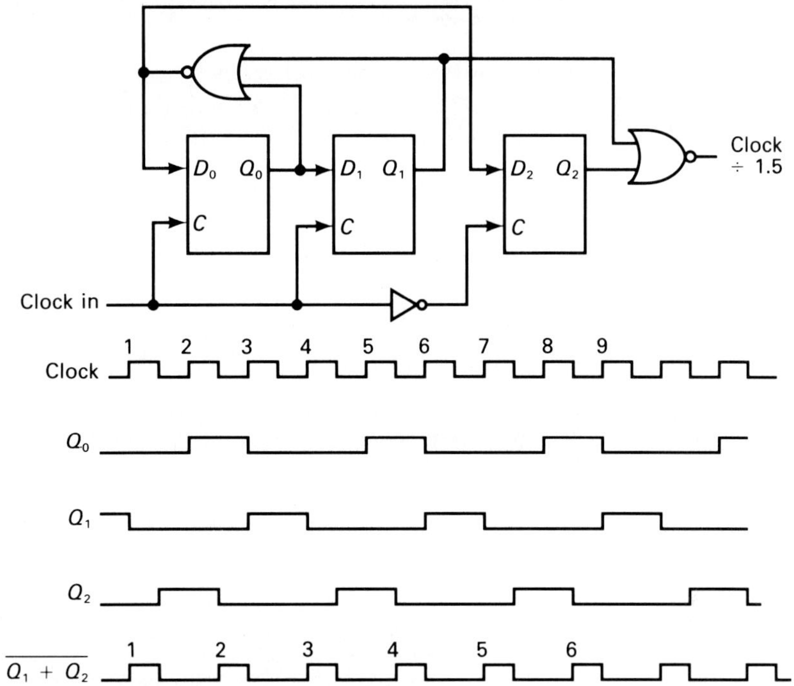

Fig. 6-24(*a*) Divide-by-1.5 scaler.

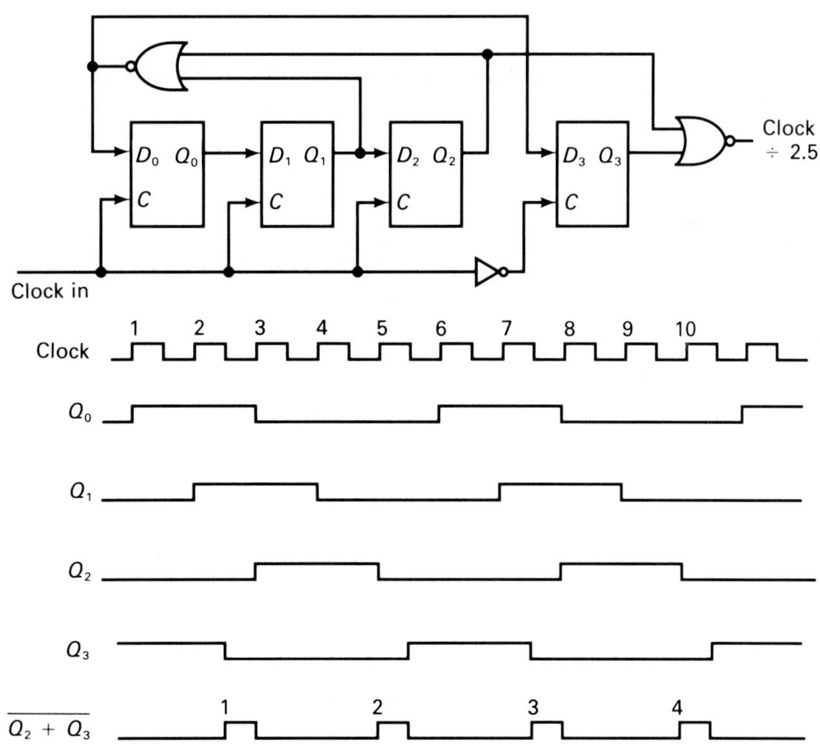

Fig. 6-24(*b*) Divide-by-2.5 scaler.

REVIEW 6-6

1. **Q.** What types of counter operation may be implemented with the '160 series packaged parts? **A.** Modulo-*n*, *n*-and-only-*n*, and variable modulus counters can be implemented using the '160 series.

2. **Q.** How does the '162 differ from the '163? **A.** The '162 has a full count of 9, making it a decade counter; the '163 has a full count of 16, making it a binary counter. Both may be configured to produce a lower modulus.

3. **Q.** How do the '162 and '163 counters differ from the '160 and '161? **A.** The lower-numbered parts have asynchronous reset inputs.

4. **Q.** Which state is internally decoded in all '160 series parts? **A.** the final count, which is 1001 in the case of the '160 and '162, and 1111 in the case of the '161 and '163

5. **Q.** Can the TC output be used to reset the '160? **A.** No, because the TC may contain glitches which could reset the '160 at the wrong time.

6. **Q.** Can the TC output be used to reset the '162? **A.** Yes, because the '162 has a synchronous reset.

7. **Q.** How can the '160 series be used to produce *n*-and-only-*n* operation? **A.** The *n*th state can be decoded and used to disable further counts with one of the two count enables.

8. **Q.** How can '160 series counters be used to produce divide ratios greater than 16? **A.** They can be cascaded by using the TC output of one package to enable the next. One enable of every stage, after the first, should be driven by the first-stage TC.

REVIEW PROBLEMS

1. Determine the maximum clock frequency of each of the synchronous scalers shown in Fig. 6-11. Assume that HC112s are specified with the parameters given in Fig. 6-2. Assume that any logic gates have a maximum propagation delay of 20 ns.

2. Use the HC112 parameters as specified in Fig. 6-2 to compute the maximum clock frequency of (a) a modulo-8 binary ripple counter, (b) a modulo-16 binary ripple counter, and (c) a modulo-32 binary ripple counter.

3. Find a ring code for a modulo-5 shift counter. Use a state table and Karnaugh map to implement this counter with D flip-flops.

4. Using the ring code given in Fig. 6-9, design a modulo-12 shift counter with D flip-flops.

5. Use the ring code given in Fig. 6-25*a* to design a mod-9 JK scaler.

6. Use the ring code given in Fig. 6-25*b* to design a mod-11 JK scaler.

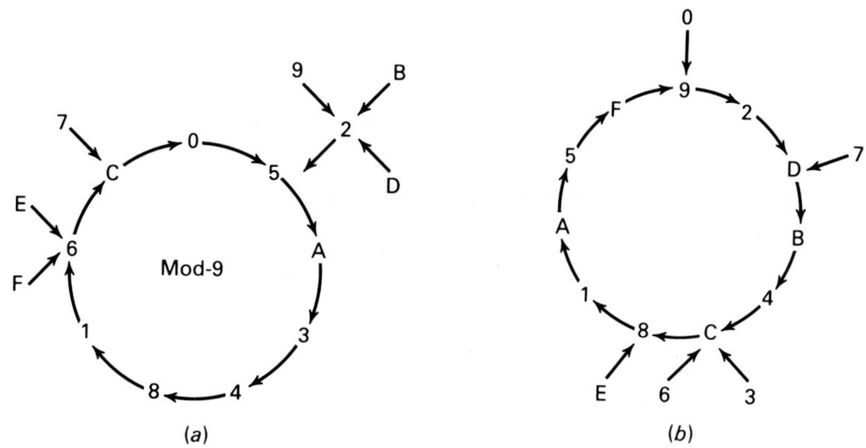

Fig. 6-25 JK ring codes for Probs. 5 and 6.

7. Show how a modulo-12 JK scaler could be built using relative primes.

8. Design a modulo-14 JK scaler using relative primes.

9. Find a modulo-6 Gray code using a Karnaugh map, and design the corresponding counter.

10. Design a modulo-6 Gray-code counter using the twisted-tail method.

11. Design a variable modulus divider which will produce the following divide ratios: 2, 3, 4, 5, 6, 8.

12. Sketch the timing waveforms for the two modulo-7 counters shown in Fig. 6-17.

13. Show where glitches might occur in the decoded reset for the mod-13 counter in Fig. 6-19. Could a '161 be used here?

14. Use a '163, a '164, a '374, and any necessary gates to construct a synchronous byte receiver circuit. Data, clock, and synchronization pulse are supplied by another system. This circuit will reset on the synchronization pulse and shift in 8 bits of serial data. When all 8 bits are received, they are to be transferred to the '374. On the same clock that the transfer occurs, the first bit of the new word should be shifted into the '164.

15. Design an 8-and-only-8 counter using the '163. This counter will be used to supply a gated clock to another circuit as shown in Fig. 6-26*a*. The decoded output should gate the clock so that 8-and-only-8 rising edges are supplied to another circuit. Assume that a proper, synchronous start pulse is supplied. Redesign this circuit so that only one gate is needed as shown in Fig. 6-26*b*.

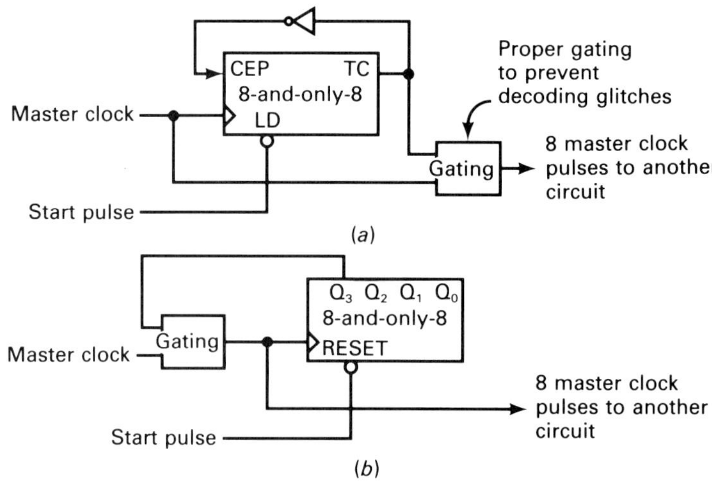

Fig. 6-26 Block diagram for Prob. 6-15.

16. Design a variable modulus scaler using the '163 which will divide by 13, 14, or 15. Draw a table which indicates which codes produce which divide ratios.

17. Show how two '163s, with proper decoding, can be used to make a modulo-96 binary-coded counter.

18. Design a synchronous counter which will produce output frequencies of 1200 Hz (cycles per second), 600 Hz, 300 Hz, and 150 Hz from an input frequency of 4800 Hz.

19. Design a circuit, using '162s, which will divide the wall outlet frequency of 60 Hz into 1 Hz. Use the 1 Hz signal to enable a 60-s counter.

20. Design a 12-hour clock counter using the circuit from Prob. 19 to drive a minute counter and an hour counter.

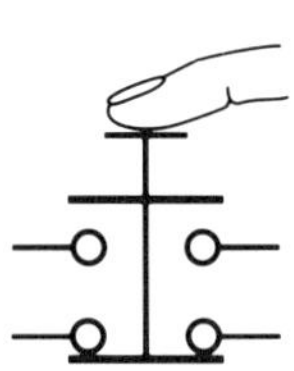

21. Devise a method of setting the 12-hour clock designed in Prob. 20 using three push buttons. The push buttons should be of the type shown at left.

22. Design the counter portion of a circuit which will measure the frequency of an unknown clock pulse by counting the number of rising edges which occur in a 1-s period. Use two 162s, a JK flip-flop, and the wall outlet frequency to produce a 1-s window in which the frequency is measured. The window counter should start, count for 1 s, and stop.

Use four 162s to count the rising edges of the unknown frequency. The general idea is presented as a block diagram in Fig. 6-27. What is the fastest frequency that can be measured with this circuit?

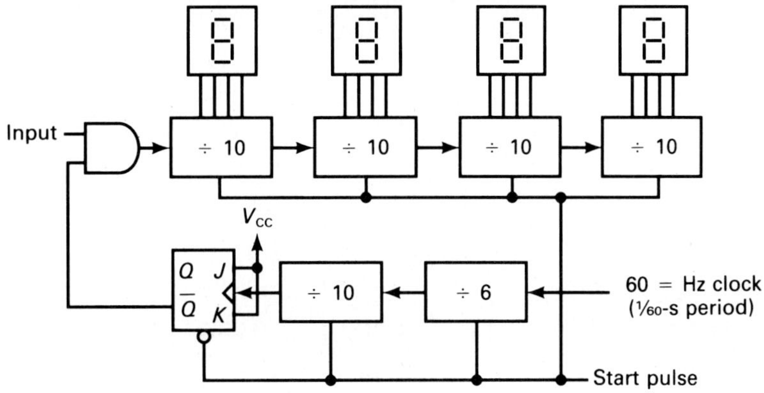

Fig. 6-27 Block diagram for Prob. 6-22.

23. Design the counter part of a digital stopwatch with start, stop, and rest inputs as shown in Fig. 6-28. The circuit is supplied with a 25.6-KHz clock and should measure minutes, seconds, and hundredths of seconds.

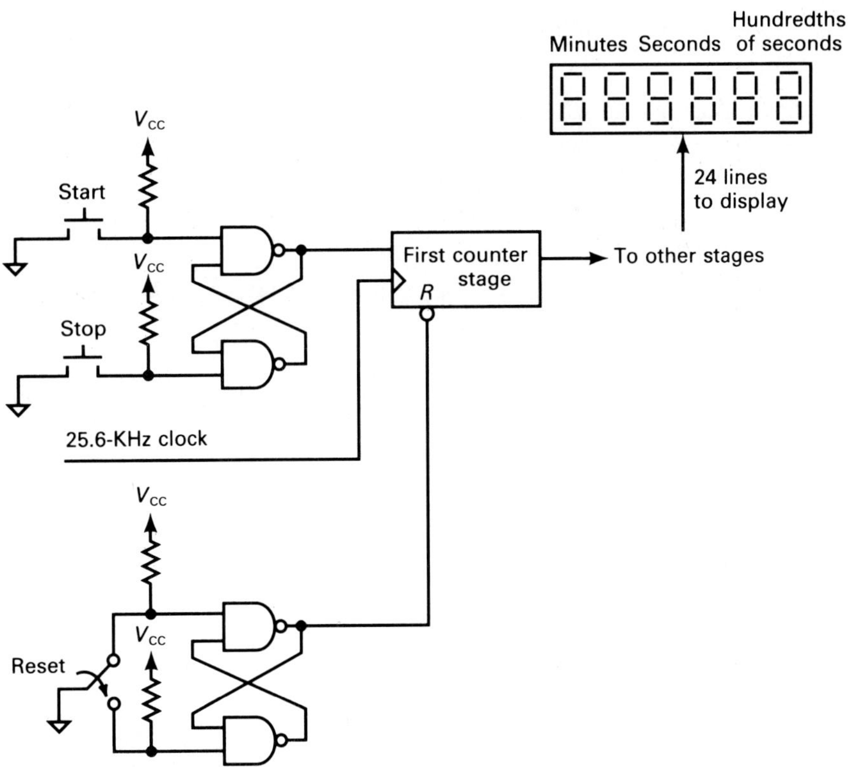

Fig. 6-28 Block diagram for the digital stopwatch in Prob. 6-23.

24. A serial *protocol* (an agreed-upon convention) separates incoming characters by two or more marks (1s). When at least two marks have been detected, a space (0) indicates the start of a new 8-bit character. Design a circuit which operates at twice the data clock rate and detects the beginning of a character. When the circuit detects the beginning of a new character, it will reset the counter. The counter will provide exactly 8 data clocks and then restart the beginning-of-character circuit. Use the state diagram and timing waveforms shown in Fig. 6-29. Show the state table, Karnaugh maps, and circuit implementation.

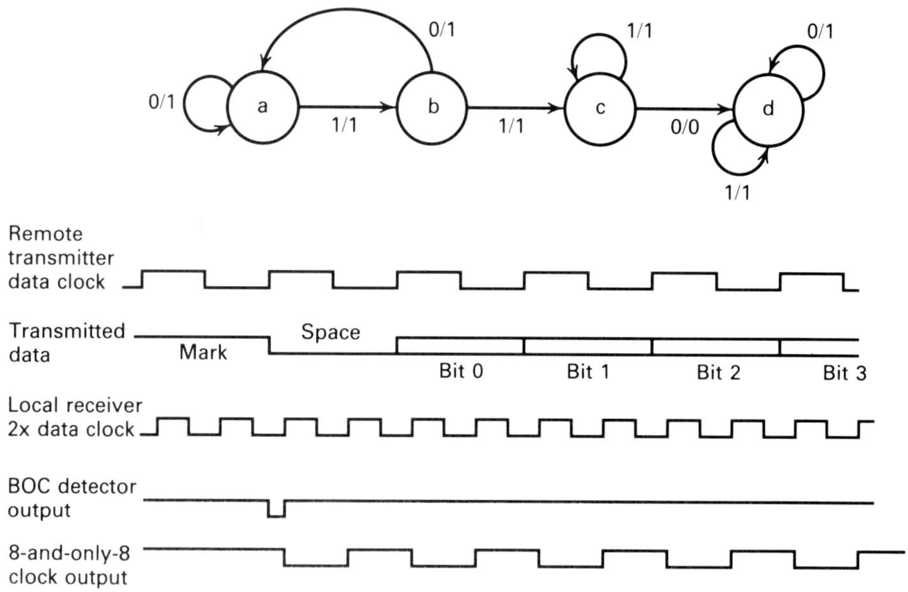

Fig. 6-29 State diagram for Prob. 6-24, beginning-of-character detector.

7 ARITHMETIC LOGIC

We naturally expect calculators and computers to perform arithmetic operations since computation is their primary function. But these are not the only systems which employ computational circuitry. Adders, subtractors, multipliers, and computational dividers are built into a variety of electronic subsystems to perform specific tasks involving numbers. Recently, high-speed computational circuitry has been used in filtering and tone detection applications. This was once the exclusive province of analog electronics, but op amps, resistors, and capacitors have, in many cases, given way to more compact, power saving digital ICs. *Digital filters* are implemented with shift registers and adders. *Digital signal processors (DSPs)* are general-purpose, high-speed state machines that shift and add numbers repeatedly to perform the very fast multiplication needed to process analog waveforms. If this type of logic becomes fast enough, it will replace large portions of radio and television circuitry. High-speed *routing engines* also rely on the basic arithmetic processes. These machines look for optimal placement of metal foils to connect ICs on printed circuit boards and are becoming standard drafting tools. Other machines which use arithmetic circuits include video games, military avionics, and digital modems. The purpose of this chapter is to introduce the basic arithmetic operations and show how they are implemented with simple logic. In this chapter we would like to answer the following questions:

1. How are binary numbers added in digital systems?
2. How are negative numbers represented in digital systems?
3. What are *accumulators* and *ALUs*, and how are they used to add and subtract binary numbers?
4. How are binary numbers multiplied and divided using basic adding, inverting, and shifting operations?
5. How can calculations be performed directly on decimal numbers without conversion to binary?
6. How are fractions and very large numbers represented in computers?

7-1 BINARY ADDITION

The most basic arithmetic function of all, and the easiest to understand, is binary addition. Adding two binary digits is just like adding two decimal digits. When learning to add two numbers, we first approached the operation in two ways, *counting* and *table memorization*. Counting is preferred because it is more instructive. We start with a number A on our fingers and add to it a number B by counting off B more fingers. The final number of fingers is the sum. In elementary school, a number line with tick marks is used instead of fingers since sums greater than 10 are needed. But the process is the same. The addition of integer A to integer B is the process of advancing B increments beyond A. However, the counting method, while easily implemented, is too slow. After the concept is understood, we usually move on to a faster method. We memorize addition tables which give the sums of decimal digits. Sometime in elementary school, these tables sink into our memories, and we never use fingers or number lines again. Digital logic also uses tables, and these tables are minimized with Karnaugh maps to produce fixed logic which will add binary numbers. Binary addition tables are, of course, much simpler than decimal tables because only two digits are involved. There are only four entries:

$$0 + 0 = 0$$
$$0 + 1 = 1$$
$$1 + 0 = 1$$

and

$$1 + 1 = 10 \tag{7-1}$$

The last entry requires two digits, a *sum* and a *carry*. When the sum of two digits is too large to be represented in a single digit, a carry must be added to the next position. We learned this with decimal addition, and it is no less true with binary addition. A more general table would include two carries: a carry-out to the next position and a carry-in from the previous position. A carry from a less significant position is called a *carry-in*; a carry to a more significant position is called a *carry-out*. In school we did not include the carries as part of the addition tables but instead treated them as rules governing the use of the table. In digital logic, however, the rules are incorporated into the hardware, and the carries are part of the table. The complete table, then, for binary addition would appear as follows:

Inputs			*Outputs*	
C_{in}	A	B	Sum (Σ)	C_{out}
0	0	0	0	0
0	0	1	1	0
0	1	0	1	0
0	1	1	0	1
1	0	0	1	0
1	0	1	0	1
1	1	0	0	1
1	1	1	1	1

(7-2)

The entries under the input variables indicate all possible combinations of 1s and 0s, and this table becomes the truth table needed to build the adder circuits. Two circuits are needed: The first will produce a sum (Σ), and the second will produce a *carry-out* (C_{out}). Together they form a *full-adder*. In CMOS these circuits are realized as complex gates which are actually no larger than NAND gates.

Full-Adders

The Karnaugh maps for the two outputs are shown in Fig. 7-1*a* and *b*. The C_{out} function is easy to implement with AND/OR logic. The Σ output

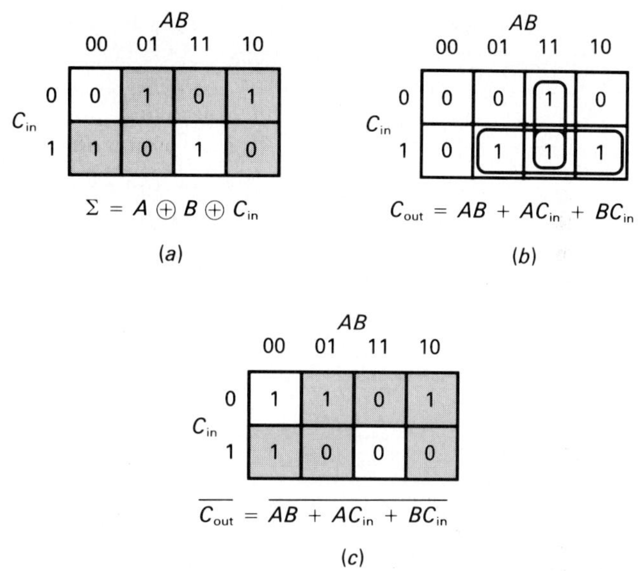

Fig. 7-1 Karnaugh maps for a full-adder.

map exhibits the checkerboard pattern indicative of EXCLUSIVE OR functions. The sum function is, in fact:

$$\Sigma = A \oplus B \oplus C_{in} \tag{7-3}$$

But implementation of the sum output in this way requires two EXCLUSIVE OR gates. A better approach, in CMOS, is to use a design technique called *bridging*, which yields a simpler and more compact layout. Bridging allows us to "bridge" between any two functions F_1 and F_2 when the Karnaugh maps are similar. In this case F_1 is the C_{out} NOT function. The C_{out} NOT map is shown in Fig. 7-1*c*. This map is obtained by swapping 0s and 1s in the original map. The inverse of the carry output map looks enough like the sum output map to make bridging useful. The entries which agree are shaded.

In order to bridge between the two maps, we find bridging functions G_1 and G_2 such that $F_2 = F_1G_1 + G_2$. The role of function G_1 is to eliminate the effect of 1s in F_1 which are not in F_2, and the role of G_2 is to add 1s which are in F_2 but not in F_1. The bridging techniques are beyond the scope of this text, and only the results are shown here. In this case $G_1 = (A + B + C_{in})$, and $G_2 = ABC_{in}$. Remembering that we chose F_1 to be C_{out} NOT, we have:

$$\Sigma = \underbrace{\overline{C_{out}}}_{F_1}\underbrace{(A + B + C_{in})}_{G_1} + \underbrace{ABC_{in}}_{G_2}$$

where, $$\overline{C_{out}} = \overline{AB + AC_{in} + BC_{in}} \qquad (7\text{-}4)$$

Using the CMOS design techniques discussed in Chapter 3, we arrive at the Σ cell shown in Fig. 7-2. One reason this cell is so compact is that the complements of A, B, and C_{in} are not required as inputs to the AND/OR structure, and this allows the polysilicon gates to be shared in a donut fashion as shown in Fig. 7-2. A full-adder is the combination of the Σ cell and the C_{out} cell. The two cells look very similar and are equally compact. They can sit one above the other in two rows, called *sticks*, which may contain as many stages as needed. One full-adder stage is represented as

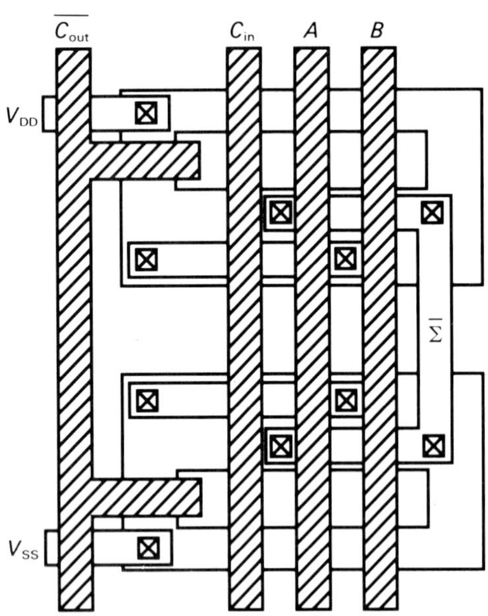

Fig. 7-2(*a*) CMOS implementation of the Σ function.

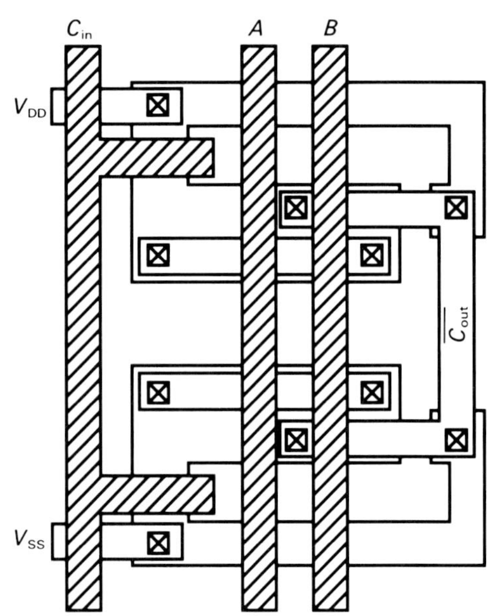

Fig. 7-2(*b*) CMOS implementation of the $\overline{C_{out}}$ function.

a box, labeled FA, with three inputs, A, B, and C_{in}, and two outputs, Σ and C_{out}.

Half-Adder

Another approach to the design of adders, which is less efficient in terms of silicon, but perhaps simpler conceptually, uses two half-adders. A *half-adder* is a circuit which accepts two inputs, A and B, and produces two outputs Σ and C_{out}. It is called a "half-adder" because two of these circuits are needed to make one full-adder (with inputs A, B, and C_{in}). The truth table for this circuit is as follows:

A	B	Σ	C_{out}
0	0	0	0
0	1	1	0
1	0	1	0
1	1	0	1

(7-5)

It is easy to see that the Σ output can be produced with an EXCLUSIVE OR gate and that C_{out} is a simple AND function. Figure 7-3 shows the half-adder (*a*) and a full-adder implementation using two half-adders (*b*).

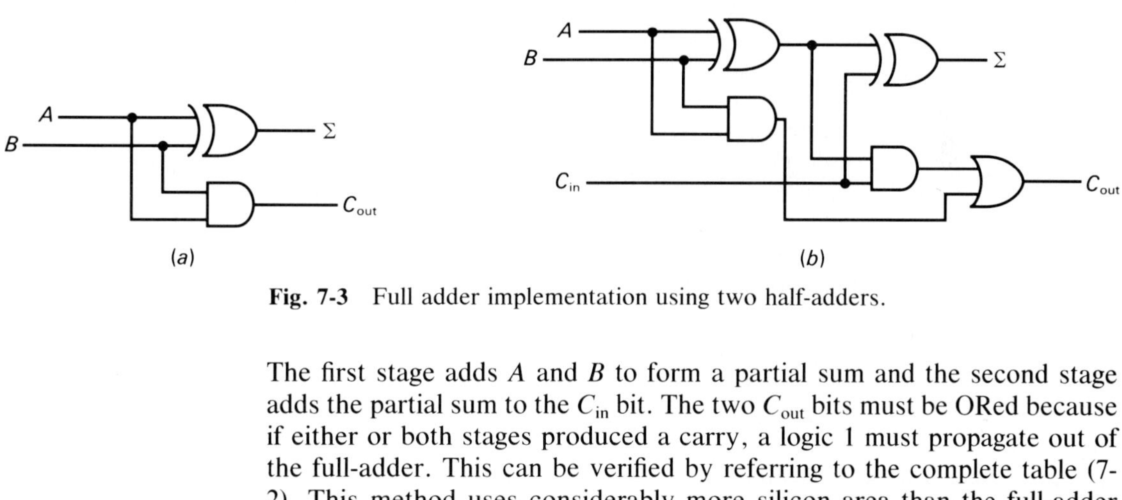

Fig. 7-3 Full adder implementation using two half-adders.

The first stage adds A and B to form a partial sum and the second stage adds the partial sum to the C_{in} bit. The two C_{out} bits must be ORed because if either or both stages produced a carry, a logic 1 must propagate out of the full-adder. This can be verified by referring to the complete table (7-2). This method uses considerably more silicon area than the full-adder approach. In the parallel addition of bytes and words, where eight or more full-adder stages are needed, the savings of the direct approach can be appreciable.

Parallel Addition

Normally, binary digits are added in multiples of 3 (octal) or 4 (hex). Microprocessors add binary numbers of 8, 16, or 32 bits each. Mainframe computers and special computational ICs typically add numbers larger than 32 bits. The basic concepts of binary addition can be understood, however, by examining a simple 4-bit adder.

The name *parallel addition* implies that each pair of digits in numbers A and B are added at the same time, yielding a total sum and final carry after one full-adder delay. This may not always be true. If a 4-bit parallel adder were designed using the FA cells developed above, it might look something like the circuit shown in Fig 7-4. Even though all digits of the

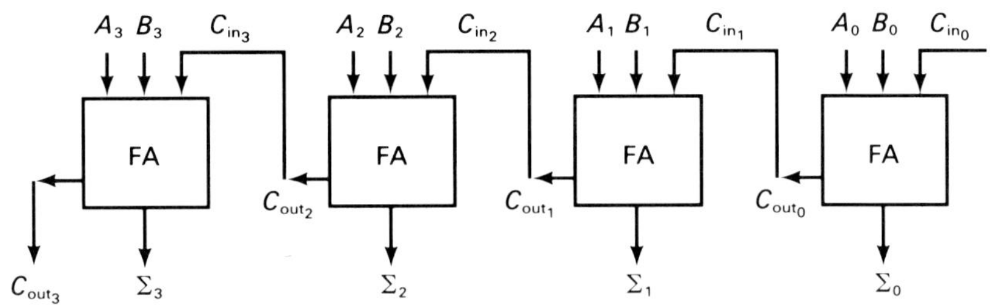

Fig. 7-4 A 4-bit parallel binary adder.

numbers A and B are applied to the inputs at exactly the same time, the resulting sums Σ_0, Σ_1, Σ_2, and Σ_3 will not be ready at the same time. The carry-in to the second stage will not be available until the carry-out of the first stage has stabilized, and this delays the second stage sum and the second-stage carry-out. Each stage must wait for the carry-outs of the previous stages to ripple through before it can form a sum and carry-out. The simple *ripple-carry adder* shown in Fig. 7-4 has the disadvantage that it takes $n + 1$ gate delays to settle out when n full-adders are used, and $2n + 1$ gate delays when n double half-adders are used. For large numbers n, this worst-case propagation delay severely limits the maximum speed of operation and is therefore impractical for computer circuits.

A better method would employ a *look-ahead carry generator*. The carry-in input for each FA depends on the inputs A, B, and C_{in} of all previous stages, and these could be combined in parallel, without ripple-through. All carries could, theoretically, be generated in two levels of delay, regardless of the number of stages. But the number of gates required would be prohibitive for long adders. Instead, adders are typically implemented in 4-bit parallel blocks, and the carries within a 4-bit block are derived, in a parallel fashion, using a look-ahead circuit. For instance, the carry-outs needed for a four-stage adder with no initial carry-in would be as follows:

$$
\begin{aligned}
C_{out0} &= A_0B_0 \\
C_{out1} &= A_1B_1 + (A_1 + B_1)\,C_{out0} \\
C_{out2} &= A_2B_2 + (A_2 + B_2)C_{out1} \\
C_{out3} &= A_3B_3 + (A_3 + B_3)\,C_{out2}
\end{aligned} \qquad (7\text{-}6)
$$

Each carry-out must wait for the carry-out of the previous stage. But if we substitute the equivalent expression into each C_{out_n} term, we get an

AND/OR function which does not depend upon any previous stage outputs. For instance, in the second equation we have:

$$\begin{aligned} C_{out_1} &= A_1B_1 + (A_1 + B_1)\,C_{out_0} \\ &= A_1B_1 + (A_1 + B_1)(A_0B_0) \\ &= A_1B_1 + A_1A_0B_0 + B_1A_0B_0 \end{aligned} \qquad (7\text{-}7)$$

Making similar substitutions for each equation down the line results in boolean equations for each carry which can be implemented directly with AND/OR structures and no ripple delay. All carries within a block, C_1 through C_3, are derived directly from the nine inputs, A_0 through A_3, B_0 through B_3, and C_{in_0} in one or two levels of delay, depending on the method of implementation. The carries between blocks may then either ripple through or be computed in a look-ahead generator over four blocks. At the packaged-part level, the 74xx182 provides look-ahead carry capability across four groups of binary adders. This allows look-ahead operation over long word lengths.

Serial Addition

Numbers of any size may be added with only one FA stage if they are shifted in serially and added 2 bits at a time. This saves hardware at the expense of time. No look-ahead circuitry is needed because the final sum of two n-bit numbers is not expected to become available until one FA delay after n clock pulses. With today's level of integration and the emphasis on speed, serial addition makes little sense unless at least one of the numbers to be added is entering the system serially. In this case, something like the circuit shown in Fig. 7-5. might be employed to add an incoming number to a local number for storage as a local sum.

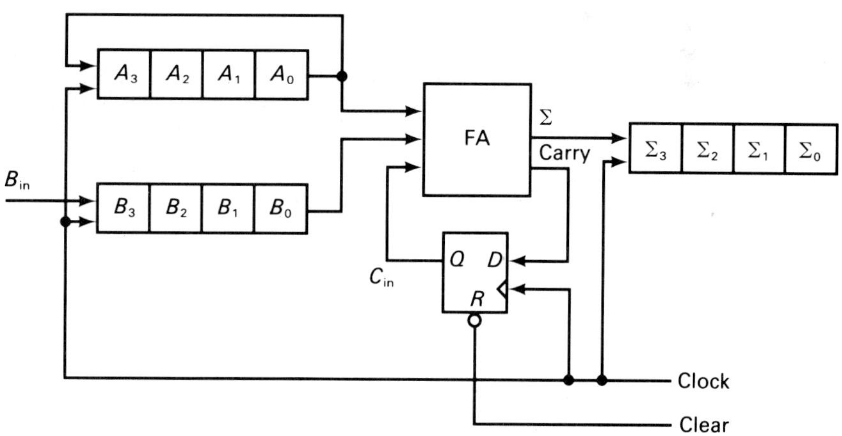

Fig. 7-5 Serial addition.

REVIEW 7-1

1. **Q.** What is the advantage of the computational table over the counting method of addition? **A.** speed

2. **Q.** How are addition tables implemented in digital systems? **A.** with boolean logic derived from the tables

3. **Q.** What three digits must be added to produce a complete sum? **A.** A_n, B_n, and C_{in_n}

4. **Q.** What is the circuit called which forms the complete sum? **A.** a full-adder

5. **Q.** What technique may be used to simplify full-adder implementation in CMOSs? **A.** bridging

6. **Q.** What is parallel addition? **A.** It is the addition process in which all digits of the two numbers to be added are available at the same time.

7. **Q.** What is serial addition? **A.** It is a process in which the numbers are added two digits at a time starting with the least significant position.

8. **Q.** What are the relative advantages of each approach? **A.** Parallel addition can produce a final sum approximately n times faster than serial addition, where n is the number of positions to be added. However, parallel addition requires n FA stages whereas serial addition requires only one FA stage for any number with n digits.

9. **Q.** What is a look-ahead carry generator? **A.** It is a combinational circuit which produces each carry-in as a function of all initial inputs. When a look-ahead carry generator is used, all carry-ins are available one or two gate delays after the inputs are stable.

7-2 SIGNED ARITHMETIC

When we learned arithmetic in school, we were taught to handle signed numbers according to certain rules. Specifically, when adding two signed numbers with like signs, we *add* the two numbers and attach the same sign to the result. When adding two numbers with unlike signs, we would actually *subtract* the smaller number from the larger number and attach the sign of the larger number to the result. This method depends upon the signs and sizes of the numbers and is therefore a data-dependent process. In digital systems we could employ rules like these, but in the time it would take to examine the numbers and determine their signs and relative magnitudes, we could have performed the original operation. It is preferable, in small machines, to treat all numbers in the same manner regardless of sign and size.

The key to understanding how negative numbers are handled in a digital machine is to realize that once it has been built, the machine is limited to a fixed number of digits. When we work out arithmetic on a sheet of paper, an extra position can always be penciled in when needed. But a machine is built with n and only n flip-flop positions. If one position in n bits is used to represent a plus or minus sign, then the range of positive binary numbers which can be represented is effectively cut in half. For instance, if a machine has 8 bits and no sign bit is used, then it has a range of 00000000_2 to 11111111_2, or 0_{10} to 255_{10}, which is 256 different positive integers. But if negative integers are allowed, and the most significant bit is used to represent the sign, then the range of numbers falls into two groups between $\pm 0000000_2$ and $\pm 1111111_2$, or $\pm 0_{10}$ and $\pm 127_{10}$. Actually, the usable numbers have been reduced by only 1, since the negative numbers are now available, and only the -0 is wasted. But the range of positive integers is cut in half, and one number, -0, must represent an undefined quantity.

The most straightforward approach to representing negative integers in digital systems is to treat the most significant bit (MSB) as a sign bit. When the number is positive, the MSB is 0; when it is negative, the MSB is 1. This approach is called the *signed-magnitude representation* and is illustrated below for an 8-bit machine:

Decimal	Positive Binary	Negative Binary	
1	00000001	10000001	
2	00000010	10000010	
3	00000011	10000011	(7-8)

The addition of two numbers in sign-magnitude representation is identical to pencil-and-paper arithmetic, and the operation depends on the signs and sizes of the numbers. This method is usually too cumbersome for small digital systems and is normally found only in larger computers.

Another approach, called *sign-complement representation*, allows the machine to treat each pair of numbers in the same manner regardless of sign. Positive sign-complement integers are identical to the positive sign-magnitude integers. A 0 in the most significant position indicates that the number is positive. A 1 in the most significant position indicates that the number is negative, but a negative number is not just a positive number with a 1 in the most significant bit. Instead, a negative number $-n$ is represented as n less than 0 in the same way that n less than zero would be represented on an odometer. Imagine a binary odometer with only 8 positions. If no miles were registered, then the odometer would display 00000000. One less than 0, or -1, would be obtained by turning the odometer back one position to produce 11111111. A -2 would be obtained by turning the odometer back another position to 11111110. The largest-magnitude negative number possible would be 10000000, which would be produced by turning the odometer back 128_{10} positions from 0. This representation of negative numbers is the most common type of sign-complement notation and is specifically called *2s complement representation*. It is the method most often used to represent signed numbers in microprocessors and computational circuits.

There are two ways to change the sign of a 2s complement number. Both of these methods eliminate the need to count backward with an odometer. The first is to invert all the bits and add the number 1. For example, to obtain a -2 in an 8-bit machine, we perform the following steps:

A $+2$ is represented as:	00000010	
Invert the bits to obtain:	11111101	
And add 00000001 to produce:	11111110	(7-9)

which is -2 in 2s complement representation.

Another way to arrive at the same result is to follow the procedure outlined below:

1. Starting with the least significant bit (LSB), copy down each bit as you find it until you reach the first 1.
2. Copy down the first 1.
3. Invert all bits thereafter.

Using the example above,

```
                              MSB             LSB
                                ↘             ↙
Positive number 2                   00000010
                                  Copy one  Copy zero
                                          ↘  ↙
                                           10
                                   Invert the rest
                                     ⏜
                                     11111110              (7-10)
```

Both procedures work equally well in reverse. Given a negative number in 2s complement notation, the positive equivalent magnitude can be obtained by using the exact same steps as outlined above. Notice that applying these rules to the number 1000000 produces itself, a special case.

When negative numbers are represented in 2s complement notation, they can be added just like positive numbers, regardless of sign. For example, the addition of $(-5_{10}) + 3_{10}$ in an 8-bit adder would take the following form:

-5 is represented as:	11111011
3 is represented as:	00000011
The binary sum is:	11111110
Which is equivalent to:	-2_{10}

Overflow

Two's complement numbers may be added directly, regardless of sign, as long as the sum can fit into the existing number of bit positions. When two numbers of opposite sign are added, there can never be a problem since the result will always be smaller than either number. However, when two numbers of equal sign are added, there is always a possibility that the result will not fit. When a machine adds 2s complement numbers and the sum does not fit, a carry either *into* or *out of* the sign bit will occur, but not both. For instance, the largest positive number that can fit in 8 bits is 01111111. Adding that number to itself produces 11111110, with no final carry-out. The problem is that the two numbers added were both positive, but the sum was negative. The same impossible situation might occur when adding two large-magnitude negative numbers. The largest magnitude negative number that will fit in 8 bits is 10000000, or -128_{10}. If we add this number to itself, the sum is 00000000 with a final carry out of the sign bit but no carry-in. Again two numbers of like sign were added, but the sum is of the opposite sign. This occurs because the sum of the two numbers is too large to fit in the given number of bits and the result is invalid. This condition is called *overflow*.

Digital adders cannot inspect each number before addition to make sure that their sum will fit. Instead, circuitry is provided to detect overflow when it has occurred so that the user can be made aware of the fact that the results are invalid. An easy way to detect undesirable overflow is to examine the carries into and out of the sign bit. If these do not match then there is overflow. This is easy to believe because if there were a carry-in with no carry-out, the numbers being added must both be positive, but the sum would be negative due to the carry-in. Likewise if there were a carry-out with no carry-in, the numbers being added must both be negative, but with a positive sum. These are both invalid situations indicating overflow. The overflow condition, then, may be detected by using an EXCLUSIVE OR gate as shown in Fig. 7-6.

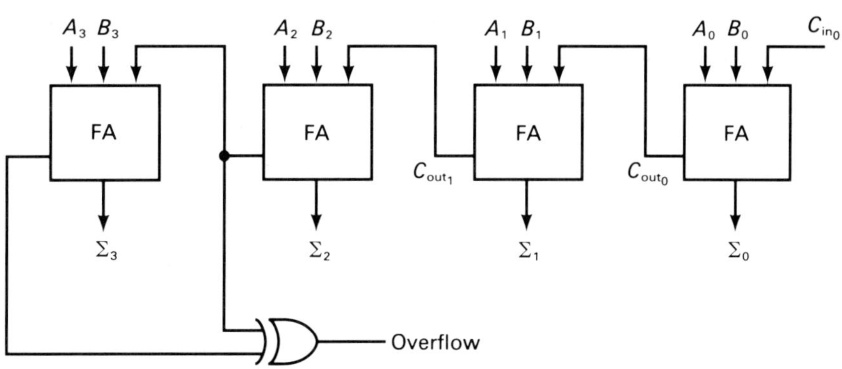

Fig. 7-6 A 4-bit adder with overflow indicator.

Binary Subtraction

Subtraction, like addition, may be approached in two ways, counting and table memorization. The counting method would indicate the difference of two integers, $A - B$, by starting with integer A and counting back B positions, either with fingers or ticks on the number line. A problem occurs, however, when B is larger than A or when A is a negative number. We have been taught to compare the two numbers beforehand, and if A is smaller than B, we actually change the problem to a different problem which is easier to solve. We subtract A from B, $(B - A)$, and attach a minus sign to the result. This changing of rules was discussed earlier in the context of addition. Previously we mentioned that the 2s complement number representation allows digital machines to bypass the examination process and treat all numbers in the same way. Using 2s complement notation, any two numbers may be added regardless of their signs and a valid sum will be produced as long as it can fit in the allotted number of bits.

Subtraction could be performed with special subtractor logic. Full-subtractors and half-subtractors can be built using subtraction tables and Karnaugh maps. The maps, however, bear a striking resemblance to the maps obtained for adders, and it is just a small step to realizing that adders may be used as subtractors by employing the rules set forth above for 2s complement representation. In order to subtract the number B from the number A, $(A - B)$, we need only to 2s complement the number B and add the two numbers with a full-adder. For instance, if we wished to subtract the number 5_{10} from the number 7_{10} $(7_{10} - 5_{10})$, we would perform the following steps:

The 2s complement of the subtrahend 5_{10} is: 11111011

The problem, then, is:	00000111	Minuend
	+11111011	Subtrahend
With a difference of:	00000010	or $+2_{10}$

But what about subtracting 7_{10} from 5_{10}?

The 2s complement of the subtrahend 7_{10} is: 11111001

and the problem is:	00000101	Minuend
	+11111001	Subtrahend
With a difference of:	11111110	or -2_{10}

The number -2_{10} will remain in 2s complement form as long as it remains in the machine. It will be converted to signed-magnitude representation only when it is time to display it to the outside world. As long as all numbers remain in 2s complement form within the machine, then subtraction is always performed by inverting all bits of the subtrahend and adding it to the minuend with a carry-in of 1. This is equivalent to 2s complementing and adding. Figure 7-7 shows how this is implemented with EXCLUSIVE OR gates and a full-adder.

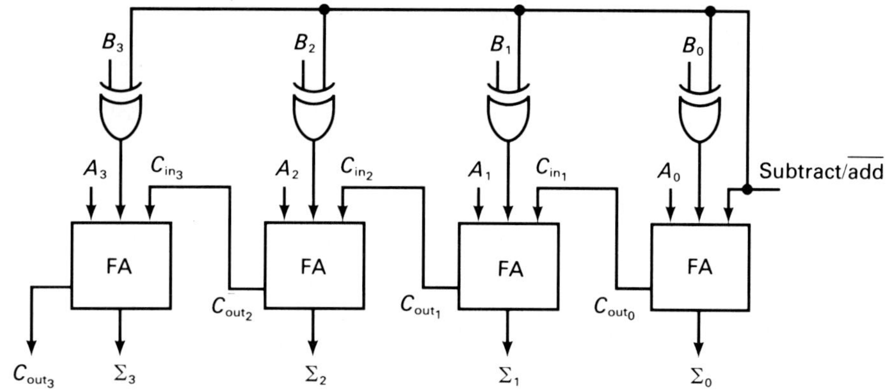

Fig. 7-7 A 4-bit subtractor-adder using 2s complement representation.

Register Operations

Digital computers always have instructions for *incrementing, decrementing*, and *clearing* registers. These are especially useful when registers are used as counters for timing events or counting occurrences of events. *Incrementing* means adding 1 to whatever number currently resides in a register, *decrementing* means subtracting 1 from the register contents, and *clearing* means resetting all register bits to 0. Incrementing, decrementing, clearing, adding, and subtracting can all be accomplished with the full-adder circuitry described above. For purposes of illustration, we will use only 4-bit registers. This means that the largest positive number which will fit into the machine is 0111 or $+7_{10}$. The largest magnitude negative number allowable is 1000, or -8_{10}. With these very small registers we can examine all the properties of larger, more useful, computing machines.

Figure 7-8 illustrates a simplified ALU. *ALU* stands for *arithmetic-logic unit*. This particular ALU lacks the logical functions which are implied in its name, but it will perform the basic arithmetic functions. The circles with numbers in them represent control inputs, and in any given clock cycle, they will be either high or low. Control input 1, for instance, is the enable input to the *accumulator* register. Both the accumulator register and the ALU latch are clocked with the same master clock. If control input 1 is low, data appearing on the four inputs of the accumulator will be accepted on the next rising clock edge. The same is true for the ALU latch when control input 5 is low. Both registers have access to a common parallel bus. Data may appear on this bus *from* the outside only when control signal 9 is low. Data on this bus may be presented *to* the outside only when control signal 10 is low.

The core of this machine is the 4-bit adder. It performs all the arithmetic tasks. The registers merely hold the results and partial results. In order for this machine to operate, there must be some control logic supplying the correct control inputs at the correct times. Control logic will be discussed in a later chapter. For now we will assume that the control signals are provided, and we will describe the ALU operation by specifying

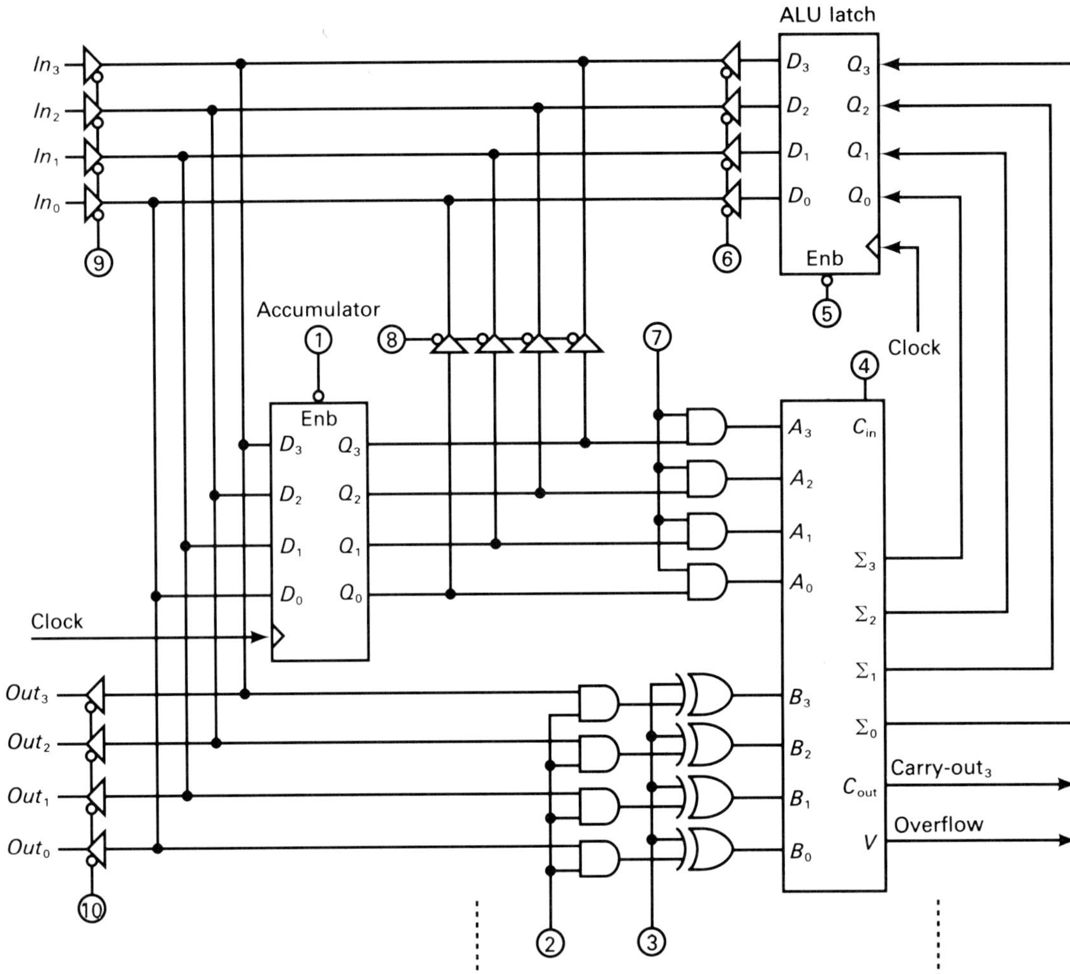

Fig. 7-8 ALU and operational registers.

which control signals are needed for each instruction. Figure 7-9 shows three instruction cycles. The first instruction loads the accumulator with a number A. This instruction, which we will call LDA, for load accumulator, consists of enabling the input drivers (control signal 9) and the accumulator inputs (control signal 1) for at least one setup time before the rising clock edge. It has the effect of bringing outside data onto the bus to be read by the accumulator. The next instruction, called ADD, will add some number B to the contents of the accumulator. Control signal 9 will again be used to enable new data onto the internal bus. This time, control signal 1 is not asserted, and the new number will *not* be latched into the accumulator. Instead, control signals 2 and 3 will gate the new number to the B side of the adder. At the same time, control signal 7 will gate the accumulator contents to the A side of the adder. Control signal 4 provides a 0 carry-in, and control signal 5 enables the ALU latch so that it may accept the sum output on the next clock edge. The new data must be stable long

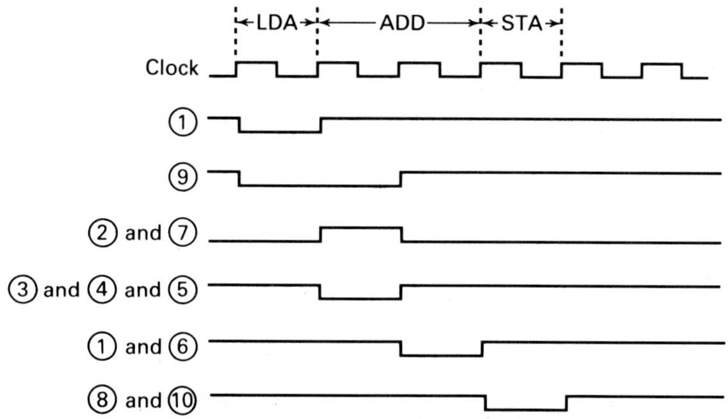

Fig. 7-9 ALU instruction cycles.

enough in advance to propagate through the logic and adder and still have time to setup the ALU latch before the rising clock edge. When the sum has been latched, the input drivers turn off and the ALU latch drivers turn on, enabling the sum onto the internal bus. While the input drivers are turning off and the ALU latch drivers are turning on, there will be a short period of contention. This means that an input driver may be still asserting a high level on a bus line while an ALU latch driver is trying to assert a low level onto the same line. This contention will not affect the operation of the circuit as long as the currents involved are limited and the contention is resolved at least one setup time before the clock edge. On that clock edge, the enabled accumulator will accept the new value. The ADD instruction required two clock periods. Finally, an instruction called STA, or store accumulator, enables the sum, stored in the accumulator, onto the internal bus (control signal 8) and external bus (control signal 10) to be presented to some other memory device on the outside. Figure 7-9 shows the timing for all three operations.

The remaining instructions will be described in less detail. In each case the ALU latch accepts a sum from the adder on one clock edge and supplies that sum to the accumulator on the next clock edge as described above. SBA, subtract from accumulator, involves inverting the incoming bits of the subtrahend B before applying them to the adder. All the bits of number B will be inverted, or *complemented*, when control signals 2 and 3 are both high. At the same time, an initial carry-in, a high at control signal 4, completes the subtraction process of $A - B$ if control signal 7 is high. INCA, or increment accumulator, is accomplished by *masking* the number on the internal bus with lows at both 2 and 3. Zeroes will appear on the B side of each FA, and a 1 will be added to the contents of the accumulator if control signals 4 and 7 are high. DECA, or decrement accumulator, requires presenting all 1s (the equivalent of a 2s complement -1) to the B side of the adder while enabling the A side with control signal 7. All 1s are produced at the B side with a low at control signal 2 and a high at control signal 3. Finally, CLRA, or clear accumulator, can be accomplished by masking both sides of the adder and supplying a low value to the initial carry-in. Zeroes will be added to 0s with a carry-in of 0 to produce a sum of 0s which is stored first in the ALU latch and then in the accumulator.

Other operations which are typically provided in real ALUs are: AND, OR, EXCLUSIVE OR, compare, shift right, and shift left. The AND, OR, EXCLUSIVE OR, and magnitude comparison ($A = B$) logic can be built right into the adder and selected with additional control inputs. The shift operations may be implemented by using a universal shift register for the accumulator instead of the simple parallel-in–parallel-out register used in Fig. 7-8.

Standard ALU chips are available which perform the functional operations described above. In the 54/74 series, the '181 provides 16 boolean logic functions plus addition, subtraction, shift A side one position, and magnitude comparison. Figure 7-10 shows a block diagram of the '181.

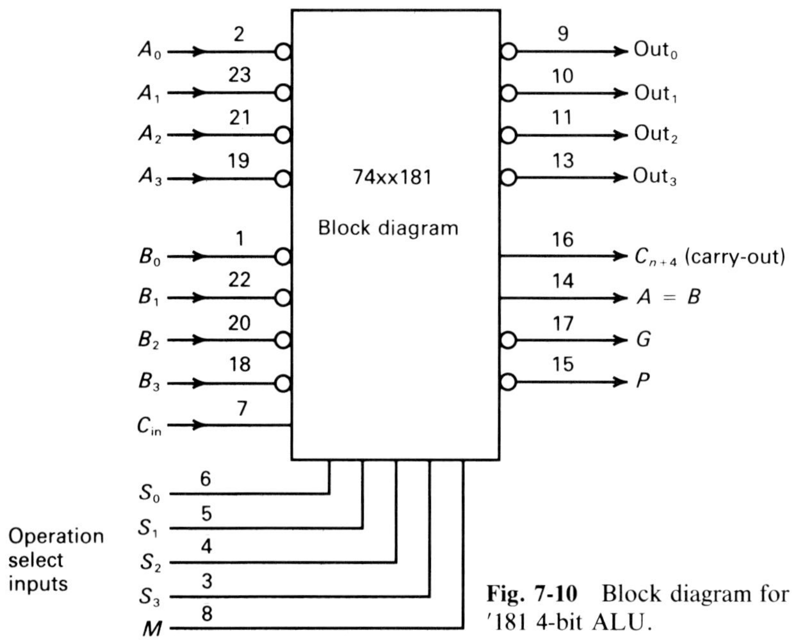

Fig. 7-10 Block diagram for the '181 4-bit ALU.

The G, P, and C_{n+4} outputs are carry generate, carry propagate, and ripple-carry output, respectively. G and P may be used with the 74xx182 look-ahead generator and four '181 ALUs to form a 16-bit unit. Or, the C_{n+4} may be used to cascade '181s when speed is not important. The mode control input M enables the internal carries when arithmetic manipulations are being performed. The '181 is an ALU only, operational registers must be supplied externally.

REVIEW 7-2

1. **Q.** Why are pencil-and-paper rules of arithmetic not suitable for addition and subtraction in small digital systems? **A.** because they

are data dependent, and smaller machines are more efficient when they can treat every number in the same manner regardless of sign and size

2. **Q.** Why is the straightforward approach of signed-magnitude representation not commonly used in smaller machines? **A.** because its rules are identical to those of pencil-and-paper arithmetic and are too cumbersome for small machine implementation

3. **Q.** Which method of signed representation is preferred for microprocessors and other small digital machines? **A.** Sign-complement representation is preferred since its rules are not data dependent. Specifically, small digital machines use 2s complement representation.

4. **Q.** What is the number 10000000 in 8-bit 2s complement representation? **A.** It is -128_{10}.

5. **Q.** How are 2s complement numbers formed? **A.** If a binary number is positive, it is represented with a 0 in the most significant position. If the number is negative, it may be converted to 2s complement by inverting all the bits and adding 1.

6. **Q.** What is overflow? **A.** It is the result of adding two numbers whose sum will not fit into the number of positions provided.

7. **Q.** How can overflow be detected? **A.** Overflow is indicated by unequal carries into and out of the sign bit during addition. It may be detected with an EXCLUSIVE OR gate.

8. **Q.** What is an ALU? **A.** An arithmetic-logic unit consists of adders and other combinational logic needed to perform computational and logical functions.

9. **Q.** What is an accumulator? **A.** An accumulator is an operational register which is used in conjunction with an ALU to store partial results of arithmetic and logical operations.

10. **Q.** What are control signals? **A.** Control signals are high and low pulses, synchronized with the master clock, which enable the various registers and combinational logic associated with computational circuitry.

11. **Q.** What does it mean to increment and decrement a register? **A.** To increment means to add 1 to the register contents; to decrement means to subtract 1 from the register contents.

12. **Q.** How can we decrement by adding? **A.** Adding an all-1s pattern is equivalent to subtracting 1.

13. **Q.** What is an instruction? **A.** An instruction is a group of sequenced control signals which will perform a specified task.

14. **Q.** How do logic devices sharing a common bus avoid interfering

with each other? **A.** The control signals maintain order by allowing only one set of outputs onto the bus at a time.

15. **Q.** Why doesn't the adder in Fig. 7-8 not need a clock input? **A.** Adders are combinational, or flow-through-type logic and have no memory. The adder outputs must be stored in a register which is sequential type logic.

7-3 MULTIPLICATION

Like addition, multiplication can be reduced to a table. Decimal multiplication tables are extensive since each digit between 0 and 9 can be multiplied by itself and every other digit between 0 and 9. Binary multiplication, however, involves only two digits, 0 and 1, and the table is quite simple:

a	b	P
0	0	0
0	1	0
1	0	0
1	1	1

(7-11)

It is easy to see that the table can be implemented with a single AND gate. The AND gate is a 2-bit binary multiplier. While binary addition requires three inputs and two outputs for each stage, binary multiplication requires only two inputs and one output. Multiplying two binary digits can never produce a carry-out, and a carry-in makes no sense in single digit binary multiplication. However, multiplying binary numbers greater than 1 involves binary addition with carries in and out. For instance, consider the product of two binary numbers a_1a_0 and b_1b_0.

$$\begin{aligned} a_1a_0 \times b_1b_0 &= (a_1 \times 2^1 + a_0 \times 2^0) \times (b_1 \times 2^1 + b_0 \times 2^0) \\ &= a_1b_1 \times 2^2 + a_1b_0 \times 2^1 + a_0b_1 \times 2^1 + a_0b_0 \times 2^0 \end{aligned}$$

This is often represented in the following form:

$$\begin{array}{rrr} & a_1b_0 \times 2^1 + & a_0b_0 \times 2^0 \\ a_1b_1 \times 2^2 + & a_0b_1 \times 2^1 & \\ \hline P_2 \times 2^2 + & P_1 \quad \times 2^1 + & P_0 \quad \times 2^0 \end{array} \qquad (7\text{-}12)$$

The numbers are grouped into columns in order to simplify the addition of like powers.

EXAMPLE 7-1

Multiply 2 *times* 3 in binary.

$$\begin{aligned} 10_2 \times 11_2 &= (1 \times 2^1 + 0 \times 2^0) \times (1 \times 2^1 \times 1 \times 2^0) \\ &= 1 \times 2^2 + 1 \times 2^1 + 0 \times 2^1 + 0 \times 2^0 \end{aligned}$$

$$\begin{array}{r} 1 \times 2^1 + 0 \times 2^0 \\ \text{or} \qquad \underline{1 \times 2^2 + 0 \times 2^1 \qquad\qquad} \\ 1 \times 2^2 + 1 \times 2^1 + 0 \times 2^0 = 110_2 = 6_{10} \end{array}$$

The easy part of this operation was the multiplication; the complicated part was the addition of partial products. Implementation of this 2-bit multiplier is shown in Fig. 7-11. The powers of 2 notation, 2^2, 2^1, and 2^0,

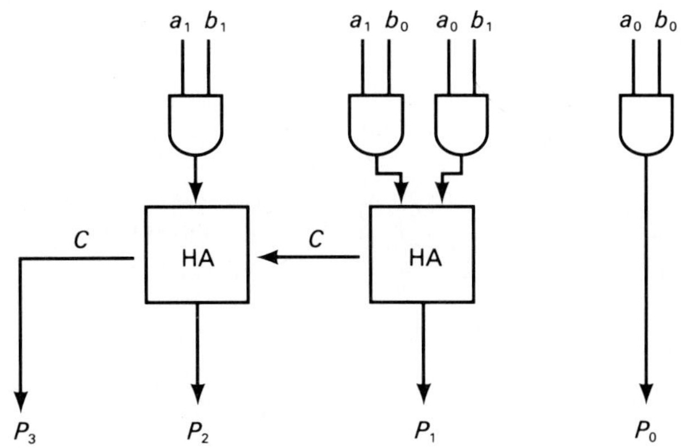

Fig. 7-11 A 2-bit multiplier.

is not actually necessary since the positions of the coefficients indicate their weight. It is used here only to show how and why the numbers are arranged in shifted positions. The same example without the powers of 2 notation would appear as follows:

$$\begin{array}{r} 10 \\ \underline{\times\ 11} \\ 10 \\ \underline{10\ \ } \\ 110 \end{array}$$

Unsigned Multiplication

When we learned to multiply in school, we were taught to first write down the *multiplicand*, then write down the *multiplier* beneath it and underline the multiplier. Each digit of the multiplicand was multiplied by the first digit of the multiplier, and the carries were added after each operation. The answer was recorded directly under the line. Then the operation was repeated using the second digit of the multiplier, and the answer was recorded beneath the first answer but *shifted* one digit to the left. The process was repeated, as shown below, until we ran out of multiplier digits. Then the columns of numbers were added to form the final answer:

$$\begin{array}{rl} 117 & \text{Multiplicand} \\ \underline{\times 105} & \text{Multiplier} \\ 585 & \\ 000\ \ & \\ \underline{117\ \ \ \ } & \\ 12285 & \text{Product} \end{array}$$

Each digit of the multiplicand was multiplied by a digit of the multiplier using tables stored in our memory. However, notice that when the multiplier digit was 0, the operation was reduced to writing down 0s and when the multiplier digit was 1, the operation was simplified to writing down the multiplicand. Binary multiplier digits are always 0s and 1s. The entire operation of binary multiplication consists of writing down 0s when the multiplier digit is 0 and writing down the muliplicand when the multiplier digit is 1. An example using two 8-bit numbers is given below:

```
               01110101
               01101001
               --------
               01110101
              00000000
             00000000
            01110101
           00000000
          01110101
         01110101
        00000000
        ---------------
        010111111111101 = 12285₁₀
```

We usually perform this operation by rote and never give any thought to why it works. However, as we indicated above, there is good reason for the shifting and adding. Figure 7-12 shows how the shifting keeps equally powered products in the same columns for an 8-bit product.

				$a_3 \times 2^3$	$a_2 \times 2^2$	$a_1 \times 2^1$	$a_0 \times 2^0$
				$b_3 \times 2^3$	$b_2 \times 2^2$	$b_1 \times 2^1$	$b_0 \times 2^0$
				$a_3b_0 \times 2^3$	$a_2b_0 \times 2^2$	$a_1b_0 \times 2^1$	$a_0b_0 \times 2^0$
			$a_3b_1 \times 2^4$	$a_2b_1 \times 2^3$	$a_1b_1 \times 2^2$	$a_0b_1 \times 2^1$	
		$a_3b_2 \times 2^5$	$a_2b_2 \times 2^4$	$a_1b_2 \times 2^3$	$a_0b_2 \times 2^2$		
	$a_3b_3 \times 2^6$	$a_2b_3 \times 2^5$	$a_1b_3 \times 2^4$	$a_0b_3 \times 2^3$			
$P_7 \times 2^7$	$+ P_6 \times 2^6$	$+ P_5 \times 2^5$	$+ P_4 \times 2^4$	$+ P_3 \times 2^3$	$+ P_2 \times 2^2$	$+ P_1 \times 2^1$	$+ P_0 \times 2^0$

Fig. 7-12 Binary multiplication of two 4-bit numbers.

This procedure can be implemented in hardware using an ALU and two accumulators similar to the one described above. There are three important differences, however, between performing multiplication on paper and performing multiplication in accumulators. *First*, accumulators have a fixed number of bits in which to store the product. If the number of bits in the product exceeds the number of bits available in the registers, the answer will not be valid. The product of two n-bit unsigned numbers will at most require a register $2n$ bits wide. Therefore, if our data registers are 8 bits wide, we will need a 16-bit register to hold the product of two numbers contained in those registers. *Second*, instead of storing shifted copies of the multiplicand and adding them all together at the end, an accumulator sums only two numbers at a time and *successively accumulates* the *partial products*. This is illustrated below with the product of two 4-bit binary numbers:

	0101	
Add zero	0110	
	0000	*First partial product*
	+0101	←Shift
Add multiplicand	01010	*Second partial product*
	+0101	←Shift
Add multiplicand	011110	*Third partial product*
Add zero	+0000	←Shift
	0011110	*Fourth partial product*
	00011110	←*Final product represented in 8 bits*

(7-13)

Successive accumulation is preferred in mini and microcomputers because the same logic which performs other functions may also be used for multiplication. Since only two binary numbers are added at a time, one n-bit adder could perform the complete multiplication of two n-bit numbers. If the addition were to be performed all at once, at least $(n - 1)$ n-bit adders would be required.

Early, special-purpose computers designed by Seymour Cray used the pencil-and-paper or, *flow-through*, method. The advantage was speed. The complete multiplication process was completed in the time required for the longest propagation path through the AND gates and the full-adder stages. This would typically be shorter than one clock period. A flow-through multiplier was used in Control Data Corporation's CDC 7600 supercomputer to produce one 48-bit × 48-bit product every 275 ns. The disadvantage was that forty-eight 48-bit adders were required. However, with VLSI technology constantly pushing the density limits of single-chip multipliers, variations of flow-through techniques have become more attractive for high-speed computational machines. Figure 7-13 shows how binary multiplication is implemented using the flow-through method. The products $a_n b_n$ are formed with AND gates not shown in the figure. The operation is asynchronous in the sense that no clock is required to produce the final product.

The *third difference* between pencil-and-paper multiplication and successive accumulation is that instead of shifting the multiplicand to the left after each addition, the partial product is shifted to the right. The result is the same, but the hardware implementation is simpler. Figure 7-14*a* shows how two unsigned 4-bit numbers may be multiplied using two 4-bit registers and one 4-bit adder. This direct *hard-wired* logic can perform only multiplication and only with unsigned numbers. Figure 4-14*b* shows the same operation using two 4-bit accumulators and one ALU. This is how it would be done in a microprocessor. The ALU is not shown in Fig.

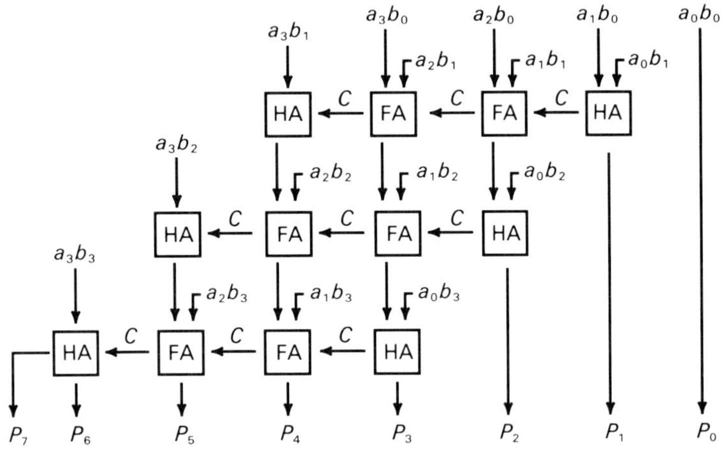

Fig. 7-13 Flow-through method of multiplication. Not shown are the AND gates which form the products $a_n b_n$.

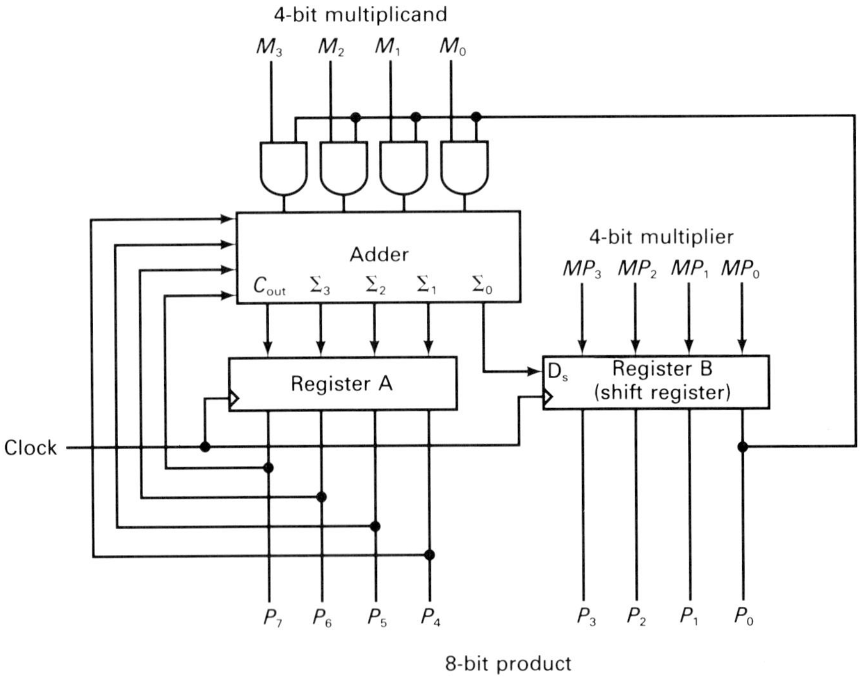

Fig. 7-14(*a*) Direct hardware implementation of an unsigned 4-bit multiplier. Initially, register A is cleared to all zeroes and register B is loaded with the multiplier. Four clock edges will then produce the unsigned product at outputs P0–P7.

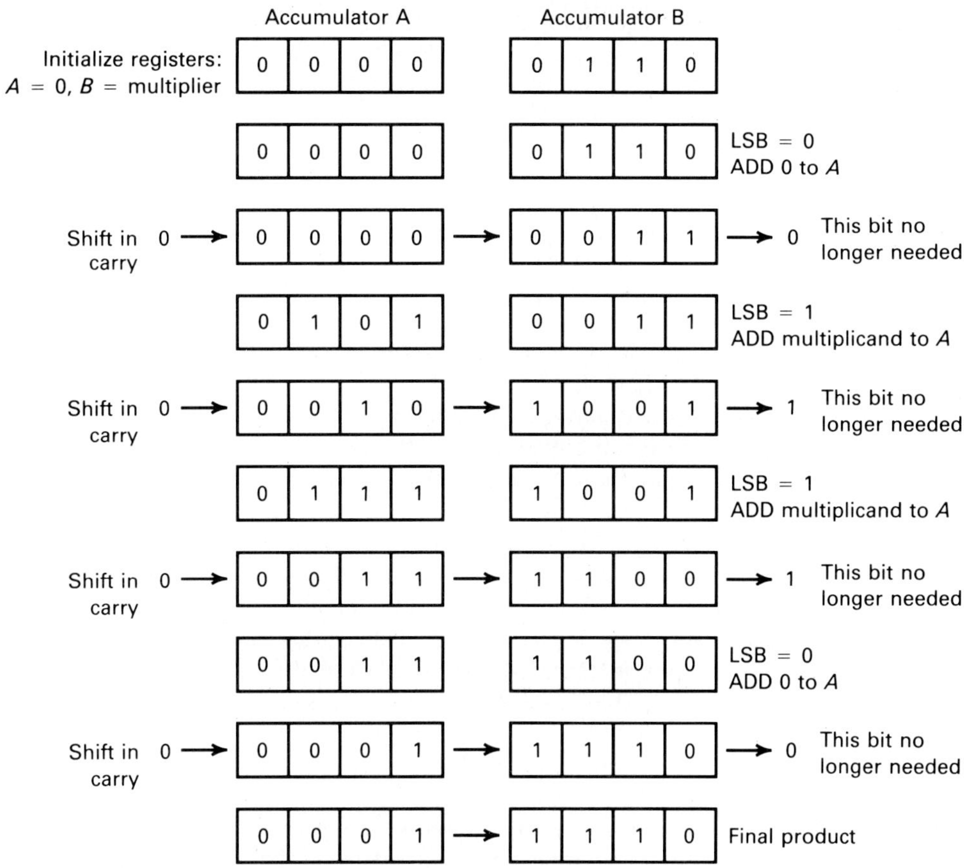

Fig. 7-14(*b*) Multiplying two 4-bit numbers with accumulators. The multiplier is 0110_2 (6_{10}) and the multiplicand is 0101_2 (5_{10}).

7-14*b* but may be configured in a manner similar to the ALU in Fig. 7-8. In microprocessors, *instructions* tell the machine when to add and when to shift. When instructions are used to sequence the accumulators, the operation is called *software multiplication*. This is because the instructions may be changed so that the accumulators will perform tasks other than multiplication. *Soft-wired logic* can perform varied tasks. *Hard-wired logic* can perform only fixed tasks. Early soft-wired logic consisted of hardware components connected through a patch panel, similar to the old-fashioned telephone switchboard. The logic was programmed with plug cords. Computers are soft-wired logic programmed with instructions stored in the computer's memory. The actual work, however, is done by digital hardware responding to control signals on clock edges.

Signed Multiplication

Multiplication of signed numbers may be performed in the same way as described for unsigned numbers if we are willing to convert both numbers to their absolute values and save the signs. Actually the procedure involves four steps:

1. EXCLUSIVE OR the sign bits of the two numbers and save the result.
2. Convert the numbers to positive values if they are negative.
3. Shift and add as described above for unsigned numbers.
4. Convert the product to a negative value if the EXCLUSIVE-OR of the sign bits was 1. (7-14)

Unsigned multiplication may be performed in two accumulators as described above. This approach will work equally well in signed-magnitude or signed-complement machines.

Another approach, which is faster and easier to implement, requires a positive multiplier but will work with either a positive or negative multiplicand. Binary multiplication is performed by adding shifted versions of the multiplicand, and negative multiplicands may be added to the partial products with valid results. In the case of unsigned multiplication, the most significant bit represents magnitude, and the shift right operations bring in the final carry-out from the adder. In signed multiplication the most significant bit represents sign, not magnitude, and the adder carry-out is not used. Instead, we shift in 0s for positive multiplicands and 1s for negative multplicands after the first 1 has been encountered in the multiplier. This guarantees that the sign extension (leading 1s or 0s) is preserved. If the multiplier is negative, then it must be 2s complemented before starting the multiplication. When the multiplier is complemented, the multiplicand must also be complemented in order to preseve the correct sign in the product. Specifically, the procedure is as follows:

1. If the sign bit of the *multiplier* is 1 (negative multiplier), then 2s complement both multiplier and multiplicand; otherwise, just go to 2.
2. If the sign bit of the *multiplicand* is 0 (positive multiplicand), go to 4.
3. Shift in 1s from the left after the first 1 is encountered in the least significant bit of register B (the shifted multiplier) and proceed as in Fig. 7-14. Stop.
4. Always shift in 0s from the left and proceed as in Fig. 7-14. Stop. (7-15)

This will yield a product in 2s complement form and is fairly easy to implement in hardware. If software-successive accumulation were used, the leading 1s or 0s would be shifted via a shift-in control signal after each addition, and no extra logic would be required. The more direct, hardware approach would require two extra logic gates as shown in Fig. 7-15. The flow-through method would require additional adders for the sign extensions of the partial products. Using pencil and paper, an example would appear as follows:

```
+5 × +7                                            +5 × −7

                        Leading bits
      0 1 1 1           are extensions          1 0 0 1
    ×0 1 0 1            of the sign           ×0 1 0 1
[0][0][0] 0 1 1 1       bit.            [1][1][1] 1 0 0 1
[0][0] 0 0 0 0                          [0][0] 0 0 0 0
[0] 0 1 1 1                             [1] 1 0 0 1
0 0 0 0                                 0 0 0 0
0 0 1 0 0 0 1 1                       1 1 0 1 1 1 0 1 = −0 0 1 0 0 0 1 1
                                                                  (7-16)
```

Standard parts are available which perform 4 × 4, 8 × 8, and 16 × 16 bit binary multiplication. In applications such as video games, one of these parts controlled by an 8-bit microprocessor is usually adequate. High-speed computers, speech processors, and digital recording systems, however, require larger operands. The smaller units must be cascaded to accommodate the larger formats. The 74S558 Schottky technology multiplier is a cascadeable 8 × 8 flow-through type multiplier which cascades into a 56 × 56 bit unit capable of handling the IBM long-format numbers. This format is used in IBM 360/370, Amdahl 470, and Data General Eclipse computers.

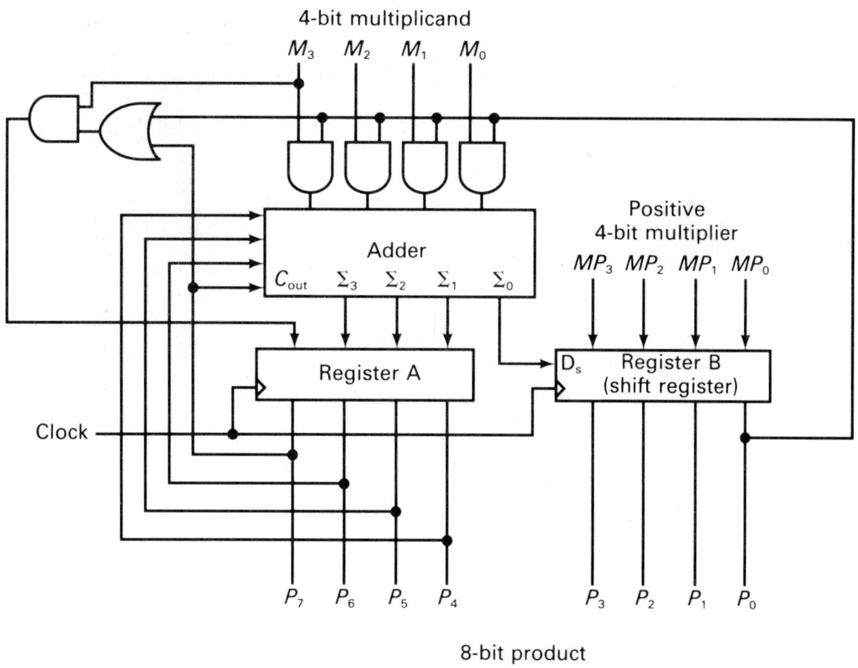

Fig. 7-15 Signed 4-bit multiplication with positive multiplier and either positive or negative multiplicand.

REVIEW 7-3

1. **Q.** Which logic gate may be used to multiply single-digit binary coefficients? **A.** the AND gate

2. **Q.** What operation must be performed when the numbers to be multiplied exceed single digits? **A.** addition of partial products

3. **Q.** What is a partial product? **A.** It is the accumulated sum of shifted multiplicands.

4. **Q.** Explain how the successive-accumulation method is used to multiply unsigned n-bit binary numbers. **A.** The multiplier is placed in the lower half of a $2n$-wide register and examined 1 bit at a time, beginning with the least significant bit (the multiplier). When the least significant bit is a 1, the multiplicand is added to the most significant half of the $2n$-wide register and the contents shifted to the right. When the least significant bit is 0, the shift occurs without the multiplicand being added first. In both cases, the carry-out from the addition is shifted in from the left. This is done n times.

5. **Q.** Explain how the successive-accumulation method is used to multiply signed binary numbers. **A.** The procedure is the same as given in Prob. 4 except that 0s are shifted in from the left until the first 1 is encountered in the multiplier.

6. **Q.** Explain the flow-through method of multiplication. **A.** Use a diagram to show how this pencil-and-paper method is implemented with AND gates and adder stages. Figure 7-13 shows this without the AND gates.

7. **Q.** Explain the difference between hard- and soft-wired logic. **A.** Hard-wired logic performs fixed tasks; soft-wired logic can be programmed to perform varied tasks.

8. **Q.** Explain the difference between signed-magnitude and signed-complement multiplication. **A.** Signed-magnitude multiplication requires two extra steps: EXCLUSIVE ORing the signs and storing the result, and returning the product to negative form if the signs were different.

9. **Q.** How many bit positions are needed to hold the product of two 8-bit numbers? How about the product of an 8-bit number and a 4-bit number? **A.** 16 and 12

10. **Q.** What is sign extension? **A.** Sign extension means filling the leading digits of a positive number with 0s and the leading digits of a negative number with 1s. These leading digits are just extensions of the sign bit and maintain the leading positions for purposes of addition.

11. **Q.** Why is sign extension necessary when negative multiplicands are added? **A.** If the nonsignificant leading positions of a negative 2s

complement number were left blank, 0s would be assumed and the number would be mistaken as positive.

7-4 DIVISION

Multiplication is performed by repeated addition and shifting to yield a product larger than the multiplicand. Division is performed with repeated *subtraction* and shifting to yield a *quotient* which is smaller than the dividend. In general, any integer A is divisible by another integer B if there is a third integer q such that:

$$A = Bq$$

A is called the *dividend*, B is called the *divisor*, and q is the *quotient*. When both sides of the equation are divided by B, we get:

$$\frac{A}{B} = q$$

However, the more general case involves a third number r such that:

$$A = Bq + r$$

where r is the *remainder*, or the part left over when the division is complete. When both sides are divided by B, we get:

$$\frac{A}{B} = q + \frac{r}{B}$$

The result comes in two parts: a quotient and a fraction of B which would not divide into A evenly. Division has few applications outside of data processing and "number crunching" and will not be described in such detail as multiplication. Only the division of two positive binary numbers will be discussed. This will be sufficient to illustrate the general approach.

The process of dividing positive numbers begins by comparing the n-bit divisor with the most significant n-bit portions of the dividend. If subtraction yields a negative result, the operation begins and the divisor is shifted one position to the right. If a subtraction yields a positive result, the quotient is longer than n-bits, indicating an overflow condition, and the operation is stopped. Thereafter, the n-bit divisor is compared with the most significant $n + 1$ bits directly above it, and a subtraction is performed only if it will produce a positive result. If a negative result occurs, the subtraction is inhibited. In either case, the divisor is then shifted one position to the right, and the procedure is repeated. For every successful subtraction, a 1 is entered into the quotient. When subtraction is inhibited, a 0 is entered. The process is illustrated below with the division of the binary equivalent of 37_{10} by the binary equivalent of 6_{10} ($37_{10} \div 6_{10} = 6_{10}$):

	0 1 1 0	← Quotient
0 1 1 0	0 0 1 0 0 1 0 1	
	0 1 1 0	No subtraction
	0 0 1 0 0 1 0 1	
	0 1 1 0	Shift right, no subtraction
	0 0 1 0 0 1 0 1	enter 0 in the quotient
	0 1 1 0	Shift right, subtraction
	0 0 0 0 1 1 0 1	performed, enter 1
	0 1 1 0	Shift Right, Subtraction
Remainder →	0 0 0 0 0 0 0 1	performed enter 1
	0 1 1 0	Shift right, no subtraction,
		enter 0

(7-17)

This procedure may be implemented by placing the dividend into two 4-bit accumulators (or two 4-bit shift registers) and applying the most significant outputs to a 4-bit adder so that they are added to the 2s complement of the divisor. The accumulators or shift registers containing the dividend are shifted to the left to simulate the right shift of the divisor in the example above. The sign of the result may then be inverted and used to serve the dual purpose of enabling the output of the full-adder to be stored on the next clock pulse (subtraction performed) and supplying the quotient with its next less significant bit.

After the initial test for overflow, the accumulators are shifted. Positive full-adder outputs result in the full-adder outputs being stored; negative full-adder outputs cause the storing operation to be inhibited. Shifting occurs in either case for a total of *n* shifts, where *n* is the number of bits in the divisor. The final result left in the most significant 4-bit accumulator represents the remainder. Figure 7-16 shows an example using $37_{10} \div 6_{10}$.

Very often, the ICs which are available to perform multiplication will also perform division. Monolithic Memories 74S508 Multiplier/Divider will divide a 16-bit binary number by an 8-bit binary number or multiply two 8-bit binary numbers. The 74S516 will divide a 32-bit binary number by a 16-bit binary number or multiply two 16-bit numbers.

REVIEW 7-4

1. **Q.** What is a quotient? **A.** A quotient is an integer which represents the number of times a divisor will "go into" a dividend.

2. **Q.** What is a remainder? **A.** A remainder is the part of the dividend which is left over after division.

3. **Q.** If multiplication can be described as shifting and adding, how would you describe division? **A.** shifting and subtracting

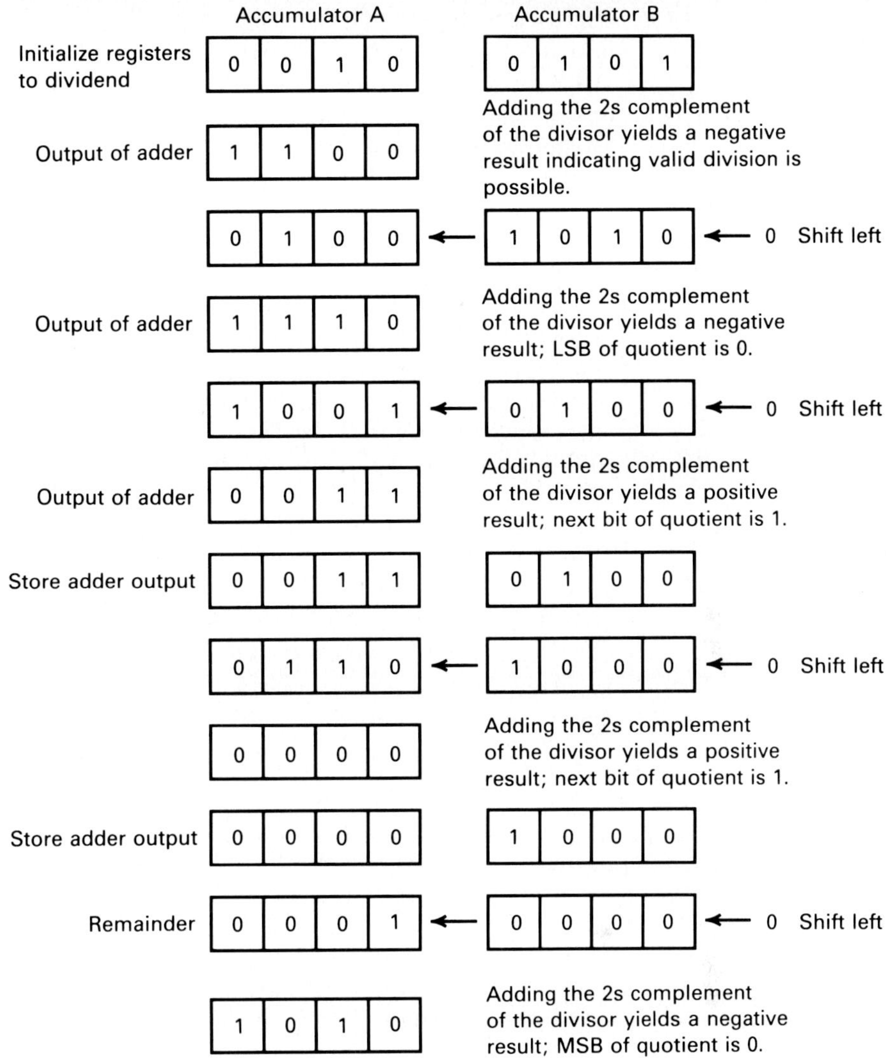

Fig. 7-16 Division using 4-bit accumulators (37 ÷ 6 yields a quotient of 6 and a remainder of 1).

7-5 DECIMAL ADDITION

Addition of decimal numbers in BCD representation can be implemented using binary adders. Binary numbers and BCD numbers are equivalent for values between 0_{10} and 9_{10}. As long as the binary sum is less than 10_{10}, the BCD sum and the binary sum are the same. If the binary sum is 10 or greater, a correction must be made to the result. Intuitively, we might come to the conclusion that, since 4 bits in binary can represent 6 more combinations than are possible in BCD, the number needed for correction might be 6. BCD numbers range from 0000_2 to 1001_2 and then roll over into another bit position. Hexadecimal numbers (binary numbers grouped into 4-bit clusters) range from 0000_2 to 1111_2 before rolling over. Some modification must be made to compensate for the fact that 6 combinations are not used in BCD. The obvious modification is to add the number 6 to results greater than 9. The implementation of a BCD full-adder stage may be approached by using two 74xx83 binary full-adders together with some combinatorial logic which would compensate for the fact that 6 of the 16 possible combinations of 4 bits are not used.

A more methodical approach would be to construct a truth table. The maximum sum of two BCD digits is $1001_2 + 1001_2 + \text{(carry-in)} = 10011_2$ since each BCD digit can be no larger than 9_{10} (1001_2) and the carry-in can be no larger than 1. A truth table, then, would show each possible binary sum between 0 and 19 together with its corresponding BCD sum. The truth table in Fig. 7-17 indicates that if the two BCD digits are added with a 4-bit binary full-adder, the sum will be a binary result which either corresponds to the proper BCD result or differs from the correct result by 1010_2 (10_{10}). In order to compensate for the excess when it occurs, we must subtract 1010_2 from the uncorrected sum. Subtracting 1010_2 is equivalent to adding the 2s complement of 01010_2, which is 10110_2. The extra leading bit is provided to accommodate the sign but is not actually necessary in this case since we will perform the subtraction only when we know that the uncorrected result is greater than 9. This condition will always guarantee a positive result. The question then becomes when to add the number 0110 to the uncorrected sum.

The table in Fig. 7-17 shows the logical expression needed to determine

Possible Outputs of the Binary Adder with BCD Inputs					Corresponding Corrected Sum				
C_{out_3}	Σ_3	Σ_2	Σ_1	Σ_0	C_{out_3}	S_3	S_2	S_1	S_0
0	0	0	0	0	0	0	0	0	0
0	0	0	0	1	0	0	0	0	1
0	0	0	1	0	0	0	0	1	0
0	0	0	1	1	0	0	0	1	1
0	0	1	0	0	0	0	1	0	0
0	0	1	0	1	0	0	1	0	1
0	0	1	1	0	0	0	1	1	0
0	0	1	1	1	0	0	1	1	1
0	1	0	0	0	0	1	0	0	0
0	1	0	0	1	0	1	0	0	1
0	1	0	1	0	1	0	0	0	0
0	1	0	1	1	1	0	0	0	1
0	1	1	0	0	1	0	0	1	0
0	1	1	0	1	1	0	0	1	1
0	1	1	1	0	1	0	1	0	0
0	1	1	1	1	1	0	1	0	1
1	0	0	0	0	1	0	1	1	0
1	0	0	0	1	1	0	1	1	1
1	0	0	1	0	1	1	0	0	0
1	0	0	1	1	1	1	0	0	1

Fig. 7-17 Truth table for 1-digit BCD full-adder.

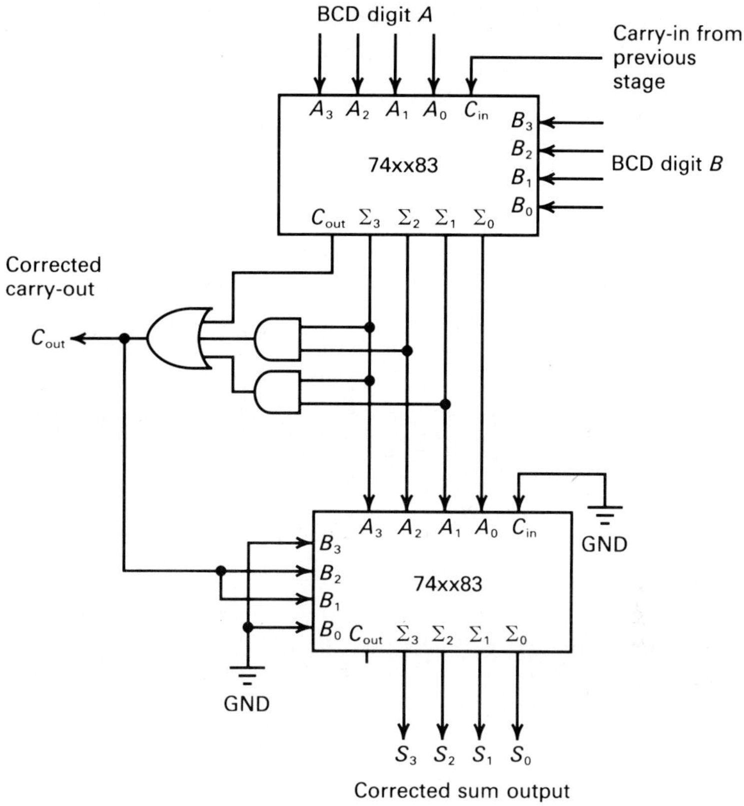

Fig. 7-18 BCD full-adder stage.

when the correction should be applied. We can see that whenever $\Sigma_3\Sigma_1 = 1$, $\Sigma_3\Sigma_2 = 1$, or $C_{out_3} = 1$, the number 0110 must be added to the uncorrected sum, and a carry-out to the next stage must occur. This may be implemented by applying the output from the expression

$$f = C_{out_1} + \Sigma_3\Sigma_2 + \Sigma_3\Sigma_1 \tag{7-18}$$

to a second 74xx83 at the b_2 and b_1 inputs while the b_3 and b_0 inputs are grounded. This is equivalent to adding 0110_2 when $f = 1$ and correction is needed, and adding 0000_2 when $f = 0$ and no correction is needed. The design is illustrated in Fig. 7-18. In IC design some economy is gained by noting that Σ_0 will in every case be added to 0 since correction is never needed for the least significant bit. Therefore, the second adder need not be a full 4-bit adder. A HA/FA/HA combination performs equally well.

REVIEW 7-5

1. **Q.** Intuitively, why would we want to add the number 6 to binary sums greater than 9 in BCD addition? **A.** We would do so because 4-bits can represent six more numbers in binary than are possible in BCD

code. These extra combinations must be cut out of the result for a sum greater than 9.

2. **Q.** What is the largest sum possible when two BCD digits are added? Why is it important to know this? **A.** 19, 9 plus 9 plus a carry-in of 1. We must show all possible BCD sums in the truth table.

3. **Q.** What does the truth table tell us about corrections? How does this justify our intuitive conclusion? **A.** When the binary sum is 1010, the equivalent BCD digit is 0000, and a carry is made to the next position. This means that the binary number 1010 must be subtracted from binary sums greater than 1001. Subtracting the number 1010 is equivalent to adding the number 0110.

4. **Q.** How do we know when to make corrections? **A.** We examine the truth table to find a function which will detect sums between 10 and 19.

5. **Q.** How many 74xx83 full-adder chips would be needed to add the numbers 25 and 15 BCD? **A.** Four 74xx83 ICs would be needed.

7-6 FLOATING-POINT REPRESENTATION

One of the limitations of machine computation is the necessity of using words of fixed length. The sum of two positive or two negative binary numbers will not always fit into the given word length. When the sum doesn't fit, overflow occurs, and the result is invalid. We know that the product of two n-bit numbers *will* fit into two n-bit registers, and this could be provided for in the hardware. However, if we have an n-bit machine, what do we do with a $2n$-bit product when further multiplication or addition is required? A second operation could cause the result to balloon into a $3n$-bit or a $4n$-bit product if the first result were not rounded back to n-bits. A variable word length is one solution, but this causes problems with the detection of the sign bit and requires excessive amounts of hardware. A more common approach is to use *floating-point representation*.

Floating-point representation is similar to scientific notation in that it represents a number in two parts, an integer or fractional part, called the *mantissa*, and an *exponent* or *characteristic* which determines the actual position of the *binary point*. Previously we have ignored radix points in discussing computer arithmetic. A *radix point* tells us which part of a number is whole and which part is fractional. In decimal notation the radix point is called the *decimal point*. In binary notation it is called the *binary point*. We have not mentioned binary points before because we have dealt exclusively with whole numbers. We have always assumed a fixed binary point to the right of the least significant digit. Now we will look beyond that fixed binary point to examine the positional representation of binary fractions. Just as with whole binary numbers, each position indicates a power of 2. If we allow the letter b to represent a binary coefficient or digit, 0 or 1, then the positions have the following meaning:

$$\ldots b_4 \times 2^4 + b_3 \times 2^3 + b_2 \times 2^2 + b_1 \times 2^1 + b_0 \times 2^0 \,.\, b_{-1} \times 2^{-1} + b_{-2} \times 2^{-2} \ldots \tag{7-19}$$

The relative weight of each position is either a positive or negative power of 2 depending upon which side of the binary point it sits.

As an example, a floating-point representation of the binary number 0.1101001×2^4, or 13.125_{10}, might be represented in two registers as follows:

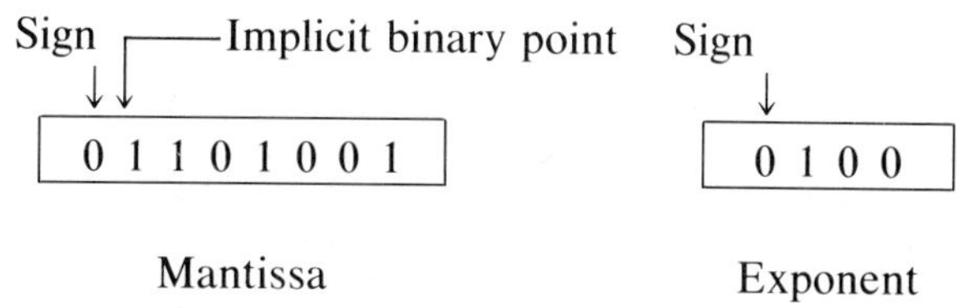

(7-20)

Some computers represent the mantissa as a fraction (fixed binary point just in front of the leading digit), some represent it as an integer (binary point after the right-most digit), and some as both with the implied binary point between the first and second digits. The binary point has no actual physical existence in the computer, but it is implicit in the organization of the data.

When a floating-point result contains n leading 0s, it must be *normalized* by shifting the mantissa n positions to the left and decreasing the exponent by n. This preserves the significant portion of the number. When floating-point numbers are added and subtracted, they must be realigned so that the exponents are equal, and this is done by shifting one mantissa and adjusting its exponent until it equals the other exponent. These two operations alone illustrate the fact that floating-point computation is more complicated and requires more complex hardware than integer arithmetic.

Truncation Errors

After each computation with n-bit words, only the most significant n bits can be saved in the mantissa, and the increased magnitude is reflected in the exponent. This will often result in the *truncation* of less significant bits. The following 8-bit example illustrates this process. Suppose the numbers 105_{10} and 69_{10} are to be multiplied. The binary equivalent is 01101001_2 multiplied by the number 01000101_2. The product 0001110001001101 may be returned to a register as a mantissa of 11100010 with an exponent of 0101 (2^5). To put this in perspective, the decimal equivalents are 7245_{10} for the product before truncation and 7232_{10} for the truncated result. You might argue to keep the entire double word. Fine, but what happens at the next multiplication? At some point, the word length must be limited or the numbers in the system will grow without bound. What we actually do is preserve a given level of resolution, or number of significant digits. In the above example only eight significant binary digits were preserved, and the result was in error by the amount 13_{10}. This type of error is called a *truncation error*.

Another approach to fitting products and sums into fixed-length registers is called *rounding*. Truncation is almost automatic. The product or sum is normalized and the result is stored, left-justified, into a register. Any trailing digits which do not fit are simply lost. Rounding requires more steps but often reduces errors by incrementing the truncated number if the lost bits represent more than half the value of the least significant surviving bit. Specifically, it involves examining the portion of a number which would be thrown away and adding 1 to the next most significant portion when the throwaway digits represent more than half of the least significant surviving bit. For example, given the decimal number 25461_{10}, if the two least significant digits will not fit, then the rounded equivalent would be $255_{10} \times 10^2$. The number 4_{10} was rounded to 5_{10} because 61_{10} represents more than half of what is possible to represent in two decimal digits.

Conversion Errors

Another source of error is inherent in the conversion of decimal numbers to binary. While any decimal integer can be represented in binary, given a long enough register, decimal fractions may or may not be representable by powers of 2. For example, while the decimal fraction 0.5_{10} is easily represented in binary as 0.1_2, the decimal fraction 0.4_{10} would require an infinite number of bits to be represented exactly:

$$0.4_{10} = (0.0110011001100110011001100110 \ldots)_2$$

The exact binary equivalent of 0.4_{10} requires an infinite repetition of the binary digit pattern 0110. Therefore, some approximation, either truncation or rounding, is necessary in order to store such a number in a computer. Assuming that an 8-bit register is provided, the truncated version of 0.4_{10} stored in binary would be 0.01100110_2 (01100110 mantissa and 2^{-8} exponent). This is equivalent to $\frac{1}{4} + \frac{1}{8} + \frac{1}{64} + \frac{1}{128} = 0.3984375$. The error in converting to 8-bit binary representation is 0.0015625_{10}. One way to avoid conversion errors of this type is to store all decimal numbers in BCD form, but this requires large machine formats.

One of the unfortunate consequences of using digital machines for computation is the introduction of errors. Truncation, rounding, and conversion errors are unavoidable and will accumulate as the computations proceed. In introductory texts, computer scientists devote much space to the discussion of errors and methods for reducing them.

REVIEW 7-6

1. **Q.** What is floating-point representation? **A.** It is a two-part representation of numbers where the first part indicates significant digits and the second part indicates the exact position of the radix point.

2. **Q.** What is a mantissa? **A.** A mantissa is the portion of a floating-point number which contains the significant digits.

3. **Q.** What is an exponent? **A.** An exponent is the portion of a floating-point number which determines the position of the radix point.

4. **Q.** What is a binary point? **A.** A binary point is the radix point for binary numbers. The digits to the left of the binary point represent the integer part of the number while digits to the right represent the fractional part of the number.

5. **Q.** What is normalization? **A.** Normalization is the process of eliminating nonsignificant leading 0s in floating-point numbers.

6. **Q.** What is truncation? **A.** Truncation is the effect of trying to fit an *n*-bit number into a register with fewer than *n* bits. Floating-point numbers are normalized before being stored so that only the less significant portion of a number is lost.

7. **Q.** What is rounding? **A.** Rounding is an attempt to minimize errors by incrementing the truncated number if the lost bits represent more than half the value of the least significant surviving bit.

8. **Q.** What are conversion errors? **A.** Conversion errors are errors inherent in the conversion of a fraction from base 10 to base 2.

REVIEW PROBLEMS

1. Add the following 8-bit unsigned numbers:

(a)	(b)	(c)
01011100	10110001	11110000
01100101	01000111	10001010

2. Design a 4-bit adder using half-adder stages and ripple carry. Assuming that each logic gate has an equal propagation delay of 5 ns, how long will it take to add two 4-bit numbers with this adder?

3. Find the maximum clock rate for a system using the adder designed above. Assume that an extra 5 ns is required to gate the numbers to the adder and another 5 ns is needed for setup to latch the sum.

4. Expand the logic equations given for C_{out} in Eq. (7-6). Note that one is already expanded as an example in Eq. (7-7). Implement these equations with AND/OR logic to provide look-ahead carry operation.

5. Using the same assumptions as in Prob. 3, find the maximum clock rate for the same adder with the look-ahead carry designed in Prob. 4. How long will it take to add two 4-bit numbers?

6. If one of the full-adder stages designed in Prob. 2 were used in a serial adder such as the one shown in Fig. 7-5, and the setup time for the shift registers were 5 ns, what would the maximum clock rate be for the serial adder? How long would it take to add two 4-bit numbers?

7. Add the following 8-bit signed-magnitude numbers and give the decimal equivalent of the sum.

(a)		(b)		(c)	
	01010110		11001010		10011101
	+10011101		+01001001		+11000010

8. Convert the following 2s complement numbers to their decimal equivalents. Convert the same numbers to signed-magnitude binary equivalents.

(a) 11101010 (b) 11110100 (c) 10000010

9. Add the following 8-bit 2s complement numbers and give the decimal equivalent of the sum.

(a)		(b)		(c)	
	01010110		11001010		10011101
	+10011101		+01001001		+11000010

10. Which of the following summations of 8-bit 2s complement numbers will produce overflow?

(a)		(b)		(c)	
	01010000		10110010		11111011
	+01101100		+11111111		+01111111

11. Perform the following subtractions using 8-bit 2s complement binary numbers.

(a) $13_{10} - 17_{10}$ (b) $-78_{10} - 1_{10}$ (c) $127_{10} - 5_{10}$

12. Sketch the control signal timing waveforms that would be required to decrement the accumulator in Fig. 7-8.

13. Multiply the following unsigned 4-bit binary numbers showing the partial products, binary final product, and decimal equivalent of the final product.

(a)		(b)		(c)	
	1110		1010		1111
	×0111		×0101		×1111

14. Multiply the following signed 4-bit binary numbers showing the partial products, binary final products, and decimal equivalents.

(a)		(b)		(c)	
	1110		1010		1111
	×0111		×0101		×1111

15. Design a 4-bit flow-through multiplier using Fig. 7-13 as a guide. Use block diagrams of the 74xx83 together with 74xx08 AND gates. The circuit will be for unsigned multiplication only.

16. Justify the conclusions reached in Fig. 7-17 and Eq. (7-18) by constructing two 4-input Karnaugh maps, one for $C_{out} = 0$ and one for $C_{out} = 1$. The minimized results of the two individual maps may be ORed to produce the final boolean equation.

17. Find the decimal equivalents of the binary numbers represented in

floating-point notation below. Assume that both the mantissa and exponent are represented in 2s complement notation.

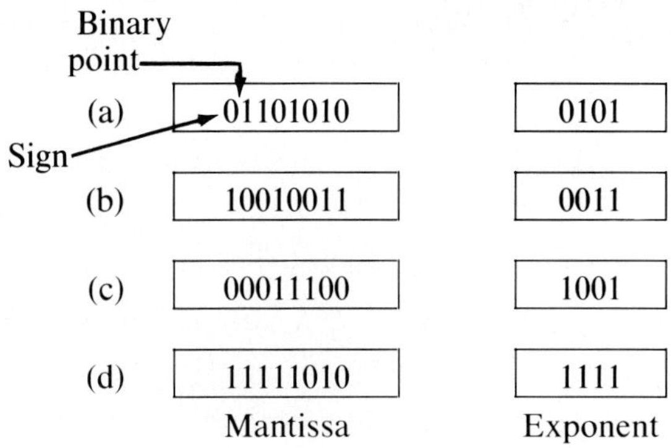

18. Repeat Prob.17 assuming a signed-magnitude mantissa and exponent.

8 SEMICONDUCTOR MEMORY

One of the first problems to be addressed in the early days of computer development was how to store instructions and data. Various mechanical, electrical, electromechanical, and optical methods were tried. Some of the more unique solutions included punched cards, neon-bulb flip-flops, shift registers implemented with wire delay lines, glass delay lines, and waves in pools of mercury. Information was also stored in magnetic fields using small toroids of wire twisted around metal cores (core memory) and small rectangles of nickel-iron-cobalt alloy (magnetic-film memory). Later developments improved on these methods by providing larger, denser, and more economical forms of storage in semiconductor, optical, and magnetic media. This chapter will describe current methods for storing and retrieving large quantities of information in semiconductor materials. It differs from the previous discussions of data storage in that the registers described in Chapter 5 represent only a departure point, and other methods for storing larger quantities of data are introduced here. This chapter will acquaint the reader with the following forms of semiconductor memory:

1. Register files.
2. Semiconductor random access memory (RAM)
3. Semiconductor read-only memory (ROM, PROM, and PLAs)
4. Erasable and electronically alterable forms of read-only memory (EPROMs, EEPROMs, EAROMs, and NOVRAMs)

In addition, memory organization and addressing will be introduced.

8-1 REGISTER FILES

One of the most straightforward approaches to storing data is to assemble individual flip-flops into word-sized registers and group the registers into conveniently sized files. A ''convenient size'' means a number which is a power of 2. Two registers can be addressed with a single bit. A 0 would enable register 0, and a 1 would enable register 1. Four registers can be addressed with 2 bits, eight registers with 3 bits, and so on. As long as the number of registers is a power of 2, every combination of n bits represents a valid address. *Addressing* a register means decoding the address inputs into a single control signal which selects that register. The *address* of a register is unique and differentiates it from all others.

Figure 8-1 shows how 4-bit registers are organized in the 74xx170 *register file*. Data is written into a particular 4-bit register by addressing that

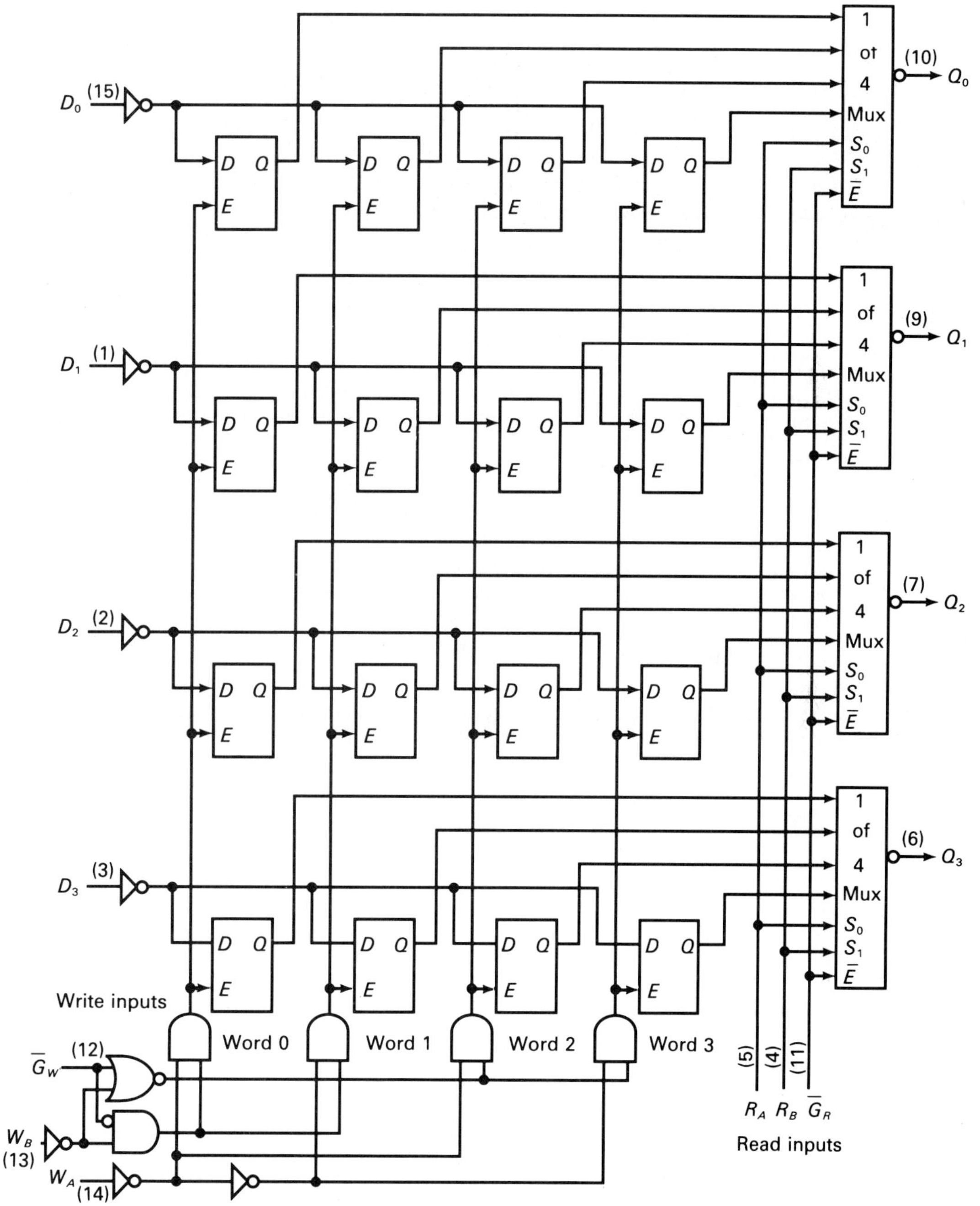

Fig. 8-1 74xx170 register file.

register with the write-address input pins W_A and W_B and strobing the write-enable input G_W low. The register elements are latches rather than flip-flops, so that whatever appears on the data inputs will also appear on the Q outputs while the write-enable input is low. In order to read a particular register, that register must be addressed with read-address inputs R_A and R_B while the read-enable input G_R is low. Data-entry addressing is separate from data-read addressing, and this allows simultaneous writes to one location while reads are being made from another location.

Two concepts invite review at this time, multiplexing and output enabling. In general, *multiplexing* means channeling data from more lines onto fewer lines. This implies a selection process. The AND/OR logic used to select between inputs for shift register and counter stages represents the most common form of multiplexing. If four possible inputs are available, then one of the four may be selected with an AND/OR structure. The AND/OR structure, introduced in Chapter 4, will always present one of the four inputs to the output and will always force the output either high or low. The 4-bit register file uses this type of multiplexer (MUX) to select bits from one of four registers.

The second concept, *output enabling*, arises in this discussion because memory devices normally sit on buses with other memory devices, and there must be some means of sharing the common bus. As discussed in Chapter 3, there are basically two types of output structures which facilitate bus sharing; open-collector/drain and tristate. *Open-collector* (*oc*) and *open-drain* (*od*) structures are quite simple. An AND gate, connected to an *npn* bipolar transistor or *n*-channel MOSFET, may pull the bus line low but cannot force the bus line high. Figure 8-2*a* illustrates this configuration. Pull-up resistors are used on each bus line to establish a high level when no output is actively pulling low. Any output may be disabled with a low input to its AND gate. The disadvantage to this approach is that pull-up resistors are not active devices and cannot pull high as quickly as transistors. The tristate approach uses transistors in both the upper and the lower connections to supply. This allows equal drive high and low. Isolation is achieved by turning off both transistors at the same time. Figure 8-2*b* and *c* show how both upper and lower transmission devices of an output driver may be turned off at the same time with logic gates. This effectively removes the output driver from the bus. Figure 8-2*d* shows how the output can also be removed from a bus with a *transmission gate*. Transmission gates are not possible in bipolor technology because bipolar transistors cannot conduct bidirectionally. However, in MOS circuits they provide an easy method for isolating a logic output under the control of an enable signal. The methods shown in *b* and *c* are best suited to outputs which must drive large off-chip loads. Transmission gates are typically used when small on-chip loads are driven.

Register files are provided in the 54/74 series with the following configurations:

8 words × 2 bits	'172 (tristate)
4 words × 4 bits	'170 (oc), '670 (tristate)
Dual 16 words × 4 bits	'870 (tristate)

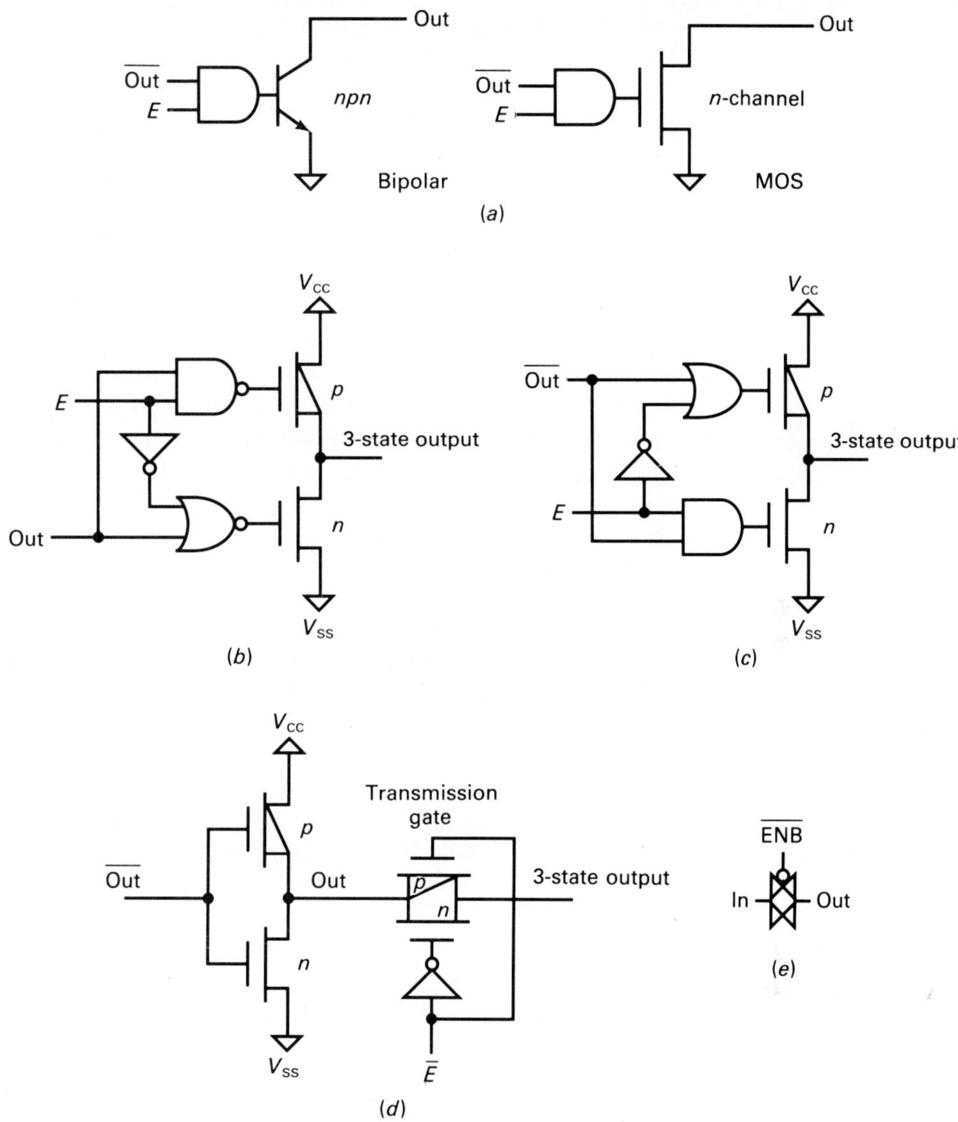

Fig. 8-2 Open-collector bipolar, open-drain MOS, and tristate MOS output devices. (*a*) Open-collector and open-drain outputs; (*b*) tristate MOS (noninverting); (*c*) tristate MOS (inverting); (*d*) tristate MOS using transmission gate; (*e*) schematic symbol for transmission gate.

REVIEW 8-1

1. **Q.** What are register files? **A.** Register files are collections of flip-flops organized into 2, 4, or 8 bit registers.
2. **Q.** How is a particular register selected for reading or writing? **A.** Read-address inputs and write-address inputs are used to select a particular register.

3. **Q.** What type of combinational circuit is used to select between register file outputs? **A.** a multiplexer

4. **Q.** How are register outputs and inputs isolated from the external buses? **A.** Outputs are isolated using open-collector or tristate drivers. Inputs are isolated with input enables to enabled D flip-flops.

5. **Q.** What is a transmission gate? **A.** A transmission gate is a bidirectional switch which may be turned on or off with an enable signal.

8-2 RANDOM ACCESS MEMORY

The register file may be classified as *random-access memory (RAM)* because access is always available and because it takes no longer to access one memory location than to access any other location. *Sequential-access memory*, on the other hand, stores data in such a way that information is available only during certain intervals of time, and the *access time*, the time it takes to read or write the data, depends on where the data is stored. Magnetic tape is an example of sequential-access memory. Although each location does eventually pass the read and write heads, some locations are available sooner than others. The term RAM is normally associated with large blocks of memory, even though it can apply to units as small as the 4-word × 4-bit register file described above.

Static RAM

A more useful form of random access memory which provides much higher densities than register files is called *static RAM*, or *SRAM*. The individual memory elements which maintain single bits of information are called *RAM cells* or *bit cells*. RAM cells are actually simplified latches. Figure 8-3 shows two examples. One uses transistors as upper transmission devices; the other uses resistors. Early RAMs used resistors. These were later replaced by transistors for economy of size. However, recent

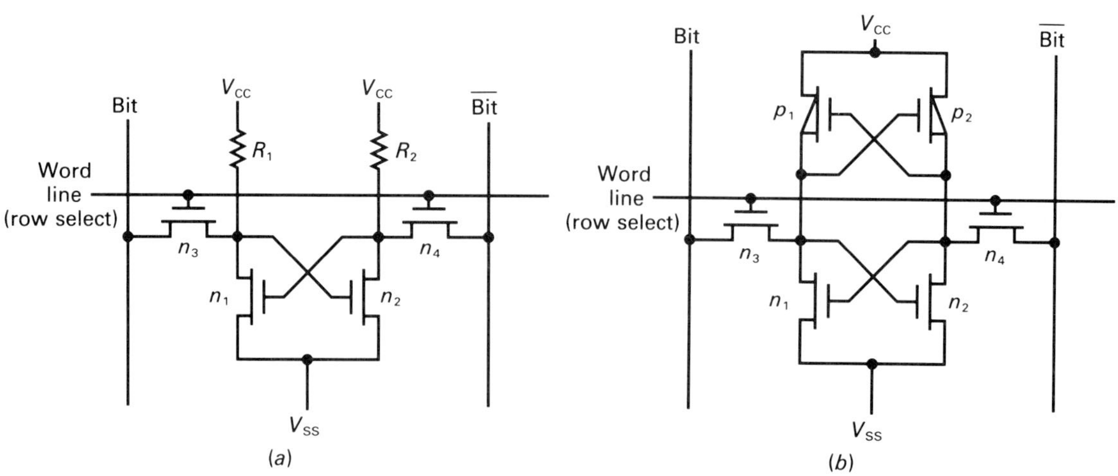

Fig. 8-3 Typical SRAM cells.

developments in the fabrication of buried, undoped polysilicon resistors have returned the size advantage to resistors. In both cases a *write* is performed by driving the bit line to the desired logic level while driving the bit NOT line to the opposite value. When the *row select*, or *word line*, is high, the value on the bit line will be forced on the cell. When the row-select line returns low, the cell is removed from any further influence by the bit lines and the latched value is maintained until another write operation occurs or until power is lost. The term *static* refers to the fact that data is retained as long as power is applied, even with no clocks running.

A *read* operation involves strobing the row-select line without forcing the bit lines either high or low. Transmission gates can be used to isolate the bit lines from the write drivers while a read is being performed. The value held in the RAM cell provides an input to a *sense amp* which, in turn, drives the external buffers. Figure 8-4 shows a CMOS sense amp.

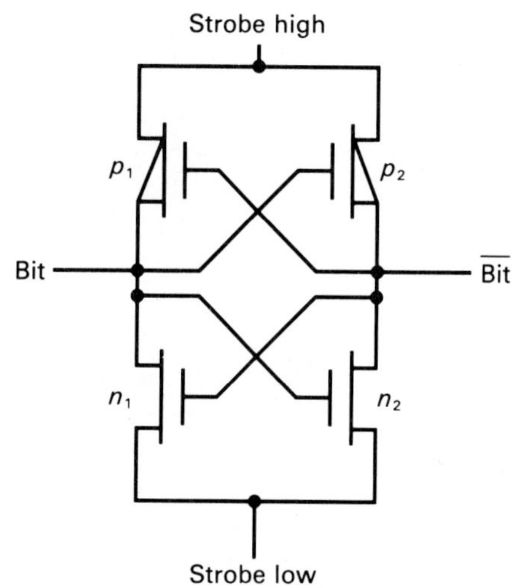

Fig. 8-4 CMOS sense amp.

The device actually looks just like a static RAM cell. However, the *n* and *p* transistors of the sense amp are larger than those of the RAM cell and are capable of driving the bit lines to full supply in a small amount of time. The idea is to make the RAM cells as small as possible since they are repeated many thousands of times and to drive the external buffers with the larger devices in the sense amp. The RAM cell transistors are too small to drive the large capacitance of the bit lines and output buffers in a reasonable amount of time. Instead, they must provide only enough difference in voltage between bit and bit NOT to set up the sense-amp inputs to the proper state. The sense amp has no power or ground connections during this setup time, so it is not active and will not oppose the change. A delayed strobe provides power and ground to the sense amp after its inputs have been established. The sense amp then guarantees sufficient drive to bring the bit lines to full V_{cc} or V_{ss} levels so that the output buffers will change correctly. One sense amp serves all the RAM cells connected to a given pair of bit lines. In this way, a very dense array of small devices may be constructed in a block, with larger sense amps along the periphery. The sense amp guarantees that the state of a cell is not lost during a read operation.

The most basic form of address decoding used in random-access memories is shown in Fig. 8-5. It consists of AND or NAND gates with

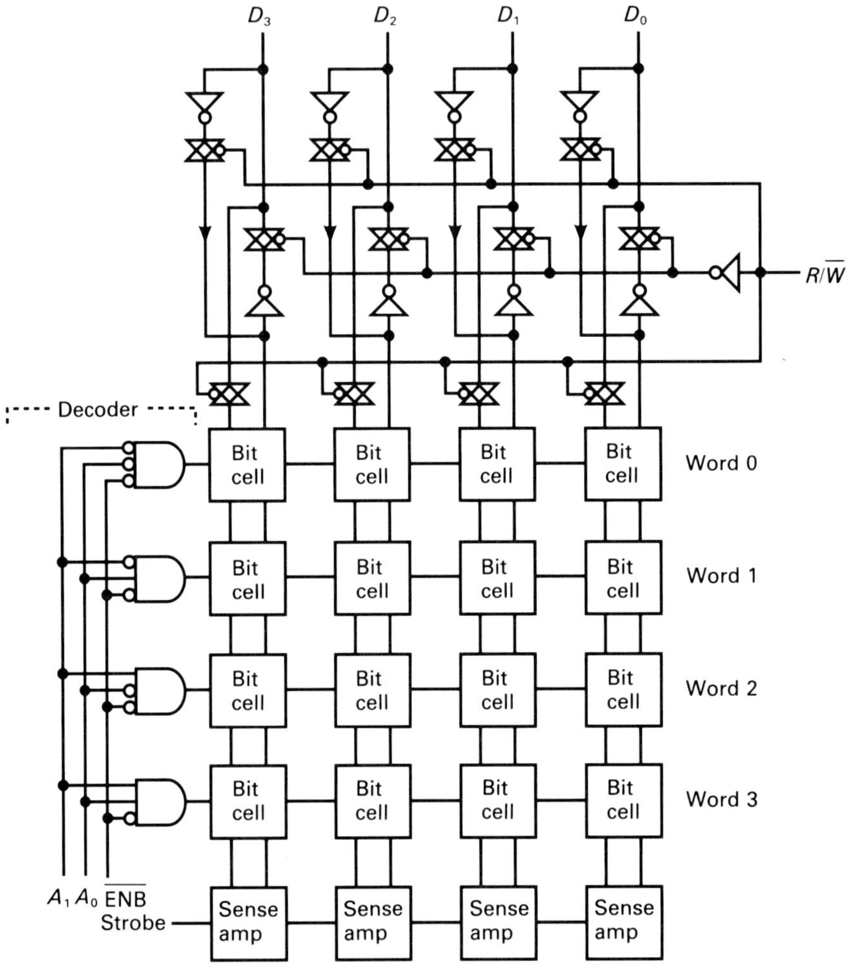

Fig. 8-5 Four-word by 4-bit SRAM array.

inverters at the inputs. A decoder will provide 1 of 2^n possible outputs for each combination of n inputs. Any n-bit address on the inputs will cause an asserted output on one and only one of the n word select lines. In Fig. 8-5, the *address bits* A_1A_0 are decoded into one of four control signals which enable *rows* of bit cells. An equivalent type of circuitry is found in the 74xx138, 1-of-8 decoder. This part is used for external address selection between RAM chips. Figure 8-6 shows the '138 logic symbol and logic diagram.

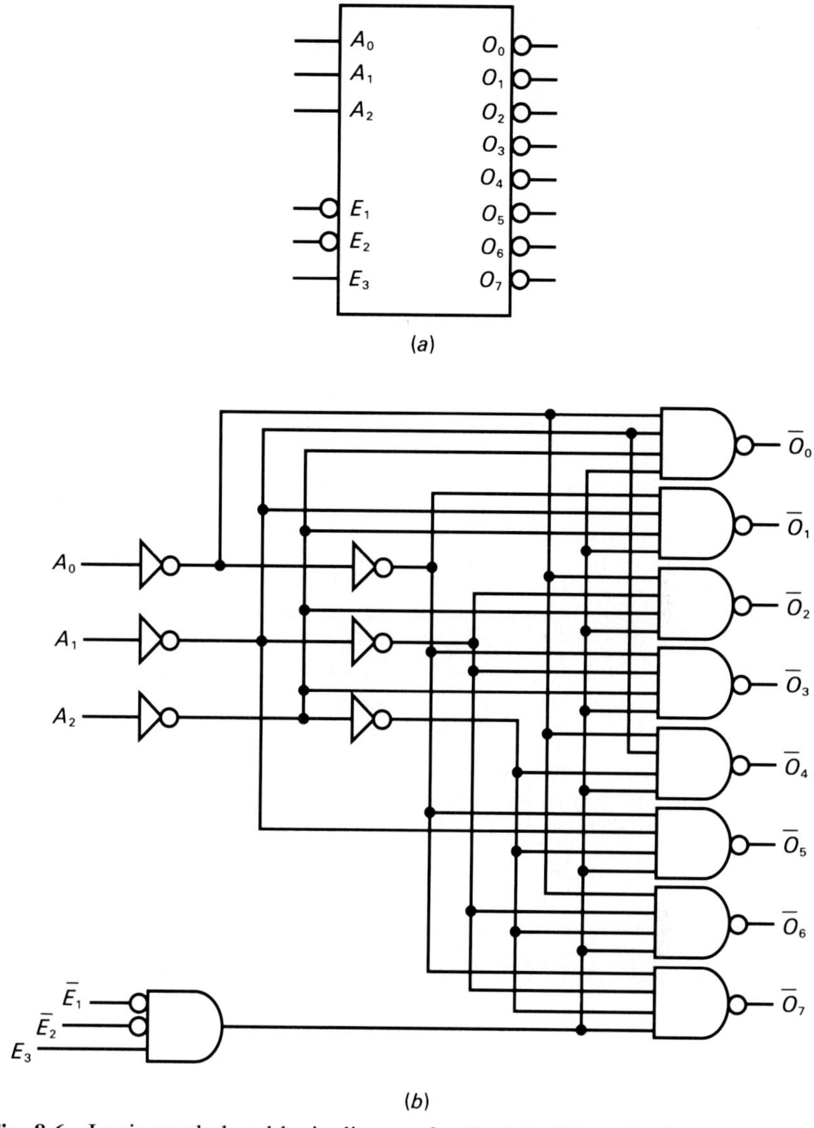

Fig. 8-6 Logic symbol and logic diagram for the 74xx138 1-of-8 decoder.

The word-length groups of bit cells in Fig. 8-5 represent *rows* of data while the individual bit positions are represented as *columns*. This, you may notice, is just the opposite of the organization used in the '170 register file. In Fig. 8-5 the rows correspond to words and the columns to individual bit positions within a word. The address decodes to a *row-select* control signal which enables the proper word. A single bit of data is uniquely identified by specifying its row and column. For instance, by knowing the address of the word (the row) and the bit position within that word (the column), we can read or write every bit cell. The direction of the data depends on a control signal called *read-write*. A high value on this line causes a read, and a low value causes a write operation.

Dynamic RAM

Another type of random-access memory, called *dynamic RAM*, or *DRAM*, uses a small capacitor to store the bit sense. Figure 8-7 shows the circuit

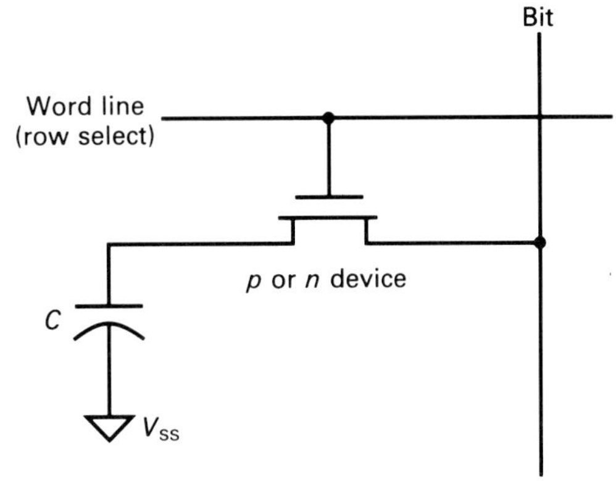

Fig. 8-7 DRAM cell equivalent circuit.

for one bit cell. Like the SRAM cells, DRAM cells are addressed with the word line. A write consists of asserting the bit line either high or low and driving the capacitor high or low through the gating transistor. A read consists of connecting the capacitor to the bit line through the gating transistor and detecting the change in charge on the line with a sense amp. Unlike SRAM, however, DRAM cells do not maintain their bit sense throughout a read operation. The state of the cell is lost when the capacitor discharges onto the bit line. This is referred to as a *destructive read.* Restoring circuits are required to write back the original charge before the read cycle is completed. Moreover, because of junction leakage currents, the cell will eventually discharge anyway if it is not read often enough. Dummy read and restore operations must be performed on a routine basis in order to maintain the correct stored data. These periodic restoration cycles are called *refresh cycles* and must be performed at least once every 2 to 4 ms. This is the primary reason for calling this type of cell *dynamic*.

RAM, both static and dynamic, is available in 1-bit, 4-bit, and 8-bit words. Eight-bit word lengths lend themselves to the 8-, 16-, and 32-bit data organization found in microprocessors. Four-bit word lengths may be used for 12-bit data buses, and 1-bit organization allows the flexibility for odd word sizes. For instance, 9- and 10-bit word sizes are commonly used in the telephone industry where 7 or 8 bits are required for digital voice storage, 1 bit is needed for signaling between telephone switches, and 1 bit is used for parity. Economics and availability are also factors in the choice of word size. Sometimes a 1-bit word size will prove cheaper or more accessible than a corresponding 8-bit organization. Economic considerations, however, must always include future projections of RAM costs since the price of memory drops very quickly when technological advances are made in design and production. As an example of how different word lengths may be assembled to achieve the same memory space, consider an application where 256K × 8 bits of RAM are needed. Figure 8-8*a*

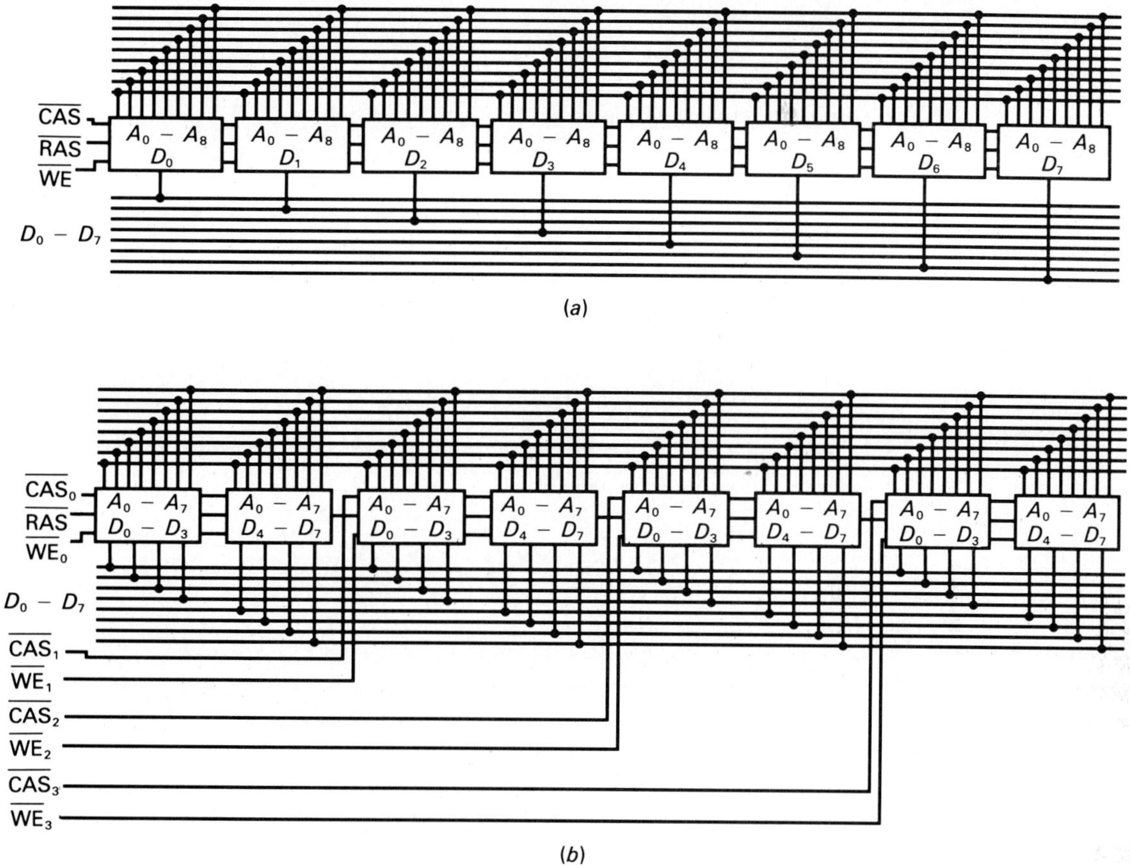

Fig. 8-8 Examples of RAM organization. (*a*) 256K × 8-bit memory using 256K × 1-bit DRAMs; (*b*) 256K × 8-bit memory using 64K × 4-bit DRAMs.

shows how eight 256K × 1-bit RAMs provide the requirement using 16-pin DIPs. Or eight 64K × 4-bit RAMs in 18-pin DIPs may be used as shown in *b*.

It is important to realize that for practical RAM sizes, the association of words with rows as presented in Fig. 8-5 is not realistic. If the word-row correspondence were maintained for a 256K × 1-bit RAM, for instance, the silicon layout would be extraordinarily long and skinny. It is desirable, instead, to keep the perimeter of the array as small and square as possible. The architecture of a 256K RAM may be visualized externally as a 512-row × 512-column memory cell array. The individual 1-bit words are selected by specifying the row with a 9-bit row address and the column with a 9-bit column address. Although an 18-bit address is required to access one of 256K bits, only 9 address inputs are provided on the IC. The 9 address inputs are shared, or time multiplexed, in order to reduce pin count. *Time multiplexing* means that two or more sets of inputs are sequentially impressed upon one set of bus lines. Access is made by first latching the 9 row address bits with the *row address strobe (RAS*)* and then presenting the 9 column address bits along with the *column-address* *strobe (CAS*)*. A third signal, *write enable (WE*)*, determines the direction of data. All these signals are active-low as indicated by the asterisks. Figure 8-9 is typical of the timing waveforms associated with reads and writes to dynamic memory.

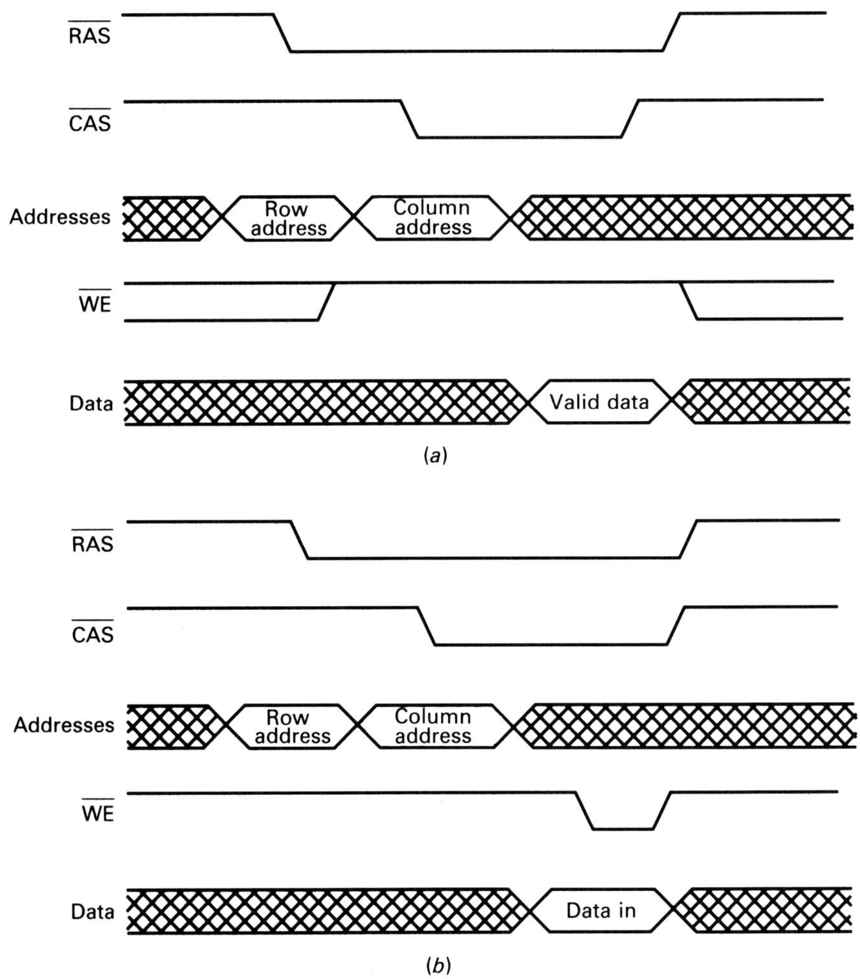

Fig. 8-9 Read and write cycles for DRAM. (*a*) Read cycle timing; (*b*) write cycle timing.

Refresh Cycles

Dynamic memory devices must be refreshed at least once every 2 to 4 ms in order to maintain correct data. Data is not actually being accessed during refresh, so only row addressing is necessary. Each row address cycle will restore the data in all the bit cells of the addressed row. In the 256K × 1-bit memory described above, for instance, there are 512 row addresses which must be read and restored every 4 ms. The internal organization of most DRAMs permits two rows to be refreshed in a single cycle, allowing the full memory of a 256K DRAM to be refreshed in only 256 cycles. For any size RAM, the number of refresh cycles needed can be computed by dividing the square root of the total memory capacity by 2. A 4K RAM, for instance, contains 4096 bits, and the square root of 4096 is 64. The 64 rows are divided into *two* 32-row sections. Since two rows may be refreshed in a single cycle, only 32 refresh cycles are needed. A 256K RAM contains $2^{18} = 262114$ bit locations, and the square root of 2^{18} is $2^9 = 512$, so only 256 refresh cycles are needed.

The memory refresh requirement can be satisfied by one of three methods: distributed refresh, burst refresh, or hidden refresh. If *distributed refresh* is used, refresh cycles are spaced at even intervals, and normal memory access is permitted in between. Using *burst refresh*, all refresh cycles are performed consecutively until completed, and the remaining time is left for normal access. *Hidden refresh* is a form of distributed refresh in which refresh cycles are attached directly behind normal memory cycles. Hidden refresh is possible when access is regular and there is a known *dead time* after each access. The first two methods are illustrated with timing waveforms in Fig. 8-10. The choice of methods is based on the time available and the ease of integrating the refresh scheme into a particular system. In all cases the normal computer cycles are interrupted by an *arbitration circuit* which must choose between memory requests

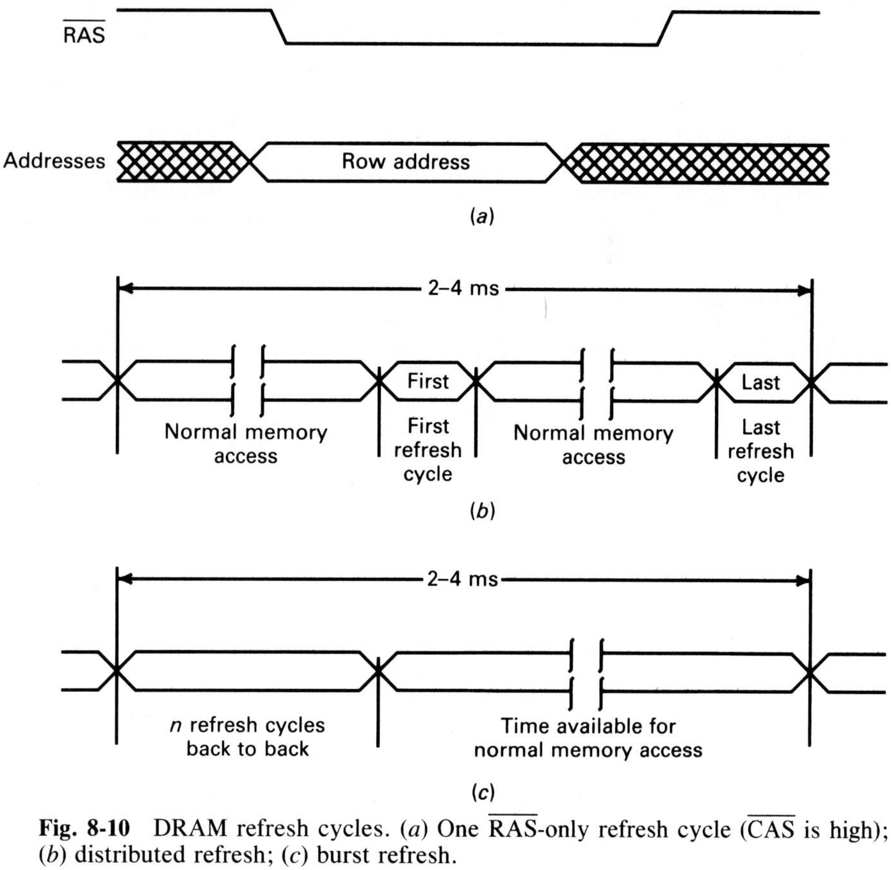

Fig. 8-10 DRAM refresh cycles. (*a*) One $\overline{RAS}$-only refresh cycle ($\overline{CAS}$ is high); (*b*) distributed refresh; (*c*) burst refresh.

made by the computer and requests made by the memory refresh circuitry. The memory refresh circuitry has priority since failure to refresh in time would mean loss of valid data.

It is easy to see that dynamic memory requires additional, complicated hardware to implement refresh. Controller chips are available to satisfy refresh requirements. The SN74S409 is a good example. Static memory, of course, is easier to use and is the obvious choice in systems where memory requirements are small, 16K bytes or fewer. However, DRAM is much denser and less power hungry than SRAM, and, when the added cost of refresh circuitry is spread over a large number of devices, the cost per bit can be substantially lower in both dollars and watts. And while SRAM is getting denser with smaller geometries, it is always a generation behind DRAMs because of the number of transistors required. CMOS SRAMs have a power advantage over NMOS DRAMs, but that advantage is lost once again to CMOS DRAMs. In many cases, cost and power considerations alone will make DRAM the obvious choice.

On the other hand, memory refresh operations do degrade system performance since data cannot be accessed during the periodic refresh intervals. Even in normal access, DRAMs are slower, and some higher-speed system requirements may preclude their use. In addition, DRAMs are more susceptible than SRAMs to soft errors. *Soft errors* are random errors in single cells due to noisy systems, poor system design, or the ionizing radiation of alpha particles. *Alpha particles* are emitted from natural radioactivity in the packaging material. When alpha particles strike the die, they generate ions, which collect in the capacitor and change the charge. The scaling of DRAM cell size has, to a great extent, been limited by soft-error rate considerations. The problem is serious enough in NMOS DRAMs to warrant additional error detection and correction circuitry. The CMOS process reduces the incidence of soft errors because the capacitor is placed in an *n* well which tends to absorb the ions.

In summary, DRAMs have higher densities, moderate speeds, and low power consumption. SRAMs are faster, require no refresh, and are less susceptible to soft errors. Systems with memory requirements of 64K bytes or fewer typically use SRAMs. Systems with memory requirements larger than 128K bytes usually opt for DRAM because of density and power considerations. But individual system requirements will often blur these guidelines.

REVIEW 8-2

1. **Q.** What is random-access memory? **A.** It is memory which is always available and characterized by the fact that any location may be accessed as quickly as any other location.

2. **Q.** What is SRAM? **A.** Static random-access memory is memory which requires no refresh and will maintain its data until either power is lost or new data is written.

3. **Q.** What is an address decoder? **A.** An address decoder is a combinational circuit which selects between 2^n rows, columns, or memory chips based on an n-bit address presented to the input.

4. **Q.** How are decoders implemented? **A.** with AND or NAND gates

5. **Q.** What is DRAM? **A.** Dynamic random-access memory is memory which requires periodic refresh of the stored data.

6. **Q.** What is memory density? **A.** "Memory density" refers to the number of bit cells per unit area.

7. **Q.** Why is SRAM always one generation behind DRAM in density? **A.** because SRAM implementation requires four to six transistors per bit cell while DRAM requires only one transistor and one small capacitor

8. **Q.** How are we able to address 64K, 4-bit words using only eight address lines? **A.** Sequential row access strobes and column access strobes allow us to latch in 8 bits at a time.

9. **Q.** What is time multiplexing? **A.** It is the operation of taking two or more sets of values over one set of input lines.

10. **Q.** What is an RAS* signal? **A.** An RAS* signal is a row address strobe. This signal latches the row portion of an address.

11. **Q.** What is a CAS* signal? **A.** A CAS* signal is a column address strobe. This signal enables the column address.

12. **Q.** Define the following terms: destructive read, soft error, and refresh cycle. **A.** Destructive read is the loss of the bit cell state during a read operation; soft errors are random errors caused by noise or alpha particles; and refresh cycles are periodic read operations needed to restore the full charge in DRAM bit cells.

13. **Q.** Describe the two methods for implementing DRAM refresh. **A.** Distributed refresh implies that one row at a time is refreshed, leaving time in between for normal access. Burst refresh implies that all the rows are refreshed sequentially.

14. **Q.** What is arbitration? **A.** It is the process of determining who gets access to memory. In the case of DRAM, an arbitration circuit must decide whether the computer may access the RAM based on whether a refresh cycle is about to occur.

15. **Q.** What are the primary reasons for using DRAM? **A.** density and power consumption

16. **Q.** What are the primary reasons for using SRAM? **A.** speed, circuit simplicity, and resistance to soft errors

8-3 READ-ONLY MEMORY

The primary difference between read-only memory (ROM) and random-access memory (RAM) is the design of the bit cell. Both types of memory are random access in nature, and both are addressed with address decoders. However, read-only memory may be written only once. This once-only writing is called *programming,* and it is achieved in one of two ways. *Mask programmable ROM*, commonly referred to simply as ROM, is programmed by the manufacturer at the foundry. When the user has decided exactly what must go into the ROM, he or she completes a form indicating the value of each bit cell. A mask is then made for the layer which completes the bit cells, and the ROM contents are then fixed. *Field programmable ROM,* commonly called PROM, is programmed by the user. Each bit cell is initially set to some default value, and the user may change any bit cell to the opposite sense by "blowing" a tiny metal fuse. A variety of *PROM programmers* are available to perform this task.

The major advantage of read-only memory is that it is *nonvolatile,* which means that, once programmed, it retains its memory contents even when the power is removed. This nonvolatility is advantageous when the memory contents represent a computer program, a set of numbers or characters for table look-up, a truth table, or microinstructions for a sequential state machine.

ROM

The basic ROM bit cell is shown in Fig. 8-11. It consists of only one transistor, an *npn* bipolar or *n*-channel MOSFET. A 0 is programmed by

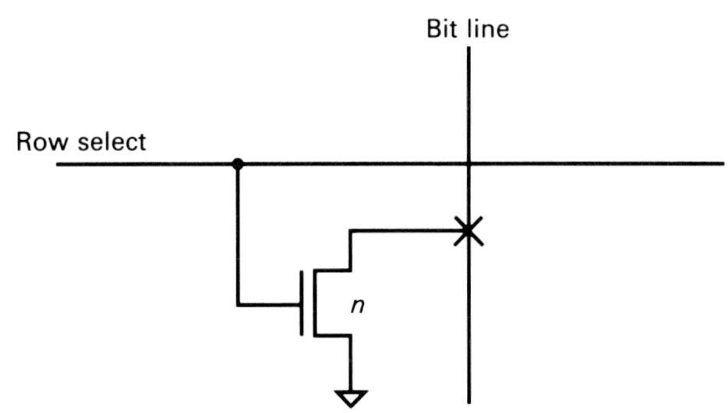

Fig. 8-11 One ROM bit cell. The X indicates a programmed contact formed by completing the transistor.

completing the transistor. A completed transistor will pull the bit line low when it is addressed by the row select. If the device is left uncompleted, a row select will have no effect on that cell and the bit line will remain high. In MOS ROM, the default high level is established either with depletion-type pull-ups or precharge devices which charge the bit lines just before a read.

Figure 8-12 shows an array of 4-bit words. As with RAM, it is important to make the array perimeter small and squarish. The decoding provides

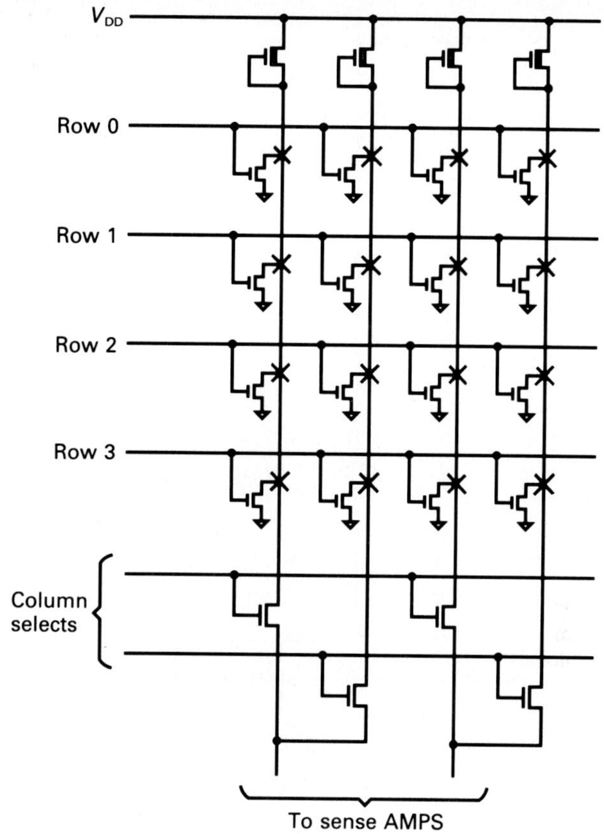

Fig. 8-12 ROM device array.

for this by breaking the address up into row address and column address. In a 2K × 8-bit ROM, for instance, seven address lines would be used to select 1 of 128 rows. Each row consists of 128 bit cells. This provides a square memory matrix. The desired 8-bit word is selected from among the 16 8-bit words in the row with a 4-bit column address. Figure 8-13 shows how, to the outside world, the row and column addresses appear as one continuous 11-bit address. The chip select input is used to choose between two or more ROM chips in the system or between a ROM chip and some other memory device. Two or three chip selects may be provided on a single memory package. In addition, if a ROM uses precharge logic to pull the bit lines high, then a separate input would be needed to provide the precharge strobe.

PROM

Field programmable ROMs are organized in the same way as mask programmable ROMs. The only difference is in the design of the individual bit cell. There are two basic types of PROM bit cell, *fusible-link* and *diffused eutectic aluminum process* (*DEAP*). Fusible-link PROMs are less expensive and more widely used. Their bit cells consist of diodes with

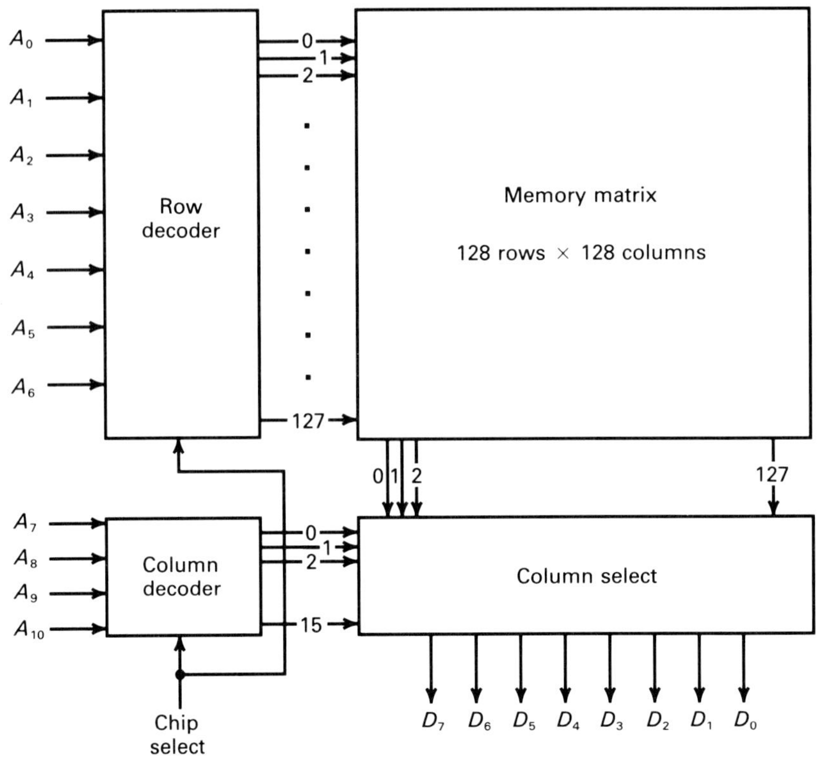

Fig. 8-13 2K × 8-bit ROM organization.

series fuses as shown in Fig. 8-14. A connection between the row select and the bit line is made during fabrication with a small metal or polysilicon link. Programming is performed by selectively blowing the fuses open. *Titanium-tungsten* (*TiW*) alloy is currently the fuse material of choice since it provides high reliability and meets Environmental Protection Agency standards. The older *nichrome fuses* (NiCR) were abandoned because of environmental hazards. The fusible links sit on the surface of the silicon

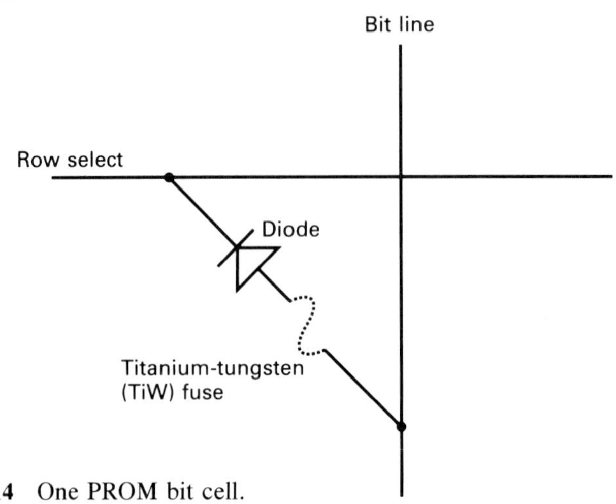

Fig. 8-14 One PROM bit cell.

and occupy a considerable amount of area. Because of their bulkiness, fusible-link PROM cells are typically 10 to 15 times larger than ROM cells. Larger cells mean more silicon area, lower yields, and greater cost.

The basic problem with this technology is that when the fuse is blown, metal is splattered around the surface. On some PROMs, the fragments may actually rupture the passivation layer which normally protects the die from contaminants. On other PROMs, an opening is deliberately designed into the passivation layer directly above the fuse. These openings are potential reliability hazards. In addition, the fuse gaps must be relatively narrow to achieve acceptable densities, which means that the possibility of growback exists. *Growback* is the partial reconnection of the blown fuse links. This may occur after many hours of operation because of the high electric field across the gap.

The newer type of PROM technology is called *diffused eutectic aluminum process,* or *DEAP*. This process uses a vertical subsurface element which is much smaller than the horizontal surface fuse links but still larger than standard ROM cells. When the DEAP memory element is programmed, it causes a *short circuit* rather than an open circuit. The DEAP technique does not splatter or exhibit growback.

Programmable Logic Arrays

Programmable logic arrays, or PLAs, are very much like ROMs. They can be mask programmed (PLAs) or field programmed (FPLAs). The only difference between a ROM and a PLA is in the decoder. RAMs, ROMs, and PROMs have fixed AND arrays which decode every combination of n address inputs into 1 of 2^n distinct select lines. There are cases, however, in which not every address combination is needed, and it would be wasteful to provide corresponding select lines and memory rows for these nonexistent addresses. For example, if read-only memory were used to implement a large truth table in which some states never occur, the don't-care states would represent nonessential addresses. Another example is when read-only memory contains control information (instructions) for sequential state machines. A combination of address inputs would produce control signal outputs, but not every combination of address inputs would necessarily represent a valid instruction. Both of these cases will be discussed in detail in Chapter 9. It is sufficient here to say that PLAs are partially decoded ROMs in which n address inputs *do not* decode to 2^n separate words of memory.

Field programmable logic arrays are fuse link memories like PROMs. They allow the user to program the memory by blowing links in the bit cells. But since there are fewer locations than the n address inputs could support, the user must also be able to program the decoder. Figure 8-15*a* illustrates the method of representing multiple input AND gates, as introduced in Chapter 3. Figure 8-15*b* shows a four-input PLA using this convention. The address inputs could see 16 possible combinations of 4 bits. But if only 10 of those combinations are valid, then only 10 AND gates and 10 memory word lines are required. The user decides which combinations decode into valid memory locations by blowing fuses in the AND array.

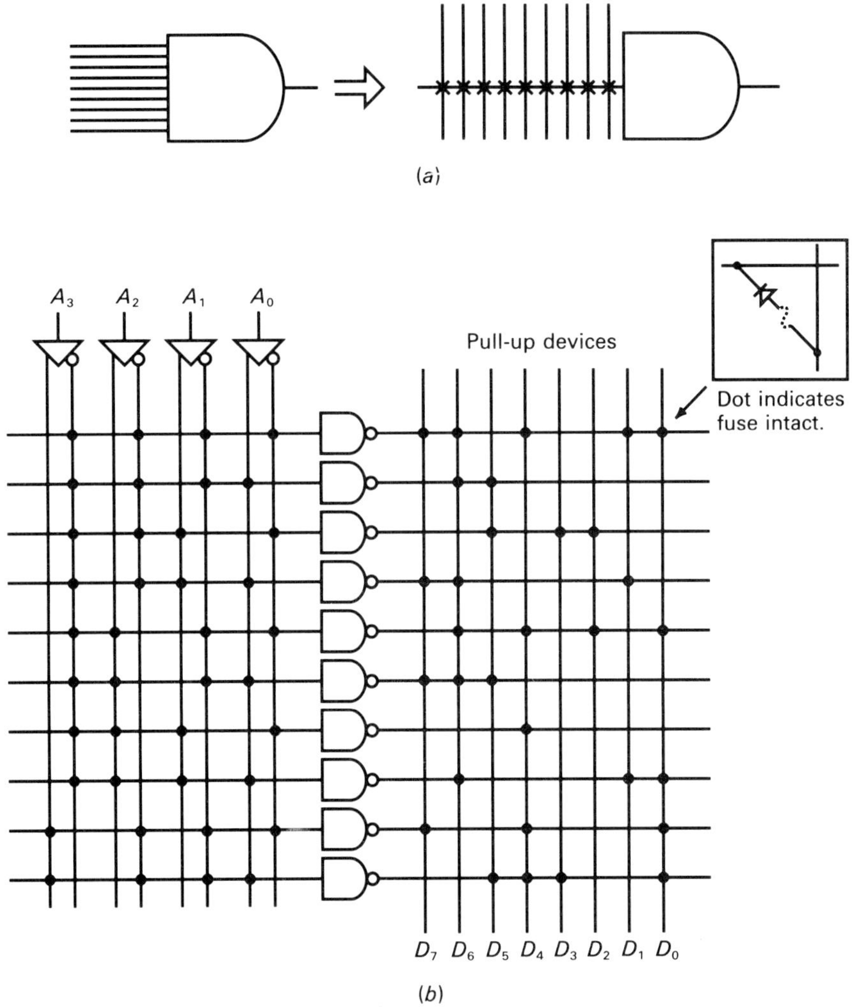

Fig. 8-15(*a*) Schematic equivalences; (*b*) field programmable logic array (FPLA) with 10 decoded select lines (10 × 8-bit).

REVIEW 8-3

1. **Q.** Is read-only memory (ROM) random access? **A.** Yes.
2. **Q.** How does ROM differ from RAM? **A.** RAM bit cells may be written any number of times. ROM cells may be written only once.
3. **Q.** What are mask programmable ROMs? **A.** They are ROMs which are programmed at the factory to users' specifications. The programming is part of the fabrication process.
4. **Q.** How are PROMs programmed? **A.** Field programmable ROMs, or PROMs, are programmed either by blowing tiny metal fuses or shorting transistors.

5. **Q.** How do ROMs and PROMs compare in size and price? **A.** Fuse technology PROM cells are about 10 to 15 times larger that ROM bit cells. DEAP cells are smaller than fusable-link cells but larger than ROM cells. Larger cells mean more silicon, lower yields, and greater costs.

6. **Q.** How does a PLA differ from a ROM? **A.** A PLA is a partially decoded ROM. Not all combinations of PLA address inputs yield a programmed output.

8-4 REUSABLE ROM

The primary advantage of ROM and PROM is that they are nonvolatile. Once programmed, the information is virtually unalterable. The primary disadvantage of ROM and PROM is the same: it is unalterable. If after programming, a mistake is found, the device must be discarded. Other types of read-only memory are available, however, which allow reprogramming. These include EPROMs, EEPROMs, and EAROMs.

Erasable Programmable ROM (EPROM)

The EPROM bit cell is illustrated in Fig. 8-16. It looks like and is used as an *n*-channel transistor. It is different from a normal *n*-channel device in that it has two polysilicon gates, one above the other. The upper gate is

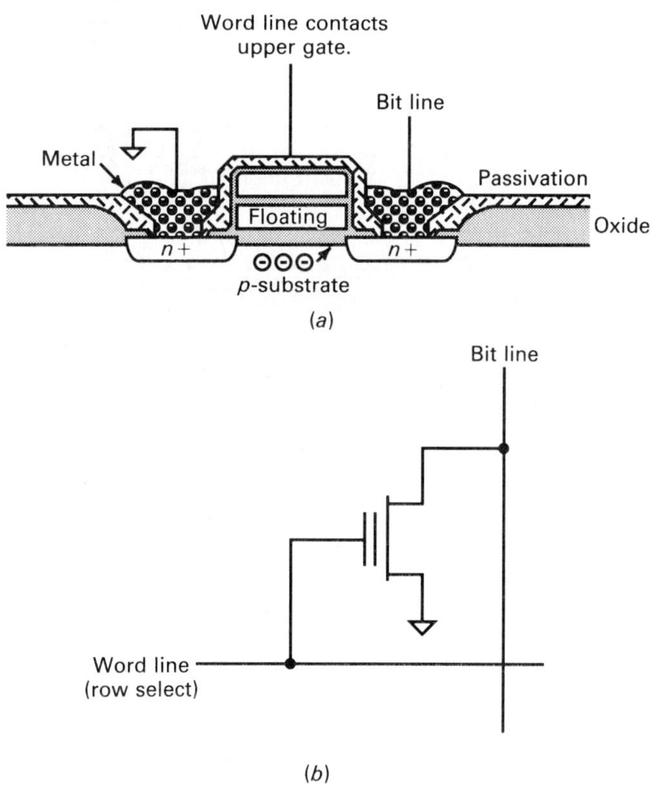

Fig. 8-16 EPROM bit cell. (*a*) EPROM transistor; (*b*) schematic representation.

connected like ROM, but the lower gate remains unconnected to any voltage. The lower gate is *floating*. This *double-poly transistor* is called a *floating-gate avalanche MOSFET,* or *FAMOS FET.* It is used in memory arrays in the same way that *n*-channel devices are used in ROM arrays (Fig. 8-12). Writing, or programming, is performed in PROM programmers by pulsing the row-select line with a high programming voltage V_{pp} of about 20 V for 10 to 20 ms while raising the corresponding drain bit line to 18 V as shown in Fig. 8-17. The nonselected rows and columns are held

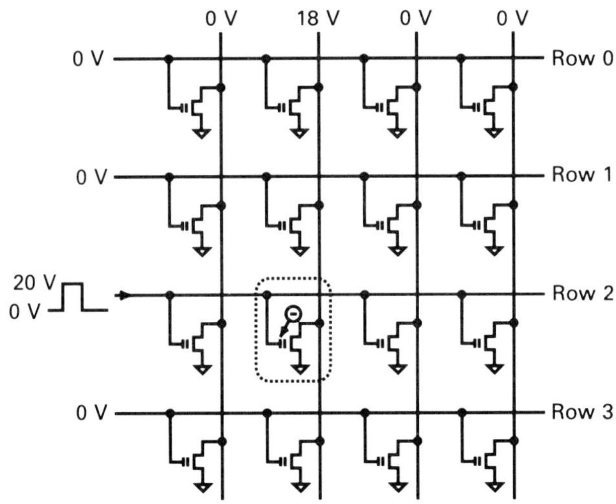

Fig. 8-17 Writing to an EPROM cell.

at 0 V. Under these conditions, electrons in the substrate gain sufficient energy to penetrate the thin oxide layer and flow into the floating gate. These *injected* electrons cause a 5- to 10-V shift in the threshold voltage of the device so that during a normal read cycle the programmed devices will not turn on. The 5-V row select will be insufficient to turn on the programmed devices, and their bit lines will remain high. The floating gates can maintain their injected charge for 10 to 100 years.

Special sockets, called *zero-insertion-force sockets,* are often used with EPROMs so that they may be easily removed for erasure and reprogramming. Erasure involves exposing the die to ultraviolet (UV) light for about 15 to 20 min. A transparent quartz window is built into the package for this purpose. The UV causes the trapped charge in the floating gate to penetrate the oxide above and below and dissipate upward into the select gate and downward into the substrate material. The complete process of writing and erasing may be performed hundreds of times before the material wears out.

Because EPROM cells consist of single transistors, they provide density close to that of DRAM. Their higher cost is determined, to a great extent, by the packaging. The need for a quartz window and a hermetic seal has traditionally made EPROMs more expensive than ROM. Initially, ceramic packaging was required in order to achieve proper bonding for a hermetic seal. The seal was needed because the passivation material used to protect the die did not resist moisture. This increased the cost and made EPROMs unsuitable for automatic insertion. However, newer passivation techniques eliminate the need for the hermetic seal, and EPROM devices have been successfully packaged in plastic. Because of the quartz window, however, EPROMs will always be more expensive than ROM. Their advantage is reusability, which makes them ideal replacements for ROM during system development.

Electrically Erasable PROMs

Electrically erasable PROMs (*EEPROMs*) use the same floating-gate technology as EPROMs. The difference is that the trapped charge can be added or removed while the EEPROM remains in the circuit. Figure 8-18 shows part of the EEPROM bit cell. In one small area at the drain, the

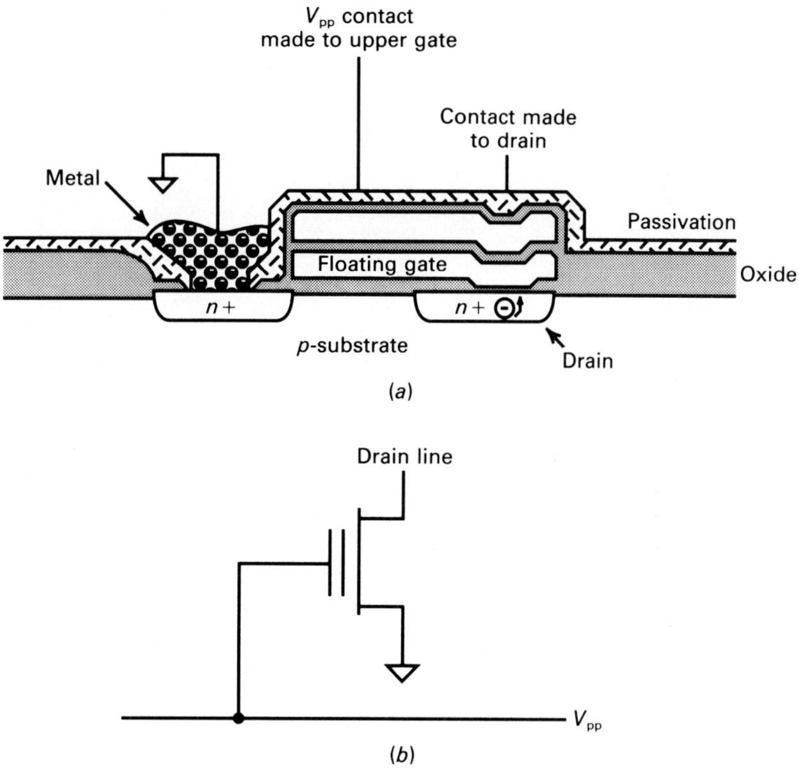

Fig. 8-18(*a*) EEPROM transistor; (*b*) schematic representation.

gate oxide is very thin, which permits electrons to *tunnel* back and forth under high-voltage conditions. As with EPROMs, the injected charge raises the gate threshold by 5 to 10 V and prevents the transistor from turning on during a read. But tunneling in EEPROMs is bidirectional and can be forced while the IC remains on the PC board. Another difference between EPROM and EEPROM is that two transistors are required. Figure 8-19 shows part of an EEPROM array during write and erase cycles.

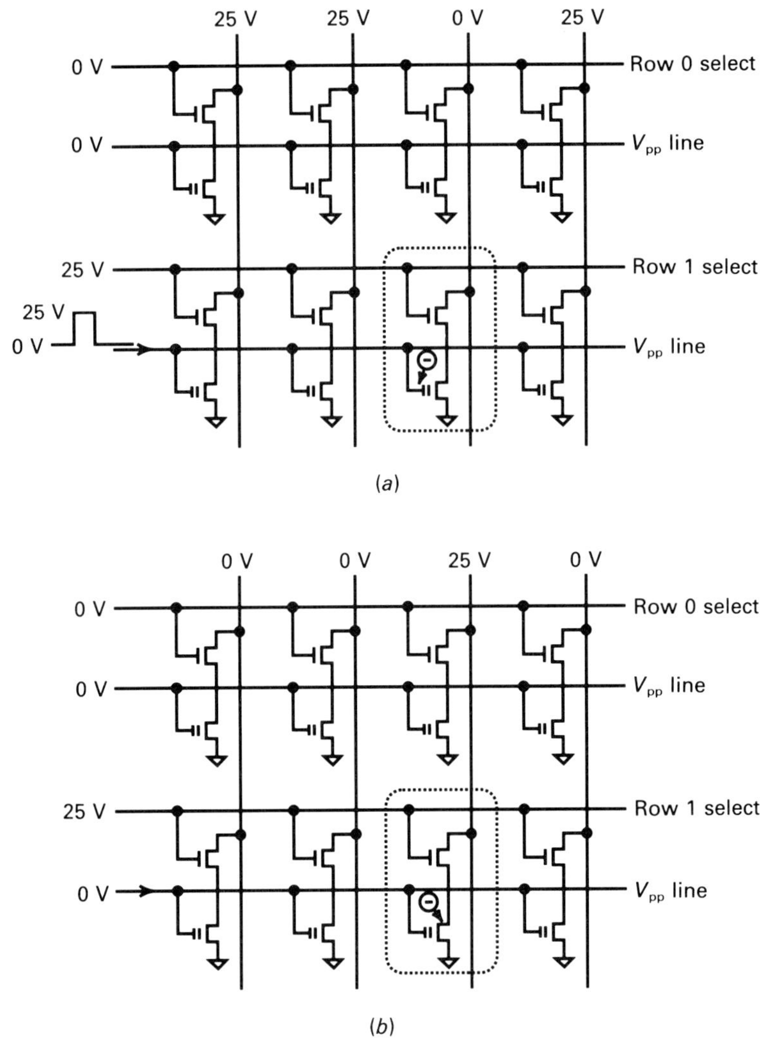

Fig. 8-19(*a*) Writing to an EEPROM cell; (*b*) erasing the same cell.

Writing is performed by applying 25 V to the row-select line while grounding the bit line. Electrons will tunnel to the floating gate if a 10-ms, 25-V pulse is then applied to the V_{pp} line. Other cells in the same row will not be programmed if their bit lines are raised to 25 V. During normal read operations, the V_{pp} line is maintained at 5 V, and a cell is read when its row select is pulsed high. The upper transistor turns on, but there will be no path to ground if the floating-gate device has been programmed. A nonprogrammed device will pull the bit line low. This is essentially the same operation as discussed earlier for mask programmable ROM. Erasure is performed by applying 25 V to both row select and bit line while keeping V_{pp} at 0 V. This causes the trapped electrons to tunnel back from the floating gate to the drain.

The high voltages needed to write and erase EEPROM bit cells may be developed on-chip with high-voltage charge pumps and analog-wave-shaping circuitry. Newer EEPROMs which support this feature can be modified, in circuit, using TTL-level signals and single 5-V power supply. Writing to this type of EEPROM is similar to writing data to a static RAM. A TTL-level signal causes address and data to be internally latched and the 10-ms write operation occurs internally without slowing down the rest of the system. One byte may be changed, or the total chip may be erased in one write operation. In addition, single-supply EEPROMs are operationally and pin compatible with older EEPROMs, which require the additional high-voltage power supply for programming.

Electronically Alterable ROMs

Electronicaly alterable ROMs (*EAROMs*) are similar in operation to the EEPROM but use a different type of bit cell. Figure 8-20 illustrates the

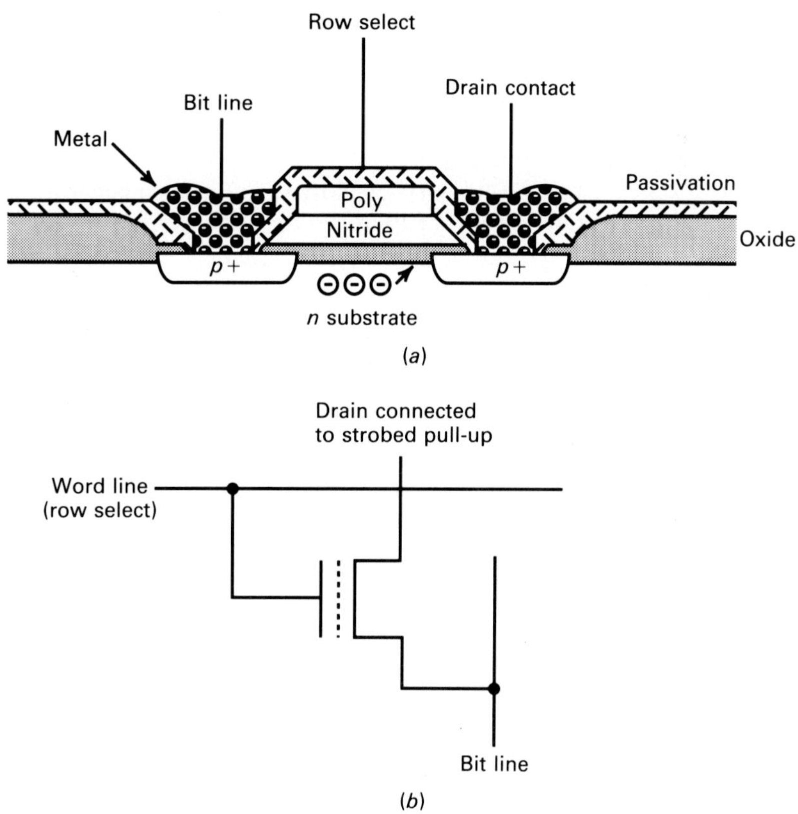

Fig. 8-20 MNOS transistor used in EAROM arrays. (*a*) MNOS memory cell; (*b*) schematic representation.

EAROM transistor. A layer of nitride (silicon nitride) is sandwiched between the poly gate and the thin gate oxide. The gate oxide is, in this case, very thin (typically less than 50 Å as opposed to less than 200 Å for EEPROM transistors and about 800 Å for ordinary *n*-channel MOSFETs). In EEPROM, electrons tunnel through the thin oxide into a floating gate when high voltages are applied. In EAROM there is no floating gate. Instead, the electrons tunnel in the nitride layer. The injected electrons then alter the threshold of the device but in an opposite manner to EEPROM programming. EAROM cells are *p*-channel devices, and the trapped charge in the nitride layer causes the device threshold to be *lowered*. The lowered threshold state is referred to as the *erased state*. Opposite polarity voltages between the gate and the substrate cause the electrons to tunnel back into the substrate, and this is referred to as the *written state*.

The EAROM bit cell is actually just a modification of the basic *p*-channel MOS transistor, but it is different enough to warrant a new name. The device is called a *metal-nitride-oxide-semiconductor*, or *MNOS*, transistor. Figure 8-20*b* shows the schematic symbol for this device. Like EPROM and EEPROM, EAROM is organized like ROM and may be used as a ROM replacement during system development.

Nonvolatile RAM

Nonvolatile RAM (*NOVRAM*) consists of a standard static RAM which overlays EEPROM bit cells as shown in Fig. 8-21. NOVRAM is faster than either EEPROM or EAROM because data can be repeatedly written to the RAM with typical RAM access times. The EEPROM acts as a shadow memory. A store operation transfers a copy of the data from the RAM to the EEPROM with either a +22-V pulse (older technology) or a

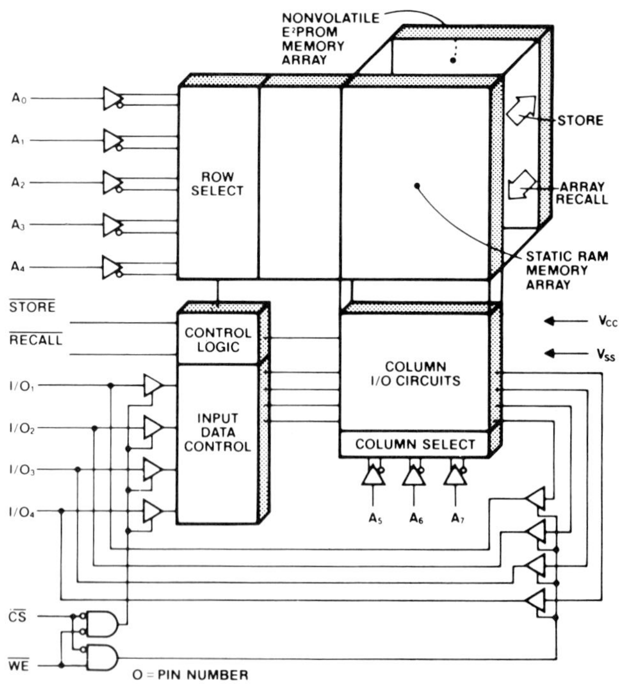

Fig. 8-21 NOVRAM functional block diagram (Courtesy XICOR).

TTL-level store signal (newer technology). The recall operation performs a reverse function by copying the data stored in the EEPROM back into the RAM. Data can be transferred back and forth between RAM and EEPROM at least 10,000 times. Typically, storage to EEPROM is initiated during a power failure. Voltage-sensing circuitry detects power loss and quickly stores away the RAM contents so that it can be recalled when power returns. NOVRAMs may, in many applications, eliminate the need for battery backup.

Standard Parts

Memory chips, in general, have not achieved the degree of standardization enjoyed by other logic devices. However, the programming requirements of EPROMs has helped to bring about some standardization. The 27xx EPROM family is the leading example. Figure 8-22*a* provides a list of 27xx device numbers and sizes available from major IC manufacturers. Figure 8-22*b* provides an example of how pinout compatibility is maintained for EPROMS of differing memory capacity. The lower portion of each chip

EPROM Model	Number of Bits	Number of Bytes	Configuration
2708	8K	1K	1K × 8
2716	16K	2K	2K × 8
2732	32K	4K	4K × 8
2764	64K	8K	8K × 8
27128	128K	16K	16K × 8
27256	256K	32K	32K × 8

(*a*)

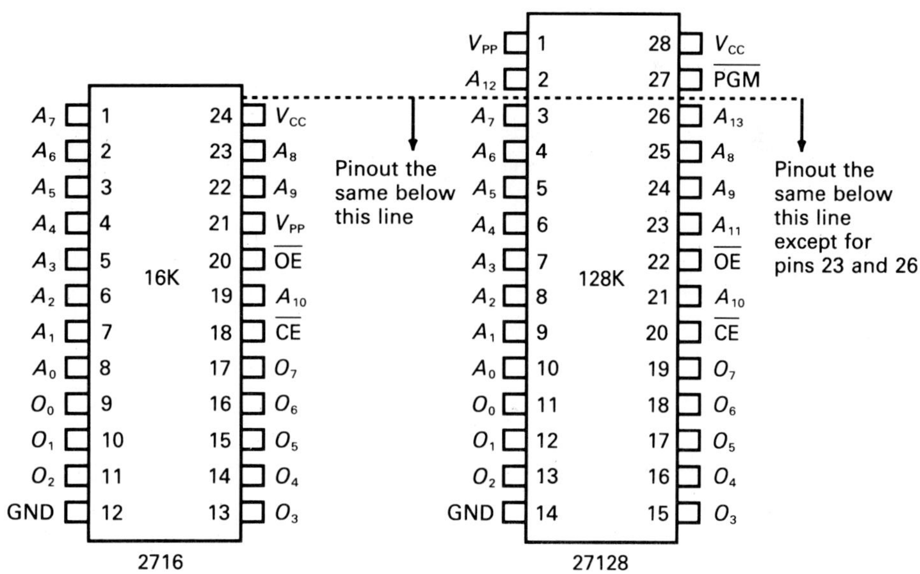

(*b*)

Fig. 8-22(*a*) Table of 27xxEPROM family parts; (*b*) example of compatible EPROM pin configuration.

is pin compatible with the two exceptions noted. Note that physical compatibility is maintained even though pin numbering shifts. The 27128 requires three additional address inputs and uses an extra input, PGM NOT, during programming. A TTL low-level pulse applied to the CE NOT input of the 2716 with V_{pp} at 21 V programs the byte selected by the address bits. The 27128 requires a TTL low-level pulse at the PGM NOT input as well. This effectively provides two chip enables.

The standardization of EPROMs has extended to a JEDEC-approved standard for all byte-wide memory parts. Figure 8-23 shows the JEDEC standard pinout with an example of compliance. RAMs, ROMs, EPROMs, EEPROMs, and NOVRAMs are now available which comply with this standard.

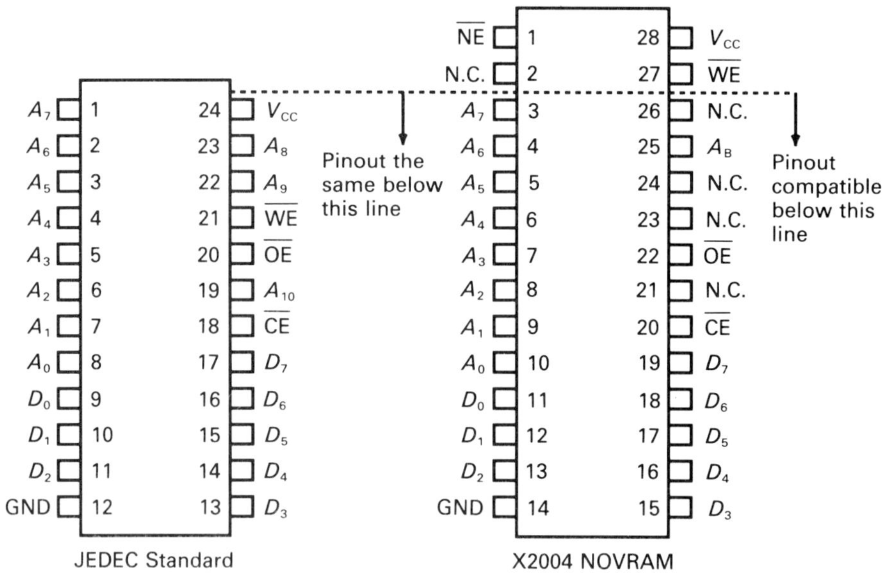

Fig. 8-23 The JEDEC-approved standard for byte-wide memories and an example of compatibility. The X2004 is a 512 × 8 nonvolatile RAM. The *Nonvolatile enable* ($\overline{NE}$) control line is used to transfer RAM to EPROM and back.

REVIEW 8-4

1. **Q.** How do reusable ROM cells differ from standard ROM cells? **A.** Reusable ROM cells are programmed by raising or lowering the gate threshold. Standard ROM cells have fixed thresholds and are permanently programmed by the presence or absence of the device itself.

2. **Q.** How are EPROMs programmed and erased? **A.** EPROMs are programmed with special high-voltage programming equipment and erased with 15 to 20 min of exposure to UV light.

3. **Q.** How does EPROM density compare to ROM? To DRAM? **A.** EPROMs provide densities close to both ROM and DRAM.

4. **Q.** What is the difference between EPROM and EEPROM? **A.** EPROMs must be removed from the circuit for programming and erasure; EEPROMs can be programmed and erased in circuit.

5. **Q.** What is the difference between EEPROM and EAROM? **A.** Functionally there is no difference. However, EEPROMs use floating-gate *n*-channel MOSFETs as bit cells while EAROMs use *p*-channel MNOS FETs.

6. **Q.** What is NOVRAM? **A.** Nonvolatile RAM is a combination of SRAM and EEPROM. Normal access is made to the RAM, but the entire RAM contents may be saved in nonvolatile EEPROM during a power failure.

7. **Q.** What advantage does NOVRAM have over EEPROM? **A.** Access times are faster for repeated writes to memory.

8-5 MEMORY ADDRESSING AND DECODING

All the forms of memory discussed so far can be mapped directly into a computer system's memory space. The concept of *memory mapping* is very important to understanding the functions of practical systems. Perhaps the best way to approach this is to start with very simple address decoding and work up to more complex and practical examples. Figure 8-24*a* shows a four-location address decoder. The corresponding memory map is also shown. Each location in the memory map contains one word of data. The actual length of the word depends on the system. It may be 1 byte for small microprocessor applications, 16 or 32 bits for larger micros, or 56 bits for mainframe computers. The memory map shows only the locations of the words, not the word length. Figure 8-24*b* shows how the addition

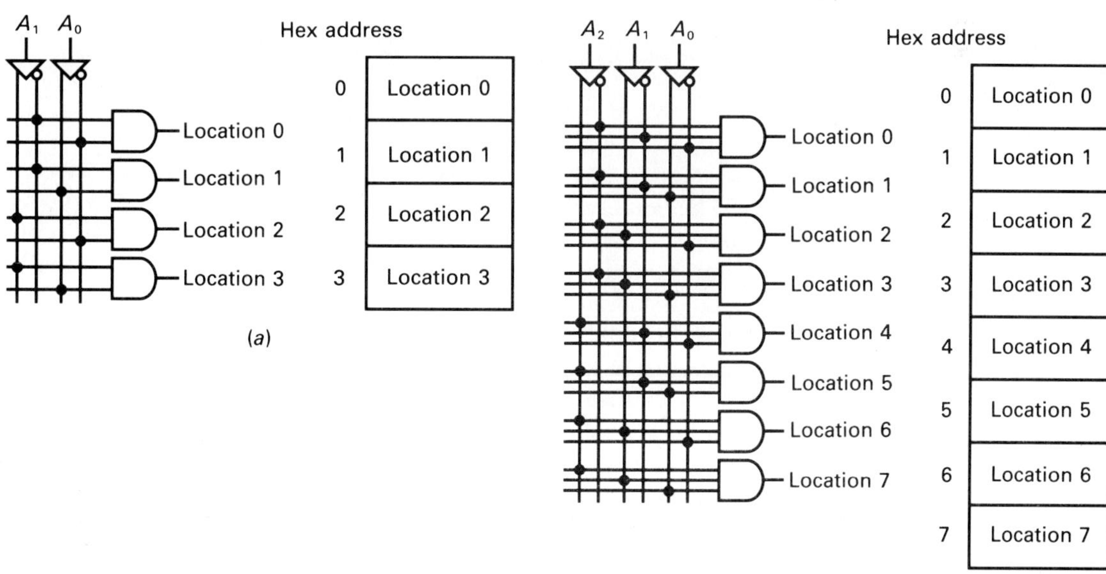

Fig. 8-24 Address decoders and memory maps. (*b*)

of one address input doubles the memory space. The number of AND gates needed to decode the address also doubles. Each additional address input will double the number of possible locations.

Figure 8-25 shows 1K of memory space divided into 256-word blocks. Note that hex notation is used so that the number 100_{16} represents 256_{10}

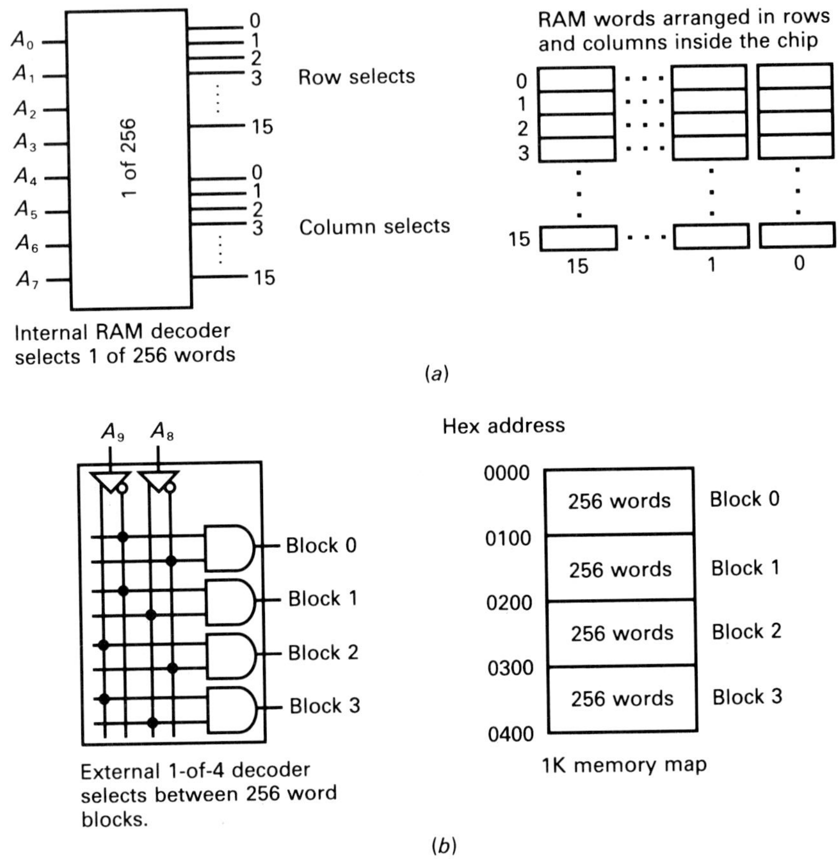

Fig. 8-25 Memory address decoding for 1K of memory space.

locations. The hex number 400_{16} represents the 1025th location in the map since numbering starts at 0000. We refer to 1024 locations as 1K because rounding down to an even thousand simplifies the conversion between hex and decimal. When we say that the system has "1K of memory space," we mean that 10 address lines are needed to decode the full memory space. Full decoding of the 10 address lines will provide individual access to 1024 separate memory locations.

In Figure 8-25*a*, eight of the address lines, A_7 through A_0, are decoded inside a 256-word RAM. The internal 1-of-256 decoder provides the row and column selects needed to access each of the 256 words. In Fig. 8-25*b* there are four 256-word RAMs, and the selection between them is made with a 1-of-4 decoder. Each RAM has a chip select which is wired to one of the decoder outputs. The write enables and output drivers of a RAM will turn on only when the chip has been selected. In SRAMs there will be one or more inputs labeled CS*; in dynamic RAMs the CAS* signal is used as the chip select.

Figure 8-26 shows an 8K memory space divided into 1K blocks. The individual 1K RAMs and 1K EPROMs use 10 address inputs to internally

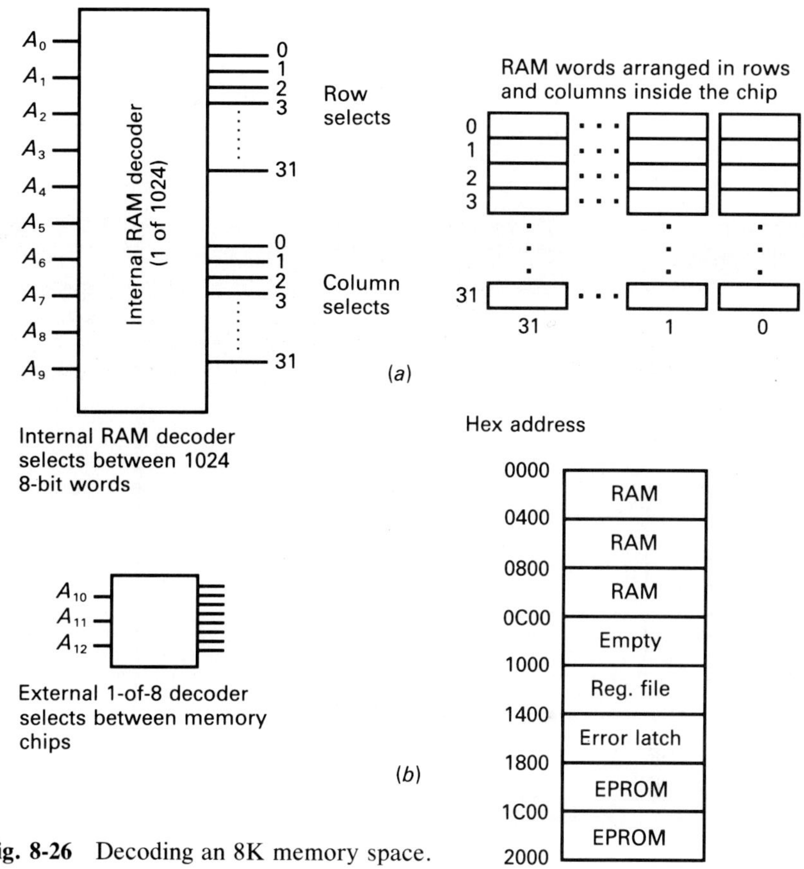

Fig. 8-26 Decoding an 8K memory space.

decode their own memory space. Externally, a 1-of-8 decoder selects between the 1K blocks. Fully decoding the 13 address lines provides access to 8192 (8K) memory locations. This does not necessarily mean that all the locations exist physically. RAM, ROM, or other memory devices may not actually be on the PC board at those locations. A memory map tells us where the memory devices are, what type of devices they are, and where the memory space is empty. In this example, memory locations 0C00 to 0CFF are empty. Memory locations 1000 to 13FF are reserved for a register file containing only 4 words and the following block is occupied by a single 1-bit latch.

The blocks are selected on the basis of only three address inputs: A_{12}, A_{11}, and A_{10}. Locations within the 4-word register file at address 1000_{16} are selected only on the basis of the first two address inputs, A_1, A_0. All the other address inputs in between do not connect in any way to the register file. This means that the address 1002_{16} will select word 2 of the register file, but $111A_{16}$ will also select that word. In fact, there are 256 different addresses which will select a given word in the register file. This phenomenon is called *aliasing*, and it is due to the fact that the decoded block contains more addresses than there are memory elements. In this case there are only four registers in a 1K decoded block. Aliasing could be prevented by dividing that block into 256 smaller blocks using a 1-of-256 decoder. However, if that block will never contain more than one 4-word register file, the additional hardware needed to prevent aliasing is unnecessary. Normally we consider the unused address inputs to be low and forget about all the extra addresses.

Reading Schematics

There are four small hurdles in the path of reading bus-type schematics. The *first hurdle* is understanding what is contained in a bus. A bus is nothing more than a bundle of wires represented on the schematic as a group. Figure 8-27 shows four common methods of representing buses.

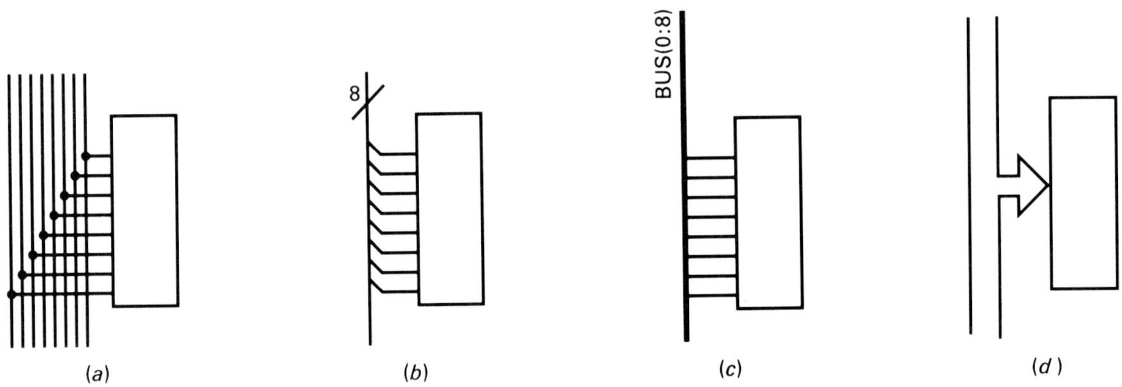

Fig. 8-27 Ways of showing buses on schematics. (*a*) All wires shown; (*b*) using one line; (*c*) CAD representation; (*d*) block diagram.

The first actually shows the wires, but this is seldom practical in large schematics. The second replaces the bundle of wires with one line. The fact that this line represents a bus is denoted with a slash and the number of wires contained in the bundle. In CAD systems the slash is not used but the line is thicker than lines representing single wires. The number of wires is indicated, in CAD drawings, with a label describing the contents of the bus. The fourth representation is appropriate for block diagrams.

The *second hurdle* is associating the number of address bus lines with a given memory space. The question is: How many address lines are needed to access X bytes of memory? Perhaps the best way to approach this is to pick one correspondence and extrapolate the rest. For instance, it is convenient to remember that 10 address lines (A_9 through A_0) are needed to address 1K of memory space. Then for each additional address line, double the memory space. Eleven lines will be needed to access 2K, 12 for 4K, etc.

The *third hurdle* is the association between the number of bytes of memory and the actual hex addresses. The hexadecimal numbering system must become familiar to anyone who works with computer systems. One approach is to remember that 1K equals 400_{16} memory locations, numbered 000 through 3FF, and extrapolate as before. Another is to use a table as shown in Fig. 8-28. For each column of this table that corresponds to one

Hex Char 3		**Hex Char 2**		**Hex Char 1**		**Hex Char 0**	
Hex	**Decimal**	**Hex**	**Decimal**	**Hex**	**Decimal**	**Hex**	**Decimal**
0	0	0	0	0	0	0	0
1	4096	1	256	1	16	1	1
2	8192	2	512	2	32	2	2
3	12288	3	768	3	48	3	3
4	16384	4	1024	4	64	4	4
5	20480	5	1280	5	80	5	5
6	24576	6	1536	6	96	6	6
7	28672	7	1792	7	112	7	7
8	32768	8	2048	8	128	8	8
9	36864	9	2304	9	144	9	9
A	40960	A	2560	A	160	A	10
B	45056	B	2816	B	176	B	11
C	49152	C	3072	C	192	C	12
D	53248	D	3328	D	208	D	13
E	57344	E	3584	E	224	E	14
F	61440	F	3840	F	240	F	15

Fig. 8-28 Table for hexadecimal to decimal conversion.

hex digit, the decimal weights are given for the corresponding hex character. Convert from hex to decimal by adding the weights. Convert from decimal to hex by subtracting the highest decimal value in the table which is lower than or equal to the decimal number to be converted. The corresponding hex character is the most significant digit. Repeat the process for the remaining characters. Eventually, with use, the hex numbering scheme will become familiar.

Finally, the *fourth hurdle* involves becoming familiar with the types of ICs used to decode addresses in bus logic systems. The most commonly used decoder chip is the 74xx138 shown in Fig. 8-6. It may be used singly, in parallel with other '138s, or cascaded. Figure 8-29 shows examples of these three approaches in a 64K-memory space. Figure 8-29*a* shows the lower memory space of a 64K system divided into 4K blocks using one '138. The entire 64K space may be broken into 4K blocks using two '138s in parallel as shown in *b*. The chip enables on the '138 facilitate this approach. The third example *c* shows how the lower half of the available space might be decoded using one '138 and one block subdivided further using a second '138 cascaded with the first. At one time, memory chips had two or three chip select inputs to facilitate memory decoding. However,

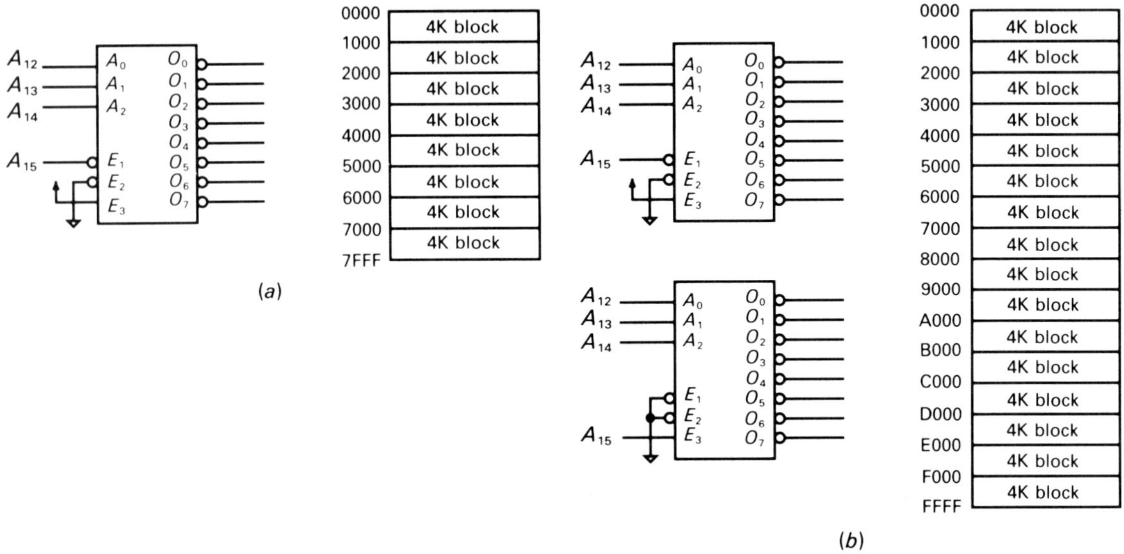

Fig. 8-29 Using '138s to decode memory space. (*a*) Decoding the lower half of memory space into 4K blocks; (*b*) decoding all memory space into 4K blocks.

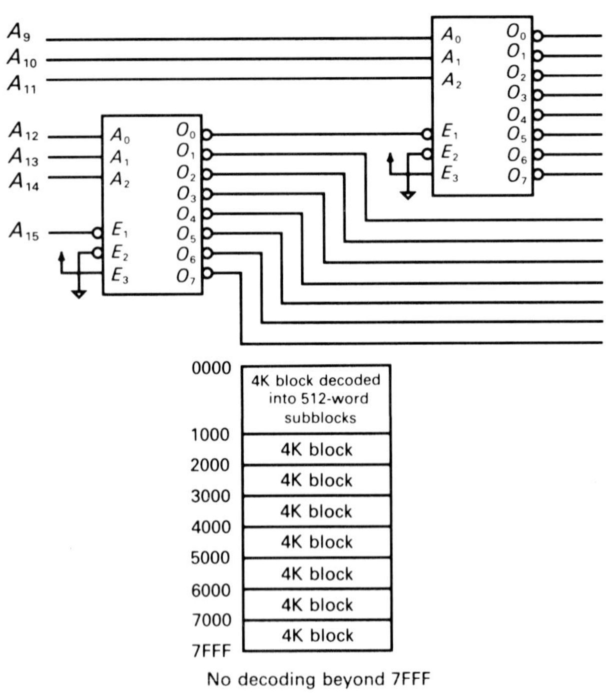

(*c*)

Fig. 8-29(*c*) Using half the memory space and cascading '138s.

with the size of systems continually growing, the use of pins for this purpose has become uneconomical. Instead, selection between memory chips is more often done externally with each chip having only one chip select input.

EXAMPLE 8-1

Consider the system illustrated in Fig. 8-30. Draw a memory map for this system showing the size of blocks decoded, the locations of each memory chip, and the address range of each chip.

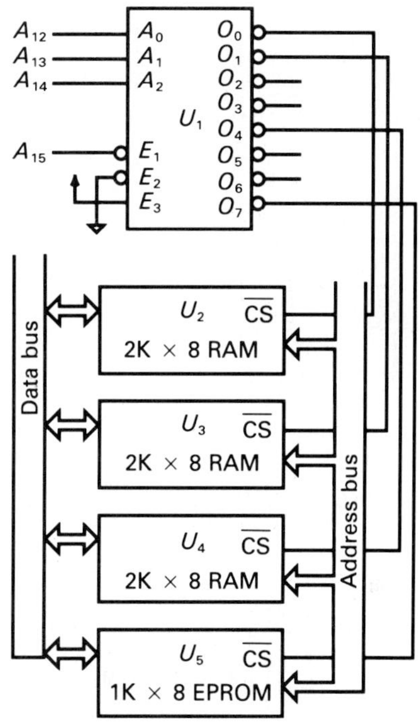

Fig. 8-30 Schematic reading example.

SOLUTION

When first approaching this type of problem, a worksheet similar to the one shown in Fig. 8-31 may be useful. Wherever an address line is decoded internally, within a chip, a letter I is written. Address inputs which are decoded externally are marked with 1s and 0s to indicate what is needed to decode that particular chip. Address inputs which are unused for a particular chip are marked with X's. Once this has been recorded, the binary representation is converted to hex. The X's are regarded as 0s in the final address, and the I's represent a *range* of locations. The range of internally decoded addresses is obtained by considering that the lowest address occurs when each I equals 0 and the highest occurs when each I equals 1. The hex equivalents of those two cases determine the range.

SYSTEM LAYOUT WORK SHEET

	MPU ADDRESS LINES A15–A0																ADDRESS	
DEVICE	15	14	13	12	11	10	9	8	7	6	5	4	3	2	1	0	FROM	TO
RAM #1	0	0	0	0	X	I	I	I	I	I	I	I	I	I	I	I	0000	07FF
RAM #2	0	0	0	1	X	I	I	I	I	I	I	I	I	I	I	I	1000	17FF
RAM #3	0	1	0	0	X	I	I	I	I	I	I	I	I	I	I	I	4000	47FF
ROM #1	0	1	1	1	X	X	I	I	I	I	I	I	I	I	I	I	7000	73FF
ROM #2																		
ROM #3																		
PIA #1																		
PIA #2																		
PIA #3																		
ACIA																		

Fig. 8-31 (Courtesy Motorola)

REVIEW 8-5

1. **Q.** What is a memory map? **A.** A memory map is a diagram which shows which areas of a system's memory space are used and how they are used.

2. **Q.** How many address lines are needed to address 1K of memory? **A.** 10

3. **Q.** What is the range of hex addresses in 1K of memory space? **A.** 000 to 3FF

4. **Q.** How many locations are in 1K of memory? **A.** There are 1024_{10}, or 400_{16}.

5. **Q.** What is aliasing? **A.** Aliasing is the phenomenon where more than one address will correctly access a single memory location.

6. **Q.** When does aliasing occur? **A.** Aliasing occurs when there are more addresses than physical memory locations and the decoding logic does not subdivide the space down to the smallest memory block.

REVIEW QUESTIONS

1. Name two factors which determine the relative cost of memory devices.
2. Why is DRAM called dynamic?
3. Why do DRAMs need refresh cycles?
4. Explain the concept of a multiplexed bus. Why is it used?
5. When is PROM used instead of ROM?
6. How is the direction of data controlled in memory systems?
7. What signals are needed to access SRAM? DRAM?
8. What advantages does CMOS DRAM have over NMOS DRAM?
9. How, in general, are EPROMs, EEPROMs, and EAROMs programmed?
10. How, in general, are PROMs programmed?

REVIEW PROBLEMS

1. How many address lines are required to access 1 Mbyte of memory?
2. Convert the following hex addresses to decimal:

   (a) ABCD (b) CFFFF (c) FFFF
3. Find the hex equivalent of the decimal number 32,000.
4. Sketch a memory map for a system which has 32K of continuous RAM starting at location 0000_{16}. The system used 4K × 8 SRAMs and 2K × 8 EPROMs. A total of 4K of EPROM resides at the top of memory ending at $FFFF_{16}$. Show all empty locations.
5. Sketch a decoding scheme using '138s for the system in Prob. 4. If partial decoding is used for the EPROM, how will this affect address selection?
6. How many different addresses will access any given location in RAM in the Fig. 8-30 example? How many for the PROM?
7. How many '138 chips are needed to completely decode a 64K memory space into 1K blocks?

8. Sketch a memory map for the system shown in Fig. 8-32.

9. Given a 16-bit address bus (A_0–A_{15}), show which address lines are needed to access the following block sizes: (a) 32K blocks, (b) 16K blocks, (c) 8K blocks, (d) 4K blocks, and (f) 1K blocks.

10. Show how a single D flip-flop might be selected using '138s and gate type logic. The flip-flop will occupy a 2K block in a 64K memory space.

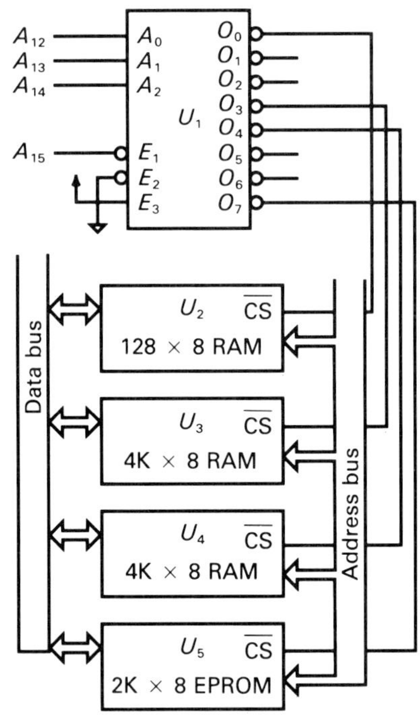

Fig. 8-32

9 CONTROL LOGIC

All the elements of a complete digital system have been discussed except two: *control logic* and interface to the outside world. The concept of control is very important to understanding digital systems because it underlies much of what is done in electronics today. Storage registers, counters, arithmetic circuits, and memory are all needed to implement electronic systems, but they must be directed by some type of intelligent logic. For many years, control systems were built around analog computers, yielding a body of science called *control theory*. Much of that circuitry has been replaced with digital logic, and in particular, with microprocessors. Control logic consists of any circuitry which provides control signals needed to perform some operation. The microprocessor is a general-purpose control machine which can be *soft-wired* with instructions stored in memory devices. Because it is general purpose, it is slower than *hard-wired* control logic, but it provides a convenient general-purpose VLSI package with space, power, and cost advantages. Recent developments in user-programmable and late mask programmable logic devices have made customized control systems more competitive relative to the microprocessor. The purpose of this chapter is to introduce control logic and show how it relates to microprocessors and other forms of soft-wired logic.

The topics covered in Chapter 9 are:

1 Direct implementation of truth tables in read-only memory
2 Control state counters and sequencers
3 Instruction registers and program counters
4 General-purpose state machines
5 User programmable and late mask programmable control logic

9-1 DIRECT IMPLEMENTATION OF TABLES

As discussed in Chapters 3 and 4, logic problems can be represented either as sum-of-products expressions or as simplified boolean equations. The sum-of-products expression comes directly from the truth table. It takes the form of AND products which are summed (ORed) together; hence the term *sum-of-products*. For instance, if there are four input variables, then the AND terms (*product terms*) will each contain four variables. All the product terms which produce an output 1 in the truth table are summed (ORed) together to produce a sum-of-products expression. This type of expression may be used directly to implement any logic function. However, when discrete logic gates are to be used, it is desirable to simplify the sum-of-products expression in order to reduce the gate count. Karnaugh mapping is the easiest way to simplify, or minimize, functions with four or fewer input variables. More complex problems can often be broken down into smaller, mappable parts.

Figure 9-1 shows a typical truth table with the corresponding sum-of-products expressions. Notice that there are four separate output variables

Inputs			Outputs			
A	B	C	f_1	f_2	f_3	f_4
0	0	0	1	0	0	0
0	0	1	0	0	0	0
0	1	0	0	0	1	0
0	1	1	0	1	0	1
1	0	0	0	0	1	0
1	0	1	0	1	0	0
1	1	0	0	1	0	1
1	1	1	1	1	0	0

Sum-of-Products Expressions

$$f_1 = \overline{A}\,\overline{B}\,\overline{C} + ABC$$
$$f_2 = ABC + \overline{A}BC + A\overline{B}C + AB\overline{C}$$
$$f_3 = A\overline{B}\,\overline{C} + \overline{A}B\overline{C}$$
$$f_4 = AB\overline{C} + \overline{A}BC$$

Fig. 9-1 Truth table and sum-of-products expressions.

representing the four different functions, but each depends upon the three input variables. Each output variable requires its own Karnaugh map. If more input variables and more output variables are added, the minimization procedure can become quite complex and tedious, and the results may involve large numbers of logic gates. If the circuit is implemented with packaged parts, this could result in tens of packages. This type of implementation is often referred to as *random logic*.

An alternative to random logic is direct sum-of-products implementation using read-only memory. A ROM (PROM, EPROM, or PLA) may be programmed directly with the desired truth table. Each address input represents one input variable, and each ROM output represents a function output variable. If a combination of input variables is expected to produce an output 1, then a 1 is programmed into the location addressed by that combination. As an example, consider the hypothetical 8 × 4-bit PROM shown in Fig. 9-2. It has been programmed with the truth table from Fig. 9-1. While this example is much too simple to warrant ROM implementation, it does illustrate the process.

The hardware needed to program PROMs, EPROMs, and FPLAs can be easily interfaced with compatible personal computers. Software is often

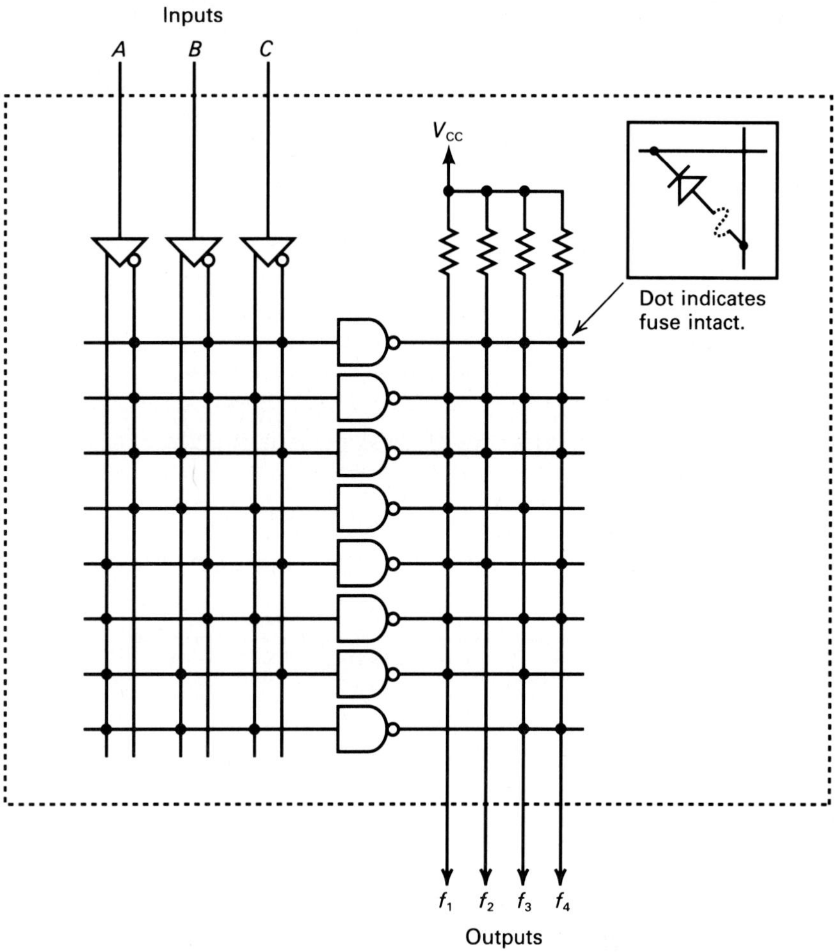

Fig. 9-2 PROM implementation of truth table.

supplied with the programmer to be executed from the personal computer keyboard. The software is sometimes menu driven so that the type of read-only memory to be programmed may be selected by the user from a list. As an example, consider a typical menu-driven programmer. After installing the disk containing the software and typing in the appropriate commands, a sequence of menus appears like those shown in Fig. 9-3. In Fig. 9-3, the user selected ⟨1⟩ from the first menu. This brought up the second menu, and ⟨1⟩ was chosen again. The second choice brought up a manufacturer's menu, and a third choice would produce a list of actual part numbers. The computer can be made to show the contents of a 256-byte "page" of *burn memory* on the screen. Figure 9-4 shows an abbreviated example of this type of display. Data can be entered either as single nibbles (hex characters) or as ASCII characters. The cursor moves to the right as the data is entered, and to the next line when one line is complete. The cursor may also be moved left or right, up or down, as needed. The data entered in the computer memory is transferred over a cable to the

```
MASTER FUNCTION MENU
=========================================
<1> Device Selection
<2> System Diagnostics
<3> System configuration
<4> Action Menu
```

```
SELECT DEVICE TYPE
=========================================
<1> PROM's
<2> EPROM's
<3> EEPROM's
<4> Microcontrollers
```

(*a*)

```
SELECT MANUFACTURER - PROMs
=========================================
<1> Advanced Micro Devices
<2> Fujitsu
<3> Harris
<4> Monolithic Memories
<5> National Semiconductor
<6> Signetics
<7> Texas Instruments
```

(*b*)

Fig. 9-3(*a*) Programmer menus. (*b*) Select manufacturer menu.

programmer and burned into the device which has been inserted into the appropriate socket. This marriage of hardware and software makes logic and memory programming very simple.

Look-up Tables

In addition to organizing random logic, read-only memory may be used to perform code conversions and arithmetic operations. For instance, BCD-

```
Memory Editor                                                  AMD 27S45/A:
ADDR: 00 01 02 03 04 05 06 07 08 09 0A 0B 0C 0D 0E 0F:   ASCII EQUIV.
--------------------------------------------------------------------------
0000  01 00 00 00 00 00 00 00 00 00 00 00 00 00 00 00    ............
0010  00 02 00 00 00 00 00 00 00 00 00 00 00 00 00 00    ............
0020  00 00 03 00 00 00 00 00 00 00 00 00 00 00 00 00    ............
....  .. .. .. .. .. .. .. .. .. .. .. .. .. .. .. ..    ............
....  .. .. .. .. .. .. .. .. .. .. .. .. .. .. .. ..    ............
00E0  FF FF FF FF FF FF FF FF FF FF FF FF FF FF 0E FF    ............
00F0  FF FF FF FF FF FF FF FF FF FF FF FF FF FF FF 0F    ............
```

Fig. 9-4 Example of memory editor display.

to-binary or BCD-to-seven-segment conversion can be implemented in a truth table burned into a PROM or EPROM. Off-the-shelf ICs are available for converting ASCII characters to dot-matrix representation for display on printers and *cathode ray tubes* (*CRTs*). These chips are basically just ROMs coded with information similar to that shown in Fig. 9-5*b*. Figure 9-5*a* shows a functional block diagram of the CRT7004 Dot Matrix Character Generator from Standard Microsystems. ROM-based chips are also used to provide ASCII-coded output from terminal keyboards. ROM is even used to convert coded English words to speech patterns for voice synthesis.

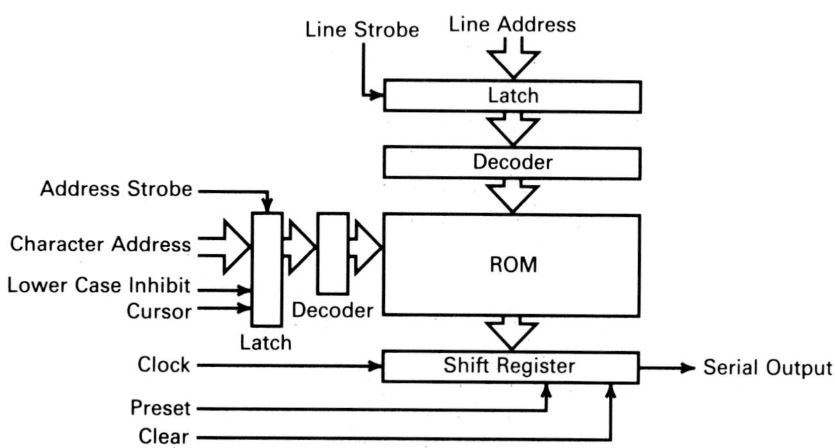

Fig. 9-5(*a*) Functional block diagram for the CRT7004.

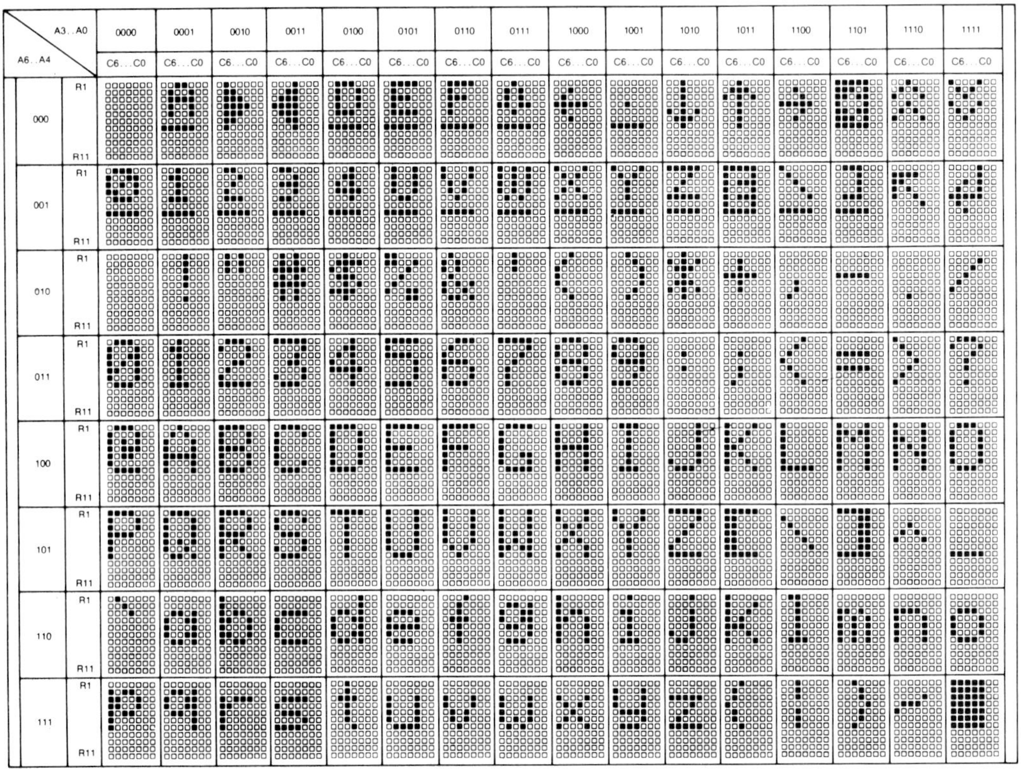

Fig. 9-5(*b*) ROM contents of the CRT7004. *Courtesy of Standard Microsystems Corporation.*

Texas Instrument's TMS 6100 16K × 8-bit ROM is a read-only memory especially designed for applications such as talking toys and annunciators.

Arithmetic operations, such as multiplication, division, and trigonometric conversion, can often be performed with table look-up at higher speeds than are possible with random logic. A simple example would be to use a 256 × 8-bit PROM to multiply two BCD digits. The lower four address inputs would represent one BCD digit while the upper four address inputs would represent the other. The PROM would be programmed so that any two valid BCD inputs on the address lines would produce a correct two-digit BCD product at the output. Figure 9-6 illustrates this approach to BCD multiplication. A little thought leads us to the conclusion that since only 100 possible combinations of two BCD digits exist, 156 locations in the PROM are wasted. For instance, the address 11001001 does not represent a valid BCD multiplication since 1100 is not a BCD digit. This suggests that a FPLA might be used instead of a PROM. At this density, however, no cost advantage would be realized by using a FPLA. At higher densities or in IC designs where the choice would influence the die size, this would be a consideration.

Another simple look-up table application would use two 1K × 8-bit

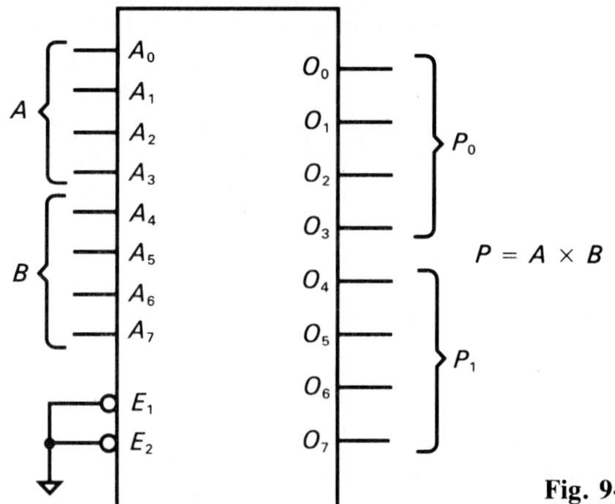

(a)

Address		Output	
A	B	P_1	P_0
0111	0110	0100	0010
0111	0111	0100	1001
0111	1000	0101	0110
0111	1001	0110	0011

(b)

Fig. 9-6 Using a 256 × 8-bit PROM to multiply BCD digits. (*a*) Pin organization; (*b*) example of relationship between data and address.

PROMs to perform trigonometric functions. The 10 inputs required to address 1K of memory could represent an angle of 0 to 360° in BCD code. The PROMs would be programmed so that the outputs represented the sine or cosine of the angle. The PROMs would receive the same input and produce a BCD output which had four significant digits. Since the largest value for a sine or cosine is 1.000, the upper 2 bits could be coded in the PROM and decoded with external gates to produce a 1 for sine 90° or cosine 0° leaving 16 outputs free for four fractional digits.

State Tables

State tables are truth tables used in sequential logic design. In the previous chapters these tables were minimized with Karnaugh maps to produce simplified boolean logic. The sum-of-products expressions were simplified in order to reduce the number of gates and to reduce the number of inputs to those gates. DeMorgan's theorem was then used to reduce the number of board-level packages. The logic function outputs were applied to the inputs of flip-flops, and the flip-flop outputs were used to drive the logic function inputs. This yielded a classic Huffman-Moore sequential state machine.

The reduction or minimization of truth tables is the best approach when small state machines are designed. However, when state machines become large and complex, ROM-based implementation has definite advantages. ROM, in this context, refers to ROM, PROM, EPROM, or any form of read-only memory. Reduction in chip count, reduction in design cycle time, and ease of circuit modification all make ROM-based logic an attractive alternative. Ease of modification is particularly important when high-priced PC boards are involved. A design change in random logic usually means a PC board modification or even a new PC board release. A change in ROM-based logic can often be accomplished with a simple reprogram.

As an example of ROM-based design, consider the traffic light state machine in Chapter 4. Very little logic was required to implement this machine, and it would hardly be worthwhile to replace the random logic with an expensive memory chip, but Fig. 9-7 shows how it could be

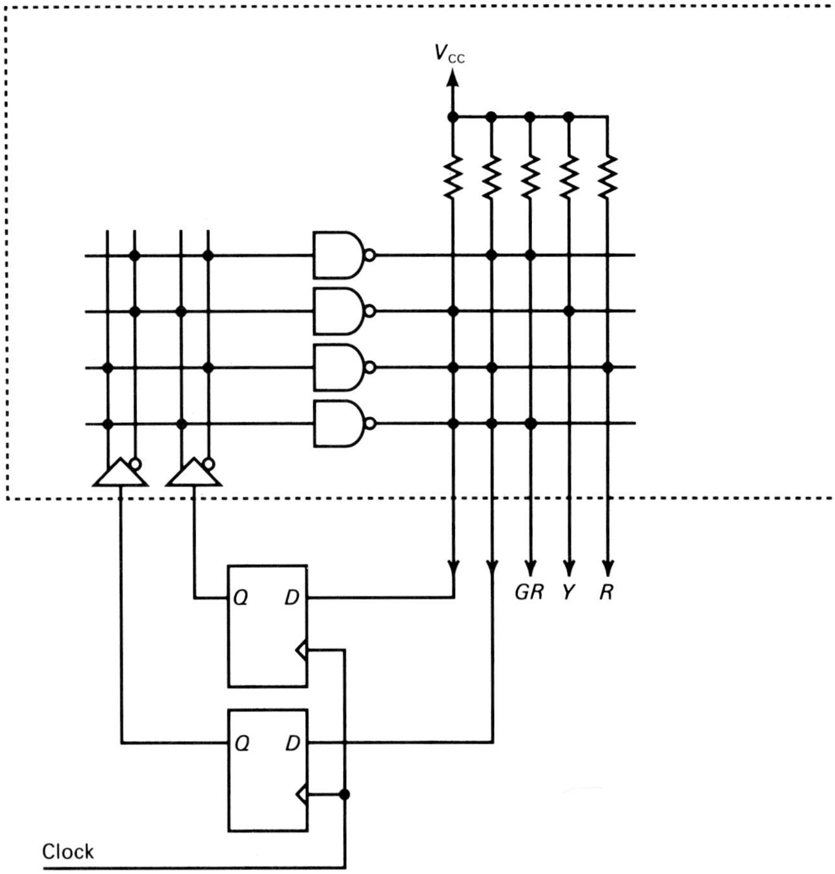

Fig. 9-7 Traffic light sequencer implemented with PROM.

implemented in a portion of PROM. Any state machine may be implemented in read-only memory provided that one or more ROM-type chips will provide enough inputs and outputs. When determining maximum clock rate,

$$\text{min. clock period} = \text{flip-flop delay} + \text{gate delay} + \text{setup time}$$

the gate delay is replaced by access time. *Access time* is the time between when the address inputs are valid to when the outputs are stable. ROMs and PROMs are available with access times down to 15 ns. EPROMs and EEPROMs are slower.

PROMs are also available with on-board flip-flops. These are called *registered PROMs*. When these are used as state machines, the registers are called *state registers*. When using registered PROMs, there are two timing parameters to consider: *setup time from address to clock* and *clock to output delay*. The first parameter represents the time needed to locate the information and present it to the internal flip-flop inputs. It is the sum of access time and flip-flop setup time. The second parameter represents the flip-flop delay of the on-board register. When the registered PROM is used as a one-chip state machine with some outputs fed back to the address inputs, the minimum clock period is simply the sum of these two parameters. This type of state machine presents a new wrinkle in the state machine model. Previously, only the *next-state* outputs were clocked into flip-flops. The decoded outputs changed when the inputs changed. When registered PROMs are used, the output lines used for next state are the same as before, but the decoded outputs are different because they do not change until the next clock edge. This type of implementation has the advantage that decoding glitches never appear on the outputs. An example of the traffic light sequencer using a registered PROM is shown in Fig. 9-8.

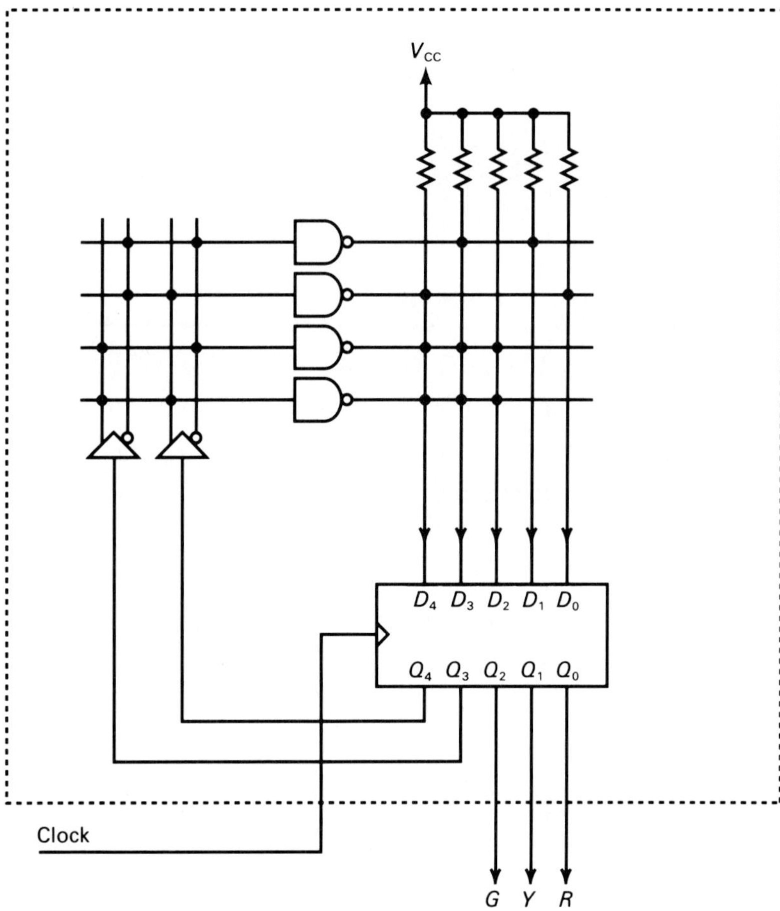

Fig. 9-8 Implementation of traffic light sequencer using a registered PROM. Note that in order to maintain the correspondence of machine state to color output, the PROM connections must shift. The outputs do not appear until the clock edge after they are decoded. This prevents decoding glitches from appearing at the outputs.

REVIEW 9-1

1. **Q.** What is a sum-of-products expression? **A.** A sum-of-products expression is the unsimplified boolean equation derived directly from a truth table with AND products ORed, or summed, together.

2. **Q.** What is minimization? **A.** Minimization is simplification and reduction of product terms.

3. **Q.** What is random logic? **A.** Random logic is gate-level implementation of minimized sum-of-products expressions.

4. **Q.** How is logic implemented in read-only memory? **A.** The truth table is programmed directly into the memory.

5. **Q.** What is a look-up table? **A.** A look-up table is a table which converts coded inputs into decoded outputs. The inputs may represent numbers or binary codes, and the outputs may represent arithmetic results or code conversions.

6. **Q.** When using ROM as logic, what parameter determines the logic delay? **A.** the access time of the ROM

7. **Q.** How is the maximum frequency of operation determined when a registered PROM is used? **A.** It is the reciprocal of the sum of two parameters: setup time from address to clock and the clock to output delay.

9-2 CONTROL STATE COUNTERS

The traffic light controller in Fig. 9-7 could be viewed as consisting of two parts. The first part is the two PROM outputs driving the flip-flops. This part is a simple modulo-3 counter, or *sequencer*. It sequences through three states, 00, 01, and 10, before repeating. The second part consists of the decoded outputs. These outputs *control* the red, yellow, and green lights, and are called *control signals*. As simple as this example is, it illustrates the two parts of a state machine which form the heart of the microprocessor. A microprocessor has more stages in its control counter, and many more control signals, but the concept is almost the same. *Almost*, because one element is missing—instructions.

Consider the expanded version of the traffic light example shown in Fig. 9-9. In the PROM block, only the memory matrix is shown, not the decoder. This machine provides a three-stage control counter allowing 8 different control states. Only 8 states are needed to provide a complete cycle of red, yellow, and green lights, but two different *sequences* are available. In the first, the red is on longer than the green with a very short yellow. The second sequence provides equal red and green times with a longer yellow. The choice between the two sequences is made with the *instruction bit*. One instruction bit is hardly adequate for a real microprocessor, but this illustrates the concept. In a microprocessor, a control counter sequences through a number of control states producing control

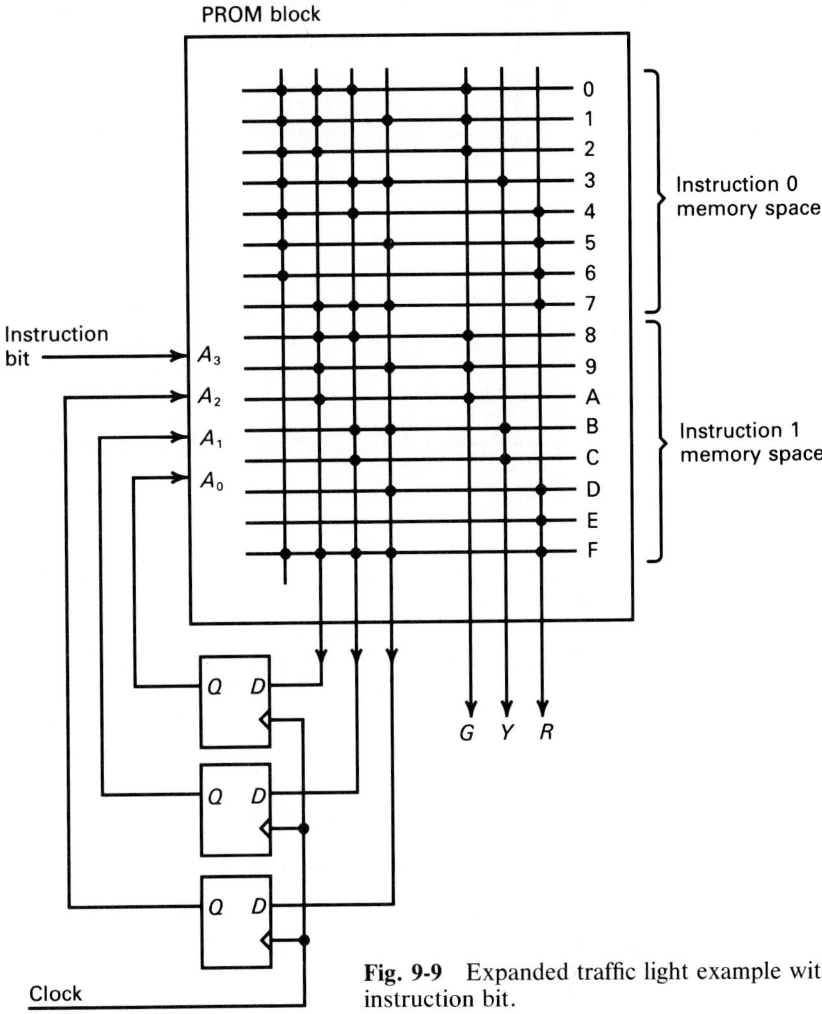

Fig. 9-9 Expanded traffic light example with 1 instruction bit.

signals. The instruction bits determine *which* sequence is used. A PLA provides the ROM-based logic, and the mask programmed connections of that PLA are called *microcode*. Programming the PLA is called *microprogramming*. One complete sequence of microcode represents an *instruction* to be executed by the machine.

If no instructions were needed, or if all instructions required the same number clock periods to execute, a simple binary counter could be used to provide control sequencing. Its outputs could be decoded in a ROM or PLA. Figure 9-10 shows a control state sequencer using two '163 binary counters decoded with a 63S280 256 × 8 PROM. The PE and TE enable inputs have been combined into one E input for simplicity. Since only 256 control states are possible from the 8 counter outputs, the 63S280 will decode every state. However, the process may require fewer states before repeating. In the Fig. 9-10 example, one control signal may be used to

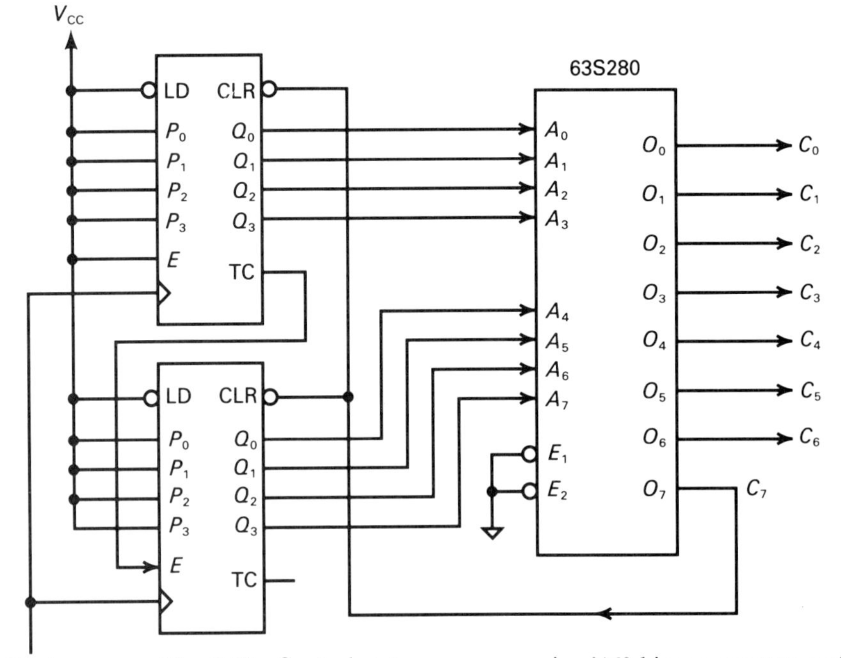

Fig. 9-10 Control state sequencer using '163 binary counters and 63S280 256 × 8-bit PROM.

reset the counter early. In this example, the synchronous clear input is used to reset the counter on the next clock after some decoded state. There are no instructions. The machine sequences and repeats in the same way every time.

Figure 9-11 shows the same counter-PROM configuration with allowance made for instructions. A 4-bit instruction may be entered as a load input

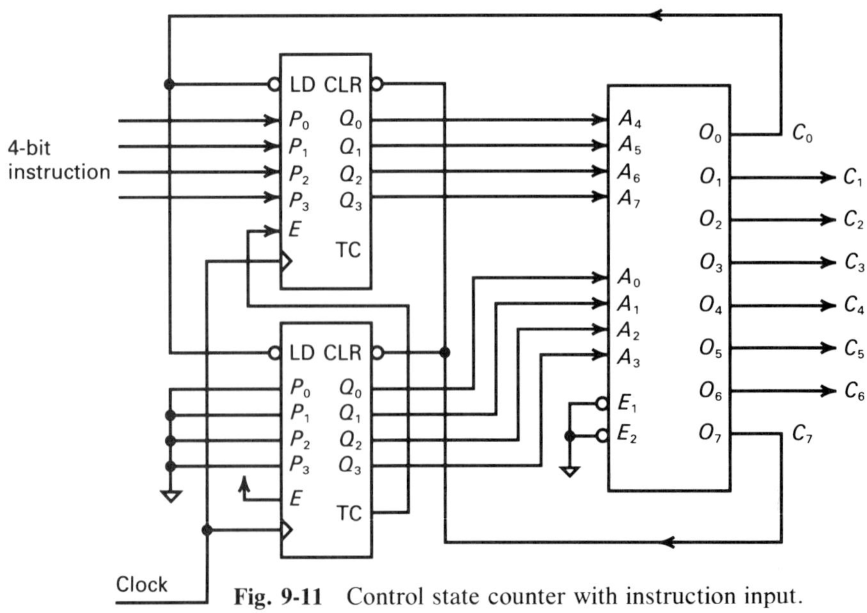

Fig. 9-11 Control state counter with instruction input.

to the upper 4 counter bits. The load occurs when one of the decoded control signals C_0 pulls low. This causes a synchronous load to the counter on the next rising clock edge. Since the PROM is not registered, glitches appear on the control outputs after every rising clock edge, necessitating the use of a counter with a synchronous load input such as the '163. The PROM is programmed so that after it completes an instruction, it loads a new instruction on the next rising edge by asserting C_0. In this way the change of instruction will occur only after the present instruction has been completed. If the present instruction is not changed, then the machine repeats the previous operation.

A new instruction causes the control state counter to *jump* to a new starting state. Each instruction may have as many as 16 clock cycles for execution, representing the 16 combinations of the 4 bits in the lower counter. Some instructions may require fewer cycles than others and occupy fewer than 16 lines in the PROM. The memory space of the PROM is shown in Fig. 9-12. There are 16 possible instructions, each occupying

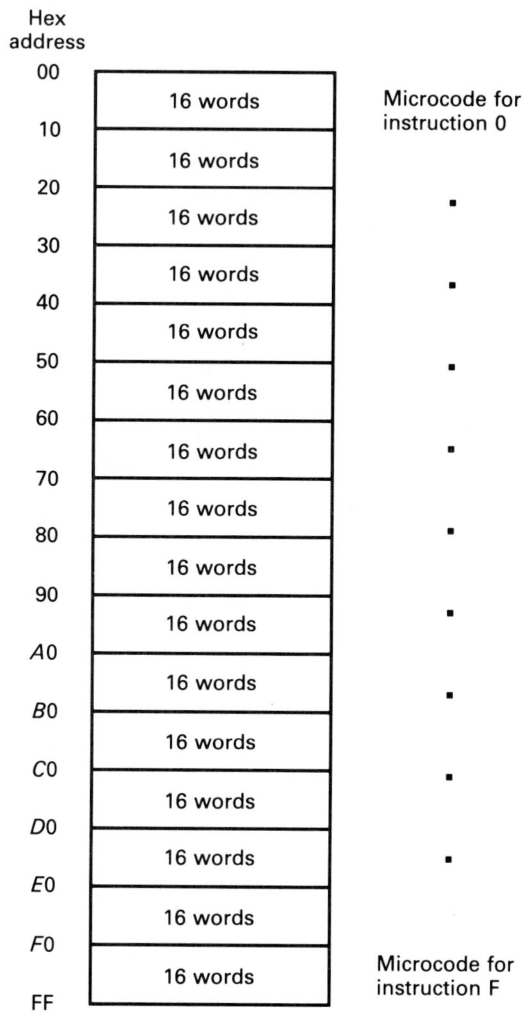

Fig. 9-12 Memory map for PROM shown in Fig. 9-11 example.

a block of 16 memory locations. The sequence of each instruction may occur only in its assigned block, and if many instructions do not require 16 locations, then quite a few memory locations may be wasted.

Another approach to control sequencing is to design the counter into the PROM itself. This was done in Figs. 9-7, 9-8, and 9-9. In those cases, some control signals were used to drive the flip-flops. They sequenced through a straight binary count, but non-binary sequences are also possible and sometimes make better use of the memory space. In each control state the flip-flop outputs driving the lower PROM addresses will select the present location in PROM, and the programmed bits in that location will determine the next control state. Additional outputs are needed for control signal outputs. Instructions may be latched and allowed to change only when the current instruction has completed. While there is no advantage to this second method at the board level, it is useful at the chip level where the ROM or PLA may have as many outputs as needed.

REVIEW 9-2

1. **Q.** What is microcode? **A.** It is the sequence of programmed ROM cells which produces control signals at the ROM output.
2. **Q.** How does the 1-bit instruction in Fig. 9-9 choose between microcoded sequences? **A.** If the 1-bit instruction is a 0, the lower portion of memory is addressed. If the bit is a 1, the upper words are addressed.
3. **Q.** How do instructions change the operation of the control state counter in Fig. 9-11? **A.** They change the operation in the same way as the 1-bit instruction. A new instruction will cause the counter to address a different part of memory.
4. **Q.** How is a new instruction loaded? **A.** The microcode for each sequence must contain a bit needed to load the counter after the current sequence has completed.

9-3 INSTRUCTION REGISTERS AND PROGRAM COUNTERS

Control logic consists of any circuitry which provides the right control signals in the right sequence to perform some operation. The operation might be a repetitive task, such as making up boxes and filling them with bolts. Or it might be a highly varied activity which relies on external inputs from the operation or operator. If very few changes are ever made to the routine process, the logic may simply be a counter sequencing through read-only memory as shown in Fig. 9-10. The control signals needed to operate a box filler are well determined, and any changes would be in the form of periodic improvements to the process. In this case, the only inputs needed from the operator would be on and off controls. On the other hand, there are processes which are constantly changing. A milling machine, for example, has several modes of operation. Each basic function, such as drilling, routing, and turning, can be combined with any other function to produce a sequence tailored to a specific job. This type of machine cannot be driven by a counter-sequenced ROM. Each function of the machine will have its own section of microcode in the ROM, and the decision as to which sequence to use next is left to the operator. All the possible functions are provided in read-only memory, and the operator may select between them.

If a human operator were to select the next operation after the completion of each task, too much time would be wasted, and the machine would be idle much of the time. Real-time choices by human operators would result in inefficient machine use. Instead, human activity occurs before the machine even starts its task. Programmers choose the tasks beforehand and store that information into an external memory. The control logic then accesses that memory and executes its own microcode on the basis of the external instructions. This is called *stored program control* and requires that each microcoded sequence finish its task by loading in a new external instruction. John von Neumann is credited with originating this concept in 1946, and stored program machines are often referred to as "von Neumann machines." From then to the present, no other computer architecture has seriously challenged the von Neumann machine.

Consider the control sequencer in Fig. 9-13. The new features are an instruction register, a program counter, and some extra gates. The machine

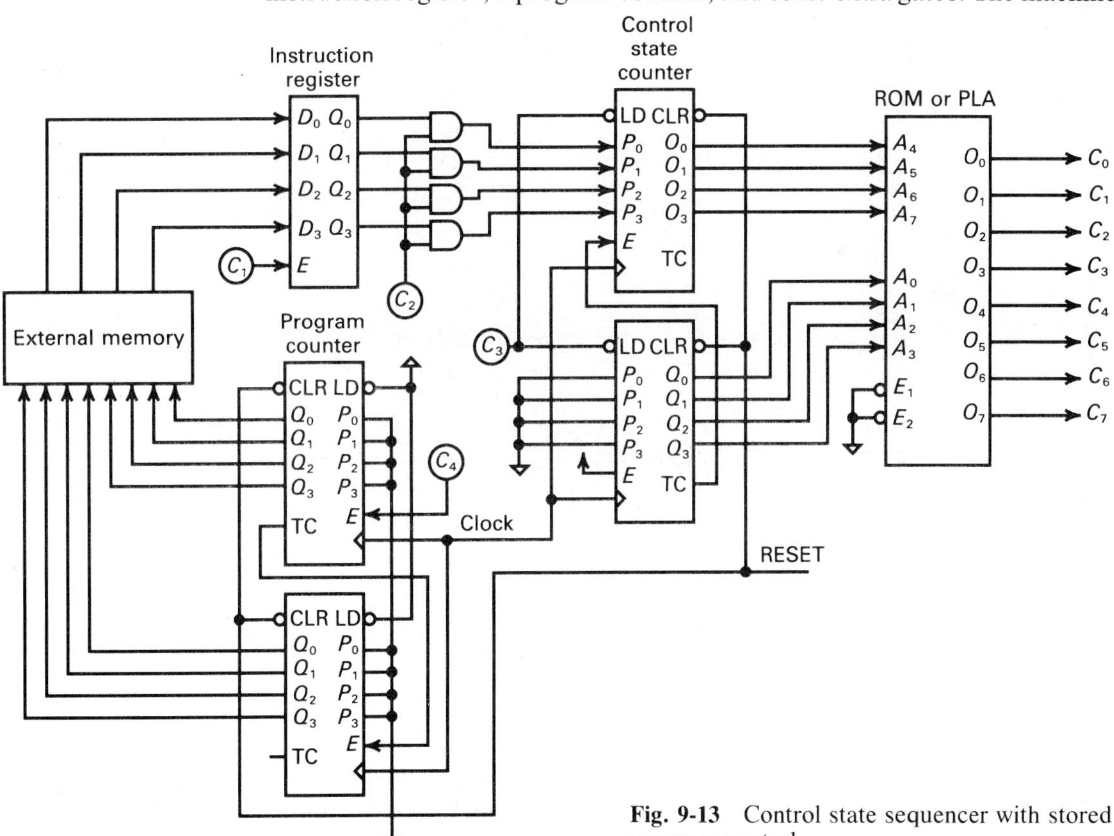

Fig. 9-13 Control state sequencer with stored program control.

clears when the operator resets the counters. This starts the microcode at location 00_{16} in the ROM (ROM, EPROM, or PLA). The control state counter outputs sequence 00, 01, 02, etc., addressing the first few locations of the ROM. The microcode at these locations causes the contents of external memory to be loaded into the instruction register. This sequence is called the *fetch cycle* because it causes an external instruction to be fetched from memory. This is accomplished with control signal C_1. The external memory is addressed by the *program counter,* which was reset to 00_{16} at the same time as the control state counter. After the contents of memory location 00_{16} have been loaded into the instruction register, control signal C_4 causes the program counter to increment. This is done by holding the enable input high for one clock period. Next, control signals C_2 and C_3 cause the contents of the instruction register to be loaded into the control sequence counter. This forces the counter to start counting from a higher value and address a different part of the ROM. The instruction, taken from external memory, in effect tells the machine to start executing the microcode in that portion of ROM. The last part of the microcode, for every instruction, produces low values at control signals C_2 and C_3, sending the control counter back to 00_{16} so that the fetch microcode will be executed. Since the program counter is incremented after each use, it is always pointing to the next instruction in external memory. The remaining control signals are left to perform actual tasks.

REVIEW 9-3

1. **Q.** What is stored program control? **A.** It is a type of control using instructions which have been written ahead of execution time and stored in external memory for access by control circuitry.
2. **Q.** What is a fetch cycle? **A.** It is a microcoded sequence which loads in an instruction from external memory.
3. **Q.** Why is stored program control necessary? **A.** If human beings loaded new instructions after each execution, much machine time would be wasted.

9-4 GENERAL-PURPOSE STATE MACHINES

The controller in Fig. 9-13 could be combined with the ALU logic described in Chapter 7 to make a simple general-purpose state machine called a *microprocessor*. Figure 9-14 illustrates the combination. Arithmetic and logical functions are needed because so many control operations require them. Even systems as crude as large rock crushers in mining complexes require multiplication to implement closed-loop servo control. The state machine in Fig. 9-14, of course, is much too simple to have practical value but it illustrates the key principles of microprocessor design. Real processors have 8-, 16-, or 32-bit registers and much more complex ALUs. However, the 4-bit processors which kicked off the micro revolution in the early 1970s were not so different from this example.

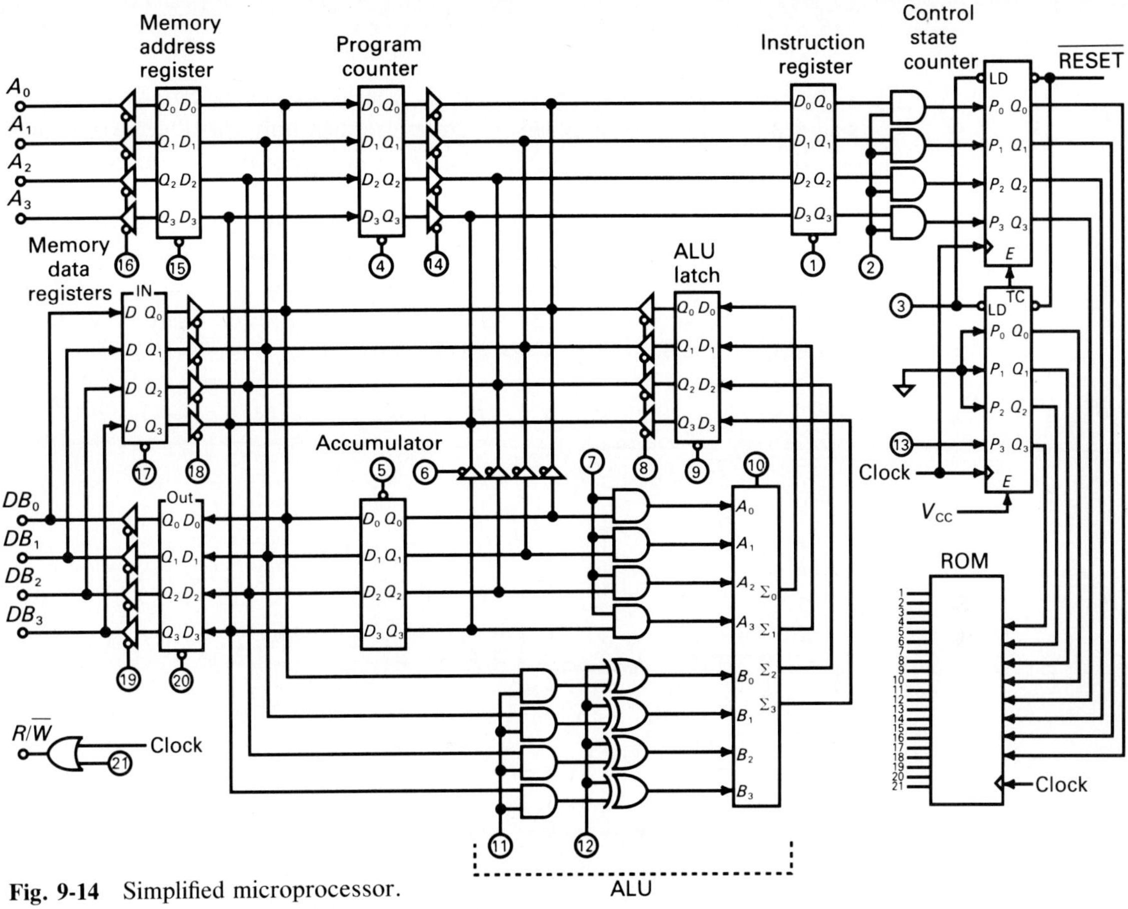

Fig. 9-14 Simplified microprocessor.

Note the following about this microprocessor:

1 The ROM has 21 outputs representing the 21 control signals needed to operate this machine. This could easily be done as a PROM or PLA on an IC, but three PROMs would be needed to implement this design at the board level.

2 The control signals, numbered 1 through 21, are connected to the corresponding circled numbers on the logic inputs.

3 The registers may be clocked flip-flops, or transparent latches, if the PROM is registered. When the control state counter changes, false values will not momentarily appear on the PROM outputs if registered PROMs are used. If clocked flip-flops are used, the clock inputs are implied.

4 The tristate buffers which isolate the register outputs from the internal bus are turned on with an active-low control signal.

5 The program counter is not actually a counter. The counter has been replaced with a register. This is common in microprocessor design because the ALU is available to load or increment the register and there is no need for the additional counter circuitry.

6 External memory is not shown in Fig. 9-14 but sits on the address and data buses labeled A_0 through A_3 and DB_0 through DB_3.

7 The control signals from PROM do not exit the machine. Control of external circuitry is accomplished through the memory data register-out.

In order to facilitate discussion of this simplified microprocessor, we use a control signal notation which references only those control signals which are important to understanding each operation. For each clock period, the asserted control signals are listed and unimportant nonasserted signals are ignored. Assertion may be active-high or active-low, depending on the function of the signal.

The first eight PROM locations are reserved for the reset sequence. A reset may be actuated with a button connected to a momentary switch at the input of an RS latch. Figure 9-15 shows how this reset circuitry may

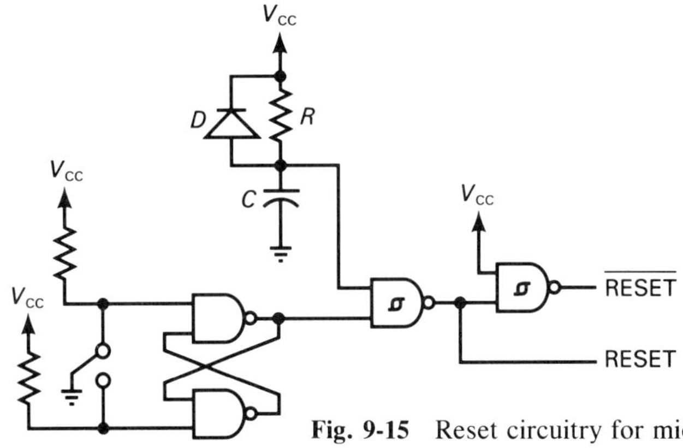

Fig. 9-15 Reset circuitry for microprocessor example.

be coupled with a *power-on reset* so that the machine may be started when the power is applied or restarted whenever someone pushes the reset button. When the power is turned off, the capacitor *C* discharges quickly through the diode *D*. When power is restored, the capacitor takes some amount of time to charge to the logic 1 threshold. In the meantime, the output of the first NAND gate is held HIGH, and the second NAND gate is LOW. During this time any circuitry whose reset inputs are connected to these lines is held reset. The reset time is determined by the resistor and capacitor values and the input threshold of the gate. One time constant *RC* is sufficient to exceed the logic threshold of any NAND gate. It is only necessary, then, that this time exceed the specified reset pulse width LOW of the logic being affected. Some logic cannot tolerate slowly changing transitions, and in those cases gates with Schmitt trigger inputs, such as the 74xx14 or 74xx132, may be used to speed the transition. Schmitt triggers were discussed in Chapter 6.

The internal registers could be reset directly with the hardware shown in Fig. 9-15. This would be called a *hard reset*. Or the reset pulse could load the control state counter to address location 00_{16} in the PROM. Then the microcode in the first few PROM locations would cause zeroes to be loaded into the registers. When coded instructions cause a reset, it is called a *soft reset*. The soft reset allows the use of smaller, nonresettable D flip-flops or D latches. The following control signals will be asserted in the start-up sequence:

Hex address		
00	⑦ ⑨ ⑩ ⑪ and ⑫ all low	Clear ALU latch.
01	④ ⑧ and ⑤ low	Clear accumulator and program counter.
02	② ③ low and ⑬ high	Start at location 08_{16}.

The last control output of the start-up sequence forces the counter to address location 08_{16} in the PROM. Location 08_{16} is the start of the fetch sequence. This sequence of microcode loads the instruction register with programmed instructions from external memory. The following sequence of control signals will load one instruction. Abbreviations are used with arrows to indicate when the contents of one register moves into another register:

Hex address		
08	⑦ ⑨ ⑫ ⑭ ⑮ all low	PC → MAR
	⑪ ⑩ high	PC + 1 → ALU latch
09	④ ⑧ low	ALU latch → PC
0*A*	⑯ ⑰ low and ㉑ high	Ext. memory → MDR
0*B*	① ⑱ low	MDR → IR
0*C*	③ low and ② high	IR → counter

The instruction causes the counter to jump to the part of PROM which contains the microcode for that instruction. Of course, when we program the external memory, we must know exactly what the instructions do. The microcoded ROM for most machines is fixed at the time of manufacture, and the *machine instruction set* is defined at that time. A few machines do allow the user to customize instructions by manipulating the microcode. For educational purposes we will assume that we can write the microcode for the processor in Fig. 9-14.

EXAMPLE 9-1

Detail the control signal sequence necessary to load the accumulator with data from external memory. The instruction code for this operation is 0001_2 and the number to be loaded follows the instruction in external memory. Show the addresses of the microcoded control signals in PROM.

SOLUTION

The following sequence of control signals will cause the number following the instruction to be loaded into the accumulator:

Hex address						
10	(7) (9) (12) (14) (15) all low	PC → MAR
	(11) (10) high	PC + 1 → ALU latch
11	(4) (8) low	ALU latch → PC
12	(16) (17) low and (21) high	Ext. memory → MDR
13	(18) (5) low	MDR → ACC
14	(2) (3) low and (13) high	Jump back to fetch the next instruction.

This instruction causes the counter to start at 10_{16} which is where we should find the microcode. The first part is identical to the fetch sequence. The program counter must be incremented by each sequence which uses it. When the actual instruction part of the sequence begins, the old program counter value will be in the *memory address register* (*MAR*), and the new, incremented value will be in the program counter waiting to be used by the next sequence. Remember that the microcode for this instruction is just a series of blown fuse links between addresses 10_{16} and $1F_{16}$. We show only the locations actually used to perform the task, and we show only the control signals which are asserted when those locations are addressed by the control state counter. *Implicit in the instruction 0001_2 is the information that accumulator is to be loaded with whatever 4-bit number follows the instruction.* Therefore the location pointed to by that instruction already contains the microcode needed to execute that particular operation.

Addressing Modes

An *addressing mode* refers to the manner in which data is obtained from memory. When discussing addressing modes, the terms *opcode* and *operand* are often used. The *opcode* is the instruction itself, and the information following the opcode is called the *operand*. In the example given above, the data to be loaded into the accumulator immediately followed the instruction. This is referred to as the *immediate mode* of addressing because the data can be processed immediately without any further address reference. In the immediate mode of addressing, the operand *is* the data to be processed.

Other addressing modes are also possible. *Direct* addressing, for example, indicates that the operand following the instruction is not data but an address which points to the data. In 8-bit processors, the opcode is 1 byte and the operand is 1 byte. The 1-byte operand contains the address needed to locate the data. Since 8 bits can address only 256 locations, it is implied that these are the first 256 locations in memory. A *page* of memory in an 8-bit system is defined as 256 locations. Direct addressing is sometimes called *zero-page* addressing because it accesses only the first, or zero, page of memory. *Extended*, or *absolute*, addressing modes require 3 bytes: 1 byte of opcode and 2 bytes of operand. As with direct mode, the operand contains an address which points to the data. But the 2-byte operand permits access to 65,536 (65K) bytes of memory.

EXAMPLE 9-2

Detail the control signal sequence necessary to implement the direct mode of addressing in the simplified 4-bit processor. The instruction code for load accumulator direct is 0010_2. Show the addresses of the microcoded control signals in PROM.

SOLUTION

The following sequence of control signals will cause the number following the instruction to be used as an address which points to the number to be loaded into the accumulator. The only difference between this sequence and the immediate addressing mode example is that the first nibble of data is stored in the *memory data register* (*MDR*), transferred to the *memory address register* (*MAR*), and then used as the address for finding the next nibble, which is then stored in the accumulator.

Hex address		
20	⑦ ⑨ ⑫ ⑭ ⑮ all low	PC → MAR
	⑪ ⑩ high	PC + 1 → ALU latch
21	④ ⑧ low	ALU latch → PC
22	⑯ ⑰ low and ㉑ high	Ext. memory → MDR
23	⑱ ⑮ low	MDR → MAR
24	⑯ ⑰ low and ㉑ high	Ext. memory → MDR
25	⑱ ⑤ low	MDR → ACC
26	② ③ low and ⑬ high	Jump back to fetch.

There are instructions which require no external data for execution. An example is the instruction INCA, or increment accumulator. The number in the accumulator is incremented in the ALU, and the result is placed back into the accumulator. This type of addressing is called *inherent* mode addressing, or *implied* addressing. In the implied addressing mode, there is no explicit operand, but rather, the operand is *implied* in the instruction code. Other addressing modes include *indirect* and *indexed* modes. The operand for the *indirect* mode of addressing is an address which points to another address. This second address, then, points to the data. This mode of addressing facilitates the storage of programs in ROM since the absolute location of the data in RAM need not be specified when the ROM is programmed. When accessing a table or list, it is convenient to increment the address to the next location. This would not be possible if the absolute address were programmed into ROM. The *computed indexed* mode of addressing requires that the processor compute the desired address by adding the operand to some already established base address contained in a register. The computed indexed addressing mode uses the operand as an *offset* from the address in the base register. This mode of addressing is also useful when accessing tables or lists of data.

Input and Output to Nonmemory Devices

So far we have explained how instructions are fetched from memory and executed and how data is brought into the machine, but we have not explained how the machine communicates to the outside world. Microprocessors use *peripheral ICs* to interface to external devices such as printers and terminals. There are three common techniques for accessing these peripheral ICs. The first method, *serial I/O*, requires serial input and output pins on the processor which can be manipulated with single instructions. An example of this can be found in the RCA COSMAC microprocessor. *Event flag* inputs sample serial data coming *in*, and a *Q* output line provides for serial data going *out*. A second technique is parallel in nature and uses *direct I/O ports* which are dedicated exclusively to nonmemory devices. In direct I/O systems, the number of addressable I/O locations is considerably less than the total memory space. An example is the Intel 8085 microprocessor, which has 512 I/O addresses, 256 in and 256 out. The total memory space for the 8085 is 65K bytes. Finally, the third approach treats external peripheral devices just like memory by accessing them and controlling them through the address register and memory data registers. Motorola family microprocessors use this method exclusively.

REVIEW 9-4

1. **Q.** How is the general-purpose state machine started? **A.** with a hardware reset
2. **Q.** What does the reset sequence of microcode do? **A.** It initializes all the memory devices by resetting them to 0.
3. **Q.** What does the fetch sequence of microcode do? **A.** It loads one instruction from external memory into the instruction register.
4. **Q.** What must be done each time the program counter is used? **A.** It must be incremented so that it will point to the next location in memory.
5. **Q.** What is the purpose of the program counter? **A.** It keeps track of where we are in external memory.
6. **Q.** What is the immediate mode of addressing memory? **A.** It is the mode of addressing in which the operand contains the data to be used by the machine.
7. **Q.** What is direct addressing? **A.** It is the mode of addressing in which the operand is an address which points to the data to be used by the machine.

8. **Q.** What is indirect addressing? **A.** It is the mode of addressing in which the operand is an address which points to a second address which points to data.

9. **Q.** What is inherent, or implied, addressing? **A.** It is the mode in which the operand is implied in the instruction.

10. **Q.** How does the machine interface with external devices? **A.** either through serial I/O, parallel I/O ports, or with the memory data registers

9-5 USER-PROGRAMMABLE CONTROL LOGIC

The key word in electronics today is *integration*. Integration means putting increasing amounts of circuitry onto a single substrate. Digital circuitry started with vacuum tube and transistor level design. During the 1950s and early 1960s one NAND gate occupied a single PC board. *Small-scale integration (SSI)* put four NAND gates into a single 14-pin package. *Medium scale integration (MSI)* put tens of gates into a single package. The JEDEC Council of the Electronics Industries Association (EIA) defines MSI as a system containing 12 or more equivalent gates. An *equivalent gate* is the unit measure of digital circuitry. It means that a circuit with a given gate equivalency could be built discretely with that number of gates. In the industry, one CMOS equivalent gate is equal to two *p* transistors and two *n* transistors, which is enough to create either a two-input NAND gate or a two-input NOR gate. *Large-scale integration (LSI)* consists of 100 or more equivalent gates according to JEDEC definitions. Today, *very large scale integration (VLSI)* dominates the industry. While no precise definition of VLSI has been given by either the JEDEC committee or the IEEE, it is safe to say that VLSI chips contain between 1000 to 10,000 or more equivalent gates.

The advantages of large scale integration are obvious:

1 Savings in board space, power, and labor
2 Fewer interconnections and therefore greater reliability
3 Proprietary advantages accruing from the difficulty in copying VLSI designs

VLSI has been available to large electronics companies for years. This takes the form of *full-custom ICs* which are designed from scratch and *semicustom ICs* which utilize standard cell libraries developed by in-house IC designers. Companies with their own fabrication facilities have led the field in LSI and VLSI. The microprocessor, to some extent, put everyone else on par with the large concerns by making VLSI accessible to everyone. Almost any low- to medium-speed application can be satisfied with the appropriate off-the-shelf microprocessor. The microprocessor is a generic, all-purpose VLSI chip which may be programmed for specific applications with instructions. The instructions are written and stored in external memory and are referred to as *software programs*. The disadvantages of microprocessors are:

1 They require a fair number of support ICs, including memory and special I/O chips.
2 Programming micros has proven to be more of a burden than anyone first imagined.
3 Verification of reliability is difficult in software. Debugging in software merely proves the presence of errors—not their absence.
4 The microprocessor solution has speed limitations. While the processors are getting faster, they will never be as fast as special-purpose hardware.

The second point requires some explanation. Originally, software programming of microprocessor logic was done by the same engineers who designed the hardware system. Even then, the time required for the task was about 10 percent hardware design and 90 percent software design. As processors became more complex and tasks more complicated, the job ballooned to monumental proportions. Software design for microprocessors is now handled by software specialists, and the overhead involved in supporting large software departments for small systems has often been underestimated.

System designers are increasingly turning to hardware alternatives. A new *application-specific IC* (*ASIC*) market has blossomed and continues to grow. Hardware alternatives include *gate arrays, standard cell libraries, silicon compilers, and EPLDs*. Each of these will be discussed in this section.

Gate Arrays

The term *gate array* originally referred to a *sea-of-gates* provided on a single die to be interconnected by the user. Now, gate arrays are not actually arrays of gates but arrays of *p* and *n* devices which may be interconnected to form NAND gates, NOR gates, flip-flops, or other types of circuitry. Because the components are standardized, the array can be produced in high volume at low cost. The user supplies the information needed to complete the last two or three metal layers which define the actual application. This is called *late mask programmable* logic because it is all prefabricated except for the last steps.

Gate arrays are structured into columns of "gates" called *diffusion columns*. An unconnected CMOS gate array is shown in Fig. 9-16. Each gate location is called a *slot* and consists of two pairs of *n* and *p* transistors. There is space between the slots to route two or three metal runs and space between the columns for a dozen or more metal runs. I/O buffers (large inverters) are provided around the periphery to give the tiny on-chip signals the boost needed to drive off-chip.

Gate array design requires computer simulation as well as breadboard testing. You can build up the equivalent circuit with standard parts and test the system for functionality. But, computer simulation is also needed to analyze the circuit's sensitivity to voltage and temperature variation and timing problems introduced by capacitive loading and parasitic effects on the silicon. Fortunately, gate array design software is available for PC compatible personal computers and other standard platforms. The design software allows schematic entry and design checks at a very low cost.

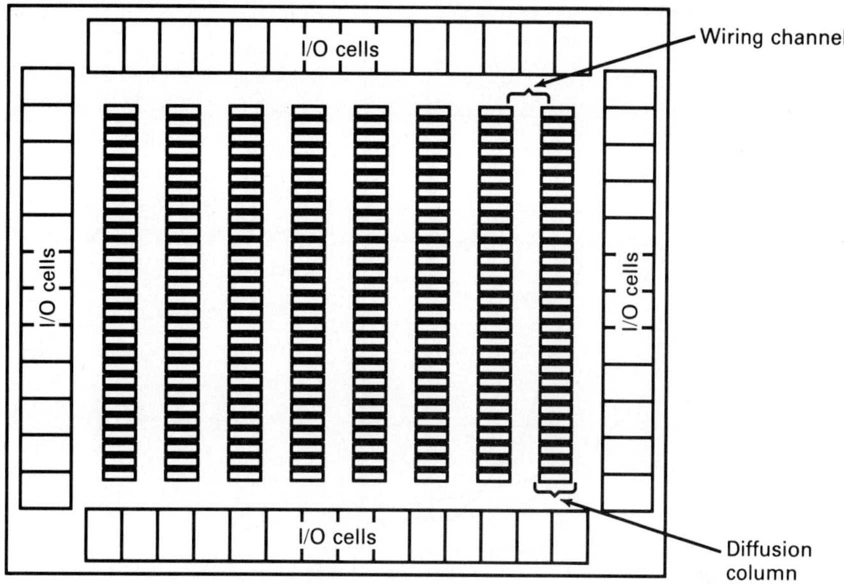

Fig. 9-16 Unconnected CMOS gate array.

Very often, the hardware needed for PC-based gate array design, the PC, extended memory, I/O ports, RGB monitor, and hard-copy output is already available to the designer. Only software packages are needed, and these are available from the companies supplying the gate array libraries as well as from independent software houses.

Standard Cell Libraries

Standard cell libraries are selections of predesigned logic cells ready to be assembled. Gates, flip-flops, I/O, and memory cells are standardized to fit together easily on a flexible floor plan. Major semiconductor houses as well as small start-up companies offer a variety of cell libraries together with the expertise needed to guide the user through the pitfalls of IC design. Software is provided for simulation and modeling. Using standard cells is very close to full-custom design and most of the work is done by the manufacturer. Savings is derived from the standardization, but high *nonrecurring engineering (NRE)* costs and lengthy turnaround times preclude their use in low-volume applications. The obvious advantages of using standard cells over gate arrays are greater densities (up to 10,000 or 25,000 equivalent gates for 3-μm and 2-μm geometries, respectively) and faster clock speeds. Also, some applications simply cannot be satisfied by gate arrays because of I/O limitations. Standard cell ICs can be made as large as needed (within reason, 300 mils on a side is considered quite large) to provide for designs with large numbers of I/O pads. Pad-bound designs may not fit on the predefined gate array structure. The question of when to use standard cell libraries instead of gate arrays is largely answered by density, volume, and cost requirements. The graphs in Fig. 9-17 provide a rough guide.

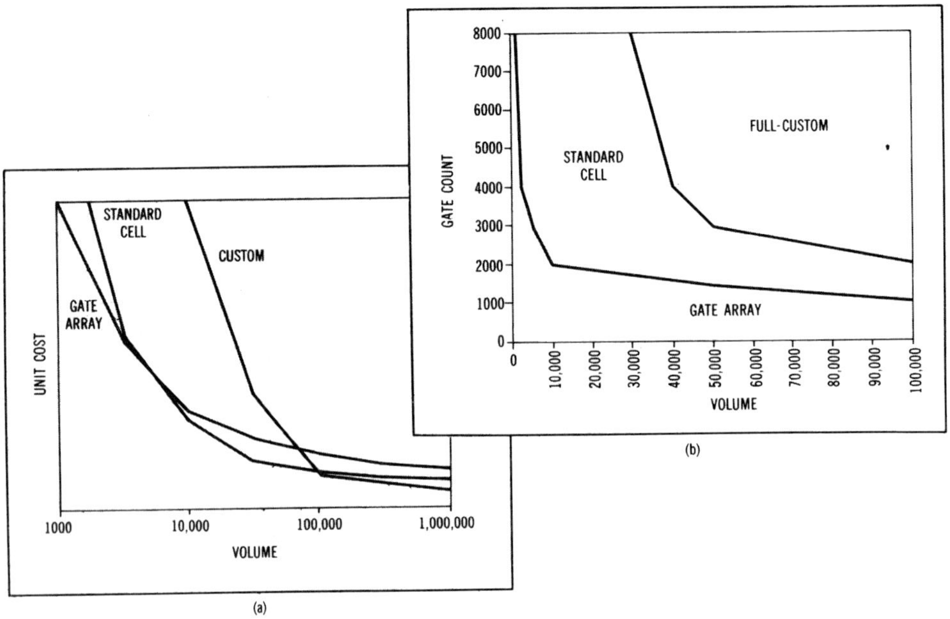

A conventional cost comparison shows that standard cell and full-custom designs have an edge at high volumes. Gate arrays are best when production volumes are low (a). Cost comparison, however, does not account for differences in chip complexity. Comparing gate count versus volume produces regions for measuring the effectiveness of each solution. Standard cells and full-custom designs are best-suited for high volume, high complexity chips and gate arrays for low volume, low complexity chips (b).

Fig. 9-17 Cost and gate count comparisons. (Reprinted with permission from *Computer Design,* March 1985 issue)

Silicon Compilation

Silicon compilation is based on techniques popularized by Carver Mead, a computer scientist at the California Institute of Technology and a founder of the company called Silicon Compilers, Inc. The basic tenet of Mead's philosophy is to use as much silicon as needed to implement your system in a PLA-based state machine because silicon is becoming cheaper every day. Traditionally, PLA-based designs were considered to be inefficient in their abundant use of silicon area. Mead maintains that the savings in design time more than compensates for this inefficiency. Moreover, he insists that with the increasing density of CMOS, the relative cost of silicon area will become insignificant when compared to labor costs. His views and techniques are elaborated in a textbook he wrote with Lynn Conway entitled *Introduction to VLSI Systems*.

A *silicon compiler* is a computer program which translates the designer's circuit requirements into customized logic cells called *bristle blocks*. These blocks fit together very closely to minimize the need for wiring channels. Full-custom and semicustom ICs often require large areas of silicon for metal routing. The bristle block design minimizes routing area and compensates for the inefficiency of PLA-based logic. The designer states his or her application in terms of state machine notation, and the compiler translates the state equations into a custom-made, microcoded PLA-type logic. Simulation and modeling are done at the workstation, and the final result appears as a set of mask layers ready to be sent to a silicon foundry chosen by the user. Tapes are made and sent to the foundry for fabrication.

The primary advantage of this approach is quick turnaround time. A key circuit in Digital Equipment Corporation's Microvax minicomputer was designed with Silicon Compiler's software in just seven months, from contract to silicon. Seeq Technology, Inc., beat Intel into the Ethernet controller market using this software even though Intel had a two-year head start. That lead helped Seeq take a large portion of the market. Silicon compilation represents the second step in IC design automation. The first step translated gate and flip-flop level logic into standard cells. The second step is a process which translates block diagram layouts into interconnected transistor circuits.

Electrically Programmable Logic Devices

Gate array, standard cell, and silicon-compiled logic are one-time programmable. There is no room for errors, cuts, or jumpers. When the design is finished and the parts return from the foundry, the designer cannot change his or her logic without incurring substantial costs. In addition, a fairly large commitment is made in terms of volume. Foundries do not start up their wafer lines for a few parts. The user must purchase thousands of parts to make the proposition viable. EPLDs, on the other hand, are prefabricated, tested, and ready to be programmed and reprogrammed. Although EPLDs do not provide the same level of integration possible in gate array, standard cell, or silicon-compiled logic, they do provide an alternative for lower-volume, medium-density products.

There are basically two types of EPLD: EPROM based and SRAM based. EPROM-based EPLD technology was developed by three small West Coast companies: Lattice Semiconductor Corp., Altera Corp., and Exel Microelectronics. As the concept grew, more interest was taken by the larger electronics companies, and Intel joined forces with Altera to manufacture and market their product. Altera uses a CMOS EPROM-based programmable logic array to implement the sum-of-products expressions. Gated logic functions are made by selectively programming EPROM cells within the AND portion of the PLA. The largest device, the EP1800, is capable of implementing functions requiring up to 2100 equivalent gates. Products from Lattice and Exel are similar. All use the two-transistor FAMOS EPROM cell. The packages come with quartz windows for UV light erasure and without windows for production use. The quartz window adds significantly to the cost of each part, and production parts presumably would not need to be reprogrammed.

Figure 9-18 shows a block diagram of the Altera EP1800. It is partitioned into four identical quadrants each containing 12 macrocells. The basic

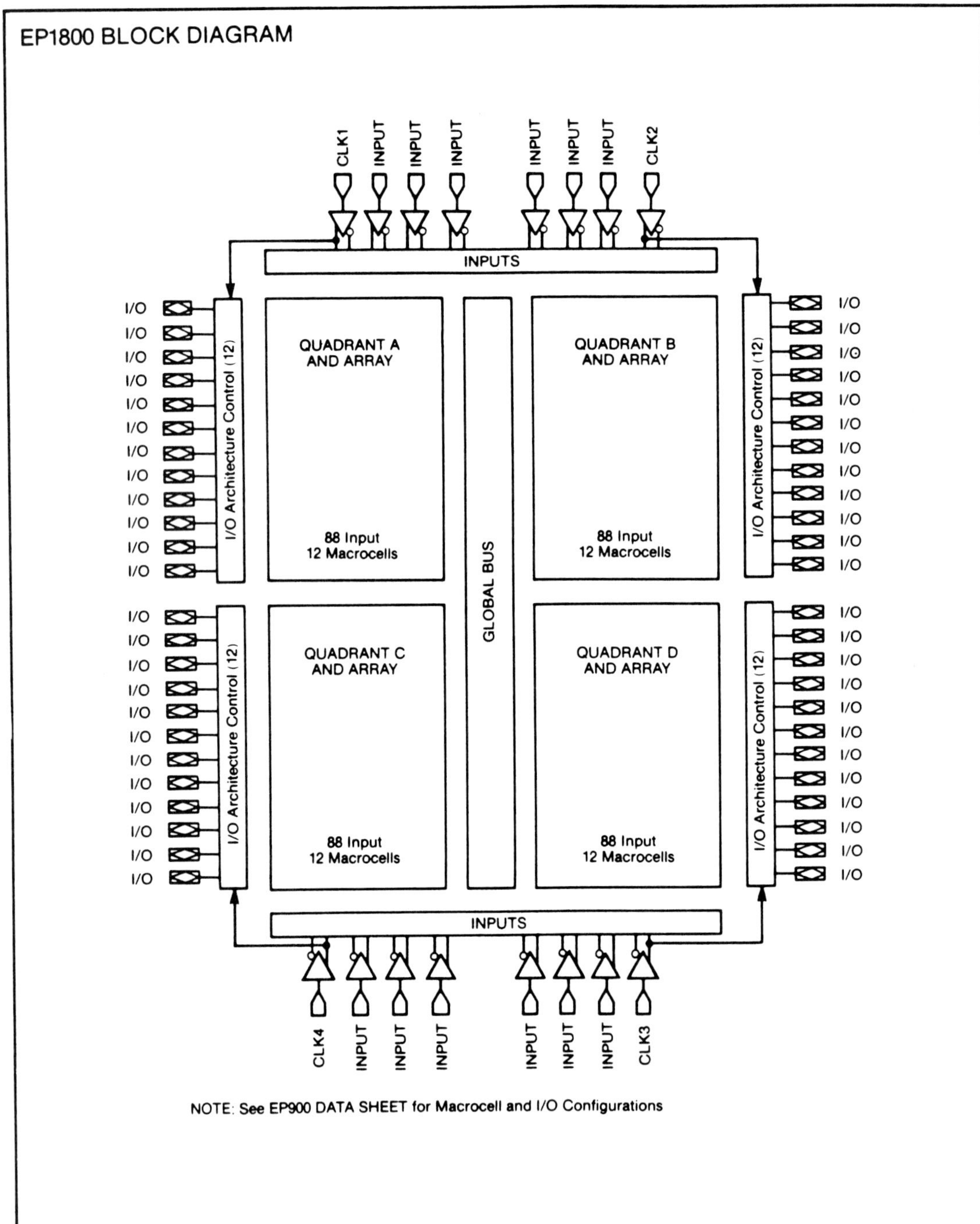

Fig. 9-18 Block diagram of Altera EP1800. (*Altera Corporation, Santa Clara, Calif.*)

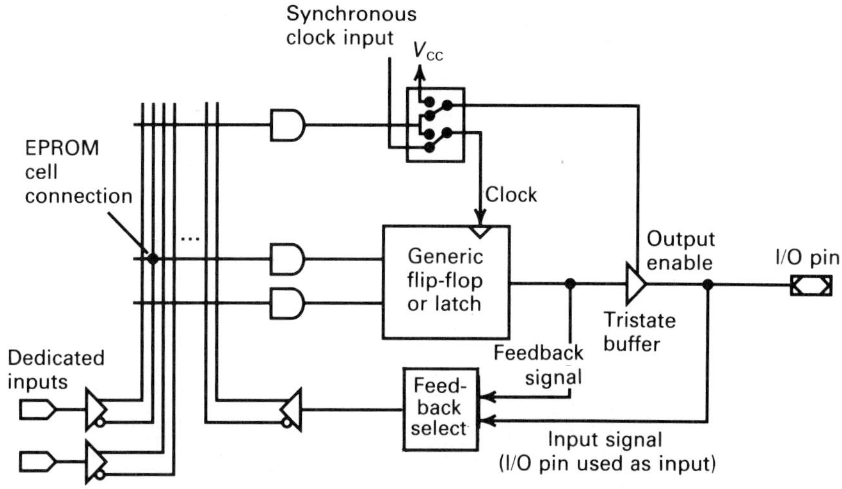

Fig. 9-19 Basic Altera macrocell structure.

macrocell structure is shown in Fig. 9-19. Inputs to the macrocells come from the local and global bus structures. The two buses comprise an 88-input AND array for each quadrant. Each quadrant may operate with a separate clock. The generic memory device in the macrocell can be configured as a D, T, JK, or SR flip-flop with sychronous or asychronous clocking. Altera provides software which runs on IBM and compatible personal computers. The logic may be entered as schematics, boolean equations, or state tables. Once the design has been entered, the software translates the design into an industry-standard (JEDEC) PLD programming file. Device pin assignments and logic minimization can be done automatically to optimize the utilization of the macrocell logic. The JEDEC output file can then be read into the programming hardware supplied by Altera.

RAM-based logic arrays were developed by Xilinx, Inc. Xilinx has trademarked the term *logic cell array* (*LCA*) to describe this type of logic. The LCA consists of a matrix of *configurable logic blocks* (*CLBs*) which are arranged in rows and columns as shown in Fig. 9-20. Interleaved between the CLBs are user-definable interconnections. Within each CLB, static CMOS RAM cells are used to directly implement combinational logic functions. Arranged around the outside perimeter are I/O blocks which contain storage elements. The device closely resembles a gate array in its organization but may be programmed in-circuit with write operations to its RAM cells.

Each CLB consists of a D flip-flop with a 16-bit SRAM look-up table to implement logic functions as shown in Fig. 9-21. Each 16-bit memory represents the implementation of a four-variable Karnaugh map. The inputs to this "function memory" are general purpose logic inputs labeled *A*, *B*, *C*, and *D*. The function memory can form any combinational logic function in four variables. External logic signals interface to the LCA through the I/O blocks. The I/O blocks may be programmed as input, output, or

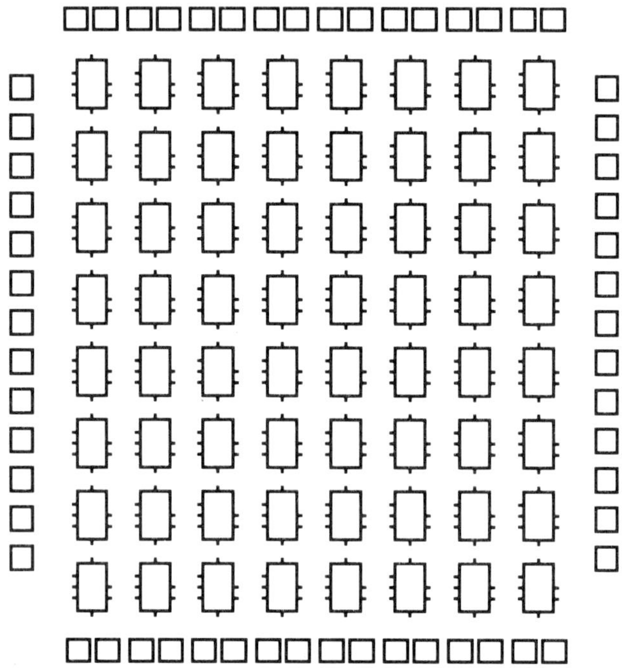

Fig. 9-20 Xilinx logic cell array (LCA) consisting of rows and columns of configurable logic blocks (CLBs).

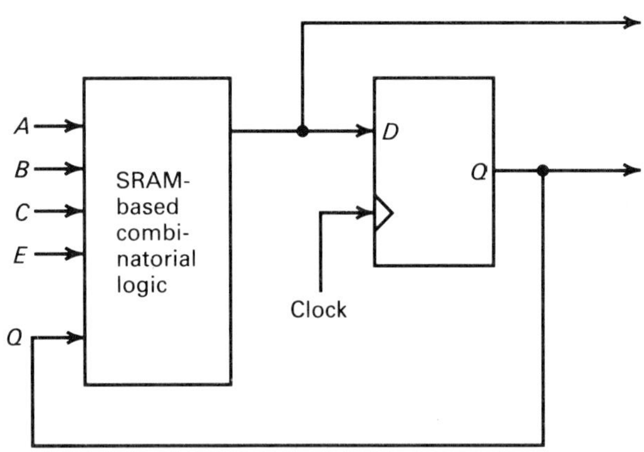

Fig. 9-21 Xilinx configurable logic block (CLB).

bidirectional. Since the logic is RAM based, it has no function when powered up. The RAM cells must be configured, or initialized, by an external PROM, EPROM, or floppy disk. The Xilinx XC-2018 logic cell array provides up to 1800 gate equivalents. Xilinx offers software packages similar in function and performance to those offered by Altera.

for bit 2 and bit 4 will allow five unit currents to be mirrored across R to produce 1.5 V on the output. The full-scale voltage is V_{max} + one unit step, or 4.8 V.

The N-channel device implementation shown in Fig. 10-2 is convenient for VLSI chips because the n-type devices are small and the DAC circuitry will not occupy a large portion of the die area. Also, the N transistors can be packed densely to improve device matching. If all devices could occupy the same exact area on the die, they would be perfectly matched. The closer they are spaced, the more closely they approach perfect matching because the effect of variations across the wafer are minimized. More closely matched transistors mean more accuracy in the DAC. DACs of this type are found in special-purpose microprocessors and in VLSI audio-processing chips.

Resistor Implementation

Resistors may also be used to measure out current in DACs, although this is seldom practical in VLSI. Resistors are quite large and are generally used only when the DAC can occupy a large part of the die area. Single-purpose DAC chips use silicon-metal combinations such as silicon chromium. When large die areas are available, highly stable silicon chromium thin film resistors can be used with great accuracy. When die area is at a premium, diffused resistors are used. Diffused resistors are formed in the same way as transistor diffusions and require no additional processing steps. However, the matching and temperature characteristics of diffused resistors are inferior to thin film resistors.

Most single-purpose monolithic, hybrid, and modular DAC chips use a weighted resistor ladder design similar to that shown in Fig. 10-3. This

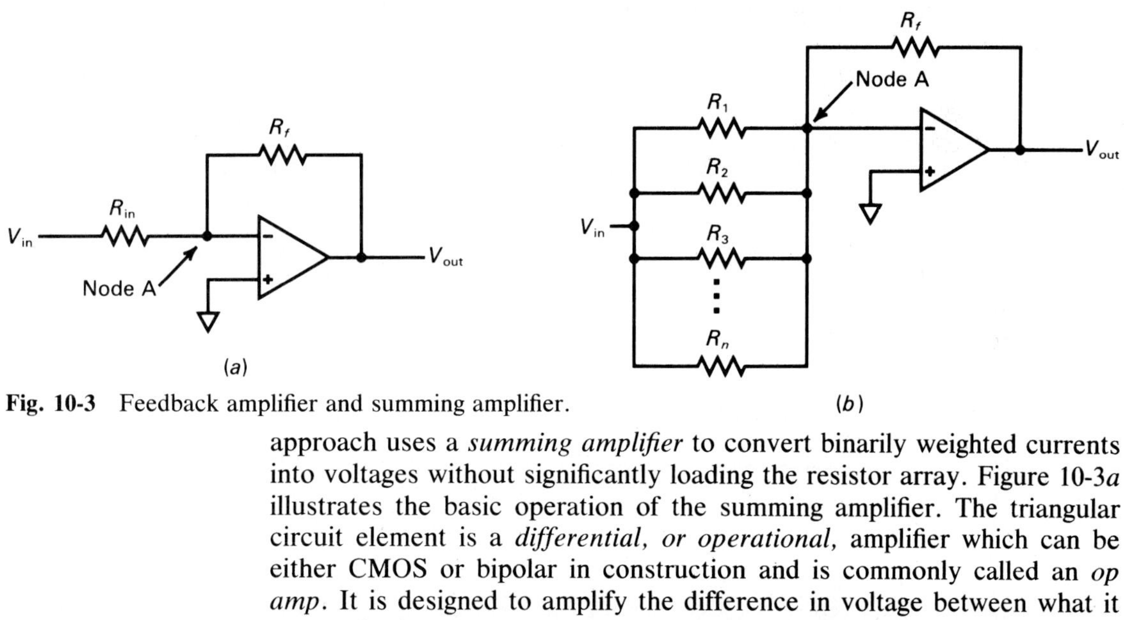

Fig. 10-3 Feedback amplifier and summing amplifier.

approach uses a *summing amplifier* to convert binarily weighted currents into voltages without significantly loading the resistor array. Figure 10-3*a* illustrates the basic operation of the summing amplifier. The triangular circuit element is a *differential, or operational,* amplifier which can be either CMOS or bipolar in construction and is commonly called an *op amp*. It is designed to amplify the difference in voltage between what it sees at its input terminals. When input and feedback resistors R_{in} and R_f are used as shown, the amplifier will produce whatever positive or negative current is needed to maintain a zero voltage difference between the input terminals.

The analysis of this circuit will assume that the positive input terminal is always connected to ground. The key to understanding this type of circuit is to keep track of the currents flowing through the resistors. This is easy because the op amp inputs do not sink any measurable amount of current. If no current is entering node A through R_{in}, then no current will be required from the amplifier output to maintain a zero voltage difference between the inputs. Both inputs and the output will rest at 0 V, and no current will flow through R_f. If V_{in} is raised above ground, current will flow into node A through R_{in}, and the amplifier will sink an equal amount of current to maintain a zero voltage difference between inputs. The amplifier does this by providing an opposite polarity voltage at its output sufficient to produce an equal amount of current *out of* node A through R_f. If the resistors R_{in} and R_f are equal in value, then equal and opposite voltages at the input node and at V_{out} will produce equal and opposite currents. If R_{in} is twice as large as R_f, then only half as much voltage is needed at V_{out} to match the currents.

Since the output voltage is referenced to ground and the negative input is also maintained at ground, V_{out} will equal the product of current and resistance as follows:

$$V_{out} = I_f R_f \qquad (10\text{-}1)$$

And the feedback current is equal and opposite to the input current, so that:

$$I_f = -I_{in} = \left(\frac{-V_{in}}{R_{in}}\right) \qquad (10\text{-}2)$$

Combining (10-1) and (10-2) gives the following relationship:

$$V_{out} = -\left(\frac{V_{in}}{R_{in}}\right)R_f = -\left(\frac{R_f}{R_{in}}\right)V_{in} \qquad (10\text{-}3)$$

Taking this a step further, if multiple input resistors were used, as in Fig. 10-3*b*, then each input current contributes to the total current that the op-amp must source or sink, and:

$$V_{out} = -R_f\left(\frac{1}{R_1} + \frac{1}{R_2} + \frac{1}{R_3} + \cdots + \frac{1}{R_n}\right)V_{in} \qquad (10\text{-}4)$$

The amplifier is designed so that its output voltage will always be large enough to produce a current through R_f which matches the sum of currents produced at the inputs. Node A is called the *summing junction,* and the configuration is called a *summing amplifier*.

In Fig. 10-4 an array of binarily weighted resistors is used with CMOS or bipolar current switches to produce a binarily weighted current at the summing junction. The switches are turned on or off with high or low voltages from logic outputs. The total current is matched by the current through R_f. The output of the amplifier not only provides the current needed to match the input currents but also provides the unbuffered voltage output of the DAC. In this example, R_f equals one-half the value of the smallest input resistor, and V_{in} is renamed V_{ref}. V_{ref} is the full-scale voltage for this circuit and must be less than the supply voltage. The analog output voltage will be a function of the binary switch inputs in the following way:

$$V_{out} = -V_{ref}\left(B_1 \times \frac{1}{2} + B_2 \times \frac{1}{4} + B_3 \times \frac{1}{8} + B_4 \times \frac{1}{16}\right) \qquad (10\text{-}5)$$

EXAMPLE 10-3

Assuming that the ground in Fig. 10-4 is 0 V and V_{ref} equals 3.2 V, tabulate the analog output voltage versus binary input code. What is V_{max}? What is the full-scale voltage?

SOLUTION

Only the first step needs to be computed; the others are incremental multiples of the unit step. Notice that this DAC, like the previous example, can never quite reach full scale. V_{max} is -3.0 V, and full scale is -3.2 V.

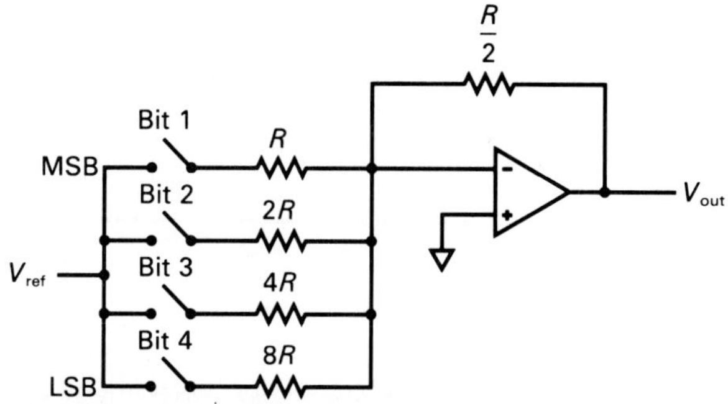

Fig. 10-4 DAC constructed with binarily weighted resistors and summing amplifier.

Binary Input B_1	B_2	B_3	B_4	Analog Output V_{out}, volts
0	0	0	0	0.0
0	0	0	1	−0.2
0	0	1	0	−0.4
0	0	1	1	−0.6
0	1	0	0	−0.8
0	1	0	1	−1.0
0	1	1	0	−1.2
0	1	1	1	−1.4
1	0	0	0	−1.6
1	0	0	1	−1.8
1	0	1	0	−2.0
1	0	1	1	−2.2
1	1	0	0	−2.4
1	1	0	1	−2.6
1	1	1	0	−2.8
1	1	1	1	−3.0

One problem in implementing higher-resolution DACs is that a wide range of resistor values are needed. An 8-bit DAC would require a range from R to $128R$. This variance can cause problems in linearity, temperature stability, and switching speed. To overcome these problems, weighted resistor arrays are used in indentical groups as shown in Fig. 10-5. The output of the lower order group is divided down by a *scaling resistor*. In Fig. 10-5*a* the scaling resistor divides the lower-order group by 16 to implement an 8-bit binary-coded DAC. Figure 10-5*b* shows how the same weighted resistor quads can be used to implement an 8-bit BCD-coded DAC by simply changing the value of the scaling resistor. It is convenient to use BCD codes when the input is received directly from a front panel. For example, when DACs are used in digitally programmed power supplies and digital motor speed controls, they are set by digit switches or front panel keypads.

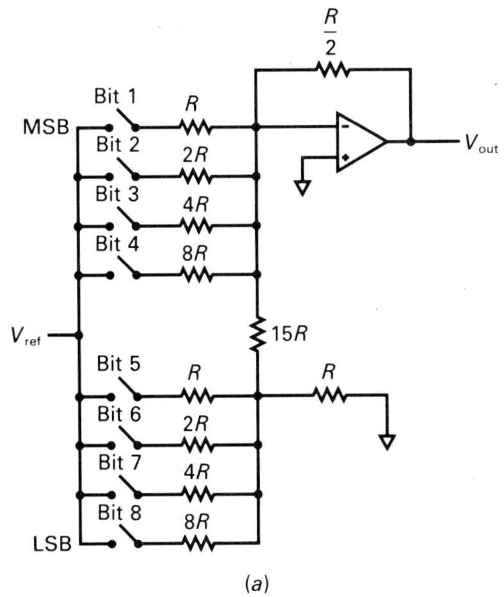

Fig. 10-5(*a*) Binarily weighted 8-bit DAC. (*b*) BCD-weighted, 8-bit DAC.

Another technique used to reduce the range of resistor values is the R-$2R$ ladder method. Figure 10-6 shows a resistor network consisting of only two resistor values, R and $2R$. The operation of the R-$2R$ network is based on the binary division of current. As current flows in from V_{ref}, it encounters a current divider (discussed in Chapter 1) consisting of equal value resistors. The shunt resistance is always $2R$, and it can be shown that the resistance to the right of any node is also equivalent to $2R$. At the input node, the current splits into two equal parts because it sees equal resistance values to ground in either direction. At the next node, the same equal current division occurs again because, again, the shunt resistance is $2R$ and the equivalent resistance of the remainder of the network is also $2R$. The remaining one-half of the input current splits in two halves. Current flowing down the ladder continues to divide into equal parts until

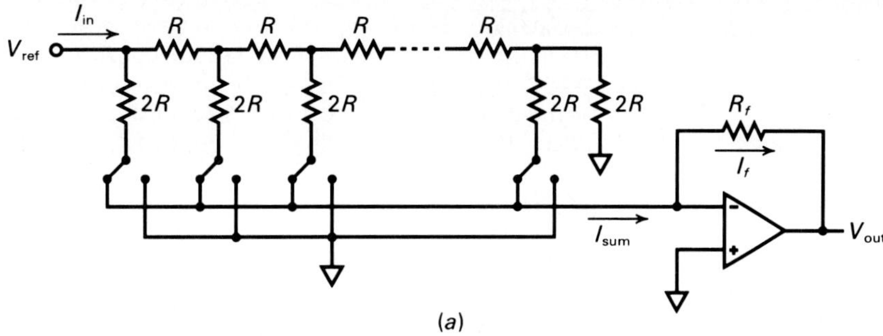

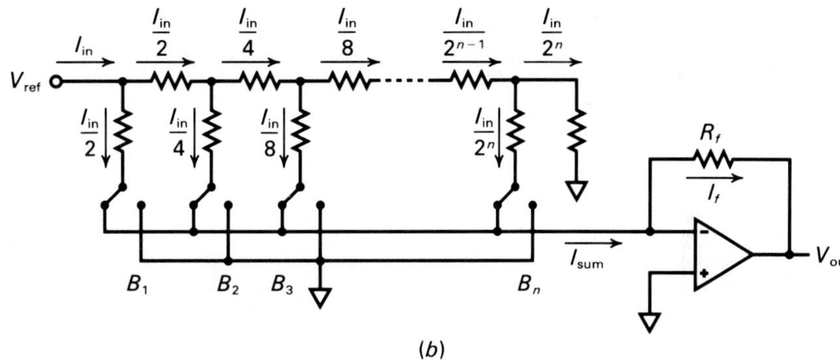

Fig. 10-6 *R*-2*R* ladder method of D/A conversion.

the final node is reached. The result is that each shunt resistor receives one half of the current received by the previous shunt resistor. Figure 10-6*b* shows how the currents split. The binarily weighted currents are switched either to ground or to the summing junction (also held at ground) so that a constant current is maintained in each ladder leg regardless of the switch positions. Since V_{ref} always sees an equivalent resistance of R Ω to ground, $I_{in} = V_{ref}/R$. As with the previous circuit, the summing amplifier converts the binarily weighted currents to an output voltage. When only 4 bits are used, the total current directed to the summing junction is given by:

$$I_{sum} = \frac{V_{ref}}{R}\left(B_1 \times \frac{1}{2} + B_2 \times \frac{1}{4} + B_3 \times \frac{1}{8} + B_4 \times \frac{1}{16}\right) \tag{10-6}$$

This converts to a voltage when I_{sum} is matched by I_f through the feedback resistor. The output voltage is the product of feedback current and feedback resistance as given below:

$$V_{out} = I_f R_f = -I_{sum} R_f \tag{10-7}$$

$$= -V_{ref}\left(B_1 \times \frac{1}{2} + B_2 \times \frac{1}{4} + B_3 \times \frac{1}{8} + B_4 \times \frac{1}{16}\right)$$

when $R_f = R$.

EXAMPLE 10-4

Assuming that V_{ref} equals 3.2 V in Fig. 10-6, tabulate the analog output voltage versus binary input code for a 4-bit version of the *R*-2*R* ladder DAC. What is V_{max}? What is the full-scale voltage?

SOLUTION

The results will be identical to those in Example 10-3.

DAC Specifications

Figure 10-1 presented an *ideal* transfer characteristic for a 4-bit DAC. The 4 bits produced 15 different output voltage steps, or quantums, between 0 and V_{max}. If 5 bits had been used, V_{max} would have been divided into 31 equal voltage steps. Figure 10-7 shows how more steps produce a finer resolution in the analog output. The term *resolution* is not equivalent to accuracy. *Resolution* refers to the step size, while *accuracy* refers to the absence of errors in the characteristic. The ramp waveform in Fig. 10-7 is more closely approximated by the 5-bit DAC than by the 4-bit DAC, but if the 5-bit DAC characteristic contained errors, its accuracy could be worse. More bits produce greater resolution and more control over the output voltage. Because the resolution of the output depends only on the number of input bits, resolution is specified in bits. Practical D/A converters range between 8- and 16-bit resolution. The ideal transfer characteristic of a 12-bit DAC would consist of 4096 unit voltage steps, or quantums.

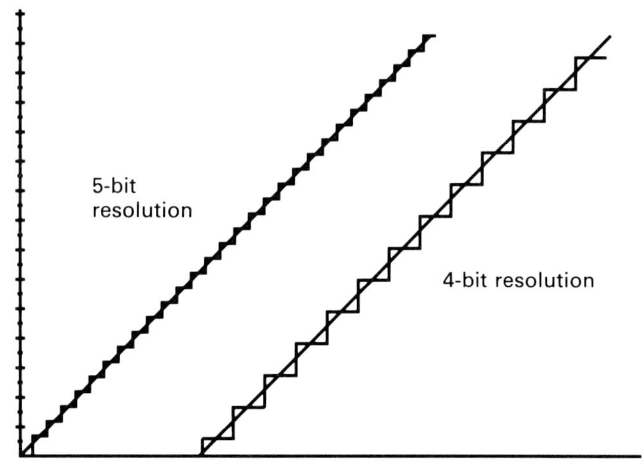

Fig. 10-7 Comparison of voltage output with 4- and 5-bit resolutions.

In an ideal DAC, the step size is equal everywhere along the transfer characteristic and starts at 0 V. Real DACs, however, exhibit departures from this ideal in the form of *offset, gain,* and *linearity* errors. Figure 10-8*a* shows a nonideal transfer characteristic which suffers from offset error. *Offset error* is defined as the voltage (in millivolts or percent of full scale) by which the transfer characteristic fails to pass through 0. Figure 10-8*b* illustrates gain error. *Gain error* is defined as the difference between the ideal V_{max} and the actual maximum output voltage of the DAC. Gain

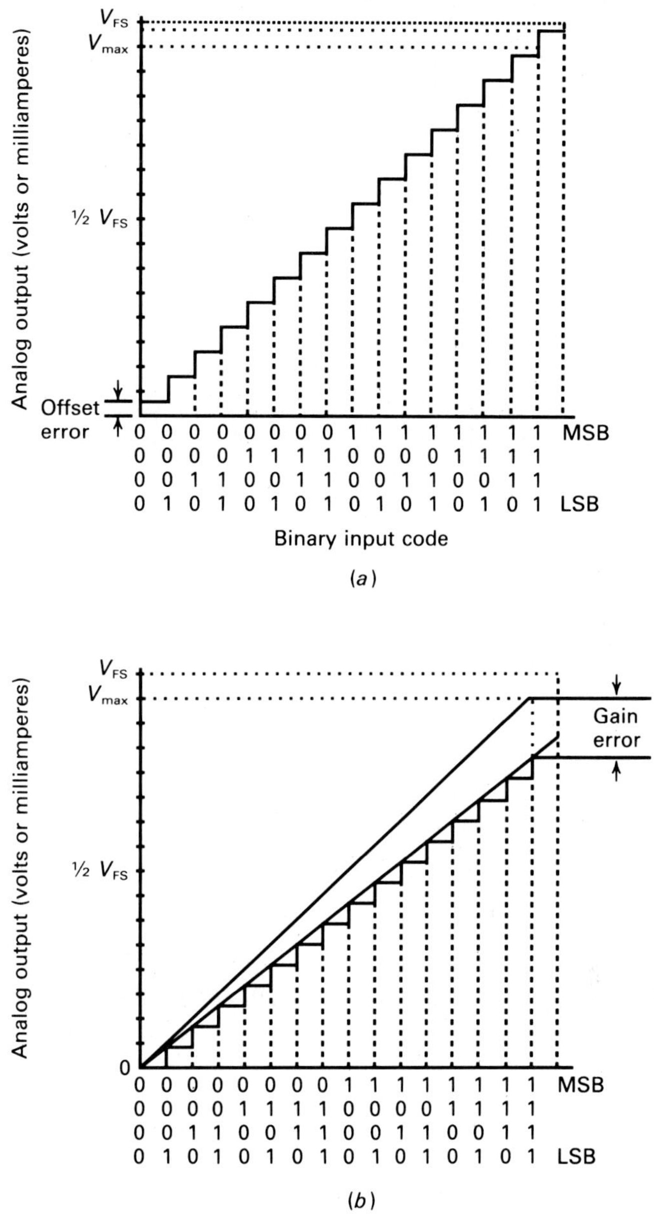

Fig. 10-8 The effects of gain and offset errors on the 4-bit DAC transfer characteristic.

error is caused by an incorrect, but uniform, step size resulting in a transfer characteristic which either undershoots or overshoots V_{max} at the maximum binary input. Gain error may be specified in percent of nominal full-scale value. The exaggerated gain error in Fig. 10-8*b* is about 15 percent. However, most DACs have provisions for trimming out gain errors, and so these errors are typically specified in terms of change per degree centigrade. Typical specifications for DAC gain errors are 5 to 10 ppm/°C.

The abbreviation ppm refers to *parts per million* of full-scale range. A gain temperature coefficient of 10 ppm/°C would be equivalent to a 0.001 percentage change in gain per degree centigrade. If the gain were adjusted to 0 at 25°C (room temperature) and the temperature rose in the equipment rack to 75°C, a 10-ppm/°C gain temperature coefficient could result in a ±0.05 percent error. In a 10-bit DAC, this would be an error of ±½ the unit step size.

Linearity errors are more serious because they cannot be reduced by adjustment. There are two types of linearity error in DACs: integral and differential. *Integral linearity errors* result in a curvature of the transfer characteristic as shown in Fig. 10-9. This type of error is specified in terms

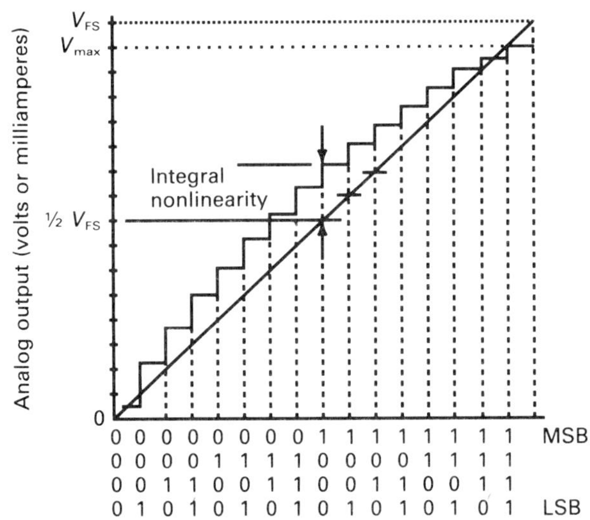

Fig. 10-9 Integral linearity error.

of the maximum deviation of the transfer characteristic from a straight line drawn between 0 and full scale. Manufacturers' specifications use the term LSB, or least significant bit, to indicate a deviation of one unit step size or quantum. The linearity error in Fig. 10-9 is exaggerated and equal to more than 2 LSBs. *Differential linearity error* is the deviation of *any one step* from the ideal. It is measured as the difference, in LSBs, between any step size and the ideal value of $V_{FS}/2^n$. If a DAC has a differential linearity error of ±½ LSB, then the actual size of any step may be between ½ and 3/2 of the nominal value. Most DACs are specified in terms of both integral and differential nonlinearity. The term *relative accuracy* refers to total error with gain and offset errors adjusted to 0. Relative accuracy is a function of both types of linearity error and is also specified in LSBs.

Another term applied to DACs is *monotonicity*. A monotonic DAC provides a voltage output which is a continuously increasing function of its input code. Figure 10-10 shows the transfer functions of two DACs with differential nonlinearity. One has differential linearity errors but is monotonic. The other is nonmonotonic because its voltage output actually

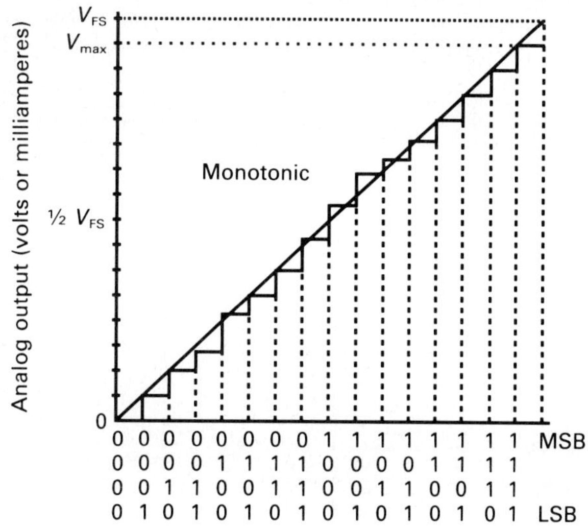

(a)

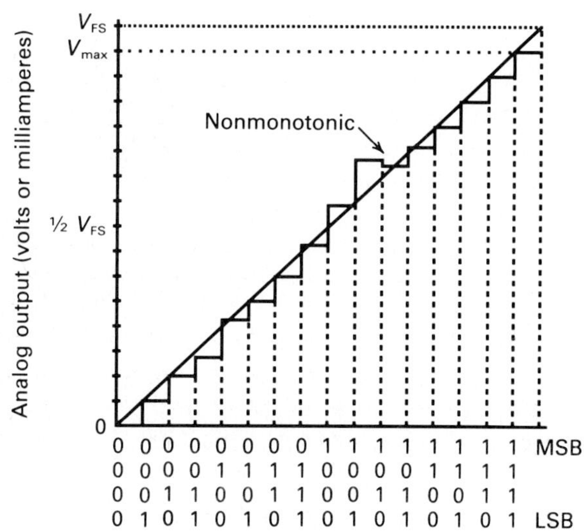

(b)

Fig. 10-10 Two forms of differential nonlinearity in DACs, monotonic and nonmonotonic.

decreases at one point. A DAC may become nonmonotonic if its differential linearity error exceeds ±1 LSB.

EXAMPLE 10-5

The 4-bit DAC in Fig. 10-4 produces the analog output values shown in Fig. 10-11*a*. Use those values to determine if the DAC exhibits gain or offset errors. Specify any offset error in terms of millivolts and gain error in terms of percent of V_{max}. $V_{FS} = -V_{ref} = +4.00\ V$.

Binary input				Analog output V_{out}, volts
B_1	B_2	B_3	B_4	
0	0	0	0	0.1
0	0	0	1	0.4
0	0	1	0	0.7
0	0	1	1	1.0
0	1	0	0	1.3
0	1	0	1	1.6
0	1	1	0	1.9
0	1	1	1	2.2
1	0	0	0	2.5
1	0	0	1	2.8
1	0	1	0	3.1
1	0	1	1	3.4
1	1	0	0	3.7
1	1	0	1	4.0
1	1	1	0	4.3
1	1	1	1	4.6

(*a*)

Binary input				Analog output V_{out}, volts
B_1	B_2	B_3	B_4	
0	0	0	0	0.00
0	0	0	1	0.26
0	0	1	0	0.60
0	0	1	1	0.88
0	1	0	0	1.18
0	1	0	1	1.50
0	1	1	0	1.75
0	1	1	1	2.00
1	0	0	0	2.25
1	0	0	1	2.50
1	0	1	0	2.75
1	0	1	1	2.85
1	1	0	0	3.13
1	1	0	1	3.35
1	1	1	0	3.55
1	1	1	1	3.75

(*b*)

$V_{ref} = -4.0$ V

Fig. 10-11 Tables for Examples 10-5 and 10-6.

SOLUTION

The first table entry tells us that the offset error is 100 mV. Beyond that, each step is 0.3 V. Since full-scale voltage is equal to $-V_{ref}$, or $+4.00$ V, the unit step is $+4.00/2^4 = 0.25$ V. V_{max} is one quantum below FS and equals 3.75 V. In order to extract the gain error, the 0.1-V offset is subtracted leaving a 4.5-V nominal full scale. The difference 0.75 V between the ideal and actual V_{max} values represents 0.75/3.75 x 100 percent, or 20 percent gain error.

EXAMPLE 10-6

Determine what, if any, linearity error exists in the 4-bit DAC output tabulated in Fig. 10-11*b*. Distinguish between integral and differential nonlinearity. Is the DAC monotonic? Again, $V_{FS} = -V_{ref} = 4.00$ V.

SOLUTION

The maximum deviation of actual values from ideal values occurs in the middle table entries and is equal to 0.25 V, or +1 LSB. This is the integral linearity error. The differential linearity errors are greatest at the low end of the table. The maximum occurs between codes 0001 and 0010 where a 0.34-V jump occurs. The ideal unit step is 0.25 V, leaving a 0.09-V difference which is equivalent to 0.36 LSB, or greater than ⅓ LSB. The DAC is monotonic.

DAC Applications

While DACs are most often used in conjunction with ADCs, they do have stand-alone applications. DACs are combined with power amplifiers for use in applications which require precise, digital control of voltage outputs. For example, the output voltage of a digital power supply can be precisely selected with an 8-bit power DAC connected to the front panel switches. DAC digital control may be initiated by either a human operator or by a computer.

Power DACs are often used in precision motor control systems. The Intersil AD7523 is an example of a low-cost, 8-bit DAC which may be used in this application. The AD7523 comes in a 16-pin DIP and uses thin-film resistors to provide a high degree of accuracy. Figure 10-12 shows a

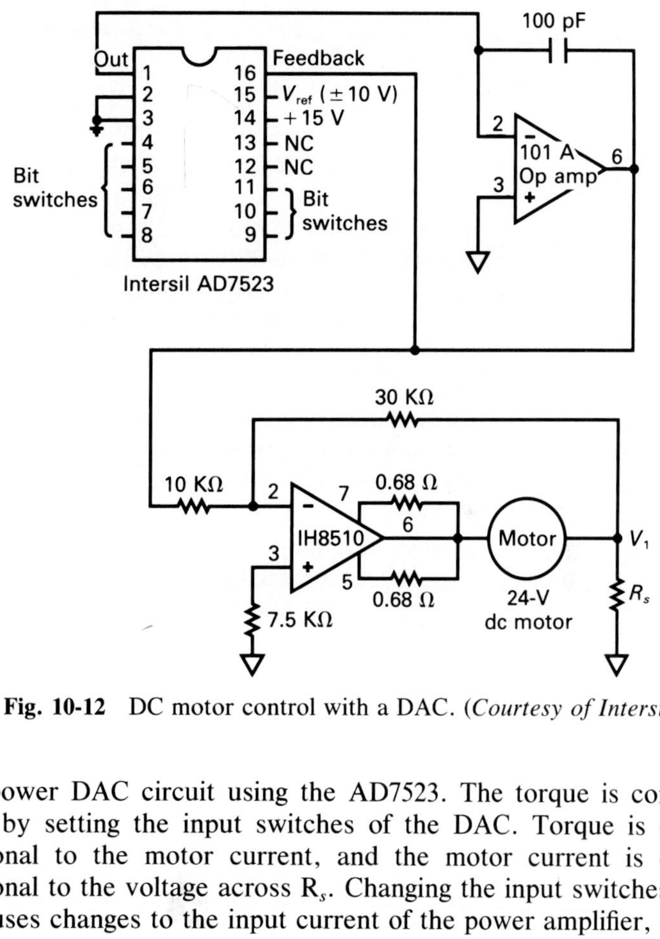

Fig. 10-12 DC motor control with a DAC. (*Courtesy of Intersil Inc.*)

typical power DAC circuit using the AD7523. The torque is controlled digitally by setting the input switches of the DAC. Torque is directly proportional to the motor current, and the motor current is directly proportional to the voltage across R_s. Changing the input switches of the DAC causes changes to the input current of the power amplifier, and the power amplifier will produce whatever voltage is needed at V_1 to match the input current with feedback current through the 30 KΩ resistor.

Another application is the function generator shown in Fig. 10-13. A *function generator* is a system which produces shaped waveforms. Typical outputs are sine waves, triangle waves, sawtooth waves, and square waves. This example uses a ROM look-up table to convert counter outputs into

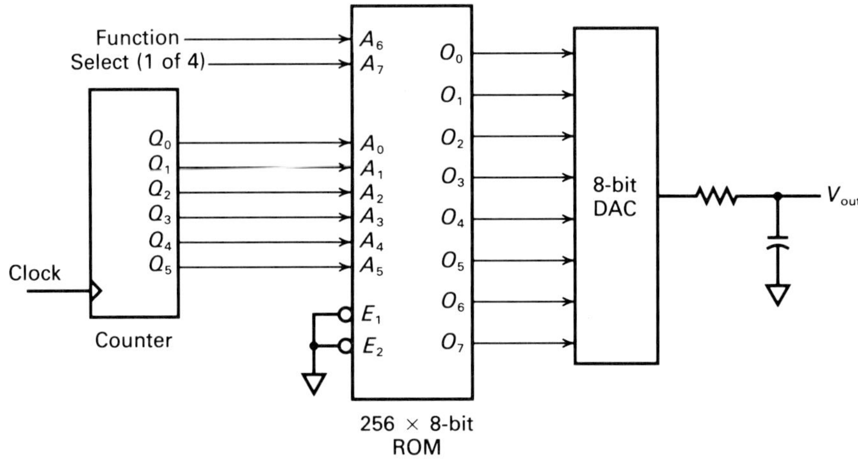

Fig. 10-13 Four-function signal generator using a DAC.

binary numbers which correspond to waveform values. The upper two address inputs determine which quarter of memory is accessed, and the lower six address inputs allow the counter to sequence the 64 locations in a given quarter of memory. A sine wave may be divided into 64 intervals, and its relative amplitude measured at each division. The same is done for three other waveforms, and the results are programmed into the ROM. The number sequence drives an 8-bit DAC which produces a stair-step output approximating the desired waveform. The resistor-capacitor combination is a low pass filter which smooths out the steps. When the voltage steps change, they change quickly. However, the capacitor requires time to charge and discharge, and the charging and discharging curves provide a smoothing effect to the waveform. Similar systems are used in speech synthesis ICs. But speech waveform values are not stored directly in ROM as described above since this would require excessive amounts of memory. Instead, compressed coefficients of speech patterns are stored in the ROM, and an ALU is used to calculate the binary numbers which drive the DAC.

REVIEW 10-1

1. **Q.** Define full-scale voltage and nominal full-scale voltage. **A.** Full-scale voltage is the quantum step voltage × 2^n. It depends only on the reference voltage and not on the number of steps. Nominal full-scale voltage is always one unit voltage step below full scale and represents the maximum voltage output attainable by a DAC. Its value depends on the unit step size. As the steps become smaller the nominal full scale voltage more closely approximates full scale.

2. **Q.** What is the unit step size of an n-bit DAC? **A.** The unit step size is $V_{FS}/2^n$, or $V_{max}/(2^n - 1)$.

3. **Q.** What do transistor array and resistor network DACs have in common? **A.** They both produce binarily weighted currents which are converted into voltages.

4. **Q.** How does an 8-bit BCD input DAC differ from an 8-bit binary input DAC? **A.** The current produced by the lower significant four switches of a binary DAC is divided by 16. The current produced by the lower significant four switches of a BCD DAC is divided by 10.

5. **Q.** What is the key concept behind the *R*-2*R* ladder? **A.** At each node in the ladder, the remaining current is split in half.

6. **Q.** What is DAC resolution and how is it specified? **A.** *Resolution* refers to the size of the unit voltage relative to full scale. It could be expressed as a percentage, but it is commonly specified in terms of input code *bits*.

7. **Q.** What is offset error? **A.** Offset error is the voltage, or percent of full scale, by which the transfer characteristic fails to pass through 0.

8. **Q.** What is gain error? **A.** Gain error is the difference between the ideal V_{max} and the actual maximum output voltage of the DAC.

9. **Q.** Explain the difference between integral and differential linearity errors. **A.** *Integral nonlinearity* refers to a curvature in the transfer characteristic. It is specified in terms of the maximum deviation from the linear ideal characteristic. *Differential nonlinearity* is the maximum deviation of any one step from the ideal step size.

10. **Q.** Define monotonicity. **A.** *Monotonicity* refers to continuously increasing output with increasing input.

10-2 ANALOG-TO-DIGITAL CONVERSION

An *A/D converter,* or *ADC,* is a circuit which converts an analog signal into an *n*-bit binary or BCD-equivalent representation. If the analog voltage, V_a is in the range $0 \leq V_a \leq V_{max}$, then each step in the binary number sequence represents $V_{max}/(2^n - 1)$. The transfer characteristic of a 4-bit ADC looks very similar to that of a 4-bit DAC. Figure 10-14 shows that

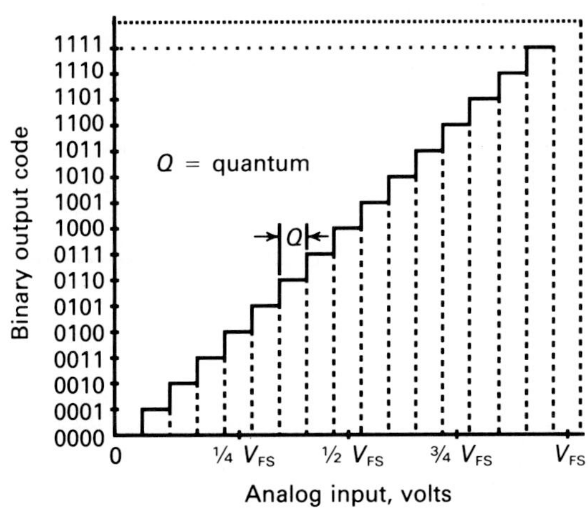

Fig. 10-14 Transfer characteristic of an ideal 4-bit ADC.

each quantum change in input results in a step or quantum change in the output. The difference between this characteristic and those presented earlier is that the analog values are *inputs* with voltage steps, or quantums, measured along the horizontal axis. Deviations from the ideal transfer characteristic exist for ADCs as well as for DACs. Offset, gain, and linearity errors are measured along the analog axis just as before, except that now the analog axis is horizontal. In ADCs, both offset and gain errors are adjustable to 0. Typical values for temperature coefficients are ± 10 to 20 ppm (of full-scale reading/°C) for offset and ± 5 to 20 ppm (of full-scale reading/°C) for gain.

Typical applications for ADCs involve converting analog signals into numbers, processing those numbers, and subsequently converting the numbers back into analog waveforms with DACs. Figure 10-15 shows one such application involving telephone service. The human voice is captured with a microphone, sampled, and quantitized into a stream of binary numbers. Those numbers are routed along with other data through a digital switching network which has made a connection with another digital switch over telephone lines designed to handle data. The receiving network routes the data to the local destination through a DAC which converts the numbers back into an analog voice signal. The box labeled S/H is a sample-and-hold circuit which holds a voice voltage level long enough for conversion. The box labeled LPF is a simple low pass filter which smooths out the steps. The conversion process is called *pulse coded modulation (PCM)*. PCM has the advantage of reducing noise in long-and medium-distance telephone links. Since only 1s and 0s are sent, any noise not large enough to change 1s into 0s (or vice versa) will not be detectable at the receiving end.

However, the process of voice-to-data-to-voice conversion does introduce some noise of its own. Even ideal A/D and D/A converters have an irreducible error called *quantization uncertainty* or *quantization noise*. Since data converters cannot distinguish an analog difference of less than one half of one unit step, their output at any point may be in error by as much as ½ LSB. Quantization occurs when the continuous analog waveform is sampled at discrete moments in time and held as discrete voltage levels. The discrete voltage levels are then assigned the quantum voltage value which they are closest to. Quantization error occurs because between two quantum steps there is an infinite number of voltage levels, and all those possible levels become one of two values. *Quantization error* is defined as the difference between the analog voltage input and the subsequent analog voltage output in an *ideal* data conversion system. *Ideal* means that there are no other sources of error. Quantization error can be reduced by using converters with greater resolution, but it can never be eliminated.

Over the long term, the *average value* of quantization error is 0, but it nevertheless distorts the received output. In order to quantify the average value of this distortion, the *root-mean-square (rms)* error voltage is used. Root-mean-square voltage is the *effective* average voltage and is defined as the constant positive voltage which would produce the same amount of

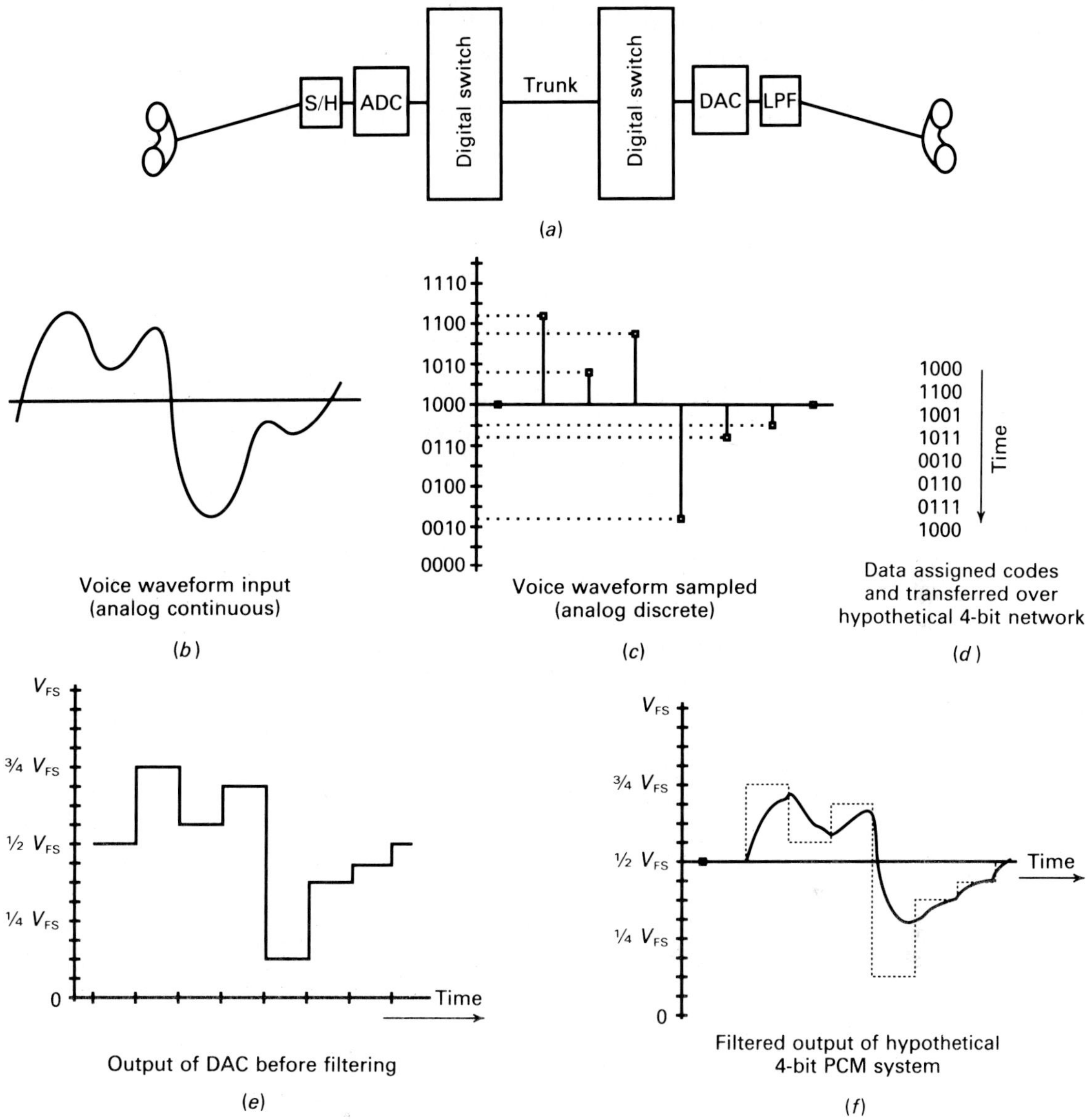

Fig. 10-15 (*a*) Voice processing and switching application using A/D and D/A converters. (*b* through *f*) Hypothetical 4-bit PCM system.

heat in a resistor as the alternating positive and negative error voltage. Basic electronic courses show that the rms voltage for a sine wave is 0.707 V_{peak}. The quantization error function looks like a sawtooth wave when the input is a ramp voltage. It can be shown that it is equal to:

$$\text{Error voltage (rms)} = \frac{Q}{\sqrt{12}} \tag{10-8}$$

where Q is the quantum step size. The rms voltage provides a measure of actual degradation and allows us to specify an important parameter of computerized signal processing systems, signal-to-noise ratio.

The *signal-to-noise ratio* (S/N) of any system is the ratio of peak-to-peak (p-p) signal over rms noise. In a data conversion system the peak-to-peak signal is equal to the number of steps multiplied by the step size, or $2^n \times Q$. The signal-to-noise ratio, then, of an n-bit conversion system is:

$$\frac{S}{N} = \frac{p\text{-}p \text{ signal}}{\text{rms noise}} = \frac{2^n Q}{\frac{Q}{\sqrt{12}}} = 2^n(\sqrt{12}) \qquad (10\text{-}9)$$

As you may have discovered in an analog electronics course, engineers are never happy with the type of representation given above. Ratios of this type must always be expressed in decibels, or dB. The signal-to-noise ratio of an n-bit data conversion system in decibels is:

$$\begin{aligned} \frac{S}{N}(\text{dB}) &= 20 \log \frac{S}{N} = 20 \log [2^n(\sqrt{12})] \\ &= 20 \log 2^n + 20 \log \sqrt{12} \qquad (10\text{-}10) \\ &= 6.02n + 10.8 \end{aligned}$$

From this we can see that the signal-to-noise ratio increases by about 6 dB for every additional bit of resolution.

Another parameter of data converters is dynamic range. The *dynamic range* of a system is the ratio of the full scale signal to the smallest resolvable signal. The smallest resolvable voltage in an ideal ADC is equal to one quantum step. Since there are 2^n quantum steps in the full-scale range, this ratio is simply $(2^n Q)/Q$, or 2^n. Expressed in decibels, this becomes:

$$\text{Dynamic range (dB)} = 20 \log 2^n = 20n \log 2 = 6.02n \qquad (10\text{-}11)$$

EXAMPLE 10-7

What signal-to noise ratio (in decibels) can be expected from an ideal 8-bit ADC? What is the dynamic range?

SOLUTION

An ideal ADC with 8-bit resolution will provide a 10.8 dB + 48.16 dB = 58.96 dB signal-to-noise ratio. The dynamic range of this converter is 6.02 × 8 = 48.16 dB.

A term used in reference to ADCs which corresponds to mononicity in DACs is *missing*, or *skipped* codes. If the differential linearity error of an ADC exceeds ±1 LSB, the output may have a missing code. This occurs when the ADC converts an analog voltage to the next highest binary code by skipping one step. Figure 10-16 shows examples of large differential linearity errors with and without missing codes.

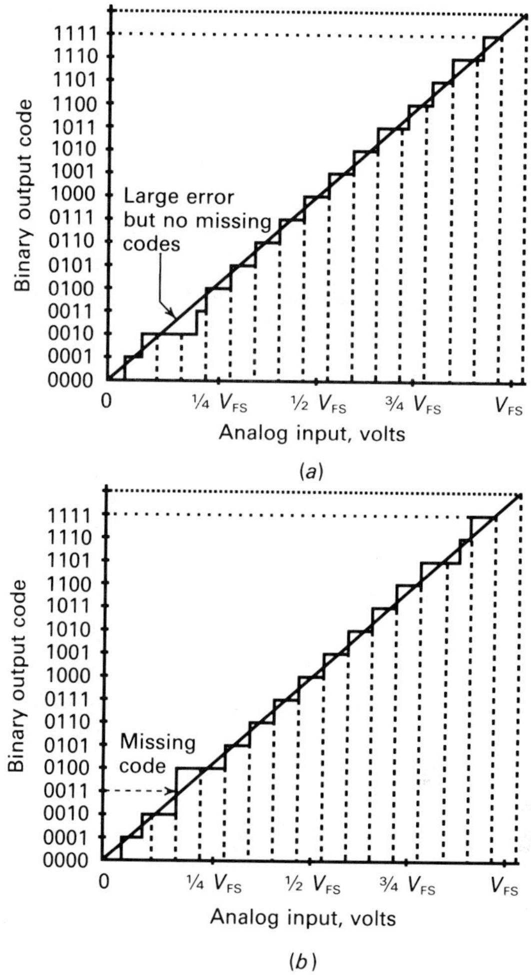

Fig. 10-16 Differential linearity errors with and without missing codes.

Sample-and-Hold

Inherent in the process of converting analog signals into binary equivalents is the need for circuitry which will *sample* and *hold* an analog value long enough for the ADC to make a conversion. *Sample-and-hold* circuits are quite simple when moderate speed and moderate accuracy are required. Figure 10-17 shows a simple sample-and-hold circuit consisting of two buffers, a CMOS switch, and a capacitor. The sine wave input in Fig. 10-17 is sampled at discrete moments by closing the switch and opening it again after the capacitor has been charged. The circuit can operate in two different modes. It can take a quick sample and then hold (*a*), or it can track the input some of the time and then hold until the next sample (*b*). Practically speaking, the latter always occurs to some degree.

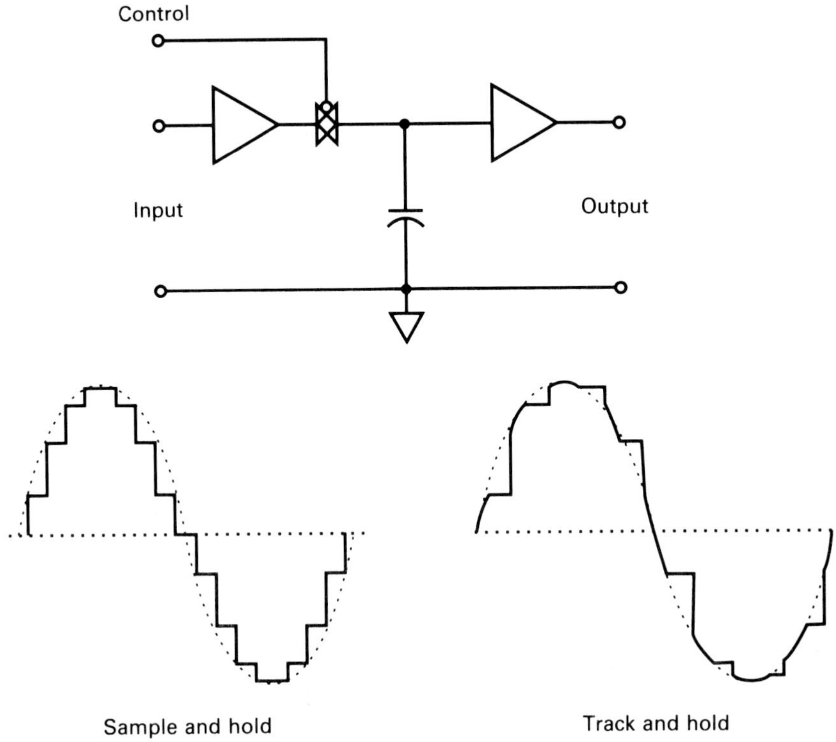

Fig. 10-17 Simple sample-and-hold circuit with two modes of operation.

Types of A/D Converters

There are five types of ADCs which account for most of the A/D conversion performed today. They are *digital ramp, successive approximation, dual-slope, flash,* and *charge redistribution.* The first two types of A/D converters, digital ramp and successive approximation, consist of an analog comparator and an "educated-guess" register connected to a DAC. The register contents, set to some initial binary number, are fed into a D/A converter. The analog DAC output is then fed into an analog *comparator* which compares the relative magnitudes of the analog signal V_a, and the guess voltage V_g. An *analog comparator* is a circuit which compares two analog voltages and produces a logic output, 1 or 0, depending upon which input was larger. The register is updated on the basis of the comparator output. The comparisons and updates continue until conversion has been completed to within ½ LSB.

Digital Ramp Conversion

Ramp-type ADCs, like the one shown in Fig. 10-18, use a counter as the guess register. In the simplest case, it is an up-counter which increments until the guess voltage V_g exceeds the analog voltage V_a. When V_g exceeds V_a, the comparator output switches low and stops the clock to the counter. The binary number left in the counter is the closest representation of the analog voltage at that time. The name comes from the fact that the DAC output ramps upward until the unknown voltage is exceeded. The worst-case conversion time is $2^n - 1$ clock pulses, but the actual conversion

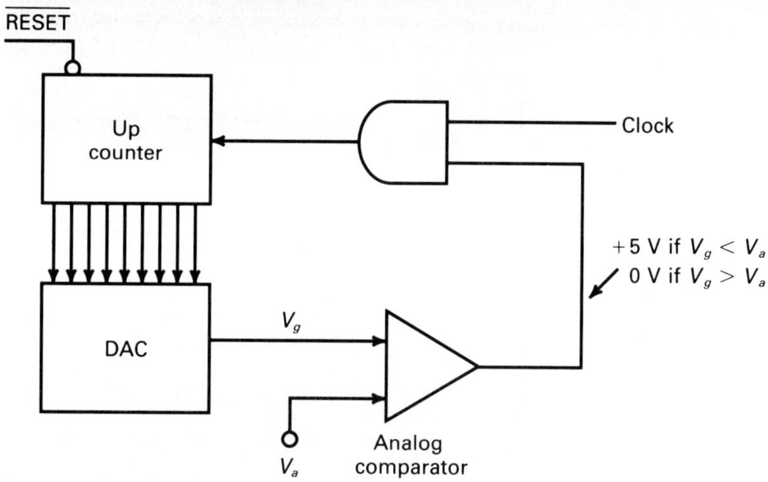

Fig. 10-18 Single ramp ADC.

time depends on the magnitude of the input voltage. The resolution of this type of A/D converter is equal to the resolution of the DAC which it uses. The final binary output will represent the analog input to within ±½ LSB. The quantitization error can be reduced only by increasing the number stages in the counter and the number of binary inputs to the D/A converter.

Successive Approximation

In successive-approximation ADCs, the guess register zeroes in on the final value by adding or subtracting successively smaller increments to or from the initial guess. The successive-approximation register (SAR) in Fig. 10-19*a* is initialized to one-half of the full-scale value by placing a 1 in the most significant bit. If the DAC output V_g is less than the analog input V_a, then the bit is left at 1; otherwise, it is set to 0. The register then moves to the next most significant bit. The DAC output is compared to the analog input, 1 bit at a time, from the MSB to the LSB. This is accomplished by shifting the initial 1 toward the LSB. After the 1 has shifted, the previous bit is either saved or cleared depending on the comparator output. Figure 10-19*b* shows how the process zeroes in on the final value with successive approximations of ±½, ±¾, ±⅝, etc. At the completion of the conversion, the bits left in the SAR represent the binary conversion of the analog voltage to within ±½ LSB of the DAC as before. The advantage of the successive-approximation method over the single-ramp method is that the conversion time is always *n* clock pulses regardless of the magnitude of the input voltage V_a. Successive approximation is the method most widely used because it is fast enough to be used in high-speed measurement systems and in PCM systems.

Dual-Slope Conversion

The *dual-slope* converters are sometimes called *dual-ramp* or *integrating* ADCs. They are more accurate than the single-ramp or successive-approximation converters, which can be only as accurate as the DAC they use. The dual-ramp converter does not require a DAC, and it cancels out any nonlinearity associated with the circuitry it does use. Figure 10-20

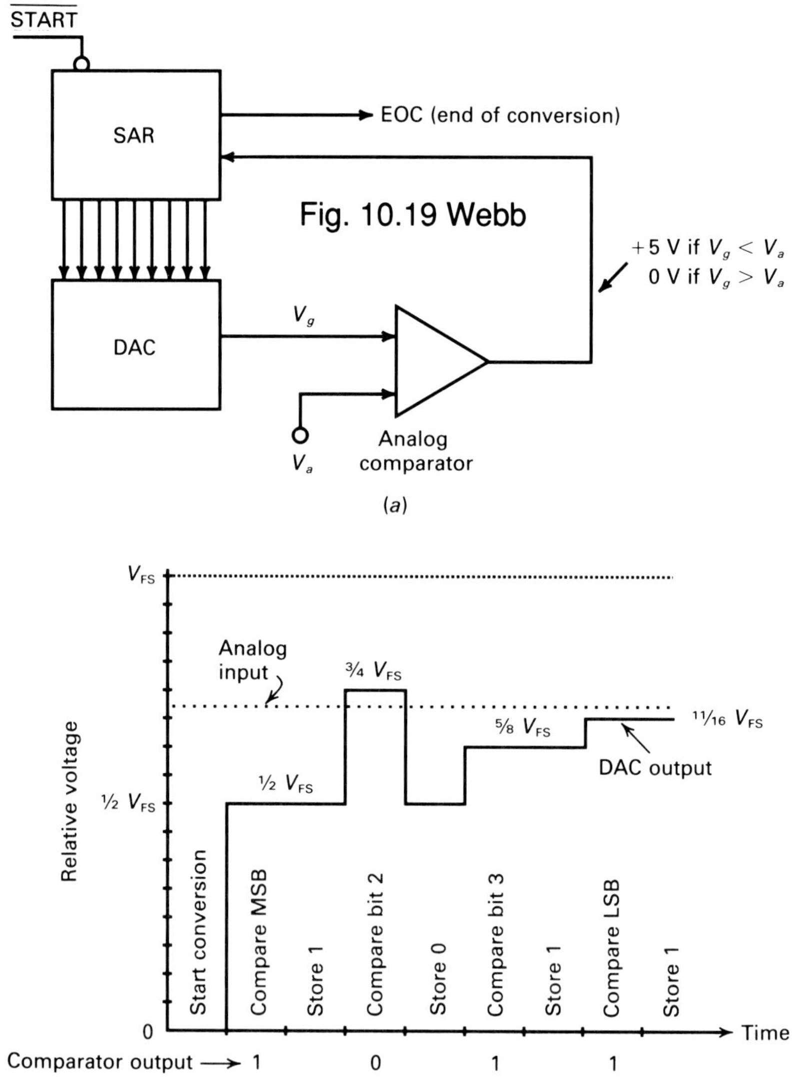

(b)

Fig. 10-19 4-bit successive-approximation ADC.

illustrates the basic circuit. The counter is allowed to count up to full value (all 1s) while the capacitor in the integrating amplifier charges up to some percentage of the analog input voltage through the transmission gate. The integrating amplifier provides a smooth upward ramp as the capacitor charges. The comparator, all the while, is comparing the ramping waveform to ground. Since the ramp is always larger than 0 V, the comparator output remains a high, and the counter clock remains enabled. After the counter has reached full count, it turns over to the all-0s state and begins to count again. However, the terminal count output of the counter sets the flip-flop on the same clock edge. This switches the integrating amplifier input to

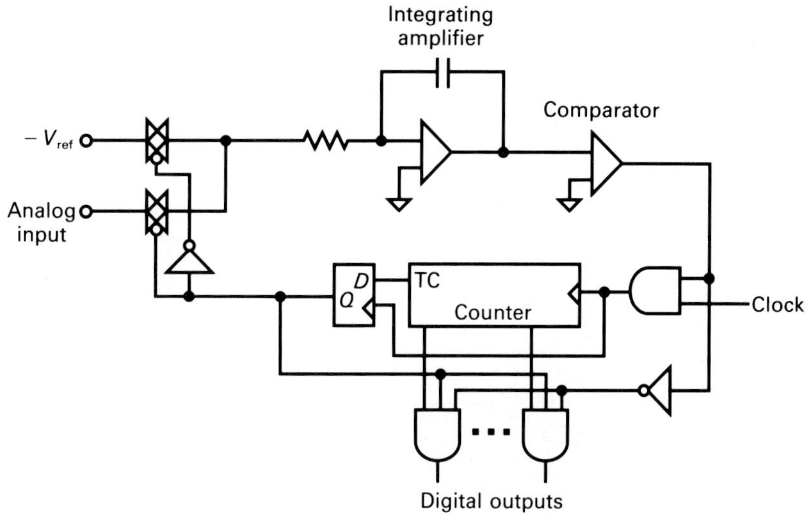

Fig. 10-20 Integrating type of ADC (dual slope).

the negative reference voltage. The counter then counts the time required to discharge the capacitor to ground. When the integrator output reaches 0, the comparator pulls the AND input low and the counter stops. The final count represents the value of the analog input relative to the reference voltage. The greater the value of the unknown analog voltage, the longer it takes for the reference voltage ramp to return to 0.

Any inaccuracy involved in producing the upward ramp is nullified by an equal inaccuracy in the downward ramp. Error sources in this type of converter are primarily due to the comparator and the voltage reference, and these errors exist in the other methods as well. Nonideal characteristics of the capacitor and the switch also contribute, but these tend to be offset-type errors rather than nonadjustable linearity errors. Because the input waveform is sampled on the integrating capacitor, this type of ADC requires no external sample-and-hold circuitry. The primary disadvantage of this type of converter is its low speed.

Flash Converters

Flash converters are the fastest type of ADC. They are sometimes called *parallel A/D converters* because they process the analog signal in a parallel fashion rather than as a serial approximation like the previously mentioned converters. Figure 10-21*a* illustrates the logic needed to make a 3-bit conversion. A separate reference level is set up for each of the seven comparators. This is done with a precision resistor divider. Each reference level is one-half step below the voltage represented by a particular 3-bit binary code. As the analog input voltage reaches or exceeds the halfway point between codes, the next comparator trips, and the next highest code is generated. If, for instance, the analog input voltage rises to just above the voltage V_{R1}, comparator *A* will switch to an output high while comparator *B*, and all the others, remain low. This will produce a high for the LSB. When the analog input voltage just exceeds V_{R2}, the LSB will

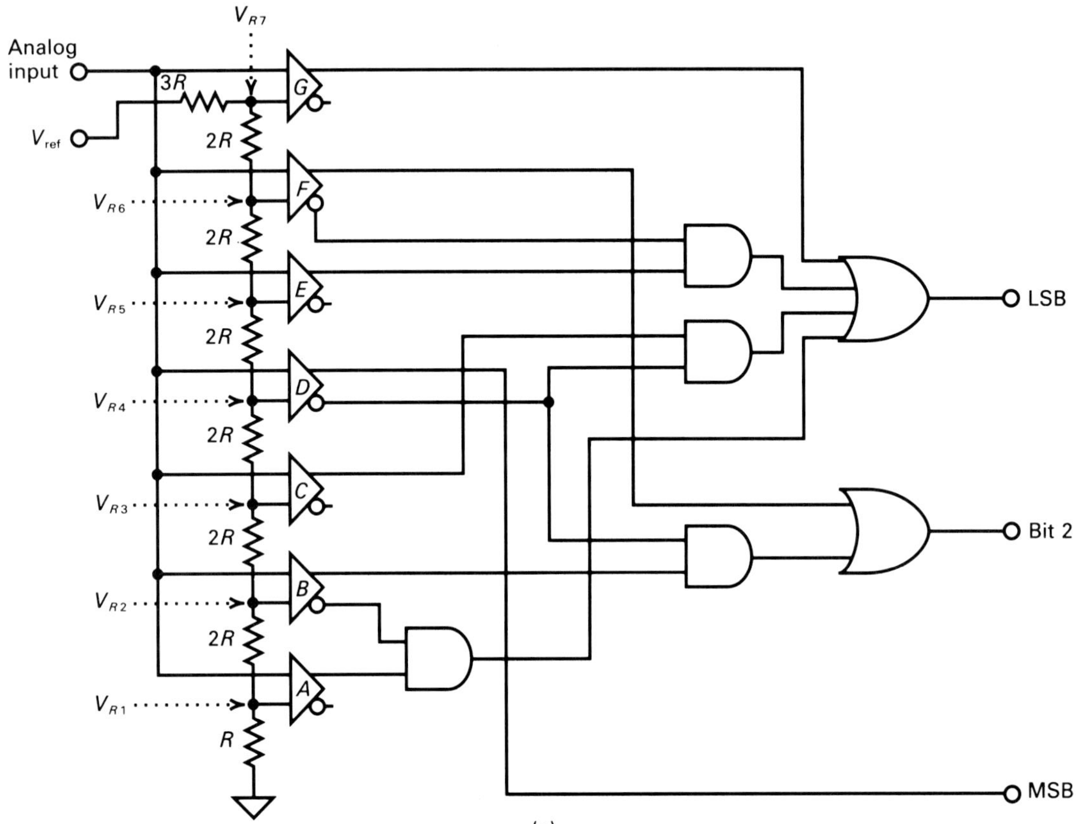

(a)

G	F	E	D	C	B	A	MSB	Bit 2	LSB
0	0	0	0	0	0	0	0	0	0
0	0	0	0	0	0	1	0	0	1
0	0	0	0	0	1	1	0	1	0
0	0	0	0	1	1	1	0	1	1
0	0	0	1	1	1	1	1	0	0
0	0	1	1	1	1	1	1	0	1
0	1	1	1	1	1	1	1	1	0
1	1	1	1	1	1	1	1	1	1

(b)

Fig. 10-21(*a*) Three-bit flash ADC. (*b*) Truth table for 3-bit flash ADC.

return to 0 and bit 2 will rise. The truth table for the decoding logic is shown in Fig. 10-21*b*. A reference table in Fig. 10-22 illustrates the relationship between the input voltage, the voltage reference steps, and the output code.

The truth table requires some explanation. Unlike other truth tables we have seen, the inputs are not binarily coded. This is because they are not actually binary inputs but converted analog inputs. As the input voltage rises and exceeds the voltage at each successive resistor node, the corresponding comparator output goes high, but none of the lower com-

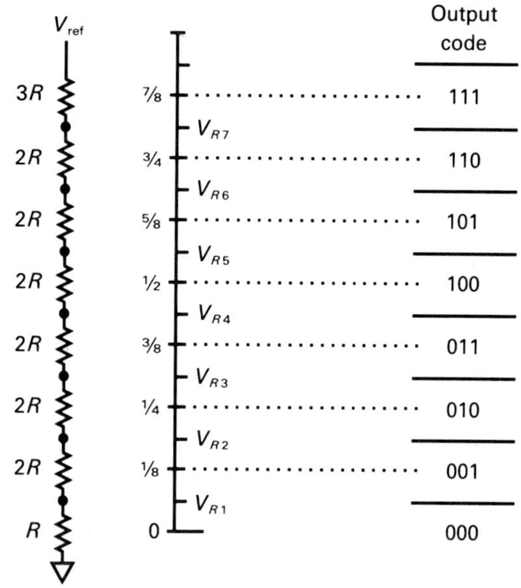

Fig. 10-22 Relationship between resistor divider, voltage levels, and output code for 3-bit flash ADC.

parators return low. This results in seven input variables but not 2^7 possible states. There are, in fact, only eight possible input states. With seven inputs, Karnaugh mapping is impractical and would produce 120 don't-care conditions. Fortunately, careful inspection of the table will provide minimization without mapping. Starting with the MSB, it is easy to see that the MSB is a 1 only when comparator D outputs a 1. Bit 2 is a 1 whenever comparator F outputs a 1 and also when comparator B is a 1 and comparator D is a 0. This same approach of finding the highest comparator with a 1 and the next highest with a 0 is used to find the expression for the LSB. The results are as follows:

$$\begin{aligned} \text{MSB} &= D \\ \text{Bit 2} &= F + \overline{D}B \\ \text{LSB} &= G + \overline{F}E + \overline{D}C + \overline{B}A \end{aligned} \tag{10-10}$$

The 3-bit flash converter requires seven comparators and several gates. The amount of circuitry needed almost doubles for each additional bit of resolution. Resolution beyond 4 bits is therefore not practical. When an 8-bit flash conversion is needed, a 4-bit converter is used in a serial fashion as shown in Fig. 10-23. The 4-bit flash ADC first determines the binary code for the four most signficant bits. This value is stored in flip-flops which drive a 4-bit DAC. The DAC then generates an analog voltage proportional to the value of the upper bits. This raises the voltage at each node of the resistor divider into a range of voltages above the level represented by the four MSBs. In this way, the analog voltage difference between the four MSBs and the next higher four MSBs is subdivided and

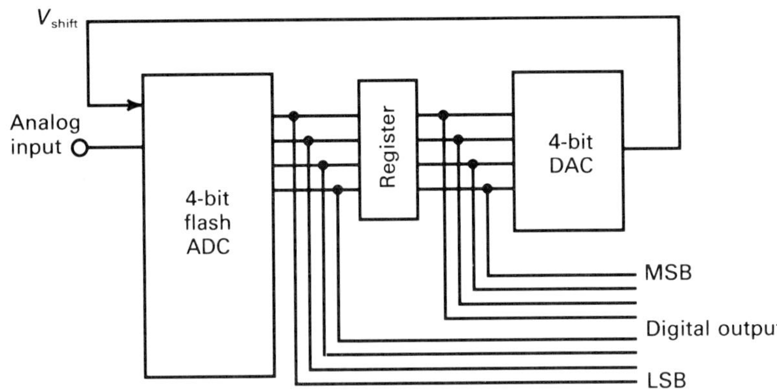

Fig. 10-23 8-bit parallel-serial ADC.

compared to the input. This two-step technique reduces conversion speed but still provides the fastest method of achieving 8-bit A/D conversion. When additional speed is essential, two 4-bit flash converters may be used in series, eliminating the need for flip-flops and the associated delay. The output of the MSB converter is fed directly into the DAC, and the DAC output drives the second flash converter.

Charge Redistribution Converters

Charge redistribution techniques are used in VLSI applications where silicon area must be shared with other digital and analog circuitry. This method has two advantages besides circuit density: It may be implemented entirely in MOS on the same substrate with MOS digital circuitry, and it provides a built-in sample-and-hold. The technique can be illustrated with a conceptual 4-bit version shown in Fig. 10-24. It uses one comparator

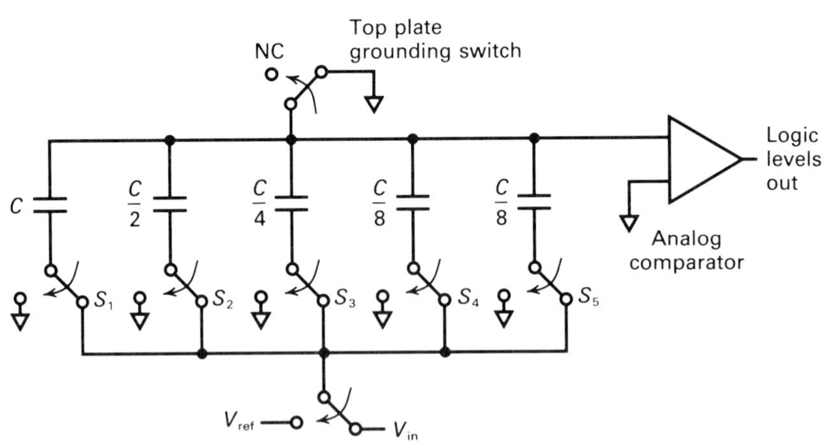

Fig. 10-24 4-bit charge redistribution ADC.

and an array of binarily weighted capacitors. The conversion consists of three steps: sample, hold, and redistribution. In the first step, the top plates of the capacitor array are connected to ground, and the unknown analog voltage V_{in} is stored on the bottom plates. The top plate grounding switch is then opened, and the bottom plates are all connected to ground with switches S1–S5. These first two operations are equivalent to the *sample-and-hold* function required by all ADCs. The result is that $-V_{in}$ is now stored on the top plates of the capacitor array.

The third step, redistribution, is actually a series of successive approximations. The bottom plate of the MSB capacitor C is switched to the reference voltage V_{ref}, with S_1. This increases the voltage on the top plates by the amount $V_{ref}/2$. The voltage V_X had been equal to $-V_{in}$ but is now equal to $(V_{ref}/2) - V_{in}$. The output of the comparator is *a logic 1 if* $V_X <$ *0 V* and is a *logic 0 if* $V_X >$ *0 V*. Therefore, if V_{in} is less that $V_{ref}/2$, the comparator will produce a logic 0.

If we assume that V_{in} is equal to a value between $\frac{11}{16}$ and $\frac{3}{4}$ of V_{ref}, we can use the graph in Fig. 10-19 to illustrate the conversion process. The voltages graphed in Fig. 10-19*b* were compared to the analog input of the successive-approximation ADC in Fig. 10-19*a*. Now the graphed voltages are added to $-V_{in}$, and the sum is compared to ground. The first comparison would produce a logic 1, which is stored in a shift register. This represents the MSB. Switch S_1 is returned to ground only if the MSB is a logic 0. In a similar fashion, the next bit is tested by switching S_2 to V_{ref}, raising V_X by $V_{ref}/4$. V_X is now equal to $(\frac{3}{4} V_{ref}) - V_{in}$ and another comparison is then made to ground. In this case, the comparison produces a logic 0 because V_{in} is less that $\frac{3}{4}$ V_{ref}. The 0 is stored, and switch S_2 is then returned to ground bringing V_X back to $(V_{ref}/2) - V_{in}$. The conversion proceeds in this fashion until all the bits have been determined. It is easy to see that this is a successive-approximation conversion and will terminate in n clock pulses where n is the number of bits of resolution. The voltage at the top plates is raised, before each comparison, by a binary fraction of V_{ref}. In the example, it was raised at each step by the amount shown in Fig. 10-19. Since $-V_{in}$ was originally stored on the top plates, the final voltage left on the top plates will be equal to the quantitizing error of the ADC. In higher-resolution versions, this voltage will be close to 0 V.

Since a great many applications involving microprocessors require some form of analog input or output, this method has obvious advantages. An 8- or 10-bit charge redistribution ADC can be patterned on the same die as an 8-bit microprocessor for low-speed applications. Motorola uses an 8-bit version of this type of converter on its MC68HC11A8 microprocessor.

Summary

Choosing an ADC design involves trade-offs in speed, resolution, and cost. The slowest and most accurate of A/D converter designs is the integrating type, or dual-slope converter. Dual-slope ADCs are relatively inexpensive, do not require sample-and-hold circuits, contain no components which require critical adjustment, and are free of linearity errors. Their one drawback is lack of speed. Ten-bit integrating ADCs require milliseconds for conversion. Successive-approximation converters are much faster and provide reasonable accuracy. A 10-bit successive-approximation ADC can complete a conversion in less than 1 μs. However, error sources are more numerous, and these designs contain components which require critical adjustment. Successive-approximation ADCs are also more expensive.

Flash ADCs provide the fastest conversions and can be used in video applications. Eight-bit flash ADCs, which consist of two 4-bit stages in series, convert in tens of nanoseconds. The major drawback of flash converters is resolution. The practical limit of parallel-type ADCs is 4 bits. Higher resolutions require multistage designs and longer conversion times. The charge redistribution design exhibits the advantages and disadvantages of successive approximation but at a much lower cost. Charge redistribution ADCs can be included with other MOS analog and digital circuitry in VLSI integrated circuits.

ADC Applications

Potential applications for A/D converters are limitless. A survey by a large manufacturer of DACs and ADCs revealed the following microprocessor-based ADC applications:

1. Monitoring of temperature, pressure, and water flow in nuclear reactors
2. Digital control of flight simulators
3. Automatic production test equipment used in manufacturing
4. Air-quality monitors
5. Medical instrumentation
6. Color correction in motion pictures
7. Process control in manufacturing

New applications are continually arising. *Digital signal processing (DSP)* is one of the most promising areas requiring ADCs. DSP involves sampling and converting analog waveforms with ADCs, filtering and enhancing those waveforms with arithmetic logic, and returning them to the analog world with DACs. *Filtering* means removing unwanted parts. Digital signal processors filter complex analog waveforms to remove unwanted noise or frequency components. Old motion pictures have been digitized, processed with digital filters, and reprinted in high-quality color without the video noise and fading associated with age and wear. The motion picture industry has also pioneered the use of computer processing for special effects.

Probably the most interesting development has been the use of ADCs in broadcast television. Analog signals from television cameras are sampled with 8-bit flash ADCs and processed digitally. The processed picture data is then reconverted into analog form with DACs before being transmitted. The digital picture data may eventually be transmitted directly to home receivers. The advantages of digitized video signals are: They are immune to noise, they can be stored in digital form for quick-access editing and read out without degradation, and they can be easily converted to different television standards. In addition, many special effects can be achieved. Digital receivers can sample and store picture data from more than one station at a time and display multiple picture windows on the screen. Video applications require conversion rates in the 10- to 20-MHz range. This means that analog signals are sampled 10 to 20 million times per second.

General-purpose digital processing is becoming accessible to PC users. Figure 10-25 illustrates a single board digital filter system which could plug into a slot of an IBM PC computer. This would convert the PC into a

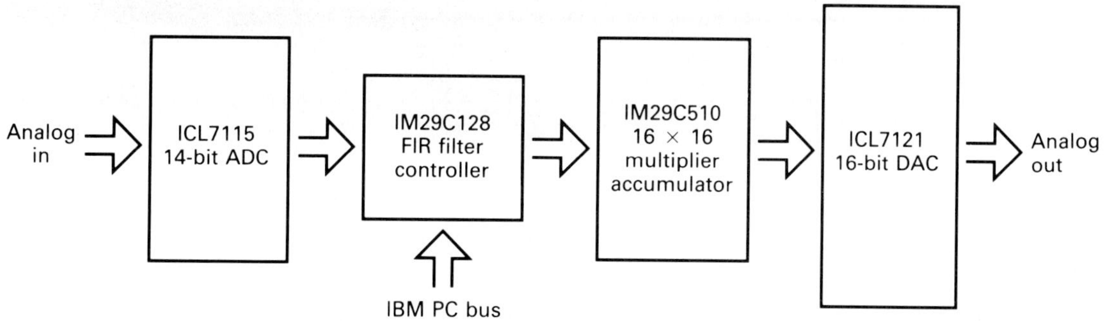

Fig. 10-25 PC-based digital signal processing system. (*Courtesy of Intersil Corporation.*)

powerful digital signal processor. The system can sample analog waveforms at rates up to and over 1.2 MHz. Memory for this system could reside in the PC or as part of the plug-in unit. PROM or RAM is required to store a look-up table containing filter coefficients. The advantage of digital filters is that they can eliminate unwanted noise without compromising the entire signal. When analog devices filter out noise they also take part of the signal with it, and in order to avoid losing too much signal, some of the noise must be left in.

A family of easy-to-use CMOS ADCs is available from Intersil in a 20-pin DIP package. The ADC0802, ADC0803, and ADC0804 are all 8-bit successive-approximation A/D converters which can be easily interfaced with all microprocessors. Figure 10-26*a* through *c* shows the pin configuration for this IC, and connection diagrams for stand-alone operation and microprocessor interface. The ADC0802-ADC0804 devices appear as I/O ports to the Intel 8080 family microprocessors and as memory locations to the Motorola family processors. Successive-approximation ADCs require a clock for conversion, and this may be derived either from the system clock or from an "on-board" clock generator. The on-board clock generator

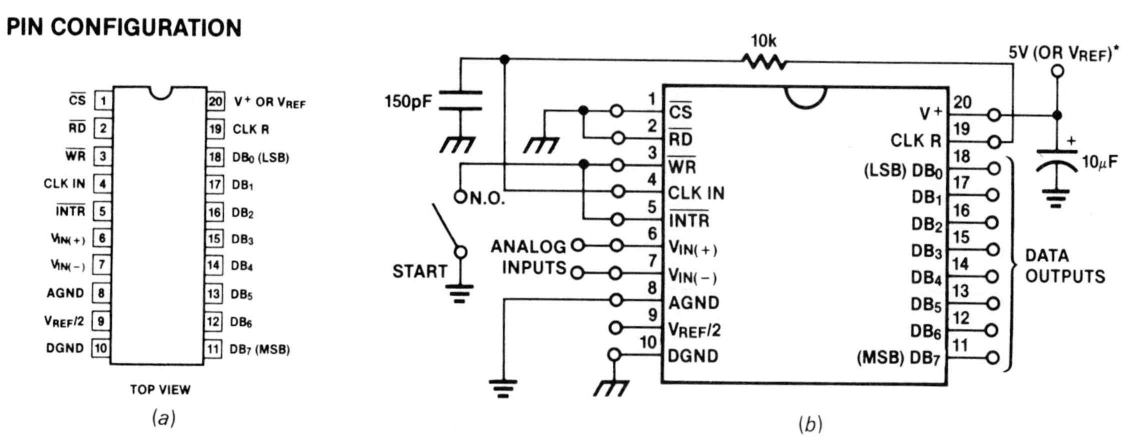

Fig. 10-26 (*a*) Pinout for ADC0802-ADC0804 analog-to-digital converters; (*b*) stand-alone configuration.

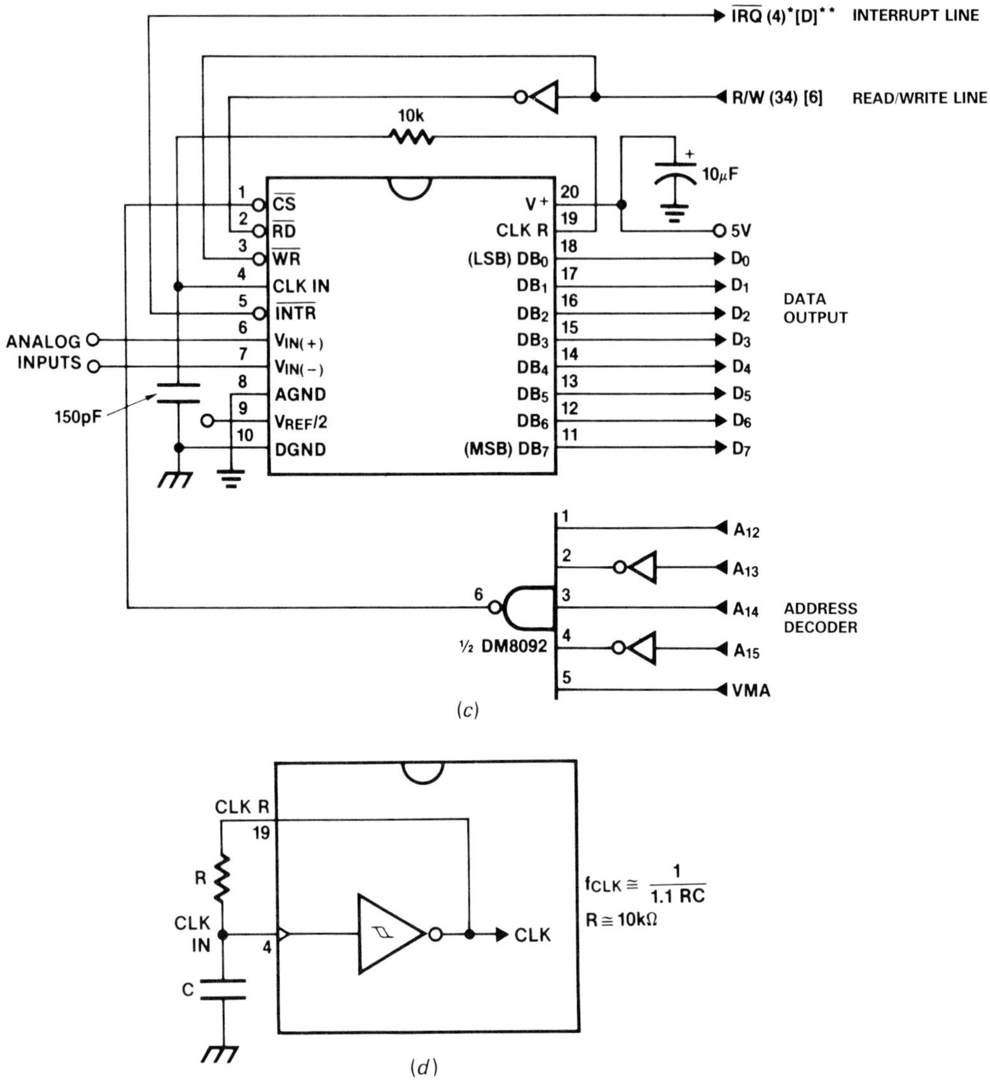

Fig. 10-26 (*c*) Interface to 6800 family bus; (*d*) connection of external *R* and *C* to internal clock circuitry. (*Courtesy of Intersil Inc.*)

uses an external resistor and capacitor combination on pins 4 and 19 as shown in Fig. 10-26*d*.

Operation is similar to the successive-approximation example in Fig. 10-19. CMOS switches in the DAC are closed in sequence by the successive-approximation register until the voltage difference between pins 6 and 7 matches the voltage derived from the DAC. The DAC uses a tapped resistor string across the *reference voltage*. The most significant bit is tested first, and after eight comparisons (64 clock cycles), the 8-bit binary result is transferred to the output latch. At this time the interrupt output, pin 5, makes a transition from HIGH to LOW, and the result may be read on pins 11 to 18.

The overall accuracy of the parts varies with increasing part number. The ADC0802 is specified as ±½ LSB of the total *un*adjusted error, and the ADC0803 provides ±½ LSB of total *adjusted* error. The ADC0804 provides only ±1 LSB of the total *un*adjusted error. The total error specification includes linearity errors, gain errors, and offset errors. *Adjustment* means adjusting the magnitude of $V_{ref}/2$ on pin 9. The accuracy of the subsequent A/D measurements will be directly affected by the accuracy and stability of voltage reference circuit used at pin 9. Voltage reference ICs are manufactured to provide constant output voltages over time and temperature. *Bandgap* reference ICs use the predictable junction characteristics of silicon diodes to provide stable outputs. Acceptable accuracy is usually in the range of 5 ppm, but specific applications may require more or less accuracy.

REVIEW 10-2

1. **Q.** What is quantization? **A.** It is the conversion of continuous multivalued voltages into discrete *n*-level voltages.

2. **Q.** What is quantization noise? **A.** It is distortion caused by differences between the analog voltage *into* and the subsequent analog voltage *out of* an ideal data conversion system.

3. **Q.** What is dynamic range? **A.** It is the voltage range of a conversion system from the smallest resolvable signal to the nominal full-scale output.

4. **Q.** What does a sample-and-hold circuit do? **A.** It stores an analog voltage and holds it long enough for an ADC to make a conversion.

5. **Q.** What is an analog comparator? **A.** It is a circuit which compares two analog values and produces a logic 1 or logic 0 output to indicate which voltage was larger.

6. **Q.** Describe successive approximation in general terms. **A.** It is a process of approaching a voltage level by adding or subtracting successively smaller increments to an initial guess.

7. **Q.** Describe the basic idea behind the dual-slope method of conversion. **A.** The dual-slope converter charges a capacitor with the unknown voltage and uses a counter to determine how long it takes to decay to a known reference level.

8. **Q.** What is a parallel converter? **A.** A flash or parallel ADC uses a multitap voltage divider and several comparators to determine the unknown voltage level. The comparator outputs are decoded with combinational logic into a binary code.

9. **Q.** Which type of ADC is best? **A.** That depends on the application. Flash converters have low resolution but are the only ADCs capable of video speed. Integrating ADCs are the most accurate but are very slow. Successive-approximation types are a compromise of speed and accuracy and are most often used.

10. **Q.** What is digital signal processing? **A.** DSP consists of sampling and converting analog waveforms with ADCs, filtering and enhancing those waveforms with arithmetic logic, and returning those waveforms to the analog world with DACs.

REVIEW PROBLEMS

1. Sketch the output of a 4-bit DAC which is driven by the ROM look-up table in Fig. 10-27. Full-scale voltage for this DAC is 4.8 V.

2. Compute the values needed for a 16 × 4-bit ROM look-up table used for sine wave generation. Sine waves actually vary between +1 and −1 but are shifted up to correspond to a range between 0 and +2. The sine wave will be divided into 16 equal intervals.

3. Tabulate the analog voltage output of a 4-bit DAC using the MOS transistor implementation illustrated in Fig. 10-2. With all switches turned on, the output voltage V_{out} equals 6.0 V.

4. How do the outputs of the MOS DAC in Fig. 10-2 differ from the 4-bit DAC in Fig. 10-4? Or from a 4-bit version of the DAC in Fig. 10-6?

5. Find the analog voltage output of an 8-bit DAC with $V_{max} = 6.0$ V and the input code equal to 11000101.

6. The DAC in Fig. 10-4 generated the output shown in Fig. 10-28. Determine what, if any, *offset* and *gain* errors exist if the reference voltage equals −8.0 V.

7. Another DAC of the same design produced the output shown in Fig. 10-29. Determine what *integral* and *differential* linearity errors are present in the DAC if $V_{ref} = -8.0$ V.

8. Find the signal-to-noise ratio of an ideal 10-bit ADC.

9. What is the dynamic range of the ADC in Prob. 8?

10. Use Fig. 10-19 as a guide to describe the path the successively larger guess voltage takes as it approximates 3.5 V in a 4-bit successive-approximation converter. Full-scale voltage is 8.0 V.

11. Explain what the graph in Fig. 10-19*b* represents to the comparator in Fig. 10-19*a*. How does the voltage in this graph relate to what the comparator sees in Fig. 10-24?

12. Explain how a 6-bit flash ADC would be constructed.

0010
0101
0111
1001
1100
1111
1101
1011
1100
1101
1010
1000
0101
0011
0000

Fig. 10-27 ROM look-up table for Prob. 1.

Binary input				Analog output
B_1	B_2	B_3	B_4	V_{out}, volts
0	0	0	0	−0.1
0	0	0	1	0.5
0	0	1	0	1.1
0	0	1	1	1.7
0	1	0	0	2.3
0	1	0	1	2.9
0	1	1	0	3.5
0	1	1	1	4.1
1	0	0	0	4.7
1	0	0	1	5.3
1	0	1	0	5.9
1	0	1	1	6.5
1	1	0	0	7.1
1	1	0	1	7.7
1	1	1	0	8.3
1	1	1	1	8.9

$V_{ref} = -8.0$

Fig. 10-28 Table for Prob. 6.

Binary input				Analog output
B_1	B_2	B_3	B_4	V_{out}, volts
0	0	0	0	0.00
0	0	0	1	0.51
0	0	1	0	1.10
0	0	1	1	1.63
0	1	0	0	2.18
0	1	0	1	2.75
0	1	1	0	3.25
0	1	1	1	3.75
1	0	0	0	4.25
1	0	0	1	4.75
1	0	1	0	5.25
1	0	1	1	5.60
1	1	0	0	6.10
1	1	0	1	6.55
1	1	1	0	7.05
1	1	1	1	7.50

$V_{ref} = -8.0$

Fig. 10-29 Table for Prob. 7.

11 DATA TRANSMISSION

Data transmission refers to the sending of information from one logic subsystem to another. The transmission path may be a long foil across a PC board, or it may be a cable across a room. Some data is transmitted for hundreds of feet and some for hundreds of miles. Up to this point we have explored the generation and storage of data but have not dealt with the practical problems involved in transferring data between digital subsystems. We have assumed that connections between logic gates and between logic subsystems do not interfere with the logic voltages. In fact, however, bit senses can be misinterpreted whenever conductors are improperly used or when the wrong choice of logic family is made for a critical application.

In this chapter we will discuss four main subtopics of data transmission:

1. General transmission line problems
2. PC board layout
3. Parallel transmission between subsystems
4. Serial transmission between subsystems

11-1 TRANSMISSION LINE PROBLEMS

A *transmission line* is any type of interconnection between logic circuits. This includes foil traces on PC boards, wire-wrapped connections, ribbon cables, and coaxial cables. The problems associated with transmitting data, *reflections, crosstalk,* and *noise,* are common to all these media. They are the primary causes of lost and damaged data.

Reflections

Sending data across a transmission line may be compared to sending a ripple across a pool of water. If the other side were infinitely far away, the waves generated near one shore would be undisturbed by reflections from the other. They would appear as continuous outbound ripples. However, if the other side were only a short distance away, the waves would be reflected back toward the sender and would interfere with the outbound waves. If the other bank were made of concrete, a very nearly identical wave would be returned. If it were made of Jell-O, there would be less reflection. This is because some of the wave energy would be transferred into the Jell-O. Water presents a certain impedance (or opposition) to the waves, and concrete presents significantly more impedance. As the bank material is changed to more closely match the impedance of water, less reflection and more absorption occurs. The same phenonemon exists when voltage pulses are sent down a wire. If the pulses encounter an impedance which differs from the *characteristic impedance* of the wire, reflections occur which may appear as extra pulses or distorted pulses in the outward-bound waves. If the wire ends abruptly (equivalent to infinite impedance), reflection is maximum. If the wire is *terminated* with a resistance to ground, less reflection occurs. As the termination more closely matches the impedance of the wire, the reflections diminish.

The characteristic impedance of a conductor is independent of its length. Characteristic impedance is, in fact, defined as the impedance (in ohms) which the conductor would present to an ac voltage if it were infinitely long. An infinitely long wire allows measurement of voltage and current in the wire without the effect of reflections from the far end. This is analogous to waves in an infinitely large body of water. Characteristic impedance is defined as:

$$Z_0 = \frac{V\,(\text{volts})}{I\,(\text{amps})} \qquad \text{for wire length} = \infty \tag{11-1}$$

This is not just a dc resistance but a complex impedance which includes series inductance and parallel capacitance. In fact, the dc resistance represents a very small part of the total impedance. Figure 11-1 shows the equivalent circuit of a wire connected to an ac voltage. The characteristic impedance Z_0 is given by the following equation:

$$Z_0 = \sqrt{\frac{R + j2\pi fL}{G + j2\pi fC}} \tag{11-2}$$

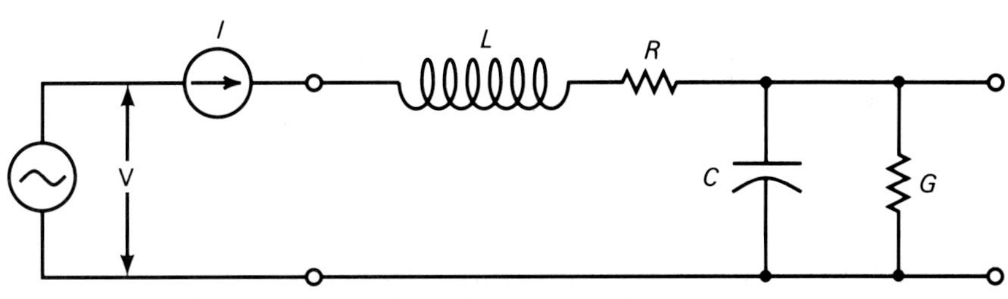

Fig. 11-1 Equivalent circuit of a cable connected to an ac voltage source.

The symbol G is the conductance through the wire insulation to ground and is normally so small that it may be neglected. The symbol j indicates that the term has a phase angle of $+90°$. At higher frequencies (hundreds of Kilohertz and above), the two terms containing f become so large that the R term may also be neglected, and the equation simpifies to the following:

$$Z_0 = \sqrt{\frac{j2\pi fL}{j2\pi fC}} = \sqrt{\frac{L}{C}} \tag{11-3}$$

If an infinitely long cable were cut to a finite length and connected to an impedance which exactly matched the cable's characteristic impedance, a pulse could be sent down the cable without any reflection. A cable may be cut to *any length* and *terminated* with its own characteristic impedance and it will behave like an infinitely long cable.

Equation (11-3) may apply even when clocks and data do not change at high frequency rates. This is because digital waveforms (square waves) contain higher frequency components. High-frequency components produce the squared edges associated with digital data. It is not the rate of the clocks or the data alone which determines the frequency content of the waveforms. The rise and fall times *(switching times)* of the waveforms are directly related to the frequency components they contain. We will see shortly that when switching times are slow, reflections are not a problem. The old CMOS metal gate 4000 series logic was trouble free in this respect.

Equation (11-3) translates into more concrete terms when we define the type of conductor we are using. For instance, if PC board traces are used, the analysis in Fig. 11-2 will apply. The inductance and capacitance of the equivalent circuit equation is translated into trace dimensions and distance from ground. The term E_r is the relative dielectric constant of the material separating the trace from the ground plane. E_r values are relative to air. The dielectrical constants for common materials are given in Fig. 11-3.

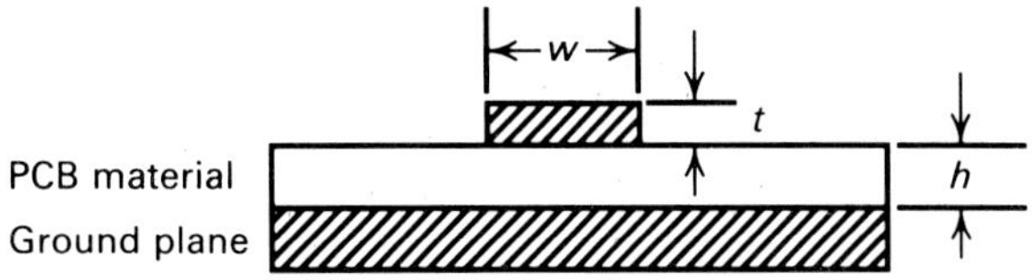

$$Z_0 = \frac{87}{\sqrt{E_r + 1.41}} \ln\left[\frac{5.98h}{0.8w + t}\right]$$

Fig. 11-2 Impedance calculation for PC board traces.

Material	E_r
Air	1.0
Polyethylene foam	1.6
Teflon	2.1
Polyethylene	2.3
Polystyrene	2.5
Polypropylene	2.6
Nylon	3.0
Silicon rubber	3.1
Polyvinylchloride (PVC)	3.5
Epoxy resin	3.6
Butyl rubber	4.0
Epoxy glass	4.7
Mylar	5.0
Polyurethane	6.5
Kynar	7.7

Fig. 11-3 Typical dielectrical constants of commonly used insulating materials.

EXAMPLE 11-1

Find Z_0 for a 12-mil (0.012-in) foil trace when the surface of the board is separated from the internal ground plane by 30 mils of glass epoxy. Assume that the strips are a uniform 0.0015 in in height. How does the length of the trace affect the impedance?

SOLUTION

The given parameters were $w = 0.012$ in, $h = 0.030$ in, $t = 0.0015$ in, and $E_r = 4.7$. Using the equation in Fig. 11-2, $Z_0 \approx 98\ \Omega$. The impedance is independent of the trace length.

Other impedances which are easy to calculate are the characteristic impedances of ribbon and coaxial cable. The impedance of a flat ribbon cable is relatively stable and usually on the order of 70 to 120 ohms. When the ribbon cable consists of twisted pairs, the impedance is even more stable and can be controlled to within ±10 percent. Coaxial cables are the most stable of the three. Figure 11-4 shows impedance equations for ribbon (flat and twisted) and coaxial cables.

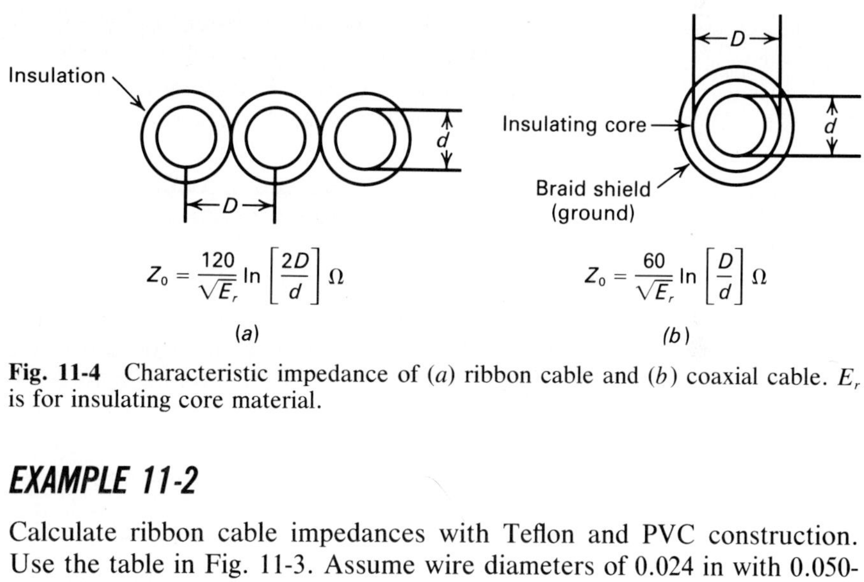

$$Z_0 = \frac{120}{\sqrt{E_r}} \ln \left[\frac{2D}{d}\right] \Omega \qquad Z_0 = \frac{60}{\sqrt{E_r}} \ln \left[\frac{D}{d}\right] \Omega$$

Fig. 11-4 Characteristic impedance of (*a*) ribbon cable and (*b*) coaxial cable. E_r is for insulating core material.

EXAMPLE 11-2

Calculate ribbon cable impedances with Teflon and PVC construction. Use the table in Fig. 11-3. Assume wire diameters of 0.024 in with 0.050-in conductor spacing.

SOLUTION

$$\begin{aligned} Z_0 &= \frac{120}{\sqrt{E_r}} \ln\left(\frac{2D}{d}\right) \Omega \\ &= \frac{120}{\sqrt{E_r}} \ln(4.17)\ \Omega \\ &= \frac{120}{\sqrt{E_r}} 1.43\ \Omega \end{aligned}$$

Teflon has a typical dielectrical constant of 2.1, so $Z_0 = 118.4\ \Omega$. PVC with a typical dielectrical constant of 3.5 provides a characteristic impedance of $91.7\ \Omega$.

It is not always possible to calculate the characteristic impedance of a transmission line. If, for instance, the PC board were two sided (no internal ground plane), the distance to ground would be unpredictable. In this case, other factors would influence the impedance more than the distance to ground. The presence of *vias* (plated-through holes in the PC board), long parallel traces, connectors, and sharp bends in the line are examples. When characteristic impedance cannot be calculated, it must be measured. This is done, at a given frequency, by measuring the impedance of a line with the far end open, Z_{oc}, and again with the far end shorted, Z_{sc}. The following equation is then used to determine the characteristic impedance of the wire:

$$Z_0 = \sqrt{Z_{oc} Z_{sc}} \tag{11-4}$$

The effect of mismatch is shown in Fig. 11-5. The ripple on the sending side is caused by reflections from the far side. This ripple may be large enough on a clock waveform to cause multiple triggers. Its height depends on the *reflection coefficient* given below:

$$\text{Reflection coefficient} = \frac{Z_t - Z_0}{Z_t + Z_0} \tag{11-5}$$

where Z_t is the line termination in ohms. The reflected signal is equal to the arriving signal *multiplied by* the reflection coefficient. When the line is open, $Z_t = \infty$, and 100 percent of the wave is reflected back toward the sending end. When the termination matches the characteristic impedance of the line, $Z_t = Z_0$, there is no reflection. Terminations having less than the characteristic impedance produce negative reflections which reduce the output voltage. Terminations having more than the characteristic impedance produce postive reflections which increase the output voltage. The *ringing*, or oscillation, following the rising and falling edges in Fig. 11-5 is caused by reflections bouncing back and forth between the near and far end.

The amount of trouble caused by a reflection depends on when the reflection arrives. This in turn depends on the length of the conductor and its intrinsic *propagation delay*. Propagation delay is the time it takes for a transition generated at one end of a line to reach the other end. This is largely a function of the type of insulation used and the distributed capacitance along the wire path. Standard wire insulations and traces on glass-epoxy PC boards provide an intrinsic delay of around 1.5 ns/ft. Coaxial cable specifications vary but are generally close to this figure. The capacitance associated with logic gate inputs will increase this delay to

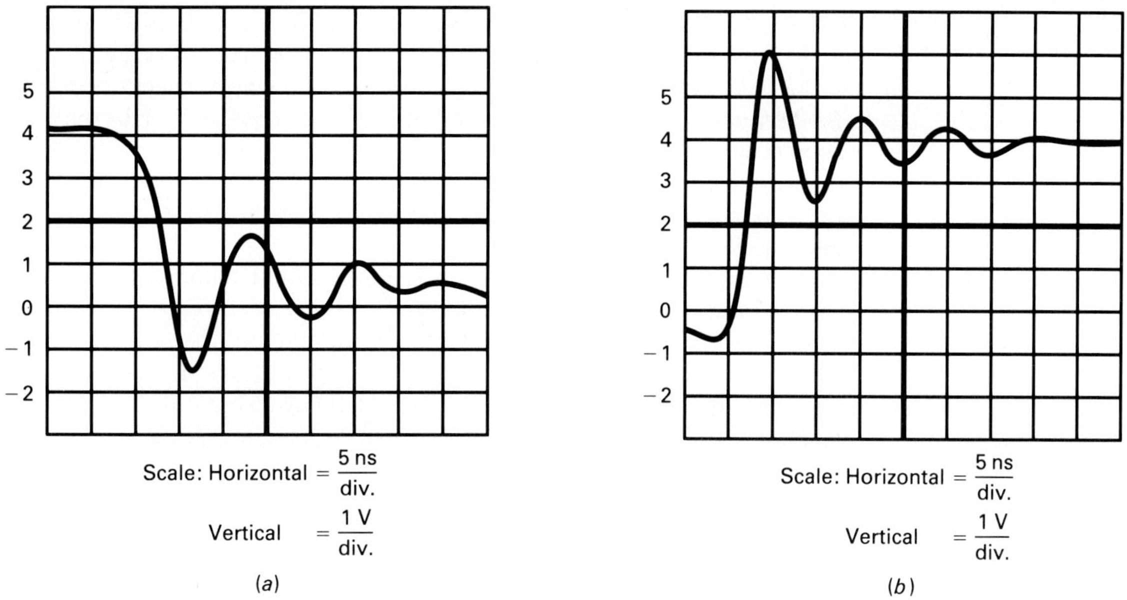

Fig. 11-5 Reflections due to line impedance mismatch.

around 2 ns/ft. The delay associated with a PC board trace with no capacitive loading is given by:

$$T_{pd} = 1.017\sqrt{0.475\,E_r + 0.67}\ \text{ns/ft} \qquad (11\text{-}6)$$

Why is the propagation delay of a line important? Because if the propagation delay were short, relative to the switching time of the driving logic, any reflections would be lost in the transition. Any reflection which returns while the driver output voltage is still changing will be "eaten up" in the transition. Although it may be noticeable as a notch in the rising or falling edge, it will not cause malfunction. A rule of thumb says that if the propagation delay of the wire is less than half the switching time of the driving logic, there is no need for termination. This is why for any given logic family, systems may be designed up to a certain physical size without worrying about reflection problems. As logic speeds increase, the maximum safe length of wire decreases. This is one motivation for integrating systems onto a single die.

Clock signals and data signals differ significantly with respect to the effect of reflections. When reflections appear on a clock waveform, a single edge can appear to a flip-flop as two edges. The system may *intermittently* go through more than one clock step. The term *intermittent* means that the problem occurs sometimes, or under some circumstances, but is not always reproducible. Intermittent problems are the most difficult to diagnose because they come and go. They may, in fact, never appear during the breadboarding and testing phase of a product design. But in production, or worse, after delivery to the customer they may appear as sporadic failures. Reflections on data signals are less critical. They serve mainly to limit the effective clock rate of a system. The propagation delay of the logic is lengthened by the amount of time which ringing occurs on the data edges. Since flip-flops cannot reliably accept new data until one setup time after the ringing has settled out, longer reflections mean slower clock rates. Reflections on data lines are generally not considered a problem unless either the maximum clock rate is being pushed to its limit or very long line lengths are used.

There are basically four types of termination used to suppress reflections. Three of these are parallel in nature and one is series. Figure 11-6*a* shows an *unterminated* line. Figure 11-6*b* shows *single resistor terminations to* V_{CC} *and ground.* Single resistor terminations draw large dc currents when used with line impedances of around 100 Ω. They also require source logic which will sink or source 25-50 mA at logic voltage levels. Remember that while the ac impedance of the line should match the dc impedance of the termination, the dc resistance of the line is actually very small. This means that most of the logic voltage falls across the terminating resistor. When terminating to V_{CC}, the voltage drops across the terminating resistor when the output is LOW. When terminating to ground, voltage drops across the terminating resistor when the output is HIGH. Single resistor termination of typical line impedances (≈100 Ω) may be used with 74F, 74S and 74AC line drivers but not with other 74xx series logic. They are not, however, recommended for CMOS because the power savings of CMOS is lost when large resistive loads must be driven.

The second type of parallel termination shown in Fig. 11-6*c* uses *split resistors* to power and ground. This produces a Thévenin equivalent of the line impedance. The advantage of this termination is that it nearly halves the dc current requirement of the single resistor termination. This method is often used in bipolar logic designs but is not recommended for CMOS inputs in bus applications. The problem is that if the line is not driven either high or low, the termination voltage will settle to about 3 V. This is below the CMOS input HIGH threshold and it is possible for both *p*- and *n*-devices to turn on and draw current. The primary advantage of CMOS, low quiescent power drain, could be nullified by this type of termination. CMOS logic with TTL compatible inputs (HCT and ACT), however, may be used with this type of termination.

AC terminations are parallel terminations which draw no dc current. A resistor-capacitor combination shown in Fig. 11-6*d* can usually be found to match the line impedance without adversely affecting the dc offset of the line. The resistor may be chosen to match the line impedance. The capacitor should be chosen large enough so that the RC time constant exceeds the loaded switching time of the signal. This type of termination may be used with any logic family but will cause some degradation to rise and fall times on the line.

Series terminations add dc resistance to the output of the driver and therefore limit the current flowing into the line. The resistance is chosen so that the series resistor value is approximately equal to the characteristic impedance of the line minus the output impedance of the driver. This causes the output voltage to divide equally so that a half amplitude pulse starts down the line. At the unterminated far end, the voltage doubles because of reflection and returns the line to its full output level. This is the type of termination most recommended for CMOS. 74AC logic provides an output impedance of 12-15 Ω so that an 82-Ω resistor may be used for 100-Ω transmission lines. This approach provides current drains which a roughly equivalent to what would be expected of an unterminated line. However, receiver inputs must be spaced very closely to the end of the line or false switching may occur as the half amplitude pulse travels down the line.

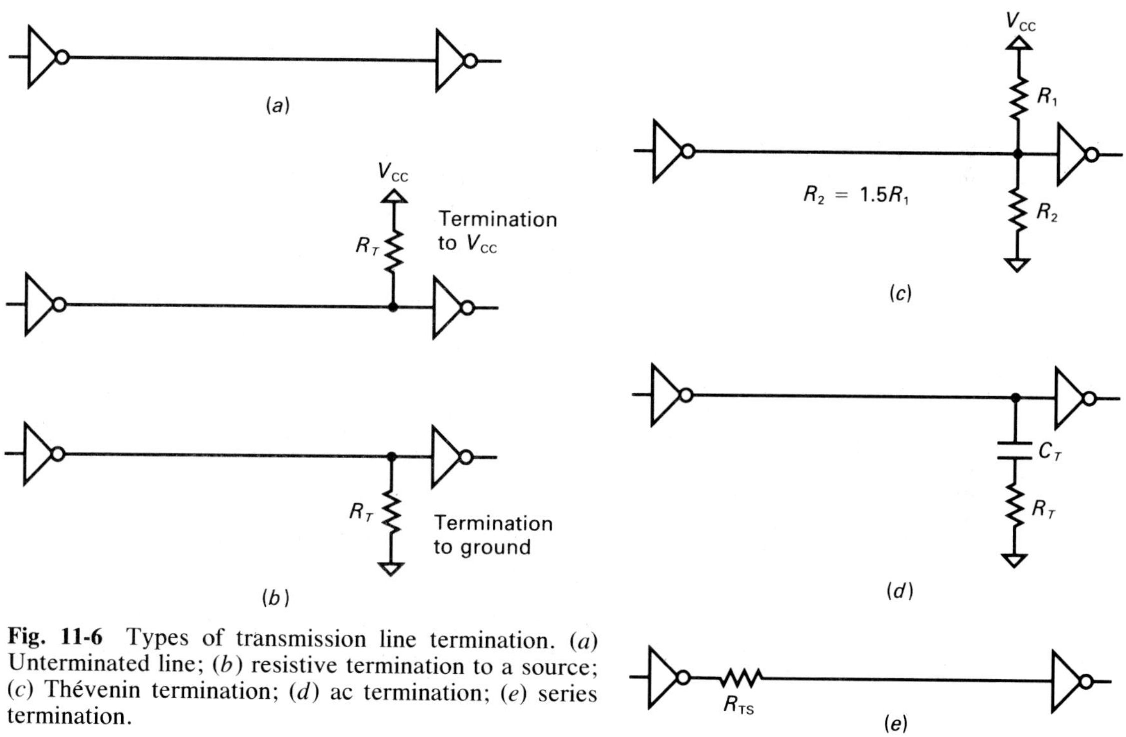

Fig. 11-6 Types of transmission line termination. (*a*) Unterminated line; (*b*) resistive termination to a source; (*c*) Thévenin termination; (*d*) ac termination; (*e*) series termination.

Crosstalk

Another potential problem which concerns PC board layout and parallel cabling is crosstalk. *Crosstalk* is electromagnetic coupling between parallel conductors. Switching currents produced in one conductor will create a magnetic field which may *couple* to a nearby conductor. Crosstalk problems increase with faster switching times and more closely spaced conductors. Figure 11-7 illustrates two forms of crosstalk, forward and reverse. Forward crosstalk occurs between two adjacent lines which are driving in the same

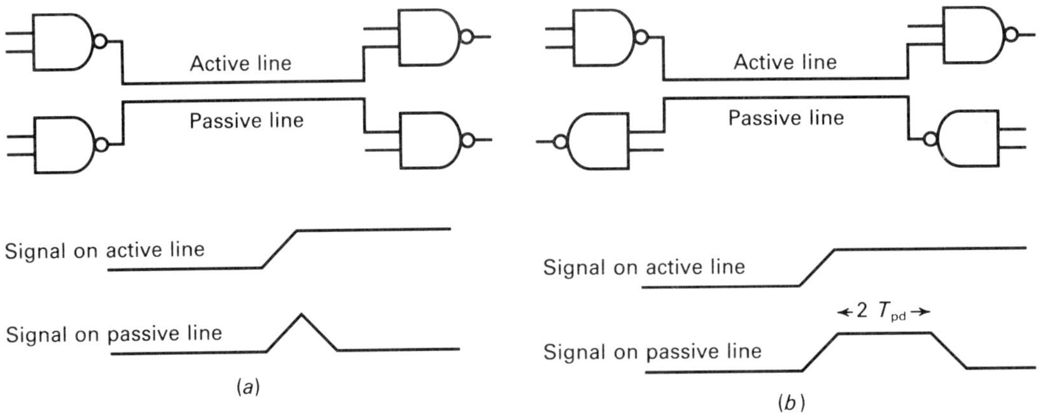

Fig. 11-7 Examples of crosstalk. (*a*) Forward crosstalk; (*b*) reverse crosstalk.

direction. Reverse crosstalk occurs when the two lines are driving in opposite directions. Reverse crosstalk is the more serious of the two because it produces a pulse with a duration of twice the propagation delay of the line. Termination of the troubled lines with the characteristic impedance will reduce crosstalk by 50 percent. General preventive measures include *maximizing* spacing between signal lines and *minimizing* the spacing between each signal line and ground.

Noise

Noise refers to any unwanted disturbance of a signal line, including crosstalk from adjacent conductors. However, aside from crosstalk, the most pervasive source of noise in digital systems is switching noise on the V_{CC} and ground lines. When logic outputs switch, they pull relatively large amounts of current from the supply. This may affect the operation of other circuitry if the V_{CC} lines and return traces cannot respond fast enough. When large amounts of current are required for short amounts of time, the resistance and inductance of the supply lines may cause noise spikes. This is particularly true when parallel buses are used on PC boards. Octal devices such as the 74xx240 series drivers and the 74xx373 register can be particularly troublesome.

REVIEW 11-1

1. **Q.** Name the primary causes of corrupted data on transmission lines. **A.** The primary causes are reflection, crosstalk, and noise.
2. **Q.** How is characteristic impedance related to conductor length? **A.** Characteristic impedance is independent of conductor length.
3. **Q.** How is characteristic impedance defined? **A.** The characteristic impedance of a conductor is the impedance, in ohms, which the conductor would present to an ac voltage if it were infinitely long.

4. **Q.** Name some parameters which affect characteristic impedance. **A.** Conductor dimensions, insulating material, proximity to ground and proximity to other conductors are parameters which affect the characteristic impedance of a conductor.

5. **Q.** Why is characteristic impedance important? **A.** Because conductors which are improperly terminated are susceptible to reflections and crosstalk.

6. **Q.** What is a reflection coefficient? **A.** A reflection coefficient is a measure of impedance mismatch.

7. **Q.** How do conductor propagation delay and logic switching speed relate to reflection? **A.** If the propagation delay of the conductor is less than half the switching time of the driving logic, reflections will not disrupt outbound pulses and termination is unnecessary.

8. **Q.** How might reflections on a clock line disrupt a sequential system? **A.** Reflections can make a single clock edge appear to a flip-flop as two clock edges. This could cause a counter or state machine to go through more than one step.

9. **Q.** What is a parallel termination? **A.** A parallel termination is an impedance connected to V_{CC} or ground, or both.

10. **Q.** What is crosstalk? **A.** Crosstalk is electromagnetic coupling between parallel conductors.

11. **Q.** What conditions contribute to crosstalk? **A.** close spacing of parallel conductors, improper terminations, and fast switching speeds

12. **Q.** What is noise? **A.** Noise is any disturbance of a signal line, including crosstalk. Aside from crosstalk, the primary source of noise in digital systems is switching noise on V_{CC} and ground lines.

11-2 PC BOARD LAYOUT

The primary goal of printed circuit board layout is to minimize electrical and mechanical problems. We will address the electrical problems of reflection, crosstalk, and noise. Mechanical problems such as warping, vibration, flexing, and heat buildup are also important but are not addressed in this text.

PC boards may be single-sided, double-sided, or multilayer. Figure 11-8 illustrates the three types. Double-sided boards allow signal traces to route across one another by means of vias and pin holes. A *via* is a hole in the PC board which provides a "feedthrough" to the other side of the board. Multilayer boards provide access to internal layers such as power and ground planes. Vias on multilayer boards interconnect traces between layers. In some cases, *buried vias* are used to interconnect only the internal signal layers, without emerging on either outside layer. This affords a savings of space on the outer layers. Figure 11-9 shows via connections

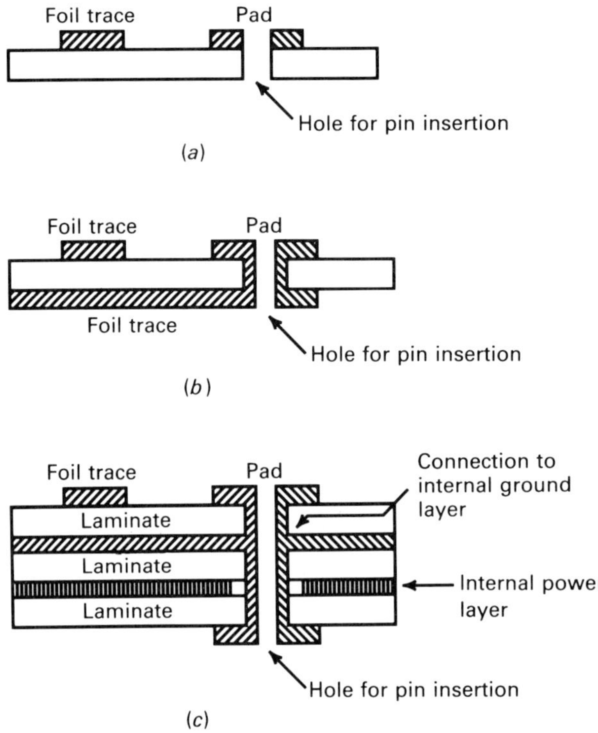

Fig. 11-8 Cross sections of PC boards. (*a*) Single-sided; (*b*) double-sided; (*c*) four-layer with internal power and ground.

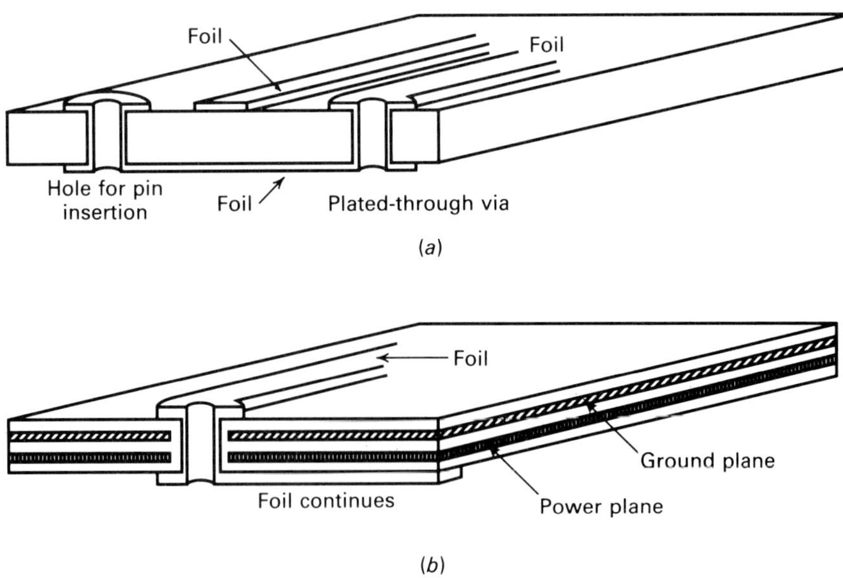

Fig. 11-9 Examples of feed throughs in PC boards. (*a*) Double-sided board with foils crossing at right angles; (*b*) plated-through-hole via in 4-layer board.

on double-sided and multilayer boards. Multilayer boards provide several advantages besides chip density. The ground and power planes provide impedance control and a reduction of noise and crosstalk on signal traces.

Terminations

The previous discussion on reflections presumed a point-to-point transmission line. This means one output connected to one input with a single transmission line. This occurs often *between* PC boards and between logic subsystems. But within a PC board this is seldom practical. Most often, on PC boards, a long line carries a clock signal to several inputs. If the clock line looks like the example in Fig. 11-10*a*, there could be trouble.

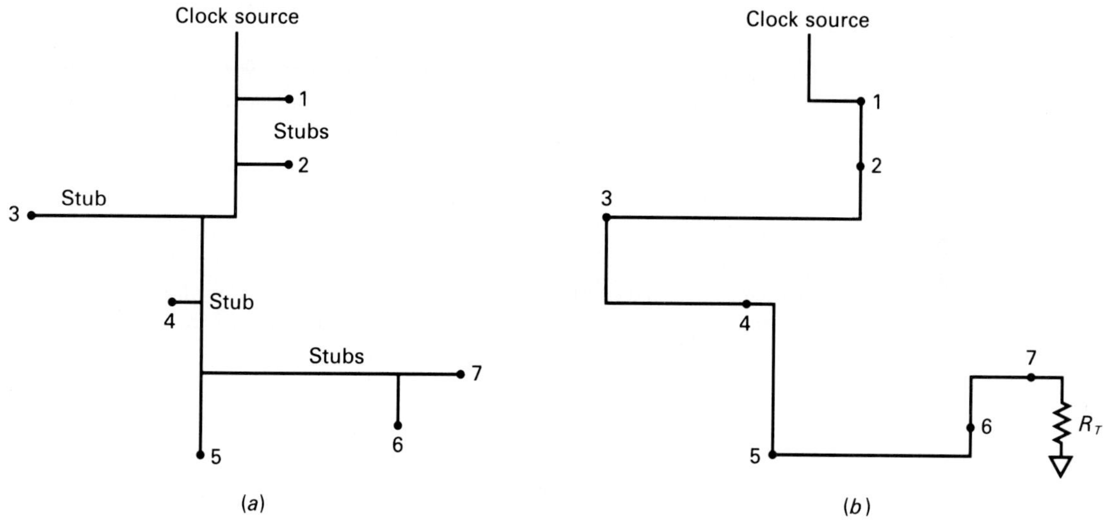

Fig. 11-10 Clock line on PC board (*a*) with unterminated stubs, and (*b*) daisy chained with termination. (● = input connection)

The unterminated *stubs* shown in Fig. 11-10*a* create their own reflections which may or may not be additive to the primary reflections from the far end. While it is possible to eliminate the effect of the stubs by spacing them widely apart, the safest approach is to avoid stubs altogether. The most common approach to this problem is to *daisy chain* the line from one input pin to another and terminate the line at the final input. Figure 11-10*b* shows how this eliminates long stubs and the possibility of additional reflections. Once again, this is necessary only when the length of the line, or stub, causes propagation delays on the order of the switching time of the driver. Any line or stub whose delay is less than half the switching time of the driving logic will cause no significant problems.

It is important to remember that logic *switching time* (rise and fall time) *is not at all the same parameter as logic propagation delay*. Propagation delay is the sum of switching time plus intrinsic gate delay. The difference is small in SSI packages because there are few stages between input and output, but in more complex packages the propagation delay is much greater than the switching time. Switching time depends on the *dynamic output current* of the logic gate and the capacitive load it must charge. Many data books do not provide the rise and fall times of the outputs, but they do provide output characteristics. Some confusion may arise concerning output current. FAST data books claim a logic 0 output drive of 20 mA for standard gates and 64 mA for buffers. This is the guaranteed minimum sink current with the output voltage no higher than 0.4 V. However, the current-voltage characteristics of a typical FAST gate show that in discharging load capacitances during an output HIGH-to-LOW transition, the circuit can absorb charge from the load capacitance at a 500-mA rate around 3.5 V. From this level it decreases steadily down to about 100 mA at 1.5 V. It is this *dynamic output current,* rather than the static low-and high-level currents, which determines actual switching speeds of the logic. The same point can be made with advanced CMOS. ACL data books claim an output source-sink current of only 24 mA for logic high and low voltages. Under typical conditions, however, a 74AC output can source 130 mA at 3.15 V and sink 150 mA at 1.35 V. Likewise, while LSTTL only promises 8 mA at 0.4 V, it can sink tens of milliamps when switching from HIGH-to-LOW.

If the output current were constant over the transition from logic low to logic high, it would be easy to calculate the rise time for a given load capacitance. The dynamic nature of the output current makes this approach too complex. Instead, we use typical switching times for test loads. Another point of confusion arises when comparing LSTTL specifications to other logic families. LSTTL switching times are specified in *most* data books for 15-pF loads while other logic families are specified for 50-pF loads. Figure 11-11 gives typical transition times for both 15- and 50-pF test loads. Since bipolar families switch faster in the HIGH-to-LOW direction, figures are given only for the fall times and not for the rise times. It makes no difference how slowly the output rises; it is the fastest transition which will determine susceptibility to reflections. When discussing transmission line problems, *switching time* refers to the shortest transition time.

74xx Family	t_s (15 pF)	t_s (50 pF)
74LS	4.5	8.5
74ALS	3.0	7.0
74HC	5.0	6.5
74AC	2.5	3.0
74F	2.0	2.5–3.0

Fig. 11-11 Typical output switching times for 74xx logic families.

It is easy to see why the 15-pF test load was abandoned when the newer logic families were developed. In a real system we must account for the capacitances associated with signal traces and the input capacitance of the receiving logic. Figure 11-12 shows capacitance values for typical 1-oz copper PC board traces over epoxy glass (G-10) material. Using this graph, we can see that a standard 12-mil trace with a 30-mil separation from the ground plane will add roughly 30 pF/ft over standard G-10 epoxy glass laminates. Additional capacitance is contributed by each gate input connection. The input capacitance of the FAST, ACL, and HC families is typically between 4 and 5 pF. LSTTL has a typical input capacitance of 5 to 6 pF. VLSI devices may have input capacitances as high as 10 pF.

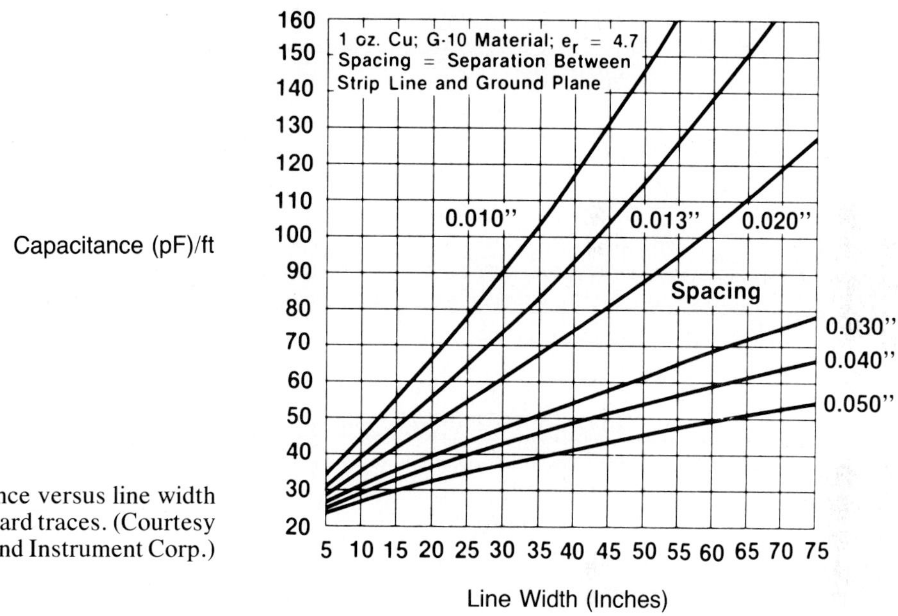

Fig. 11-12 Capacitance versus line width and spacing for PC board traces. (Courtesy of Fairchild Camera and Instrument Corp.)

EXAMPLE 11-3

How much capacitance must a logic gate drive when it is connected to five logic gate inputs with 6 in of 12-mil copper? Assume a 30-mil separation between trace and internal ground plane.

SOLUTION

The graph in Fig. 11-12 indicates that the copper foil contributes roughly 30 pF/ft or 15 pF for a 6-in trace. The gate inputs contribute 4 to 6 pF each, or 20 to 30 pF of total input capacitance. The total capacitive load would be somewhere between 35 and 45 pF.

The input capacitance of the logic also contributes to the propagation delay of the transmission line. The intrinsic propagation delay of the trace is increased by the distributed capacitance of the inputs as shown in the following equation:

$$t_{pdw'} = t_{pdw}\sqrt{1 + \frac{C_{inputs}}{C_{line}}} \tag{11-7}$$

In this equation C_{line} is the capacitance of the signal trace, and C_{inputs} is the additional capacitance provided by logic gate inputs. The additional distributed capacitance of logic inputs also lowers line impedance and increases current requirements from the driving logic.

EXAMPLE 11-4

Can the 6-in signal trace in Example 11-3 be used without termination if a 74F04 is the driver? Assume 5 pF per input and 1.5 ns/ft intrinsic wire delay.

SOLUTION

The propagation delay of the line is increased by the logic inputs in the following manner:

$$\begin{aligned} t_{pdw'} &= t_{pdw}\sqrt{1 + \frac{C_{inputs}}{C_{line}}} \\ &= 1.5\text{ ns/ft}\sqrt{1 + \frac{25\text{ pF}}{15\text{ pF}}} \\ &= 2.45\text{ ns/ft} \end{aligned}$$

In order to avoid the use of termination, the following inequality must be satisfied:

$$\begin{aligned} t_{pdw'} &< \frac{1}{2}t_s \\ \text{Length (ft)} \times \text{delay/ft} &< \frac{1}{2}t_s \\ \frac{1}{2}\text{ ft} \times 2.45\,\frac{\text{ns}}{\text{ft}} &< \frac{1}{2}(2.5\text{ ns}) \\ 1.225\text{ ns} &< 1.25\text{ ns} \end{aligned}$$

t_{pdw} is the intrinsic propagation delay of the wire, and $t_{pdw'}$ is the total delay due to line capacitance and input capacitance. With a 50-pF load, FAST logic switches in 2.5 to 3.0 ns and the worst-case value of t_s = 2.5 ns is used. The inequality shows that, in this case, a 6-in foil may be used without termination. In general, with typical parameters, 6 to 8 in is acceptable for FAST and ACL logic.

Crosstalk on PC Boards

Crosstalk on PC boards occurs whenever unterminated traces are closely spaced and running parallel for several inches. Close spacing is unavoidable for practical board densities, but long parallel runs can be avoided. The lack of termination causes voltage doubling for short periods of time and increases crosstalk problems. Even partially matched terminations make significant reductions in crosstalk. Crosstalk problems can be minimized with proper layout techniques, ground spacing, termination, and proper routing of critical signals.

Proper layout and ground spacing go hand in hand. Boards with internal ground planes provide much better crosstalk immunity than boards with ground grids. Multilayer boards with small dielectric thicknesses provide more immunity than multilayer boards with greater distances between signal and ground layers. Separation between traces should be about the same as trace-to-ground plane separation. If the spacing between traces is reduced to half the distance to ground, the crosstalk amplitude approximately doubles. Even when spacing is good, long parallel runs should be avoided. If the propagation time along parallel lengths of trace is long compared to the rise time of the signal on the active line, crosstalk has time to develop to its full amplitude. Figure 11-13 shows the dependence of crosstalk on coupled length and line delay. For delays of 1.8 ns/ft and rise times of 3 ns, crosstalk amplitudes are maximum for coupled lengths greater than 10 in. Signal lines on all two-sided and multilayer boards

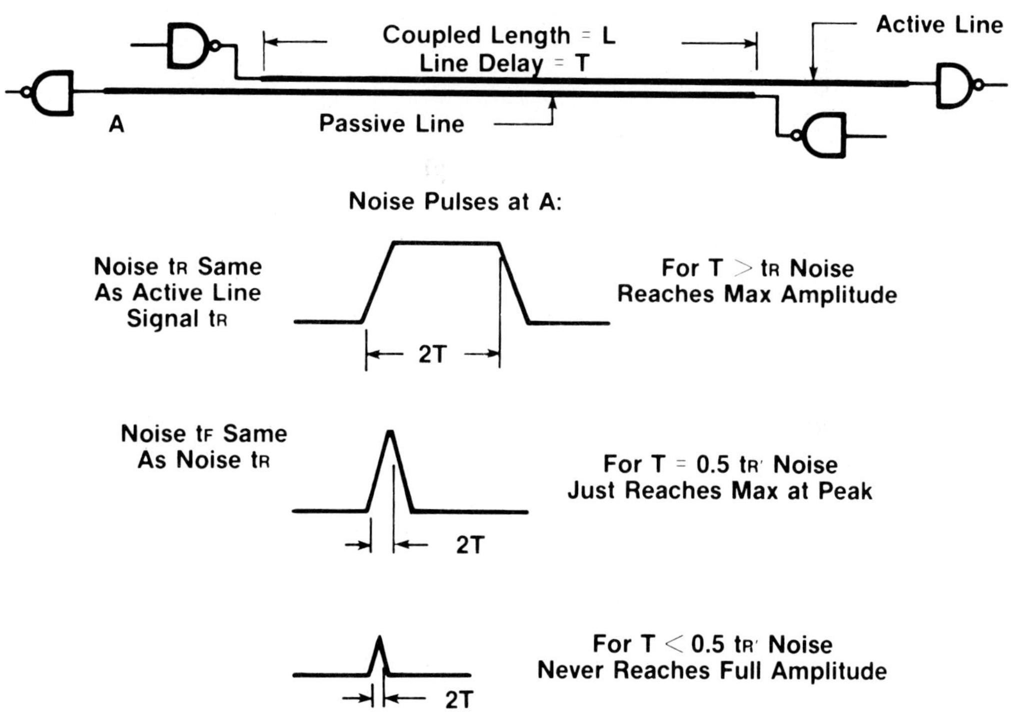

For Line Delay of 1.8 ns per Foot and FAST Rise Time of 3ns, Noise Pulse Rises to Max for Coupled Length Greater Than 10 Inches.

Fig. 11-13 Crosstalk, rise time, and coupled length. (Courtesy of Fairchild Camera and Instrument Corp.)

benefit from the standard practice of perpendicular routing on adjacent layers.

Termination and proper routing of critical signals also go hand in hand. Partial termination of parallel bus lines to twice the characteristic impedance reduces the crosstalk amplitude by one-third. Full termination of parallel bus lines is seldom necessary because the data generally changes on the same clock edge and has time to settle out before the next edge. *Critical routes* refers to clock and asynchronous signal lines. Asynchronous set, clear, and load inputs to flip-flops, counters, and shift registers are particularly susceptible to crosstalk-induced malfunction. If a clock or data line with fast transitions runs parallel to an asynchronous input line for several inches, the crosstalk could cause unexpected and intermittent sets or resets. These lines should be protected by proper spacing and termination.

Switching Noise on Power and Ground

Switching noise on power and ground lines can be controlled with good power distribution practices. Power distribution on single-sided and two-sided boards is made with power supply grids. Supply line traces on two-sided boards are thicker than signal traces but still exhibit some impedance to ac waveforms. Figure 11-14 shows impedance calculations for V_{CC} and

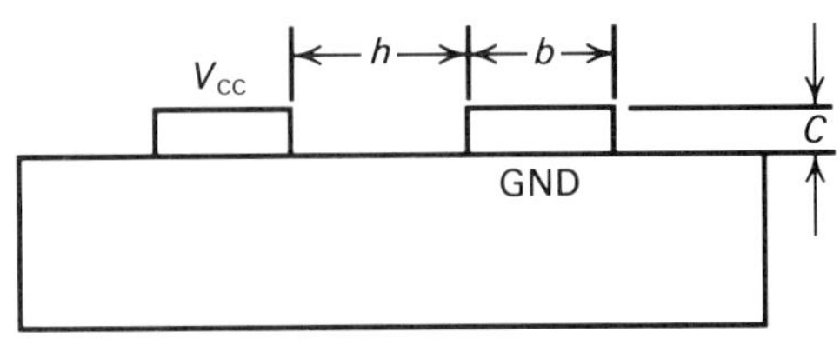

$$Z_0 = \frac{120}{\sqrt{E_r}} \ln\left(\frac{\pi h}{b + c}\right) \Omega$$

Fig. 11-14 Impedance calculation for power rails on single- and double-sided boards.

ground connections. This ac impedance prevents the supply and return lines from providing extra current as quickly as it is sometimes needed. If many outputs change at the same time, the supply line may be unable to respond quickly enough to prevent spikes on the V_{CC} inputs of some ICs. This can have the same affect as crosstalk.

The standard remedy for switching noise is the use of *bypass capacitors*. Bypass capacitors are small, 0.01 to 0.1 μF, and are connected between power and ground near the IC. They store charge when the V_{CC} line is quiet and deliver extra charge to the V_{CC} line when it is needed quickly. A good practice is to place one bypass capacitor near the V_{CC} pin of each IC. This helps to isolate one IC from another on the same supply line.

When faster, higher drive logic is used, the supply grid with bypass capacitor solution may prove inadequate. This is particularly true when 74F and 74AC octal devices are used. Current drain on the V_{CC} line for 74F240 series octal buffers can be 50 mA per buffer or 0.4 A per package. With this type of switching load, a good bypass capacitor *is required* for each package, but multilayer boards are indicated. When using faster, higher drive octal devices, the impedance between the bypass capacitor and the IC becomes important. If the impedance is too great, the extra current will not be available to the internal circuitry of the IC in time. Multilayer boards with internal power and ground planes provide much lower impedance between the leads of the bypass capacitor and the supply inputs of the IC.

Simultaneous switching of multiple outputs is common in synchronous systems but was not a major concern with the older CMOS and TTL logic families. Now however, with the emergence of the advanced logic families, 74F, 74AS, and 74AC, the analysis of simultaneous switching noise has become a major issue. Large switching noise spikes can be observed whenever octal packages drive capacitive loads. Figure 11-15 shows one

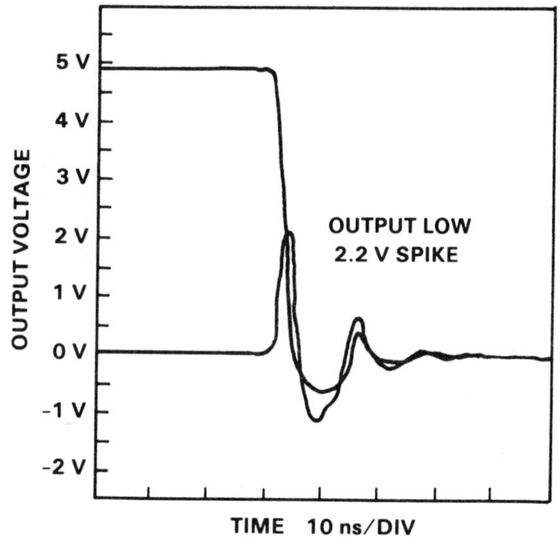

Fig. 11-15 Switching noise spike on 74AC244 output. (Courtesy of Texas Instruments Incorporated).

output of a 74AC244 octal buffer switching a 50-pF load with $V_{CC} = 5.5$ V and $T_A = 25°C$. When seven outputs switch at the same time, a spike is observed on the output of the eighth buffer in the same package even though that buffer's input is passive and is tied to ground. This crosstalk between buffers in the same package is due to a voltage spike on the V_{CC} and ground lines inside the package. This is generally acceptable with data buffers in synchronous systems since the change is not permanent and will settle out before the next clock edge. However, if the buffers are used as clock drivers, the disturbance could cause malfunction. Currently there is movement by two major IC manufacturers, TI and Signetics, to control switching noise by locating power and ground pins opposite each other in the center of the package. This *center-pin package* proposal would reduce the impedance of the supply lines and minimize output glitches.

Noise Immunity

Disturbances on signal lines can be caused by reflections, crosstalk, or switching noise. The extent to which a disturbance will affect a given logic input depends on the input thresholds of that logic. Reflection or crosstalk will not affect a logic input if the maximum amplitude is less than the logic threshold. Figure 11-16 shows input thresholds for TTL and CMOS logic.

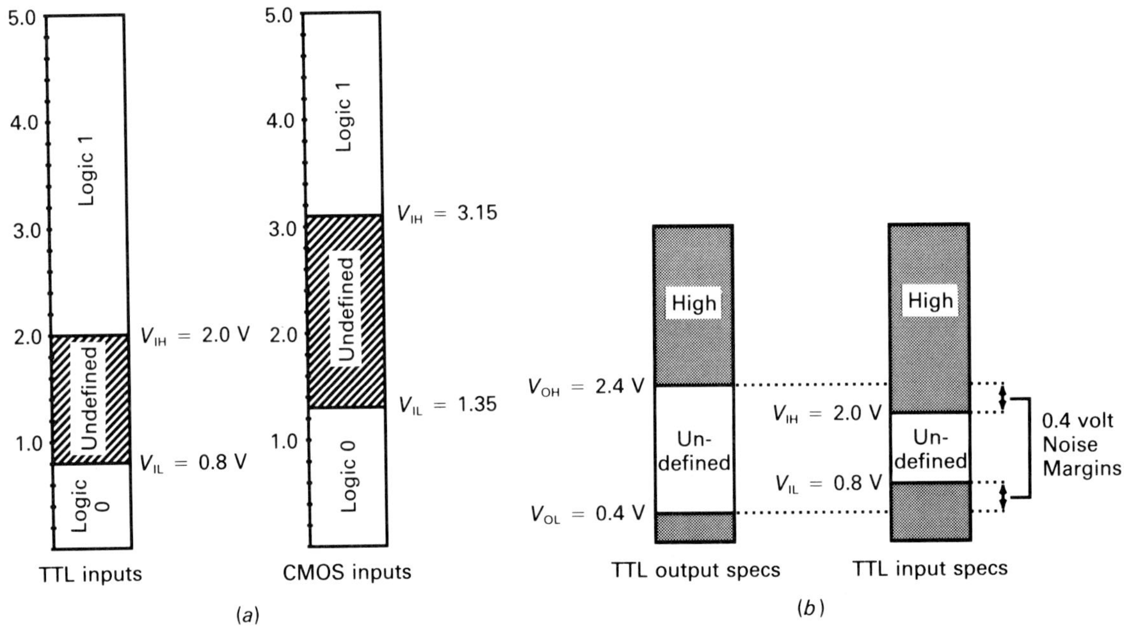

Fig. 11-16 (*a*) TTL and CMOS input thresholds; (*b*) noise margins for TTL input and output specifications.

A TTL low-level input voltage (logic 0) is defined as falling between 0.0 and 0.8 V. A TTL high-level input voltage (logic 1) is defined as falling between 2.0 V and V_{CC}. These are generic TTL specifications and apply to all *bipolar* 74xx series parts. TTL specifications are made for a nominal supply voltage of 5.0 V ± 5 percent.

CMOS 74xx parts have two sets of input specifications. One is TTL compatible and agrees with the definitions described above. CMOS parts with TTL inputs are labeled with a T, such as 74HCT or 74ACT. The T indicates TTL-compatible inputs. These parts should be used whenever CMOS inputs are driven by TTL outputs. The other specification is "CMOS-only." Since CMOS may be operated with supply voltages above and below 5 V, CMOS-only input specifications are made for three V_{CC} levels, 2.0 V, 4.5 V, and 6.0 V. In order to make a comparison with other 74xx logic families, we have used V_{CC} = 4.5 V. At V_{CC} = 4.5 V, a CMOS low-level input voltage (logic 0) is defined as falling between 0.0 V and 1.35 V. At V_{CC} = 4.5 V, a CMOS high-level input voltage is defined as falling between 3.15 V and 4.5 V. When CMOS inputs are used, ringing on a falling edge caused by an unterminated line may be as high as 1.35 V without causing malfunction. Thus, with CMOS, full impedance matching is not always necessary.

The guaranteed *input* low voltage in all cases is referred to as V_{IL} (voltage input low). The guaranteed *input* high voltage in both cases is referred to as V_{IH} (voltage input high). If the logic outputs provided exactly V_{IL} for an *output* LOW and V_{IH} for an *output* HIGH, there would be no room for error. Reflection, crosstalk, and noise could easily cause misinterpretation at the inputs. Instead, every logic family guarantees an output low voltage V_{OL} and an output high voltage V_{OH} which is different from the thresholds for high- and low-level inputs.

The low-level *noise margin* of a logic family is defined as the difference between V_{IL} and V_{OL}. The high-level noise margin is defined as the difference between V_{OH} and V_{IH}. Fig. 11-16(b) shows these differences for TTL. TTL logic provides an *output* low voltage no higher than 0.4 V for rated output current but guarantees that any voltage up to 0.8 V will be interpreted as a logic 0. This provides a 0.4-V low-level noise margin. This says that noise levels as high as 400 mV may be added to the output without misinterpretation at any input. The TTL high-level noise margin is also 0.4 V. The rated current at these voltage levels differs between 74xx families. Rated current determines how many input loads can be driven by a single output. Figure 11-17 shows input thresholds, guaranteed output levels, rated current, and noise immunity for common 74xx series logic families.

Logic Family	V_{IL}	V_{IH}	V_{OL} at I_{OL}	V_{OH} at I_{OH}	Noise Margin
74LS	0.8	2.0	0.4 at 8 mA	2.4 at 400 μA	0.4 V
74HC	1.35	3.15	0.1 at 20 μA 0.4 at 4 mA	4.4 at 20 μA 3.75 at 4 mA V_{CC} = 4.5	See note
74HCT	0.8	2.0			
74ALS	0.8	2.0	0.4 at 8 mA	2.4 at 400 μA	0.4 V
74AC	1.35	3.15	0.1 at 20 μA 0.4 at 24 mA	4.4 at 20 μA 3.75 at 24 mA V_{CC} = 4.5	See note
74ACT	0.8	2.0			
74F	0.8	2.0	0.4 at 20 mA	2.7 at 1 mA	0.4 low 0.7 high

Fig. 11-17 Comparison of noise margins for common logic families. V_{CC} = 5.0 V for bipolar, and V_{CC} = 4.5 V for CMOS. *Note:* 74HCT driving 74LS 0.47/1.84. 74LS driving 74HCT 0.4/0.7. 74HCT driving 74HCT 0.7/2.4. 74HC or 74HCT driving 74HC 1.25/1.25. For advanced CMOS, substitute AC and ACT for HC and HCT.

CMOS logic provides a higher noise immunity than TTL. The noise margin for CMOS is usually given as 1.25 V. This assumes that the CMOS logic is driving other CMOS logic. CMOS inputs are purely *capacitive*. After the dynamic output current has charged or discharged the line and the receiver inputs, current ceases to flow. There is, of course, some leakage current associated with each input, but this is very small. Typically, CMOS input leakage currents fall well below the JEDEC standard of 100 nA between $-55°C$ and $+25°C$ and 1 μA at $+85°C$ and $+125°C$. Data books specify the higher JEDEC value of 1 μA but typical I_L values are in nanoamps or tens of nanoamps. CMOS inputs have a very high impedance to dc current. After a HIGH-to-LOW transition has occurred, V_{OL} will settle to no more than 0.1 V and I_{OL} to less than a microamp.

When CMOS must drive TTL logic, the picture is entirely different. TTL inputs are *capacitive and resistive*. After the dynamic output current has charged or discharged the input capacitance, a static current continues to flow. After a HIGH-to-LOW transition has occurred, V_{OL} will settle to no more than 0.4 V if the current does not exceed the I_{OL} rating. When the outputs must sink more than the rated current, the voltage rises above 0.4 V. 74HC and 74HCT logic can sink *up to* 4 mA and still maintain the TTL V_{OL} requirement of 0.4 V. 74AC and 74ACT logic can sink up to 24 mA and still maintain $V_{OL} \leq 0.4$ V. It is important to understand that the output stages of 74HC logic and 74HCT logic are identical; only the inputs differ. The same applies to 74AC and 74ACT.

The term *fanout* refers to the number of inputs an output can drive and still maintain the V_{OL} and V_{OH} requirements. Satisfying V_{OH} is easy because all logic family inputs are high impedance in the HIGH state. Each 74LS input sinks only 20 μA in the HIGH state. This parameter is referred to as I_{IH} in the data books. Fanout is determined by how much current an output can *sink* in the LOW state and how much each input sources when it is pulled low. The latter parameter is referred to as I_{IL} in the data books. I_{IL} for 74LS is 0.4 mA when the input is pulled down to 0.4 volts. I_{IL} for 74F is 0.6 mA when the input is pulled down to 0.4 V.

EXAMPLE 11-5

Using the table in Fig. 11-17 as a guide, determine the fanout of a 74HC logic gate driving 74LS inputs.

SOLUTION

74HC outputs can sink 4 mA when pulled up to 0.4 V. The maximum current which each 74LS input will source is 0.4 mA. Dividing 4 mA by 0.4 mA tells us that each 74HC has a fanout of 10 when driving LSTTL inputs.

Obviously the fanout depends on what type of receiving logic is used. This complicates fanout specifications. There are three ways to deal with this. The specification may list only the maximum output current, leaving it up to the user to determine how many of each type of logic gate can be driven. Or, the fanout may be specified in terms of *unit loads*. One unit load in the LOW state is defined as 1.6 mA. One unit load in the HIGH state is 40 μA (all logic families provide so much output drive in the HIGH state that the LOW state is the limiting condition). A third alternative is presented in Fig. 11-18 where fanouts are listed for major logic families.

From To	HC/HCT	74LS	74ALS	74F	AC/ACT
74HC/HCT	*	10	20	6	*
74LS	*	20	40	13	*
74ALS	*	20	40	13	*
74F	*	50	100	33	*
74AC/ACT	*	60	120	40	*

Fig. 11-18 Fanout of common logic families.
*CMOS to CMOS is limited only by the maximum clock speed. Many inputs can be driven, but each input adds 5 p of capacitance.

It must be emphasized that the numbers presented so far have been for standard logic gates only. Every logic family also has special driver and buffer ICs which provide increased output drive. These will be discussed in the next section.

REVIEW 11-2

1. **Q.** What is a transmission line stub? **A.** A transmission line stub is any conductor which results in a fork in the main signal line.
2. **Q.** How can long stubs be avoided? **A.** by daisy chaining inputs along one single line
3. **Q.** What is dynamic output current? **A.** Dynamic output current is the current which a logic gate delivers at transitional voltages.
4. **Q.** Why is logic switching speed important? **A.** because the speed at which a logic gate switches determines its susceptibility to ringing and reflection
5. **Q.** Name two sources of load capacitance. **A.** Signal line capacitance and logic input capacitance combine to load a driver output.
6. **Q.** How does logic gate input capacitance affect the intrinsic propagation delay of a signal line? **A.** Input capacitance lengthens the delay.
7. **Q**. What is crosstalk? **A.** Crosstalk is the mutual coupling of parallel signal lines.
8. **Q.** How is crosstalk controlled on PC boards? **A.** with proper board layout, ground spacing, termination, and special routing of critical signals
9. **Q.** What is switching noise? **A.** Switching noise is the disturbance of the supply and return lines due to simultaneous switching of multiple outputs.

10. **Q.** What is noise margin? **A.** Noise margin is the difference between the voltage an output supplies and the voltage an input needs.

11. **Q.** What is the difference between CMOS and TTL inputs? **A.** CMOS inputs are capacitive only; once charged or discharged they do not draw any appreciable amount of current. TTL inputs are capacitive and resistive. They draw current even in the quiescent state.

12. **Q.** What is fanout? **A.** *Fanout* refers to the number of inputs a given output can drive while maintaining proper threshold specifications.

11-3 PARALLEL DATA TRANSMISSION

Parallel data transmission starts with the PC board card edge and proceeds through some transmission medium to another card edge. The transmission medium may be a *backplane*, or a cable. Backplanes bolt onto the back of equipment frames and support some type of interconnection between the PC boards in a shelf. Backplanes may be wire wrapped or fabricated with PC board technology. PC board fabrication provides better control of trace impedances, reduced noise emissions, and a higher level of reliability. *Motherboards* are small PC board backplanes which are found in personal computers and small equipment cabinets. The other transmission medium, cabling, is used whenever subsystems are too widely separated to share a backplane, or when the number of signal lines is insufficient to warrant the expense of a backplane.

PC board backplanes exhibit many of the same characteristics as PC boards. The only difference is that signal traces are longer and more susceptible to transmission line problems. Because of the increased size, backplanes are more sensitive to layout, trace length, and loading. But because there is little or no circuitry on these boards, transmission line problems are more easily predicted and controlled. Characteristic impedance can be controlled with an internal ground plane and straight, vialess connections. This makes the choice of termination easier. On the other hand, stubs can be a problem if the receivers on the PC boards are too far from the card edges. Uniform and widely spaced card slots help in this regard, but receivers should nonetheless be placed as close to the card edge as possible. Crosstalk may be controlled with wide spacing of parallel runs and the use of alternating ground and signal traces.

A major problem with backplanes and motherboards is variable loading. The characteristic impedance of the bus by itself is predictable, but when cards are added to slots along the bus, the impedance will vary. This also makes the placement of termination components less than optimal. A typical termination for LSTTL backplanes is shown in Fig. 11-19. This is an ac termination which does not affect the dc drive and load of the bus signal. The 1 = K Ω pull-up resistor provides a full 5-V logic swing which is not attainable with TTL outputs alone. Many systems keep high-speed activity on the individual PC boards and use the bus only to exchange slower data and messages between boards. For this type of system 74LS

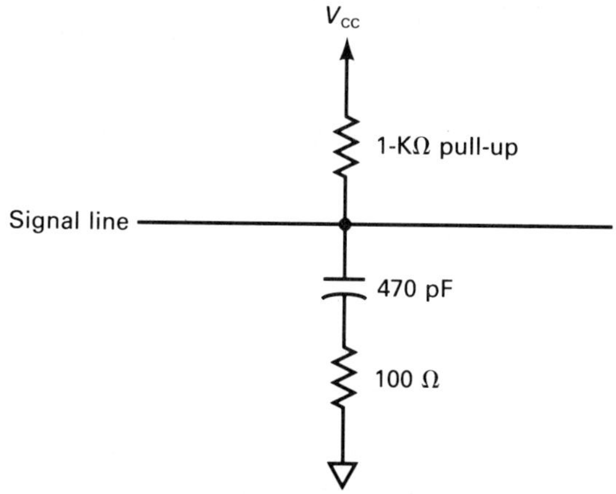

Fig. 11-19 AC termination for 100 Ω parallel bus.

bus drivers are a good choice because their slower switching speeds make stub reflection and noise emission less of a problem.

Cabling between subsystems presents more of a challenge. Parallel transmission over cables is common for distances of up to 25 ft. Round cable bundles may be used, but ribbon cables are preferred. Ribbon cables are lighter and more compact. They are more flexible, can turn corners, fit through slots, and even fold back on themselves without damage to the insulation or conductors. They also dissipate heat better and are easier to install. *Bonded* ribbon cable can be soldered, crimped, or wire wrapped. *Laminated* ribbon cable is connected with *mass termination* insulation displacement connectors. Mass termination connectors make contact to every conductor in just a few seconds without prestripping. Small forks with uniform spacing cut through the insulation and bite into the conductors. Simple termination equipment is supplied by 3M and other manufacturers.

Figure 11-20 shows the three most common types of ribbon cable: unshielded, with ground plane, and shielded. The conductors in each case

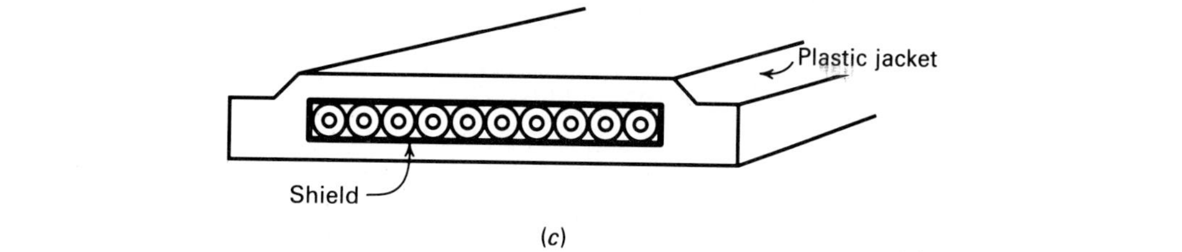

Fig. 11-20 Types of flat ribbon cable. (*a*) Unshielded; (*b*) with ground plane; (*c*) shielded.

may be flat as shown, or run in *twisted pairs*. Twisted-pair conductors provide a more uniform impedance, better noise immunity, and less noise emission. However, twisted-pair conductors exhibit substantial losses for clock frequencies above 10 MHz. Crosstalk would be a major problem in ribbon cables if preventive measures were not taken. Most flat ribbon cables are configured in alternating ground and signal arrangements. Twisted pairs are either signal-ground pairs or *differential* signal pairs. Differential signal pairs will be discussed shortly.

Types of Line Drivers

Each 74xx logic family has buffers, transceivers, and tri-state registers which provide the extra current needed to drive long parallel buses. The table in Fig. 11-21 illustrates some choices. Some provide active-low

Type	Part Number	Output Enable
Inverting buffer	74xx240	Active-low enable
Noninverting buffer	74xx241 74xx244	Complementary enable Active-low enable
Transceiver	74xx245	Active-low enable
Register	74xx373 74xx374	Active-low enable (latch) Active-low enable (flip-flop)

Fig. 11-21 Octal buffers and tri-state registers for line interface. Note that in the case of octal registers, the output enable characteristic refers only to the means of enabling the data onto a bus; it does not refer to the method of clocking.

enables which enable the data onto the bus with a LOW level. Complementary enables provide both active-low and active-high enables on the same package. These are useful for driving buses in which two sources share a common output. Transceivers allow bidirectional data flow on and off the bus.

The output drive capabilities of these devices are substantially greater than standard gate logic. 74LS, 74ALS, and 74AC/ACT families provide 24 mA, and 74F buffers guarantee 64 mA. 74HC buffers provide 6 mA. In addition to the 74xx240 series drivers, there are other non-74xx drivers for special applications. Three types of non-74xx buffers are illustrated in Fig. 11-22 and listed by part number in the appendix. They are active pull-up, active pull-down, and differential voltage drivers.

Pull-up drivers *source* current from V_{CC} through a transistor. They can drive a line high for a logic 1 output but cannot drive the line low. The line termination must pull the line low for logic 0. The amount of current sourced depends on the value of the termination resistor. This type of output is referred to as *emitter-follower* because the voltage follows the output of the bipolar emitter. It is used for IBM 360/370 buses and other general applications. The output source currents for these drivers is in the 60- to 75-mA range.

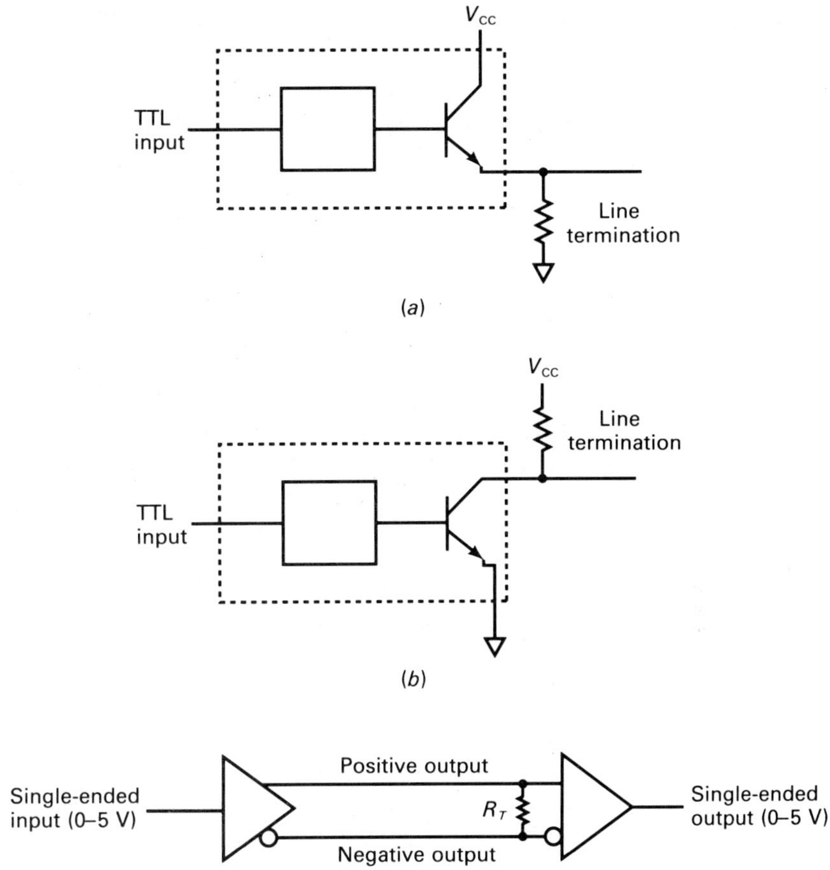

Fig. 11-22 Line driver configurations. (*a*) Active pull-up configuration; (*b*) active pull-down configuration; (*c*) differential configuration.

The advantage of the emitter follower is apparent when you consider how little voltage range there is between V_{IL} and ground in a TTL environment. If a cable connects two equipment frames which have slightly different ground reference voltages, the noise margin can disappear. It is easy for two equipment frames separated by several feet to have ground references which differ by half a volt. The emitter-follower configuration is not troubled by this difference because each system uses its own ground reference. Differences between V_{CC} levels are much less critical since there is a 3-V difference between V_{IH} and V_{CC}. Emitter-follower drivers were especially attractive in the days before CMOS had the output drive and speed needed for line driving.

Pull-down drivers *sink* current from the termination and load. The *open-collector* or *open-drain* outputs can drive the line low for a logic 0 but simply turn off for a logic 1. The line termination must pull the line back to V_{CC}. As with the emitter-follower output, the amount of output current depends on the termination. Bipolar open collector drivers can sink large currents of up to 300 mA. The main advantage of open-collector–open-drain outputs is that several drivers can share the same line without danger of device damage during contention periods. In shared-bus environments, it is easy for two or more outputs to be active for short periods of time. If driver outputs provided current in both directions, large current spikes could develop whenever two outputs were in opposite states. This could cause severe crosstalk and switching noise and even result in damage to the devices. Both active pull-down and active pull-up configurations prevent contention problems. A disadvantage of the active pull-down configuration is lack of speed. The maximum bus speed depends on the longest switching time of the bus drivers, and this is the time it takes to charge the large bus capacitance through a pull up resistor. Greater speeds can be obtained with smaller resistors, but the pull-up resistor must always be at least 10 times larger than the on-resistance of the transistor switches. This problem was discussed in detail in Chapters 1 and 2.

Differential drivers change the logic level voltages into voltages above and below ground. Differential receivers are used at the other end to convert the differential voltages back to logic levels. Differential drivers and receivers are particularly well suited to high-noise environments. Previously we considered electromagnetic coupling between parallel conductors as a major noise source. When crossing a room containing electical or electromechanical equipment, however, other noise sources become significant. Motors, relays, light dimmers, and spark plugs are major sources of noise. Their disturbances are picked up by long cables as radio waves are picked up by antennas. Conductors carrying differential signals are also affected by these spikes, but the result at the receiving end is very different. Figure 11-23 shows a long differential transmission line with

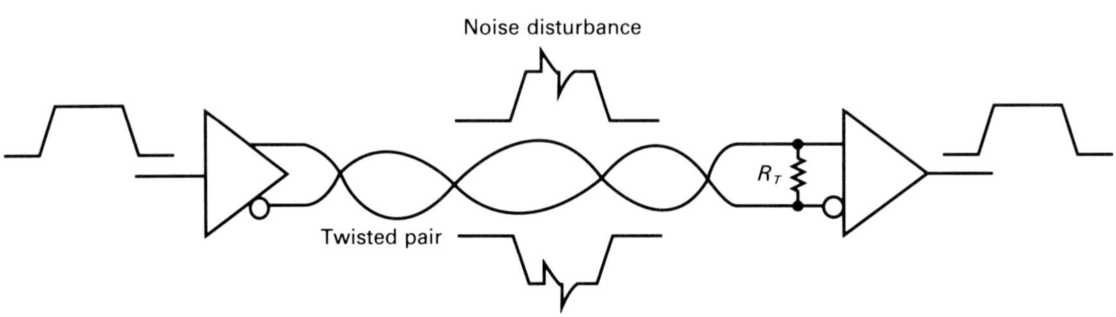

Fig. 11-23 Differential transmission with noise coupling from external sources.

noise coupling from external sources. The cable is twisted pair, which provides near identical noise coupling to both conductors. The two signals exhibit equal disturbances in the same direction, but the differential receiver amplifies only the *difference* between the voltages, and the noise is not passed on to the single-ended output. Differential drivers are the best choice for long cables and often the only choice when ribbon cables must extend for distances beyond 25 ft.

Crosstalk, Noise, and Reflections

Crosstalk on ribbon cables and backplanes can be reduced with ground planes but is best controlled with alternating signal and ground conductors. This can double the number of conductors, but that is part of the price of using a parallel bus. Alternating signal and ground is not necessary when synchronous bused data is sent on adjacent conductors and the crosstalk has time to settle out before the next clock edge. Reflections are handled by terminating each signal line to the characteristic impedance of the conductor. Terminations also help to reduce crosstalk since no voltage doubling occurs at the receiver end.

Noise is a special problem for cables. Cables *pick up* external noise and *radiate* switching noise just like antennas. Switching noise emission is especially bothersome around radio and television receivers. Uncontrolled emissions pose dangers to the public safety as well as cause annoyances. The need for regulation of digital equipment became obvious in the late 1970s when video arcade games and supermarket cash registers began to interfere with police communications and airline guidance systems. The Federal Communications Commission (FCC) has placed strict limits on the permissible levels of digital noise emissions. These emissions are referred to as *electromagnetic interference (EMI)* and are covered by FCC Docket 20780 (1980). Emission limits for the frequencies produced by most digital switching is 30 dBμV (30 dB below one microvolt) at a 30-m distance. This can be put into perspective by considering that a 2½-in length of wire, switching 1 A of current at a 30-MHz rate will emit approximately 75 dBμV at 30 m, or 45 dBμV above the legal limit. The FCC has the power to stop a manufacturer from selling any product which exceeds these limits. Parallel cabling, therefore, must not only be protected from external noise *but must also not be a source of noise.*

The solution to noise control both *into* and *out of* parallel cabling is to use shielded, flat ribbon cable with single-ended transmission or twisted-pair ribbon cable with differential transmission. Twisted-pair ribbon cable with differential drive provides good noise immunity and reduced noise emissions but cannot be used for frequencies above 10 MHz. Shielded, flat ribbon cable provides a solution for both lower and higher frequencies. At lower frequencies, less than 1 MHz, the shielding foil should be grounded only at one end. Otherwise, noise current will flow in the shield. At frequencies above 1 MHz, it becomes necessary to ground the shield at more than one point in order to guarantee that it remains at ground potential. Normally this means grounding at both ends, but for very long cables it may be necessary to provide multiple ground points along the entire length of the cable.

Bus Standards

Standard bus interfaces exist for 8-, 16-, and 32-bit microprocessor-based systems. Motorola's Exorcisor bus and Intel's Multibus I were the original 8-bit buses. Prolog introduced the 8-bit STD bus and sponsored an industry-based STD Manufacturer's Group. This organization defined the STD standard and adopted it to promote compatibility between bus-oriented products. Figure 11-24*a* illustrates applications for the STD 8-bit micro-

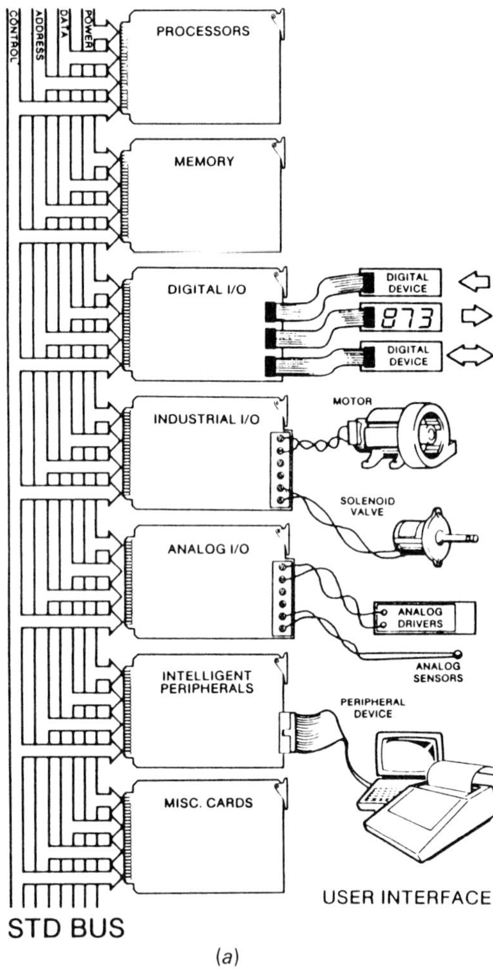

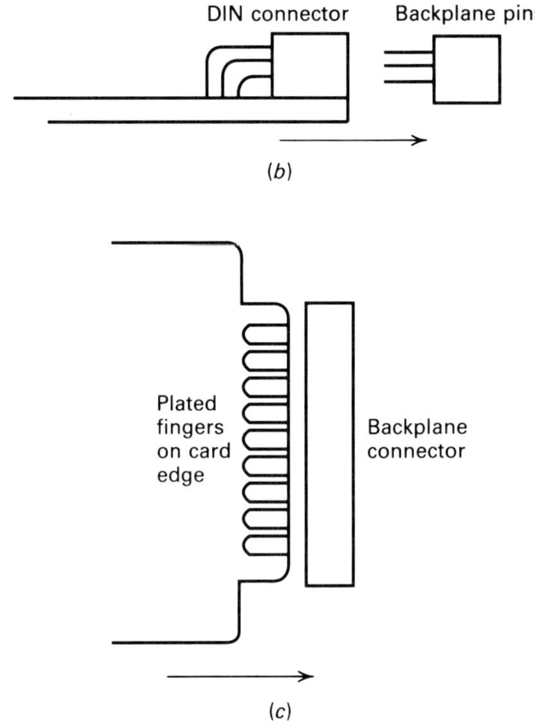

Fig. 11-24 (*a*) STD bus applications (Courtesy Prolog); (*b* and *c*) Types of connections made to backplane.

processor bus. Standard connections are specified for the card edge, and cards which adhere to the specification may plug into any slot on the backplane.

Bus standards have been proposed and adopted for 16- and 32-bit microprocessor-based systems. Motorola created and promoted the *VERSAbus* to meet the needs of advanced 16-bit microprocessors, particularly the MC68000. The *VMEbus* was introduced 2 years later in response to user inputs and market requirements. Although the VERSAbus and VMEbus are close in functionality, they have differing physical characteristics. The VMEbus conforms to a European card format called *Eurocard*. The Eurocard is smaller than VERSAbus cards and uses *DIN "pin-in-socket"* connectors as shown in Fig. 11-24*b*. The VERSAbus cards use the standard plated-finger-edge connectors shown in Fig. 11-24*c*. Both VERSAbus and VMEbus support 8-, 16-, and 32-bit data widths. Two semiconductor suppliers, Motorola and Signetics/Phillips, provide VMEbus users with off-the-shelf hardware support. Intel promotes a bus standard called *Multibus*. Multibus I supported 8- and 16-bit data widths, and Multibus II is extended to cover 32-bit applications. Multibus uses the larger-sized card format and fingered-edge connectors. Other backplane bus standards include Nubus and Futurebus. The Nubus, developed at MIT, is a low cost 32-bit standard similar to Multibus II. Nubus (IEEE-1196) uses standard TTL transceivers. The IEEE-896 Futrebus is a new 32-bit backplane system designed to take advantage of special Backplane Transceiver Logic (BTL) for higher speeds. BTL transceivers transmit data with a 1 V swing (1 V low to 2 V high).

What do these standards tell us? They provide electrical and mechanical specifications which, if adhered to, will guarantee that a product will interface with other products supporting the same standard. First-generation standards provided only physical compatibility requirements, such as drive currents, loading, and terminations. The second-generation standards went a step further to define control and timing signals needed for conversation between boards on the bus. VERSAbus and VMEbus are asynchronous. This means that instead of supplying a system-wide clock on the backplane, handshaking signals are used to maintain order on the bus. When a processor board needs access to a memory board on the bus, it initiates a request. The memory board sees that request and provides both the data requested and a "here-is-data" signal to inform the processor board that data is waiting to be enabled onto the bus. The processor then asserts a signal which enables the data onto the bus and reads it. If more than one processor shares the bus, the memory card must arbitrate between requests, assigning one or the other priority. The signals which go back and forth between the boards are called *handshaking signals,* and the set of rules governing the handshaking process is called a bus *protocol*. A protocol is a set of rules used for sending and receiving information on a bus.

Synchronous communications on backplane buses is faster, but the electrical problems associated with charging and discharging the large bus capacitances make synchronous bus design difficult. In order to understand the electrical problems associated with interfacing to parallel buses, consider a typical VMEbus backplane with an unloaded capacitance of 150 pF for each signal line. The VMEbus specification requires that no card should add more that 20 pF per line. This says that a system with eight cards would present roughly 300 pF to each signal line. If 74F240 drivers were used, the switching time would be on the order of 10 ns. If $V_{CC} = 5.0$ V this means a 10/90 percent voltage change of about 2.4 V in 10 ns. The total current needed for this transition would be:

$$\text{Switching current} = \frac{\text{capacitance} \times \text{change in voltage}}{\text{change in time}} \qquad (11\text{-}8)$$

or

$$I = \frac{C(dv)}{dt}$$

which would be 300 pF $\times$ 2.4 V/10 ns = 72 mA. When this figure is multiplied by 31 address lines and 32 data lines, a total of 4.5 A is possible between the backplane ground and the card ground for short periods of time. With this amount of dynamic current, an inductance of only a few nanohenrys in the connection could cause a 1-V difference between the two ground references. It is easy to see why parallel busing presents nontrivial problems to the designer both in terms of electrical problems and FCC emissions compliance.

REVIEW 11-3

1. **Q.** What is a backplane? **A.** A backplane is a board which supports interconnection between PC boards in an equipment rack.
2. **Q.** What is a motherboard? **A.** A motherboard is a small PC board backplane.
3. **Q.** How is crosstalk controlled in ribbon cables? **A.** With alternating signal and ground or with differential transmission.
4. **Q.** What is the advantage of using emitter-follower outputs for parallel bus applications? **A.** Emitter-follower outputs reduce grounding problems between equipment racks.
5. **Q.** Which two types of driver output may be safely operated during periods of contention? **A.** Emitter-follower and open-collector outputs.
6. **Q.** Which type of output offers the best noise immunity? **A.** differential outputs
7. **Q.** What are emissions, and why are they regulated by the FCC? **A.** Emissions are electromagnetic disturbances which leave one system and interfere with another. The FCC regulates the level of emissions because they jeopardize the public safety—police and aircraft radio—and the public convenience—broadcast radio and television.
8. **Q.** What is the purpose of shielded cables? **A.** Shielded cables protect signal lines from external noise sources and reduce emissions.
9. **Q.** How should shielded cables be grounded? **A.** When clock rates are below 1 MHz, the shield should be grounded at one end only; above 1 MHz, the shield should be grounded at both ends.
10. **Q.** What is a protocol? **A.** It is a set of rules governing the exchange of data.
11. **Q.** Define handshaking. **A.** Handshaking is the process of exchanging control signals between subsystems to indicate when data is needed or when data is ready.

11-4 SERIAL TRANSMISSION

Serial data transmission is much simpler, from an electrical point of view, than parallel transmission. When 16 bits of synchronous data are sent on a serial bus instead of a parallel bus, the number of drivers, receivers, conductors, and connectors is reduced by 16. The surge current and switching noise is also reduced by a factor of 16. In addition, the economy of using 1 wire instead of 16 makes the use of coaxial cable a practical alternative. Coax has a well-defined and well-controlled characteristic impedance which is specified by the manufacturer to ±5 percent or better. This makes termination easier and virtually eliminates reflections. Coax also allows greater cable lengths and faster clock rates. Crosstalk is not a problem because coax provides 360° shielding. The shielding also reduces noise problems both into and out of the conductor. Coax is bulky and expensive but less bulky and expensive than 16 parallel conductors in a ribbon cable.

Serial transmission paths are sometimes called *channels* or *links*. There are three basic types of channel: *simplex, half-duplex,* and *full-duplex.* The term *simplex* describes a transmission path that is one-way only. Data flows from point A to point B but *not* from point B to point A. Half-duplex transmission refers to a path in which data may flow in *either* direction but *not* simultaneously. Full simultaneous transmission in either direction is called *full-duplex*. Two other terms associated with serial transmission are *point-to-point* and *multidrop*. A *point-to-point line* starts at point A and terminates at point B. A *multidrop line* makes multiple connections to terminals between the far and near end.

The one serious drawback of serial transmission is the reduction of information bandwidth. *Information bandwidth* refers to the information-carrying capacity of the transmission path. This relates to two factors: the maximum speed of the data across a channel, or link, and the number of links used. The term *bandwidth* actually refers to the allowable data rate only. When the number of links is considered, the term should be *effective bandwidth*. If a conductor in a 16-conductor cable could operate at the same clock rate as a single conductor cable, then the 16-conductor parallel approach would provide 16 times the effective bandwidth. For short connections, parallel cable is preferred because of higher *effective* bandwidth. For runs longer than 25 to 30 ft, serial transmission is used because the expense and the problems associated with parallel transmission become overwhelming.

Considering only one link at a time, *bandwidth* refers to the range of frequencies which can pass through the link without appreciable loss, or *attenuation*. A cable begins to *bandlimit* a digital waveform by reducing or removing higher-frequency components. Remember that it is the higher-frequency components that give the digital waveform its distinctive squared edges. Figure 11-25 shows what happens when the higher-frequency components, or *harmonics*, are removed. With the removal of roughly one-half of the harmonics (the upper half), the waveform looks like Fig.

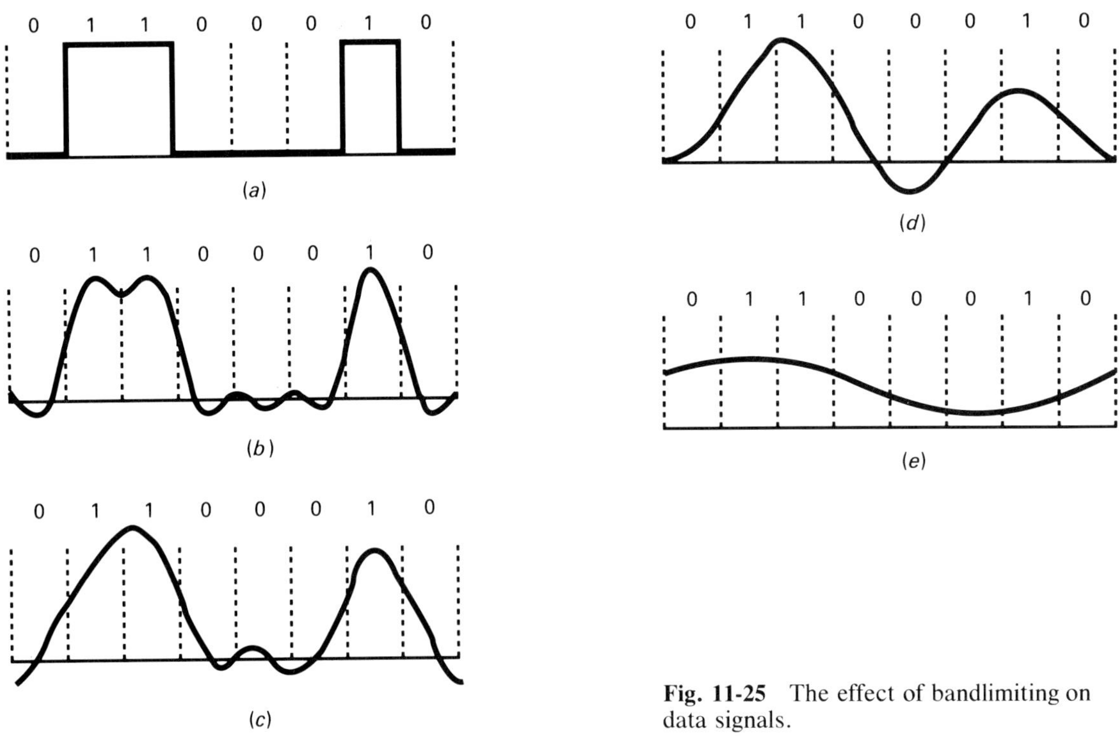

Fig. 11-25 The effect of bandlimiting on data signals.

11-25*b*. The upper harmonics contain very little energy when compared to the lower harmonics, and the waveform does not change drastically. But as more harmonics are removed, the waveform becomes less and less of a digital signal as shown in Fig. 11-25*c* through *e*.

Bandlimiting occurs when the switching speeds become too fast for the cable specification. Since switching rates are directly related to maximum clock rate ($f_{max} = 0.3/t_s$, from Chapter 2), cables can support clock frequencies only up to some maximum rate. Specification of cable bandwidth is made in one of two ways: by specifying minimum rise time versus transmission distance or by specifying the maximum-frequency harmonic which will pass unattenuated across a length of the cable. The second type of specification usually defines attenuation in terms of the half-power point. If a harmonic is reduced to 0.707 of its original voltage amplitude (which is equivalent to losing one-half of its power), the harmonic is said to be attenuated. The point at which this occurs is referred to as "the lower 3-dB point," or "−3-dB point," and the frequency at which this occurs is called the "cutoff frequency."

As cable lengths become longer, more harmonic attenuation occurs because of the increased capacitance of the line. Good coaxial cables can contribute between 15 and 30 pF/ft. This distributed capacitance, together with the very small dc resistance of the cable, will, at some point, reduce the voltage levels of all harmonics. For long lines, regenerative repeaters may be needed. T1-carrier telephone transmission lines, carrying data on twisted pairs, employ repeaters every 6000 ft. *Repeaters* are amplifiers which bring the voltage levels back to full amplitude, but they also produce unwanted jitter. *Jitter* is instability of the clock and data edges due to external noise and imperfections in the repeaters. Jitter accumulation can become so great as to prevent proper clocking of incoming data.

The effects of bandlimiting and jitter can be viewed on an oscilloscope as a closing of the data *eye pattern*. Figure 11-26 shows random data with

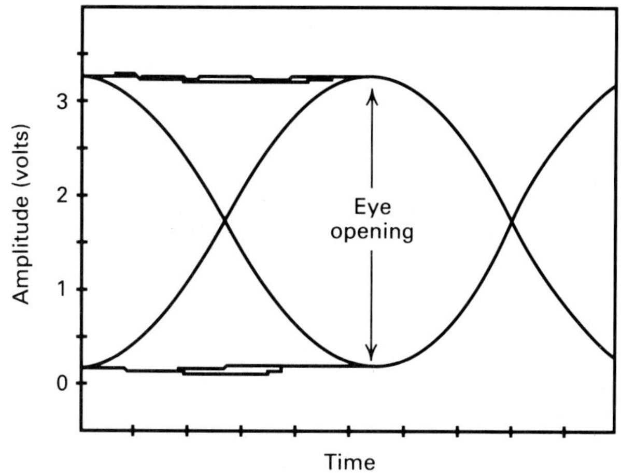

Fig. 11-26 Eye pattern produced by random data.

an open-eye pattern. Since 1s and 0s occur randomly, the data causes a pattern resembling an eye. As bandlimiting and jitter increase, the eye begins to close. As the eye closes, it becomes increasingly more difficult to correctly interpret the exact meaning of the data. Even without jitter, the limited bandwidth of the cabling can cause the rectangular pulses associated with logic 1s to produce long decaying *tails* which interfere with subsequent logic 0s. This is known as *intersymbol interference*.

There are various ways of dealing with these problems. The effects of bandlimiting can be *equalized* with special filters, and jitter can be controlled with jitter-eliminator circuitry. Also, coding schemes can be used which minimize the effects of these factors at the receiver end. *Bipolar coding,* found on telephone digital transmission lines, uses only one-half the bit time for a logic 1 and represents the logic 1 as an alternating voltage above and below ground. This reduces intersymbol interference and prevents the dc level of the line from drifting upward when long strings of 1s are encountered. Ordinary telephone lines, called *voice-grade lines,* are so severely bandlimited that they cannot support rectangular waveforms at all. These lines are purposely bandlimited to about 3 KHz to allow *frequency division modulation (FDM)* of telephone conversations. FDM allows multiple voice conversations to share the same wire pair by placing them in separate frequency channels in the same way that broadcast radio stations share the same air space by using different frequency channels. When data must be transmitted across voice-grade lines, 1s and 0s are translated into sine waves of differing frequency, amplitude, and phase. Since sine waves contain no higher harmonics, they pass through voice-grade lines unattenuated as long as their fundamental frequencies do not exceed 3 KHz. This is actually analog data transmission because it requires the translation of digital waveforms into analog signals before transmission. The equipment used to make this translation is called a *modulator-demodulator,* or *modem.* Transmission over voice-grade lines is very slow compared to digital T-carrier transmission. A single T1 line transmits data at 1.544 Mbits/s while modems transmit at rates of 300 to 9600 bits/s for single, unconditioned voice-grade lines and up to 48 kilobits/s with 12 bundled voice-grade lines.

The number of times per second that a signal changes its voltage level is measured in baud. *Baud* is the rate of change of a signal value, and the term *baud rate* means the same thing. *Baud* and *bit rate,* however, are not always equivalent terms. If signal values are binary, with voltage levels representing only 1s and 0s, then baud is equivalent to bit rate. However, if eight voltage levels are used to represent discrete values between 0 and 7, then each voltage-level change conveys 3 bits of information. In this case, the bit rate would be equal to three times the baud rate. Some modems use a combination of amplitude levels and phase shifts to produce greater effective bit rates over bandlimited telephone lines. Bauds versus bit rates for voice-grade telephone lines are listed below:

Bauds versus Bit Rates for Voice-Grade Lines (Bandwidth = 300 Hz to 3.4 KHz)

Bauds (Symbols/s)	Modulation Technique	Signaling Conditions per Symbol	No. of Bits per Symbol	Bits/s
150, 300, 600, 1200, 1800	FSK	2	1	150, 300, 600, 1200, 1800
1200	Dibit PSK	4	2	2400
1600	Tribit PSK	8	3	4800
2400	QAM	16	4	9600

Serial Transmission Standards

As with parallel transmission, there are standards for serial transmission of data. The purpose of serial transmission standards is to facilitate the interface of equipment manufactured by different vendors. One standard, EIA RS-232-C, stands out above the rest. This is not because it is inherently superior to the others but because it was one of the first and has become widely used. RS-232 standard was developed when computers were large and expensive and few in number. Companies requiring computer service would often transmit to and receive information from the computer over telephone lines. The EIA RS-232 standard defined an interface between a computer or terminal, referred to as *data terminal equipment* (*DTE*), and a data communication device (or modem), commonly referred to as *data communications equipment* (*DCE*). Since this standard was developed before the existence of small computers, it did not anticipate RS-232 connections between computer terminals and computers. RS-232 is used for data rates up to 20 kilobits/s. Both asynchronous and synchronous transmission are supported by the standard.

Asynchronous transmission refers to data sent in self-synchronized packages, called characters, without an accompanying clock. A start bit (a space, logic 0) determines when a new package begins, and one or more stop bits (marks, logic 1s) separate the packages from one another. Asynchronous data streams are sent with an LSI-shift-register-based device which converts parallel data into a serial stream. The same device receives serial data and returns it to parallel format. These are called *universal asynchronous receiver-transmitters* (*UARTs*). Asynchronous RS-232 data may be sent over voice-grade telephone lines by connecting the UART to an asynchronous modem.

When asynchronous data is received, it must be sampled in the middle of the bit period because the exact location of the edges is uncertain. This means that the local receive clock must be faster than the received data rate so that the UART can detect the edge of the start bit and then count some number of subperiods into the first bit to determine the approximate center. Normally the local receive clock is 16 or 64 times the data rate. A SPACE (logic 0) is interpreted as the start bit. If a 16-times clock is used, the UART counts 8 or 9 subperiods and then samples again to ensure that it has indeed received a SPACE. If so, the start bit is considered valid, and each succeeding bit is sampled at intervals of 16 fast clocks from that point. After the expected number of bits is received, the UART looks for one or more MARKS (logic 1s) before expecting another start bit. Figure 11-27 illustrates this scheme. All of this is unnecessary with synchronous transmission since a 1-times clock is sent along with the data.

Synchronous transmission requires that a clock signal be sent along with the data. Synchronous data streams are transmitted and received with *universal synchronous receiver-transmitters* (*USRTs*). Synchronous RS-

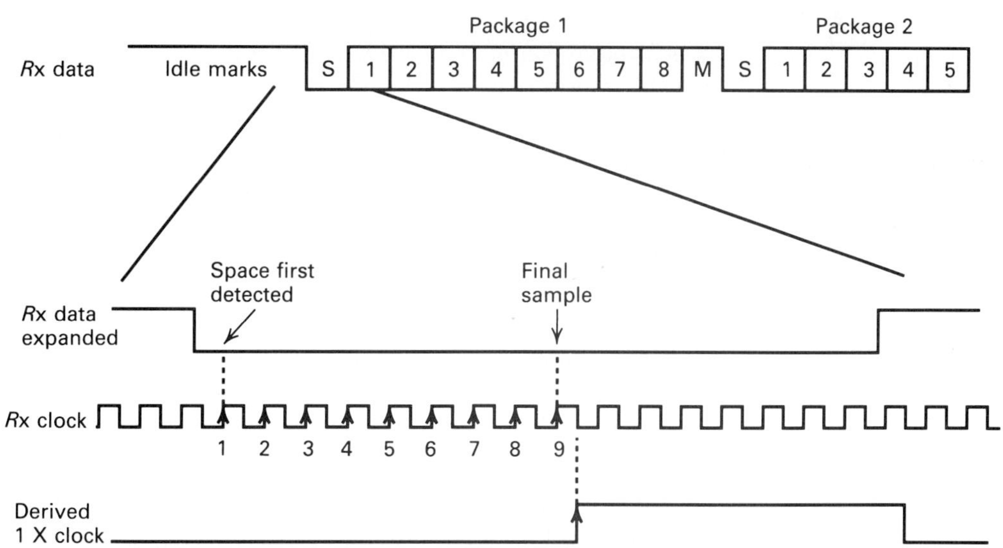

Fig. 11-27 Asynchronous clock recovery for RS-232 data.

232 may be sent over voice-grade telephone lines by connecting the USRT to a synchronous modem. Synchronous transmission is inherently faster because overhead bits (start and stop bits) are unnecessary. In addition, the center of the bit is exactly known, and maximum clock rates can be achieved. The transmitter portion of a synchronous modem must combine the data and the clock into a single encoded bit stream, and the receiver must separate clock and data from the single stream of edges. Encoding schemes with strong timing components are used for this purpose, Manchester being the most popular. Manchester encoding produces a transition at the center of each bit time. The direction of the transition determines the bit sense. Figure 11-28 illustrates Manchester encoding of data. The encoding may be viewed as an EXCLUSIVE OR function of clock and data. Since a transition is guaranteed in the middle of each received bit, the receive end circuitry can separate the clock and data by synchronizing with the midpoint edges.

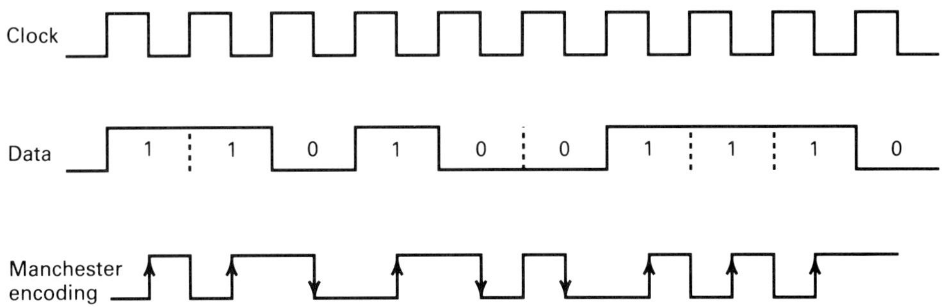

Fig. 11-28 Manchester encoding of data.

UART and *USRT* are generic terms. Manufacturers of microprocesssor family ICs often use their own names for these devices. For example, Motorola calls its microprocessor bus-compatible UART an *Asynchronous Communications Interface Adapter (ACIA)* and its bus-compatible USRT a *Synchronous Serial Data Adapter* (*SSDA*).

Compatibility with the RS-232 standard means that the electrical and mechanical characteristics of an interface do not violate RS-232 specifications. Some of these specifications are listed below:

- Output voltage levels are single ended (not differential) but are defined as -3 to -15 V for MARK (logic 1) and $+3$ to $+15$ V for SPACE (logic 0).
- Maximum switching rate is 30 V/μs.
- Load capacitance must be $\leq$2500 pF. Theoretically this limits the maximum cable length to 50 ft.
- Twenty-one electrical connections are defined in the RS-232-C interface specification, and while the standard does not specify the connector style, a 25-pin D connector (DB-25) with male shell for DTE and female shell for DCE has become the industry standard.

Deficiencies in the RS-232 standard include its speed limitation and cable distance limit. In addition, the split-supply, single-ended RS-232 line driver cannot be used in single +5-V systems without dc-dc conversion. Differential drivers are less susceptible to noise and can be powered off a single +5-V supply. Differential serial standards were developed to replace RS-232 but have not gained wide acceptance. RS-449 became an EIA standard in 1977 and was intended to overcome the deficiencies of RS-232, but acceptance has been poor. RS-449 provides a 2 Mbits/s maximum data rate with a 4000-ft maximum cable distance and its differential receiver specification minimizes noise. However, few modems and terminals support it today. RS-422A is a differential standard which supports 10 Mbits/s maximum data rate and is compatible with RS-449. It has been used by several DTE manufacturers but has not seriously challenged RS-232. The latest differential standard is RS-485.

The RS-232 interface may never be seriously challenged because of its limited area of application: low-end computer and terminal interface. With the advent of *local area networks* (*LAN*), new standards have emerged which allow a number of computers and terminals to communicate with each other on a single channel within a building or campus at data rates of 1 to 20 Mbits/s. The IEEE Standard 802 defines a family of LAN standards supported by popular vendor products. *Ethernet,* based on IEEE 802.3, is the most widely known LAN and is supported by over 100 vendors of computer and workstation equipment including DEC, XEROX, Intel, Daisy, and Cadnetix. Ethernet operates at 10 Mbits/s over 500-m coax segments. Repeaters permit extensions up to 2.5 km. Ethernet compatible products communicate with Manchester encoding of clock and data.

REVIEW 11-4

1. **Q.** What are the advantages of serial transmission? **A.** economy of hardware, ease of termination, longer cable distances, no crosstalk, less switching noise emission

2. **Q.** What is the disadvantage of serial transmission? **A.** reduced information bandwidth

3. **Q.** What is bandlimiting? **A.** Bandlimiting is the reduction of higher-frequency signal components.

4. **Q.** How does bandlimiting affect digital signals? **A.** Bandlimiting causes the edges of digital data to change more slowly and trail into adjacent bit times. This makes it difficult to differentiate between 1s and 0s at the receive end.

5. **Q.** What are harmonics? What is attenuation? **A.** Harmonics are the frequency components of a waveform. Attenuation is amplitude reduction.

6. **Q.** Explain half-duplex and full-duplex serial transmission. **A.** *Half-duplex* refers to transmission from point A to B and *then* from B to A, but not simultaneous transmission in both directions. *Full-duplex* refers to simultaneous transmission in both directions.

7. **Q.** What is jitter? **A.** Jitter is instability of signal edges.

8. **Q.** Define the terms *eye pattern* and *intersymbol interference*. How are they related? **A.** An eye pattern is the appearance of randomly changing data on an oscilloscope. Intersymbol interference is the effect of long decay tails from one pulse on subsequent bit times. Bandlimiting and intersymbol interference cause the eye pattern to close.

9. **Q.** What is a modem? **A.** It is a piece of data communications equipment which translates digital waveforms into analog waveforms for transmission over telephone lines.

10. **Q.** What is a baud? **A.** A baud is a symbol per second.

11. **Q.** What are the disadvantages of RS-232? **A.** RS-232 requires two power supplies, is slow, and operates to distances of only about 50 ft.

12. **Q.** What is the advantage of RS-232? **A.** It is widely used.

REVIEW QUESTIONS

1. What are the primary causes of lost and damaged data in digital systems?
2. What does it mean to terminate a wire?
3. How does the switching time of a logic signal relate to the frequency content of the waveform?
4. How can the characteristic impedance of a cable be measured?
5. How is the reflection coefficient used to calculate the amplitude of reflected pulses?
6. When is it permissible to ignore reflections?
7. What does it mean to say that a problem is intermittent?
8. When are reflections on data lines a problem?
9. What type of termination is preferred for CMOS? How does it work?
10. What is reverse crosstalk and how does it differ from forward crosstalk?
11. How, in general, is crosstalk avoided?
12. What is the major source of noise in a digital system, and how can it be minimized?
13. What are some of the advantages of multilayered PC boards?
14. What are stubs? What problems do they cause?

15. What value of test load capacitor is used for assessing propagation delay for 74LS logic? What value is used for 74AC and 74F logic?

16. What is a critical route? What signals require this?

17. How does crosstalk between gates in the same package occur? What type of part is most susceptible? When is this a problem?

18. Explain the difference between purely capacitive and resistive-capacitive loads. How does this relate to fanout?

19. What is one unit load in reference to fanout?

20. Where should backplane bus receivers be placed on a PC board?

21. How should 74LS bus devices be terminated?

22. Why are ribbon cables preferred over round bundles?

23. How is crosstalk controlled in ribbon cables?

24. What do bus standards tell us?

25. Explain the terms *handshaking* and *protocol.*

26. What is information bandwidth? Effective bandwidth?

27. How does a UART synchronize incoming data? Explain how it detects the beginning of a character.

28. What is the difference between baud rate and bit rate?

29. What practical significance does the difference between baud and bit rate have to data transmission over telephone lines?

REVIEW PROBLEMS

1. Determine the range of characteristic impedances of a 12-mil foil trace running across PC boards with ground plane separations of between 10 and 50 mils. Assume a uniform 0.0015-in height (t = 0.0015 in) and glass-epoxy construction.

2. Calculate the range of propagation delays for the 12-mil trace in Prob. 1 assuming between 1 and 10 input loads of 5 pF each.

3. What is the characteristic impedance of 25-mil power rails with 2 oz copper (t = 0.003) and 25-mil spacing between V_{CC} and ground? How does this number differ from the DC resistance of the traces?

4. How much current can a 74AC240 sink and still maintain an output voltage at or below the CMOS input threshold V_{IL}?

5. How much current can a 74AC240 sink and still maintain an output voltage at or below the TTL input threshold V_{IL}?

6. Determine the fanout of 74AC standard logic gates driving hypothetical logic family inputs which source 1 mA of current at $V_{IL} = 0.4$ V.

7. How much capacitance must a 74AC logic gate drive if it is connected to three 74AC logic inputs with 10-in of 15-mil-wide copper separated from the ground plane by 20 mils?

8. Use the methodology developed in Example 11-4 to determine if the line described in Problem 7 will require termination.

9. Explain how noise margin is calculated. What is the noise margin for 74LS driving 74HCT logic?

10. What is the function of the 1-K Ω resistor in Fig. 11-19?

11. Explain why bypass capacitors reduce switching noise on PC boards. Explain why they do not prevent noise spikes caused by simultaneous switching on a single octal package.

12. Explain how the bidirectional serial bus in Fig. 11-29 works.

13. What is the maximum clock rate for data on the bus in Fig. 11-29 assuming that switching from LOW to HIGH is complete in one *RC* time constant. This is a good assumption if the receiver logic has TTL inputs. Use $f_{max} = 0.3/t_s$ and a bus capacitance of 50 pF.

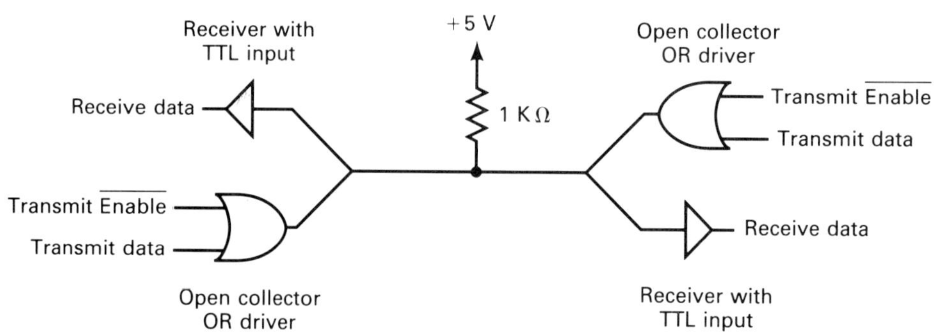

Fig. 11-29 Bidirectional bus example for Probs. 12 through 14.

14. If the receiver logic shown in Fig. 11-29 has TTL inputs, why can we not assume that switching from HIGH to LOW is complete in one *RC* time constant?

15. How many time constants are required to pull a bus from 5 V to a TTL input LOW threshold of 0.8 V?

16. How much current would be required to switch between 0.4 V and 2.4 V in 3 ns if the load were 50 pF?

17. How much dc current is required to maintain each of the terminations in Fig. 11-6 at $V_{OL} = 0.4$ V and $V_{OH} = 2.4$ V if $R_T = 100\ \Omega$ and $R_I = 220\ \Omega$ and $V_{CC} = 5$ V? Assume that the receiver inputs are high impedance.

18. When an 1800 Hz clock is used to transfer data having 4 distinct signal levels, what is the bit rate and baud rate of this data?

19. Define the voltage level requirements specified by EIA RS-232.

20. Why is a synchronous serial link faster than an asynchronous serial link using the same clock rate?

12 TROUBLESHOOTING

Troubleshooting is a difficult subject to teach. It is often said that every troubleshooting problem is unique and that experience and familiarity with the system under test are the key ingredients for success. One cannot argue with experience, and it is true that a thorough understanding of the system is an important element in fault diagnosis, but there are some basics which can be passed along. These basics may be organized into two categories: test equipment and procedures. This chapter does not presume to provide a comprehensive survey of troubleshooting equipment and procedures. It is intended, rather, to provide the beginner with some basic knowledge and a general sense of direction.

12-1 TROUBLESHOOTING TOOLS

Debugging, or *troubleshooting*, means diagnosing and correcting circuit faults. The most difficult part of the process is the diagnosis. In complex digital systems, failures in one part of the circuit often cause symptoms in other areas, and multiple faults or hazards may mask one another. It is quite common to find, after hours of probing, that you have been looking in the wrong place, or, after having found and fixed a problem, to find that another fault exists which was not apparent before. Moreover, there is always the danger of misinterpreting the findings because of the improper use of debugging tools. Even the simplest results can be misleading if the equipment is not fully understood. This section will discuss commonly used debugging tools with emphasis on their limitations and potential misuse. The questions you should ask yourself while reading this section are:

1 What is this piece of equipment?
2 How is it used?
3 What are its limitations and pitfalls?

The Logic Probe

The *logic probe* is an inexpensive hand-held tool which can sense logic levels and transitions. Some higher-priced models also sense floating (high-impedance) nodes. Connections to circuit power and ground are normally made with grabber squeeze connectors and connection to the node under test is made with a pin tip or a special accessory tip. There are generally one or two switches and one or two visual indicators. The main switch provides a choice between TTL and CMOS logic input levels, and the main indicator lamp or LED provides an indication of the logic state of the node under test. A second lamp or LED may provide an indication of a single pulse occurrence.

Fig. 12-1 The HP model 545A Logic Probe (Courtesy of Hewlett-Packard)

An example which incorporates all these features is the Hewlett Packard 545A logic probe shown in Fig. 12-1. Fig. 12-2 shows how the logic indicator lamp on the HP 545A responds to voltage levels and pulses. The indicator lamp is normally in the "dim" state and must be driven into one of its other three states—off, bright, or flashing—by the voltage levels at the probe tip. The lamp is bright for inputs above the logic 1 threshold and off for inputs below the logic 0 threshold. The lamp is dim for voltages between the logic 1 and logic 0 thresholds (and for open circuits). Pulse-stretcher circuitry in the probe converts positive pulses of ≥ 10 ns into pulses of ≥ 50 ms. Negative pulses cause the indicator to go off momentarily. Pulse train inputs cause the lamp to flash at about a 10-Hz rate.

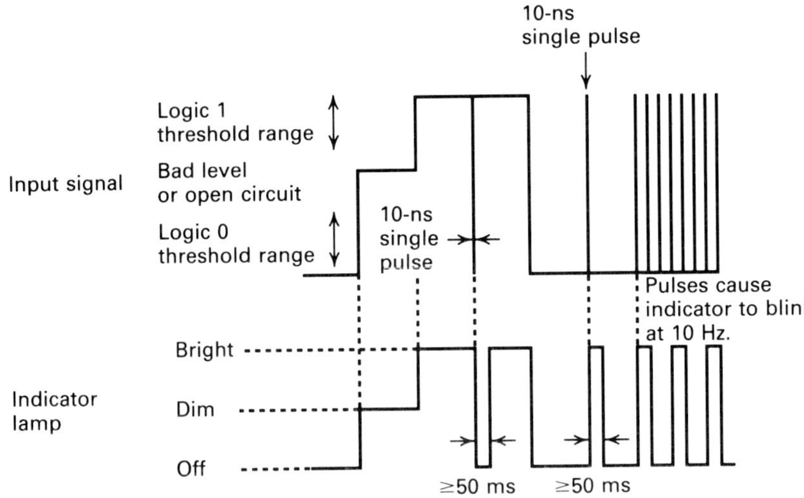

Fig. 12-2 Logic probe level indications.

The logic probe is particularly useful for monitoring control signals such as reset, start, stop, shift/load, and clock. The stages of a long counter may also be checked by moving the probe backward from the final output to the input and noting along the way if the indicator is flashing. This front-to-back examination is often useful for quickly isolating the problem to a specific area. The dim indicator feature is useful for testing tristate logic outputs. It also detects floating or unconnected gate inputs. Floating inputs are commonly caused by bent-under pins. When a dual-in-line package (DIP) IC is inserted by hand into a socket, a pin may be bent under the body of the package. This may go unnoticed until the logic probe detects a high impedance point. Some logic probes, including the HP 545A, contain a memory indicator LED which turns on when a transition (HIGH to LOW or LOW to HIGH) occurs. The LED remains on until it is cleared with the MEM-CLR reset button. This allows detection of short-duration, one-shot occurrences, such as unwanted glitches on a reset line.

A special type of logic probe, called a *logic clip* or *logic monitor*, attaches to DIPs and continuously monitors the I/O logic states of the package. The logic clip derives its power from the IC itself regardless of which pins are used. It contains its own gating logic for locating V_{CC} and ground, and it has buffered inputs to reduce circuit loading. The logic clip is useful for analyzing static systems or systems with low clock rates. High-frequency dynamic systems may be checked by either reducing the clock rate or by using a logic pulser.

The *logic pulser* is a device, packaged in a similar way to the logic probe, which allows the user to momentarily drive a node either high or low without disconnecting it from the circuit. The logic pulser output is tristateable and is normally in the high-impedance state. When the pulse button is pressed, a narrow pulse is delivered to a node by sourcing or sinking up to 650 mA of current. This overrides any other IC outputs on the node and provides a test input to a clipped or probed IC. The pulse width is limited so that the amount of energy delivered to the circuit is not excessive. Some pulsers also provide continuous pulsing and a means of putting an exact number of pulses into counter or shift-register clock inputs. The logic pulser is normally used with the logic probe or logic clip for isolating faults in a particular portion of a circuit.

Logic probes and clips cannot isolate faults caused by slow transitions, reflections, or marginal setup and hold times. They are basically static hazard indicators. If a system can be slowed down, they will do a better job of exposing faults. Many dynamic hazards, however, will not appear unless the system is run at normal speed. Moreover, probes and clips have no way of indicating timing relationships between data inputs and the system clock. Their usefulness is generally limited to isolating gross failures, bad connections, and stuck nodes. They tell us whether counters are changing and whether output signals are reaching their intended inputs. They can also detect troublesome glitches on lines which should be quiet, but only if those glitches cross logic thresholds on a regular basis. Often a glitch will occur regularly but not always with sufficient amplitude to trigger the logic probe. A reset line, for instance, may exhibit periodic glitching but only occasionally cause a false trigger. The logic probe may or may not uncover such a hazard.

The Oscilloscope

The oscilloscope allows us to actually look at a signal. If there are glitches, we can see them, whether or not they cross logic thresholds. We can also see timing relationships between clocks and data. There are other measuring devices, such as voltmeters, frequency counters, and spectrum analyzers, which provide a more accurate measure of specific parameters, but they cannot provide the overall visual representation which is so valuable in troubleshooting. The scope provides a view of signal integrity. *Signal integrity* refers to how closely the actual digital waveforms approximate the ideal. When discussing digital logic, we assume logic 0 or 1 square waves of a specific duration. Actual waveforms deviate from this ideal because of timing ambiguities (due to variations in propagation delay), transmission line effects, and faults (shorts, opens, and high-resistance points). The scope shows us a stable picture of the actual waveforms and transfers momentary events from real time into a time domain which is suitable for human observation.

Nevertheless, one of the most common problems with troubleshooting in the field is misuse of the oscilloscope. Much time and money is wasted every year because field site personnel cannot properly trigger on one signal while looking at another. Remote site diagnosis over the telephone is particularly difficult when the technician on site does not have a good working knowledge of the test equipment. In our experience, many a costly trip to the field site could have been avoided if field service personnel had correctly used the oscilloscope. For this reason, particular emphasis is placed on this piece of test equipment.

Sweep and Trigger

Periodic waveforms are the easiest to display on an oscilloscope because a period of time can be defined in which the pattern exactly repeats. Many waveforms are repetitive but not periodic. They exhibit patterns which do not always repeat within a given period of time. These non-periodic signals must be stabilized by referencing their transitions to a periodic signal such as a clock or control pulse. A misunderstanding of the *sweep* and *trigger* functions of the oscilloscope is the most common cause of error and frustration among beginners.

Basically, the oscilloscope consists of an *electron gun* which beams electrons to the screen of a *cathode ray tube* (*CRT*) in such a way that it follows the voltage levels of a waveform for a specified period of time. The electron beam is said to *sweep* across the screen. The sweep starts when the reference signal crosses a certain voltage level in either the positive or negative direction. The starting place is called the *trigger point* and is determined by the *level and slope* controls on the front panel of the scope. The exact location of these controls will vary from one scope to another, but they often appear together as shown in Fig. 12-3. The level

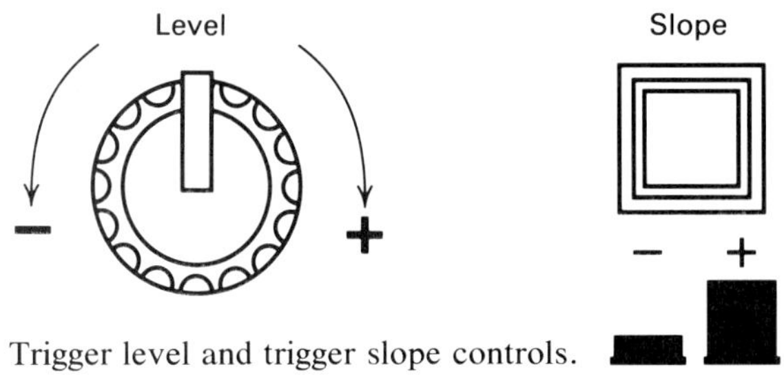

Fig. 12-3 Trigger level and trigger slope controls.

control is a potentiometer and therefore appears as a knob. The slope control is a switch and may be a push button. Trigger level determines the voltage level which starts the sweep. Trigger slope determines whether the sweep will start on a rising or falling edge. Normally, the level is set to 0 V, but it may be necessary to raise it in order to avoid false triggering on noise. After the sweep has been triggered, the beam follows the movement of the signal until it reaches the right-hand side of the screen. It then returns to the left-hand side to be triggered again at the next trigger point. The return portion of the sweep is referred to as *flyback*. During flyback the beam is turned off, or *blanked*. The visual persistance of the eye prevents us from observing this blanking when the scope is set to around 60 or more sweeps per second.

Sweep Rate

The *sweep rate* is determined by the TIME/DIV knob, sometimes labeled SEC/DIV. *DIV* refers to the major grid divisions. At 1 μs/DIV, for instance, each horizontal grid, or graticule, represents 1 μs of time. Fig. 12-4 shows typical graticule measurement markings. Also shown are the knobs which control the scale of those graticule markings. The horizontal, or *x*-axis, divisions represent time in SEC/DIV. The vertical, or *y*-axis, divisions

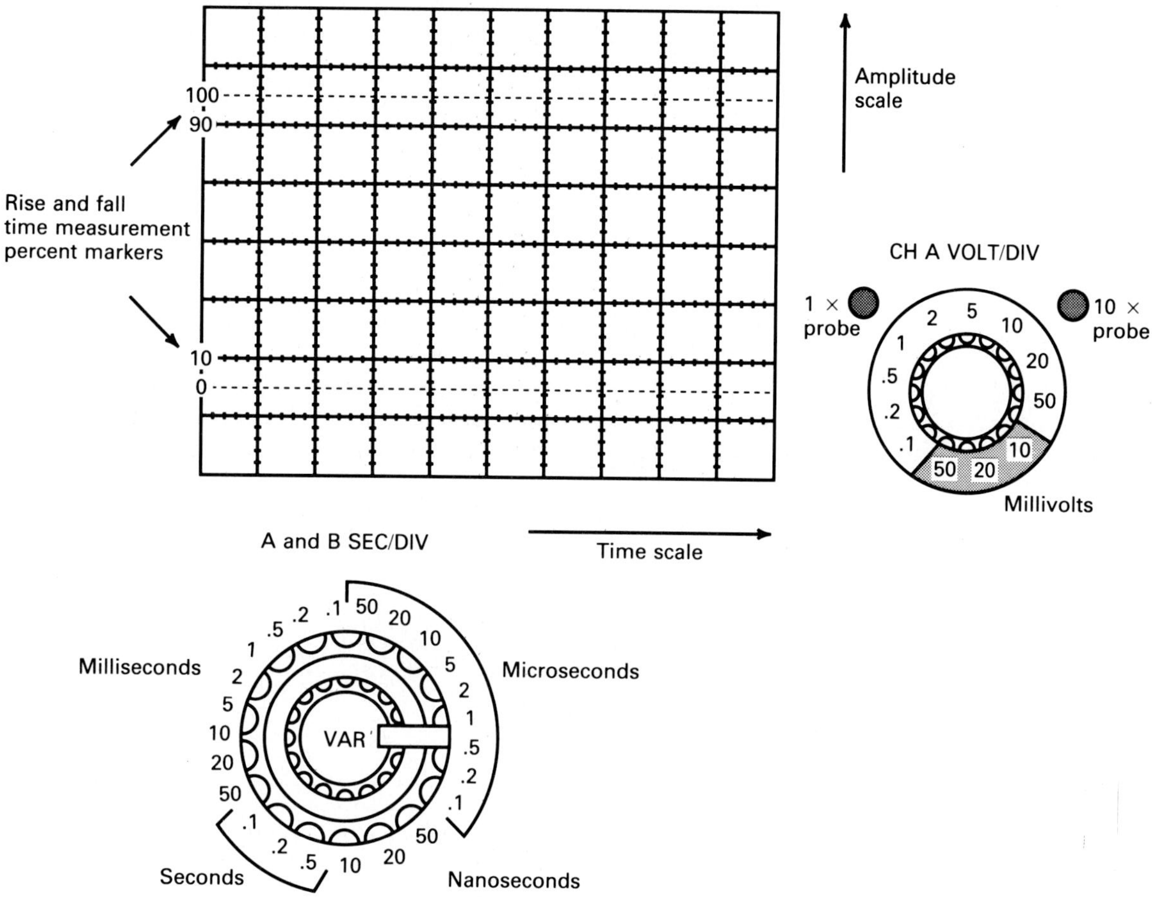

Fig. 12-4 The horizontal scale is controlled with the SEC/DIV knob and represents time. The vertical scale is controlled with the VOLT/DIV knob and represents voltage amplitude.

represent signal amplitude in VOLT/DIV. The relationships between trigger level, trigger slope, and sweep rate is illustrated in Fig. 12-5.

The choice of sweep rate depends on the type of measurement being performed. When the frequency of a signal is being measured, the most accurate method is to select a sweep rate which displays one cycle across the full horizontal scale. For example, one cycle of a 10-MHz signal can be displayed across 10 divisions with a 10-ns/div sweep. When rise times are being measured, the sweep should be chosen so that the rising edge appears with a nearly 45° slope. Percentage marks for the measurement of rise and fall times are located on the left side of the display grid. The VOLT/DIV knob can be used to enlarge the scale of the displayed waveform to fit into the 10/90 percent marks for switching time measurement. For very fast signals neither of these rules of thumb may be possible due to the frequency limitations of the scope. Most often, the sweep rate is chosen visually by expanding the time base just enough to see the time relationship between one edge and another.

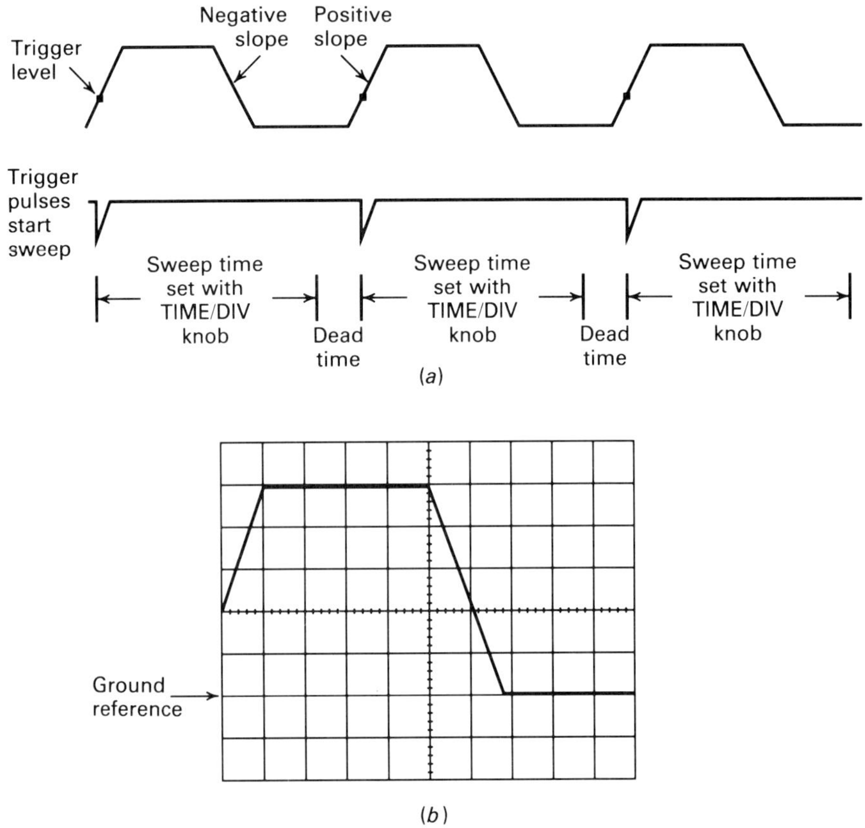

Fig. 12-5 The effect of trigger level, trigger slope, and sweep time on the displayed waveform.

Triggering Options

The front panel provides other trigger options in addition to level and slope. *Trigger source, trigger mode, and trigger coupling* are also selectable. *Trigger source,* controlled with a front panel switch, determines whether triggering occurs from one of the displayed waveforms (INT) or from an external source (EXT). Any internal input may be displayed *and* used to trigger the sweep. External inputs are used for triggering purposes only. Often a line trigger option is provided so that the display may be triggered with the power line voltage (60 Hz). This is useful when 60-Hz interference (hum) is suspected to be a culprit in circuit malfunction.

It is possible to present a perfectly stable picture only when the waveform repeats. This is the case with clocks and control signals. Data waveforms may vary with time and will usually present a more complex picture. If the data rate is very slow, as with 150-baud asynchronous serial transmission, the data changes can be observed directly as the beam wiggles across the screen. Fast data, however, appears as an *eye pattern* with both highs and lows displayed in the same time period. A periodic waveform with a fixed time relationship to the data must be used as a trigger. The trigger waveform may be displayed or it may enter the scope through the EXT trigger input. Figure 12-6 shows how serial data and clock may be viewed

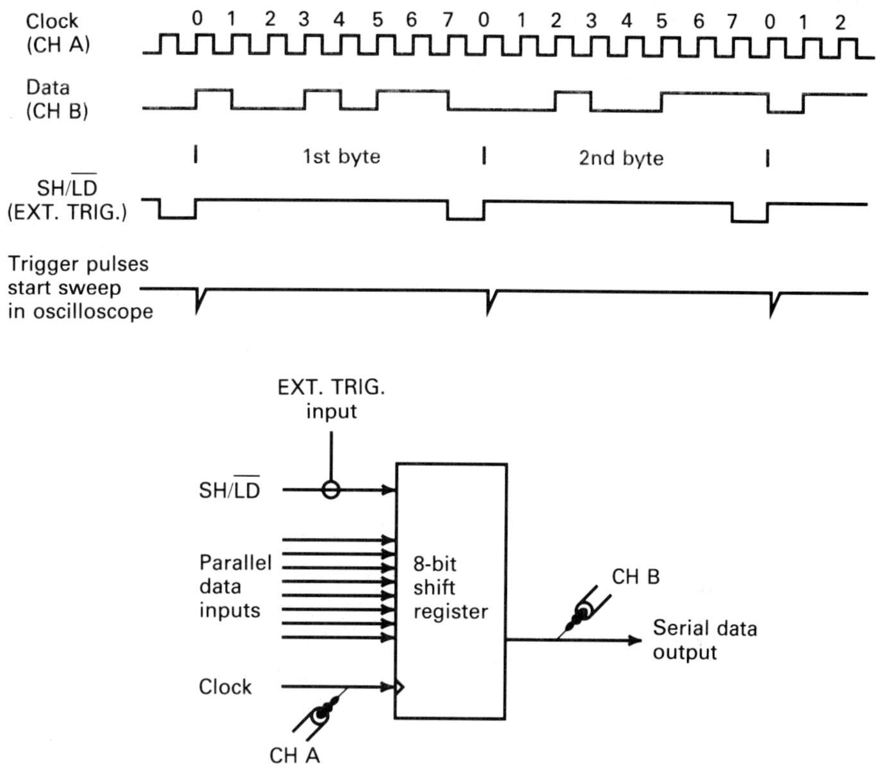

Fig. 12-6 Using the EXT trigger input to trigger on serial data and clock.

on a dual trace scope using a shift-load pulse as the trigger. The EXT trigger input is useful for stabilizing random data when a trace input cannot be spared. When displaying random data, only the edges will appear stable. This is unavoidable since each sweep displays a different data pattern and the eye cannot respond quickly enough to differentiate between sweeps when high sweep rates are used.

Even when signals are periodic, the waveforms may appear unstable because of the relationship between sweep rate and signal period. Figure 12-7 shows a series of evenly spaced periodic control signals which presents an unstable display because the sweep finishes between two double pulses. Some scopes provide a *variable hold-off* adjustment which corrects this problem. After each sweep, the trigger circuitry waits for an adjustable hold-off time before looking for another trigger voltage. This hold-off time can be varied so that multiple sweeps will coincide and a stable image is obtained. If the hold-off time is not properly adjusted, as in Fig. 12-7, the result is overlapping of noncoincidental sweeps. The dashed line indicates the proper adjustment needed to prevent this. When using a scope without the variable hold-off feature, this problem can be avoided by using the vernier adjustment on the TIME/DIV knob to select a sweep rate somewhere between the main settings. This, however, results in an uncalibrated display, and the graticule divisions no longer provide a usable time scale.

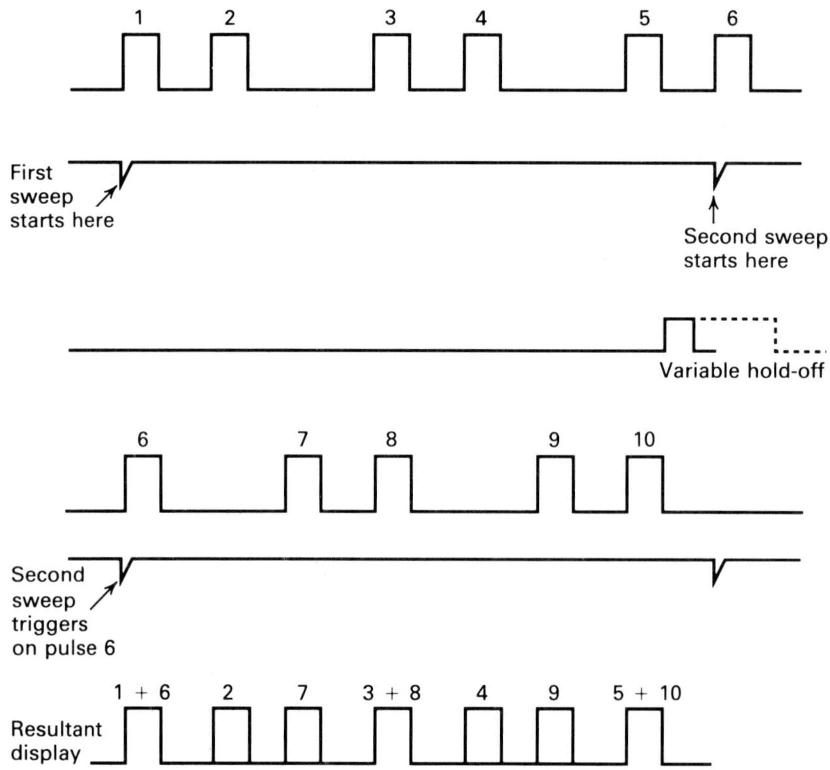

Fig. 12-7 Instability of periodic waveforms due to misadjustment of variable hold-off control. (Example from Van Erk, *Oscilloscopes,* McGraw-Hill, 1978)

Trigger mode refers to a group of front-panel controls labeled AUTO, NORM, and SINGLE. When NORM, or normal mode, is selected, the level knob determines when sweeps occur. When no triggering occurs, no sweep is initiated, and the screen appears blank. A blank screen is ambiguous: it may indicate that no signal is present, or it may indicate that the level control is adjusted beyond the range of the signal amplitude. The AUTO mode resolves this ambiguity by providing a sweep even when there is no trigger. Small voltages, too low to cause a trigger, will appear as an unstable signal in the AUTO mode. When no signal is present, a stable reference trace will appear indicating the position of ground. When a voltage large enough to cause a trigger occurs, the sweep reverts back to normal mode and provides a synchronized sweep until the trigger is removed. AUTO should be used whenever trigger repetition rates are greater than 10 Hz. Below 10 Hz, the AUTO mode may cause instability.

The SINGLE, or single sweep, mode provides only one sweep, triggered on the first incidence of the selected level and slope. Subsequent occurrences of the trigger voltage are ignored until the RESET button is pushed. Still photographs of nonperiodic waveforms may be obtained with standard oscilloscopes using the single-sweep trigger and a special camera. This is an effective way to capture hard copies of fleeting images. An alternative is to use a storage scope. *Analog storage scopes* employ long persistance phosphors and charge storage techniques to maintain a single-sweep image long enough for human viewing, but the persistant image tends to bloom and then fade with time. *Digital storage scopes* freeze nonperiodic images by sampling the waveforms with sample-and-hold circuitry and digitizing their values with a high-speed ADC (as described in Chapter 10). The digitized information is then stored in RAM and displayed, using a DAC, as stable multiple sweeps. Quantized waveforms stored in memory provide several measurement capabilities not possible with analog storage scopes. Digital storage scopes allow you to look at a waveform both *before and after* the trigger event. Digital scopes also provide unattended monitoring of transient events and signal data output to other instruments over RS-232 serial, or *general-purpose interface bus (GPIB)* parallel, data buses. Moreover, they provide unlimited storage times and clear, crisp displays of nonrepetitive events without the blooming or fading associated with analog storage scopes.

Trigger coupling determines how the signal used for triggering is coupled to the internal trigger circuitry. The trigger coupling may be dc or ac. Direct-current coupling allows triggering on all signals down to a 0-Hz frequency. Alternating-current triggering blocks dc voltages. This is useful when observing low-amplitude switching noise on the V_{CC} supply line. Some scopes provide two ac settings: one for low-frequency signals and one for high-frequency signals. The low-frequency coupling may be used to suppress high-frequency noise which may cause false triggering and interfere with the production of a stable picture on the screen. The high-frequency setting will suppress low-frequency noise, such as 60-Hz hum. The crossover frequency between the two settings is normally 50 KHz. Alternating-current trigger coupling should be used only when viewing low-amplitude disturbances on digital waveforms. Direct-current coupling will provide correct triggering and amplitude measurement for all frequencies in the absence of interference and is usually preferred for digital troubleshooting applications.

CHOP and ALT

The most commonly used oscilloscopes are *dual trace*. This means that two waveforms may be displayed at the same time. This does not mean that there are two electron guns shooting separate beams. Scopes with separate beams are called *dual beam*, and scopes with simultaneous sharing of a single beam are called *split-beam* oscilloscopes. The *dual-trace* scope uses a single-gun CRT which is alternately connected to two signals with switching circuits. Figure 12-8 shows the two modes of dual-trace operation which are selectable with controls on the front panel. The first, called the *alternate mode* (or *ALT*), provides alternate sweeps of the two waveforms. One waveform is swept through an entire period, and then the second waveform is swept. With slower repetition rates, the viewer sees the traces being displayed one after the other. With faster repetition rates, the alternating effect is not noticeable. The *chopped mode* (*CHOP*) alternates between waveforms much more often than once per sweep. These time-

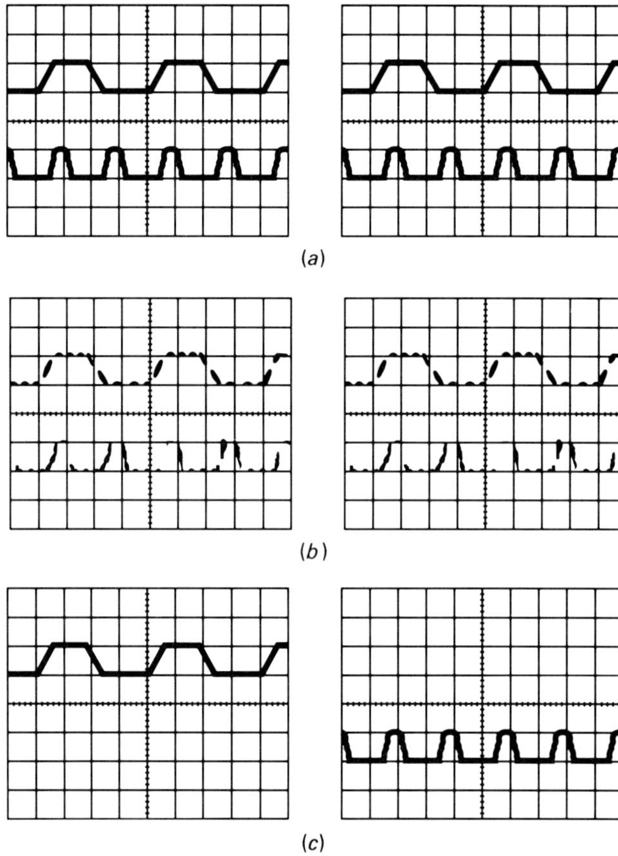

Fig. 12-8 (*a*) Dual beam; (*b*) dual-trace CHOP mode; (*c*) dual-trace ALT mode.

sharing techniques can be extended to more than two signal traces to accommodate four-channel scopes.

When troubleshooting digital faults, it is usually important to know the time relationship between signals. Triggering in the alternate mode will sometimes cause incorrect information to be displayed. Figure 12-9*a* shows two signals displayed with their correct timing relationship. Figure 12-9*b*. shows the result of triggering on CH A in the ALT mode. The incorrect result is due to the fact that the first sweep (CH A) triggers off of pulse 1 of the CH A input and the second sweep (CH B) triggers off of pulse 2 of the CH A input. The correct display would have been obtained if triggering had been made on CH B. A good rule to follow when measuring the phase relationships between two or more time-related signals is to trigger on the signal with the lower repetition rate. This type of situation is often encountered when troubleshooting counter outputs and control signals.

Delayed Sweep

When a signal voltage reaches the trigger point, as defined by the level and slope controls, a sweep is initiated, and subsequent trigger voltages are ignored until the sweep is complete. When we are interested in only a

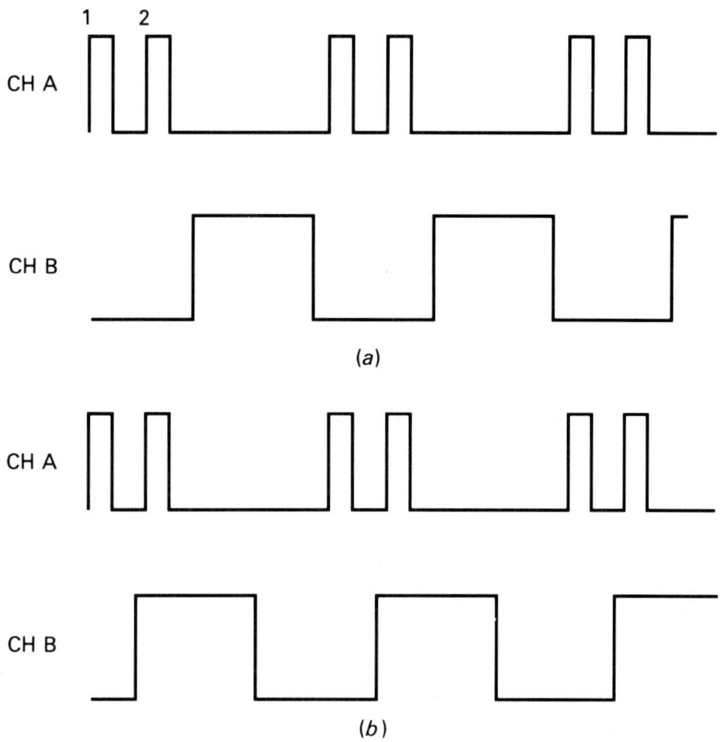

Fig. 12-9 Incorrect timing relationships displayed in ALT mode. (*a*) Correct time relationship between signals. (*b*) Incorrect time relationship caused by triggering on CH A in ALT mode. The first sweep is triggered on the rising edge of pulse 1, and the alternate sweep is triggered on the rising edge of pulse 2. (Adapted from Van Erk, *Oscilloscopes,* McGraw-Hill, 1978)

small part of the waveform, we can adjust the TIME/DIV knob to get a closer look. However, expanding the time base with the TIME/DIV knob may cause the relevant portion of the waveform to disappear. For example, if we wish to view a small glitch or spike which is triggered off a signal transition, we may not be able to view it closely enough without pushing it off the right side of the screen. This problem is solved with *delayed sweep*. Scopes with *delayed sweep* have two calibrated time bases. The second time base sweep may be started some time after the main time base trigger. This allows comparisons of the same signal at two different sweep rates. If the second time base has an independent trigger, then stable measurements on the delayed sweep are possible. Figure 12-10 illustrates the function of delayed sweep.

Input Coupling

The signal inputs to a scope are controlled with a switch which selects dc coupling, ac coupling, or GND. When ac coupling is selected, the dc component of the signal is blocked by a capacitor connected between the input terminal and the internal circuitry. This is useful when examining switching noise on the V_{CC} line. Alternating-current coupling brings the

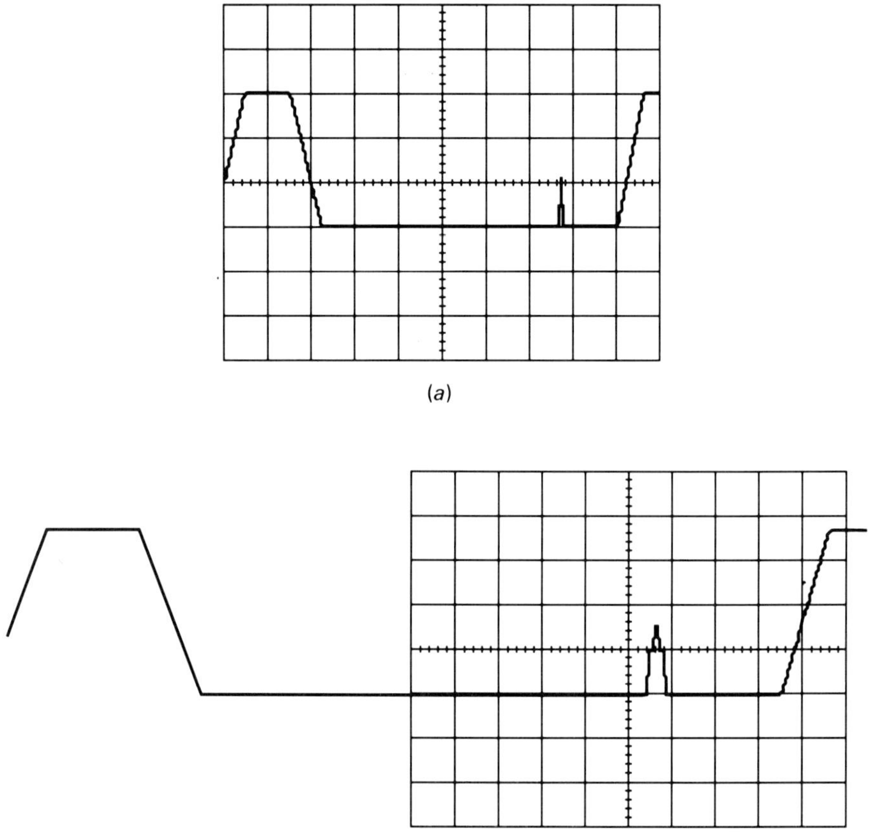

(a)

(b)

Fig. 12-10 Using delayed sweep to obtain an expanded view of hazards. (*a*) Original view of signal and glitch; (*b*) expanded view of glitch using delayed sweep.

noise down to the middle of the screen where it can be expanded for closer observation. Normally the input should be left at dc coupling for testing digital circuitry. The actual dc values of high and low outputs are of great interest when troubleshooting digital logic. The switch position marked GND causes the ground reference to be displayed so that the sweep can be positioned with the POSITION knob. The ground reference should be positioned so that it is coincident with a horizontal graticule line. This facilitates the measurement of V_{OL}, V_{OH}, reflection amplitudes, and other dc levels. After this adjustment has been made, be sure to switch the coupling back to dc.

Differential Inputs

Many scopes have an ADD function which permits algebraic summing of the two input signals on the A and B channels. This may be used with the INVERT switch to provide differential measurements. Inverting the B input and selecting the ADD function will cause an A-B output to be displayed. If the signals on the A and B inputs are identical, then no signal at all will be displayed. This is useful when there is a suspected short between signal lines. Nonshorted lines will show activity in this mode, but shorted lines will appear quiet.

Probes

Some of the most common measurement pitfalls involve the misuse of scope probes. Probes are the connecting cables between the circuit under test and the oscilloscope. Whenever a cable is involved, the effects of capacitive loading and impedance mismatch must be considered. When discussing propagation delays and switching times in Chapter 11, we noted that specifications were made for standard test loads of 15 or 50 pF. Three feet of cable in the form of a scope probe can add up to 90 to 100 pF to the load. This is in addition to the 15 to 25 pF of input capacitance in the scope itself. Moreover, if the cable is not properly terminated, the scope will see, and possibly cause, reflections which are not actually present during normal operation.

Figure 12-11*a* shows how a 3-ft-foot scope probe might appear to the output of a signal source. The signal source must charge 105 to 125 pF of

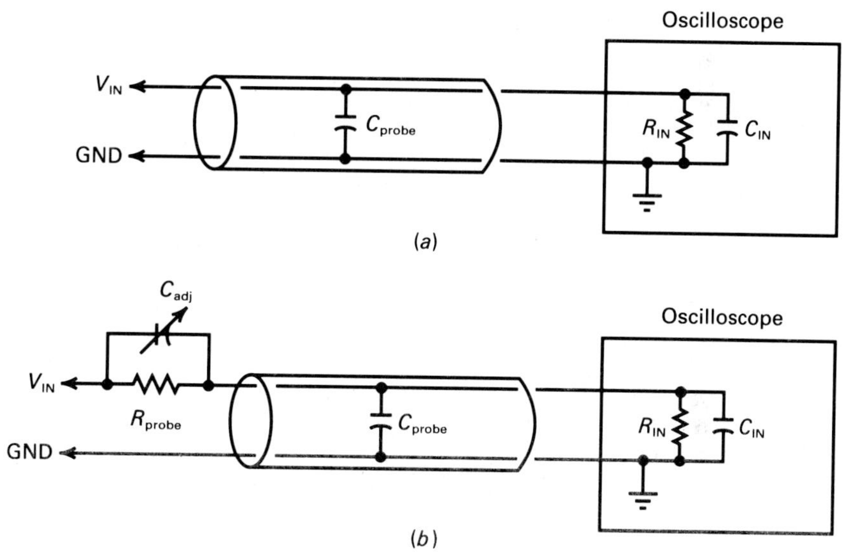

Fig. 12-11 Equivalent circuits using 1:1 and 10:1 probes. (*a*) Equivalent circuit of 1:1 probe with oscilloscope; (*b*) equivalent circuit using a 10:1 probe. C_{adj} is adjusted so that $R_{probe} C_{adj} \approx R_{in} (C_{probe} + C_{in})$. $R_{probe} = 9R_{in}$

test equipment capacitance in addition to the circuit load. This will lengthen switching times and propagation delays and may alter circuit behavior. When hold times are marginal, the circuit may work properly while the probe is attached but fail after the probe is removed. When setup times are marginal, the opposite may occur. In either case, the test equipment has become part of the circuit under test.

This problem can be reduced by using a 10:1 probe. A 10:1 probe creates a voltage divider for both ac and dc signal components and reduces the signal load by a factor of 10. Figure 12-11*b* shows the equivalent circuit when a 10:1 probe is used. Spurious ringing and reflections will also be diminished with the use of 10:1 probes. The actual dc resistance of a 3-ft cable would normally be much less than 1 Ω. However, special resistance wire is used in 10:1 probes to dampen any reflections between the probe tip and the scope input.

Proper operation of any probe depends on a short ground connection between the probe and the circuit under test. Grounding clips are provided with all scope probes, and the ground connection should be made as closely as possible to the ground lead of the signal source. Even when ground connections are kept short, a false picture of the waveform may result if the probe is not correctly adjusted. Since the input capacitance is not the same for all scopes, the divider capacitance in 10:1 probes is adjustable. The adjustment is made by placing the probe tip onto the calibration-generator output of the scope, marked CAL, and turning the adjustment screw with a screwdriver until a flat square wave is obtained. The screwdriver adjustment may be found either at the probe tip or in a box at the end of the cable. Improper adjustment will cause undershoot or overshoot as shown in Fig. 12-12.

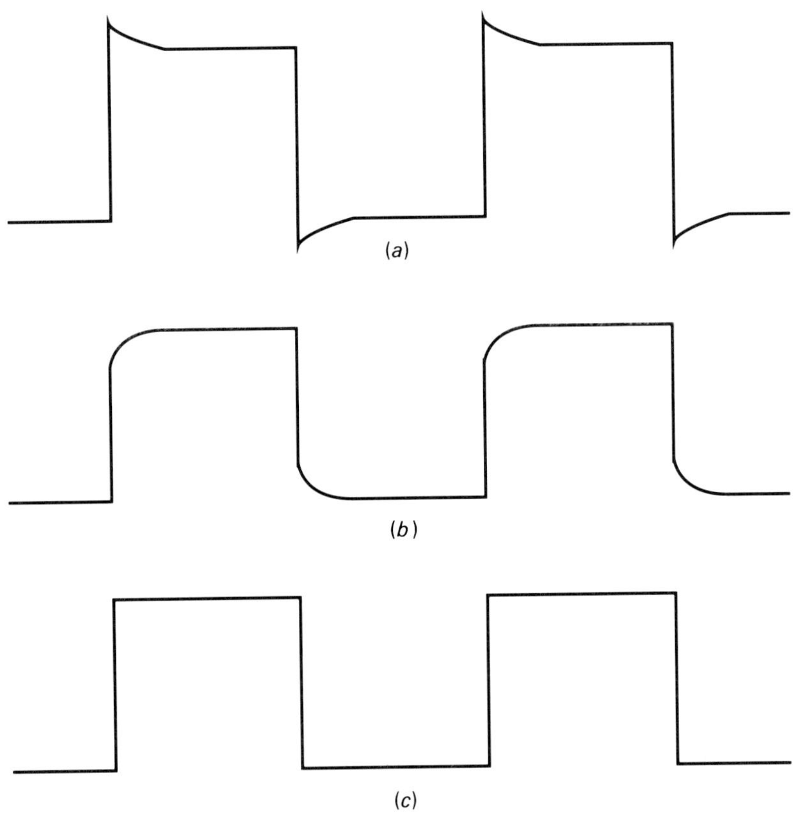

Fig. 12-12 Effects of over and undercompensation of the oscilloscope probe. (*a*) Overshoot due to loss of low-frequency signal components. (*b*) Undershoot due to a loss of high-frequency signal components. (*c*) A correctly compensated probe produces a flat response when connected to the CAL test output.

The purpose of the 10:1 probe is to provide a higher resistance and lower capacitance to the signal source output. This reduces the effects of loading, but also reduces the *sensitivity* of the scope. Sensitivity is adjusted with the VOLT/DIV knob and refers to the input signal level (in volts) needed to produce a given deflection of the electron beam (in graticule divisions). Equipment specifications for sensitivity are given in millivolts per graticule division or millivolts per centimeter. With this specification you can determine if small signals will be displayed with enough amplitude to make accurate measurements. The reduced sensitivity associated with 10:1 probes is generally not a problem when measuring digital waveforms. Sometimes, however, when high-frequency waveforms are to be observed, a 100:1 probe would be necessary to prevent loading and reflections. A 100:1 reduction of signal amplitude may very well present a sensitivity problem. In these cases an active, FET input probe should be used. An active probe contains a miniaturized differential amplifier with high-impedance FET inputs within the probe tip. The amplifier obtains its power from either the oscilloscope power supply or from a separate power unit.

Bandwidth and Rise Time

Another consideration when using an oscilloscope is *bandwidth* and *rise time*. Bandwidth is the range of frequencies that a scope can handle with less than 30 percent loss of voltage amplitude. The bandwidth specification of a scope normally appears on the front panel (100 MHz, 350 MHz, etc.). The rise time determines how quickly the scope can respond to signal transitions. Given the bandwidth, the rise time may be calculated in most cases with the following approximation:

$$\text{Rise time} = \frac{0.35}{\text{bandwidth}} \tag{12-1}$$

The switching times of digital waveforms are measured between the 10 and 90 percent voltage levels. These measurements will be in error if the scope cannot respond faster than the waveform can change. The rule of thumb for timing measurements is to use a scope with a rise time at least five times faster than the signal you wish to measure. A 5:1 ratio provides a rise time measurement with ≤ 2 percent error.

The Logic Analyzer

The oscilloscope offers the best means of verifying signal integrity. Timing ambiguities, reflections, and glitches can be viewed directly and accurately. However, when troubleshooting larger, system-type problems, the oscilloscope suffers from three shortcomings:

1. The scope can view only one or two (or sometimes four) signals at a time. When problems occur in bus-oriented systems, it is often necessary to view eight or more signals at one time.
2. The scope can trigger only on individual waveforms, not on patterns. If a control signal, such as a load pulse or terminal count, occurs at a

convienent time, then a scope will do the job. But often a problem occurs after the occurrence of a particular pattern of bits. This is common when troubleshooting microprocessor systems. The scope has no way of recognizing and triggering on patterns in the data.

3. Most scopes have no means of looking at events which have occurred *before* the trigger event. The exception, of course, is the digital storage scope.

These shortcomings have been answered by an instrument called the *logic analyzer*. Logic analyzers sample and store multiple waveforms from a system bus and display them either as signal traces or as textual or graphical information. The technology used in logic analyzers is the ADC to memory to DAC configuration discussed in Chapter 10 and used in digital storage scopes. Logic analyzers have the additional capability of referencing the signal waveforms to the software programs which generate them. In microprocessor applications, this allows the user to follow the execution of a program as it progresses through each of its steps. Any incorrect steps will be obvious when the logic analyzer output is compared with the expected results.

There are two types of logic analyzer: synchronous and asynchronous. The *synchronous analyzer* samples the signal lines with the master clock from the system under test. It accepts data in the same way that the system under test accepts the data—on the clock edge. Only changes in machine state are displayed, and for that reason it is called a *state analyzer*. The *asynchronous logic analyzer* uses an independent clock, running much faster than the system clock, to sample and store waveform information. It reflects changes *between* system clock edges and is sometimes called a *timing analyzer*. The advantage of the asynchronous analyzer is its ability to detect glitches and marginal signals. It operates much like a digital storage scope. Some logic analyzers offer both modes of operation: state analysis and timing analysis. Traditionally, the state analyzer has been considered a tool for software troubleshooting and the timing analyzer a tool for hardware troubleshooting. In either case, the waveform data is clocked into RAM and compared with a preset trigger word selected by the user. When the incoming data matches the trigger word, a preselected switch determines whether that data word will be the first or last to be displayed. Thus the trigger word can be used to start or stop the display. Trigger delays are also possible. The analyzer can be set up to start displaying information any number of clock periods after the trigger word is detected.

Once the waveforms are sampled and stored in RAM, there are three modes of display: timing waveform, logic state display, and graph mode. Figure 12-13 shows examples of screen displays for each mode. Selection of the *timing mode* turns the analyzer into a multiple-trace storage scope. Eight or more waveforms can be displayed at once. This makes a very powerful hardware troubleshooting tool. In the *logic state display mode,* all the information stored in RAM, beginning or ending with the trigger word, is displayed in binary, octal, hex, or mnemonic format. Since both

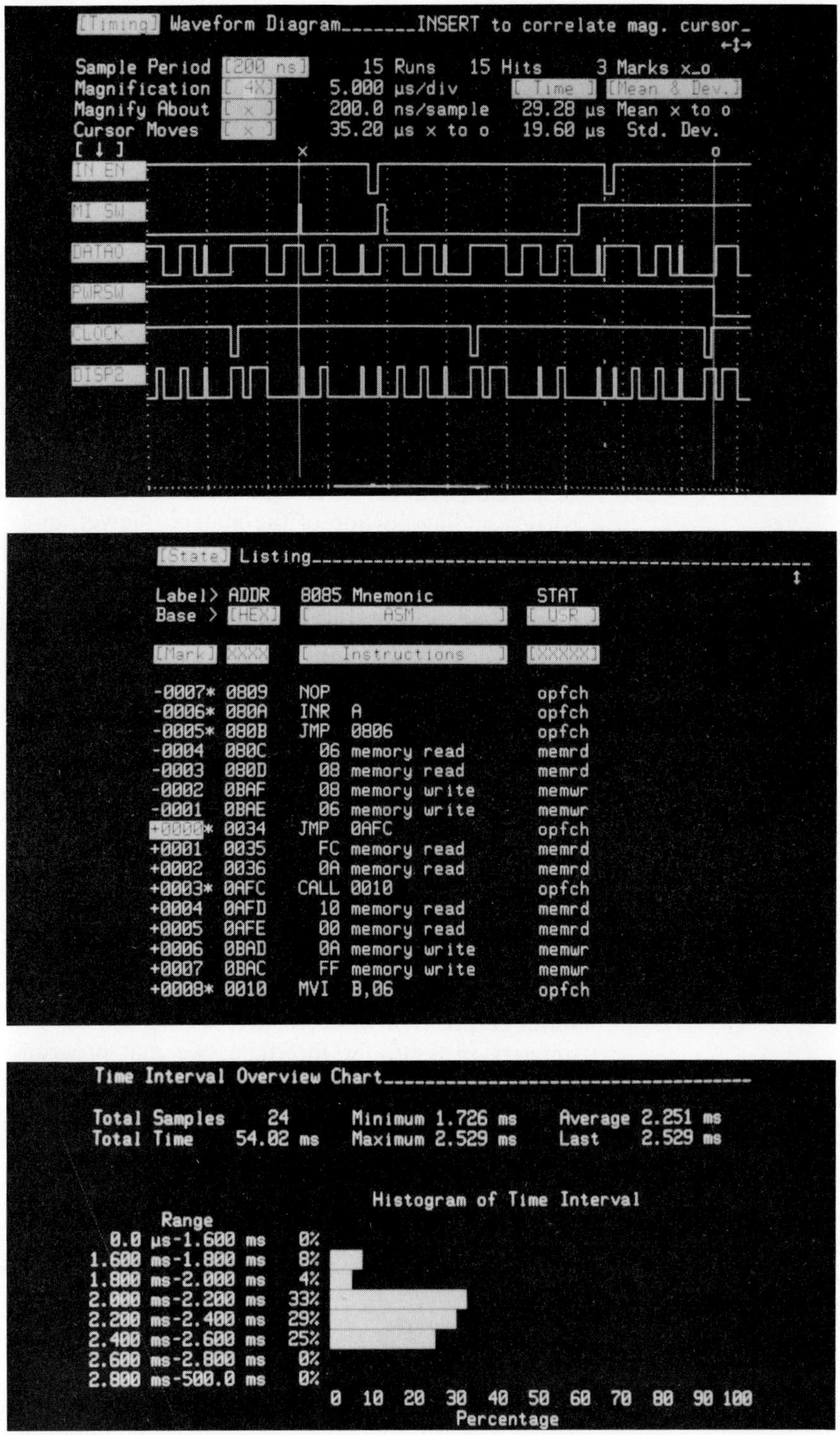

Fig. 12-13 Examples of logic analyzer display modes (*a*) timing waveform, (*b*) logic state, and (*c*) histogram (Courtesy of Hewlett-Packard)

address lines and data lines are sampled, the movement of a state machine can be followed as it sequences through its programmed states and its "trail" can be displayed on the logic analyzer CRT. When the state machine is a microprocessor with a known instruction set, the address and data waveforms can be reverse assembled (translated back) into instruction mnemonics and comments. This provides a very powerful software debugging tool.

The third mode of display, the *graph mode,* presents a picture of system activity so that tight looping and illegal memory intrusions are highlighted. *Tight looping* means that the program is stuck in a localized area of memory. This may indicate a software bug, or it may mean that a peripheral device is not responding. Older analyzers displayed a *map* of system activity on a screen matrix. The screen was divided into a 16 × 16 matrix representing a portion of address space in RAM. A graph of RAM activity was displayed by highlighting portions of the matrix. More frequently accessed portions of RAM were highlighted with brighter lines in the corresponding portion of the matrix. Newer analyzers use a *histogram* to indicate which areas of RAM exhibit the greatest activity. Several ranges of address space may be selected, and the display will show the percentage of time spent in each range.

Logic Analyzer Applications

The synchronous state analyzer is most often used to trace program flow, but it may also be used in hardware debugging. The synchronous state analyzer provides a minimum of 16 address inputs and 8 data inputs to monitor microprocessor bus activity. In addition, clock inputs, clock qualifier inputs, and trigger outputs are provided. Clock qualifier signals such as read, write, and valid memory access, allow the user to *qualify* the clock so that only specific operations will be captured and displayed. Using specific addresses as a trigger and the read or write signals as qualifiers, we can expose timing irregularities by forcing the logic analyzer to trigger an oscilloscope. This is especially useful when troubleshooting input/output interfaces to the outside world. For example, if a problem appears in a printer interface, we can program the processor to make repeated accesses to the printer while triggering the logic analyzer on either reads or writes to the printer interface address. A trigger output from the analyzer provides an EXT input to the oscilloscope so that handshaking and data waveforms traveling between printer and computer can be displayed on the oscilloscope. This allows us to view signal integrity, voltage levels, switching times, and potential hazards at the time the problem is occurring.

Interfaces to external devices, such as printers and CRTs, are particularly susceptible to signal integrity problems, but even on-board troubleshooting may benefit from the analyzer-oscilloscope combination. The state analyzer may detect that data written into RAM is not always the same as the data read back from the same address. This could indicate defective RAM or problems in the write control, read control, or bus driver circuitry. The state analyzer, used alone, may not be able to isolate the problem, but an oscilloscope, looking at the write-enable pin of the RAM during a write cycle, may show marginal values for V_{IL} caused by poor connections, bent-under pins, or shorts. Defective drivers to data or address lines can also be exposed in this manner.

The synchronous state analyzer, when used alone, cannot see timing irregularities because it samples just once for every change of state. Figure 12-14 shows how glitches may be missed when they appear between state

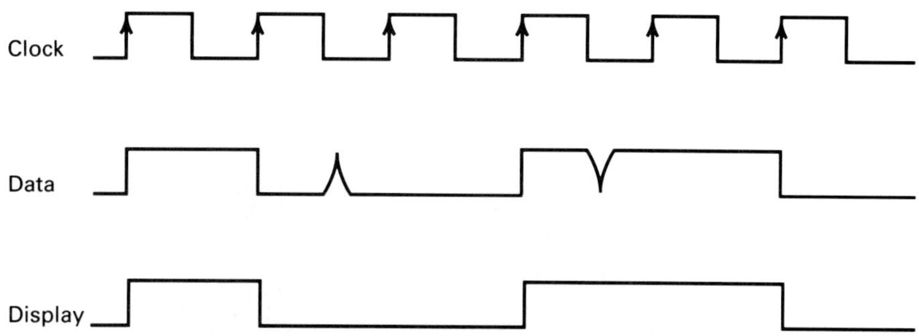

Fig. 12-14 The synchronous state analyzer samples just once on each clock. Glitches and other signal integrity problems will not be displayed.

changes. Asynchronous timing analyzers correct this problem by sampling the waveforms at rates much higher than the system clock and displaying events which occur between system clock edges. They employ glitch detection circuitry which is activated whenever a signal crosses a threshold two or more times between samples. Timing analyzers also provide a mode of operation which identifies marginal signals and slow transitions. These are often the cause of intermittent hardware problems.

The timing analyzer is used much like the state analyzer-oscilloscope combination described above. The difference is that functional verification and fault isolation may be performed with one rather than two instruments. Initially, the functional activity of the system is verified by selecting a slow sampling speed. A trace specification allows the user to specify how long a pattern must remain stable in order to qualify as a trigger. This will be a function of the particular system under test. The trigger duration selection must always be longer than a transition and shorter than the time between system clocks. The functional verification will then expose any software errors and indicate trouble spots. Higher sampling rates with glitch capture mode or marginal-signal-level mode allow the troubleshooter to focus on potential hardware problems.

The Signature Analyzer

The *signature analyzer* is an instrument which senses waveform patterns at a particular node in a circuit and converts them into a hexadecimal display. The waveforms at each test point are converted into *signatures*. The signatures of a circuit under test are compared with the signatures of a properly working system and the differences are used to isolate a faulty subsystem or component. Signature analyzers are particularly useful in microprocessor-based systems where test programs can be run to provide stimulus for the signatures.

A signature analyzer makes at least five connections to the circuit node under test. A *logic probe* senses the voltage levels at the node, and a ground lead provides a common *ground reference* between the analyzer and the circuit under test. A *clock input* sychronizes the waveform at the test node with the sampling circuitry in the analyzer. And the *start* and *stop* inputs tell the analyzer when to start and stop the measurement. Sometimes provision is made for read and write qualifiers to be used. This allows the display of read-only or write-only signatures and helps to isolate the malfunction even further. Figure 12-15 shows the Hewlet Packard 5006A Signature Analyzer.

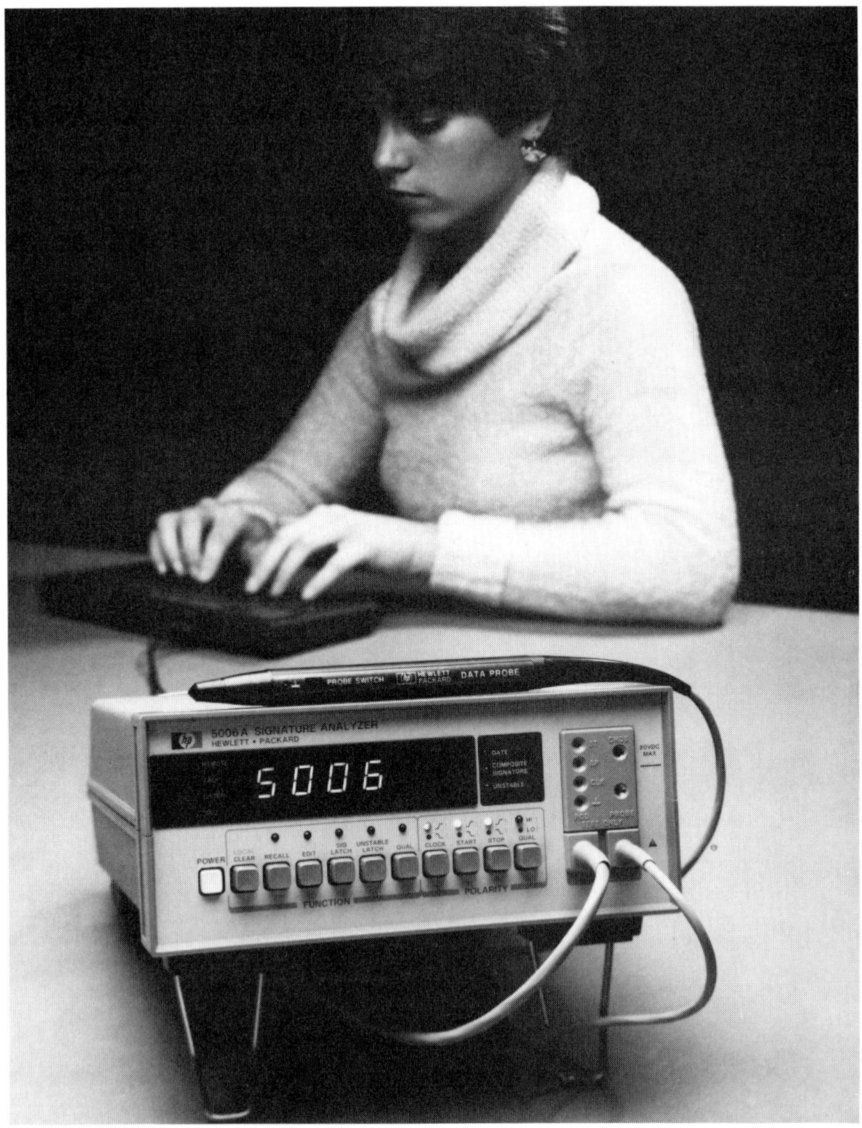

Fig. 12-15 The HP Signature Analyzer (Courtesy of Hewlett-Packard)

The signature is actually a 16-bit *cyclic redundancy code* (*CRC*) presented on four hexadecimal displays. The serial stream of 1s and 0s appearing at the test node is "divided" by a 16-bit binary number using a 16-bit shift register and several EXCLUSIVE OR gates. The code produced on the final clock edge is the remainder resulting from the binary division. The result left in the shift register represents a remainder which is not easily duplicated by a different serial bit stream. Thus, if the codes obtained from two different nodes are the same, then the chances are very high that they are operating identically. If the codes are different, then the two nodes are most certainly not operating identically.

Signature analysis is relatively slow for overall fault isolation. Its strength lies in isolating the problem to a specific node after the trouble area has been determined. On the surface it appears to offer a means for board testing by relatively low-skilled personnel, but there are pitfalls. Misleading and erroneous results can be obtained if both the analyzer and the circuit under test are not fully understood. Misleading results can be due to testing the problem circuit during a different state sequence than was used for the known good circuit. Problems may also occur because of a misunderstanding of the required relationships between the clock and the start and stop inputs. For example, most signature analyzers require a clock edge before the start pulse and after the stop pulse. If, for instance, the terminal count (TC, RC, or RCO) output of a '160 series counter is used as the stop input, and the counter's clock is turned off by the same signal, the stop edge will not be detected by the instrument. Other sources of malfunction are start and stop pulses which may be too short in duration for the clock frequency. Signal integrity problems of any kind may make the signature analyzer unusable. An in-depth discussion of signature analysis pitfalls can be found in the *Hewlett Packard Application Note Series 222*.

REVIEW 12-1

1. **Q.** What faults can be isolated with a logic probe? **A.** incorrect logic levels, bad connections, floating inputs, stuck nodes, chip failures, and sometimes glitches or spikes on reset lines

2. **Q.** What faults cannot be isolated with a logic probe? **A.** timing failures caused by setup and hold-time violations, multiple clocking caused by reflections, incorrect time relationships between control signals, and glitches which cross logic thresholds only intermittently

3. **Q.** What faults can be exposed with an oscilloscope? **A.** failures caused by marginal setup and hold times, hazards associated with reflections, glitches which may cause marginal operation, marginal logic levels, switching noise on power supply lines, shorted bus lines, and any fault which can be revealed by observing the time and amplitude values of waveforms

4. **Q.** Name some pitfalls which may be encountered when troubleshooting with a scope. **A.** improper triggering, misuse of scope probes, improper grounding, and the bandwidth limitations of a given oscilloscope

5. **Q.** How can an oscilloscope be used to find shorted signal lines? **A.** If shorting is suspected, the suspicious signal line, displayed on channel A, may be ADDed to the inverted sense of channel B while channel B is used to probe adjacent lines. An A-B display will appear quiet when the shorted signal line is found.

6. **Q.** What is the advantage of using a 10:1 probe? **A.** A 10:1 probe reduces signal loading by providing a higher resistance and lower capacitance to the signal source output. It also reduces ringing and reflection.

7. **Q.** When should a 10:1 probe not be used? **A.** When measuring high-speed signals, a 10:1 probe may not provide enough isolation, and an active—FET input—probe should be used.

8. **Q.** What is a storage scope? **A.** It is an oscilloscope which stores the picture of a transient waveform for easy viewing. Analog storage scopes use stored charge and persistent phosphors. Digital storage scopes store waveform information in RAM.

9. **Q.** What is a logic analyzer? **A.** A logic analyzer is similar to a digital storage scope in that it samples and stores logic waveforms before or after a trigger event. The waveforms may then be displayed concurrently on eight or more channels. It is different than a digital storage scope in that it can interpret and display those waveforms as mnemonic text, indicating the execution of program instruction codes.

10. **Q.** Describe the difference between a state analyzer and a timing analyzer. **A.** A state analyzer samples the circuit output synchronously with the system clock. A timing analyzer uses an external high-speed clock to sample waveform information between system clock edges. A state analyzer reflects only state changes while the timing analyzer can reflect events which occur between state changes.

11. **Q.** What is a signature analyzer? **A.** A signature analyzer is a debugging tool which captures waveforms and converts them into four-digit hexadecimal signatures. These signatures can then be compared with the waveforms extracted from a properly working system to isolate faulty nodes in a circuit.

12-2 PROCEDURES

Digital equipment has in recent years become increasingly complex. This has put greater demands on designers to provide acceptable levels of manufacturability, reliability, and maintainability. The success of a product depends greatly on all three of these qualities. If a device cannot be manufactured with a high yield (percentage of working units), the costs of manufacturing soars, which causes either low profits or noncompetitive pricing. If the units sold do not prove reliable, the cost of service contracts rise, repeat business dwindles, and word gets out to potential customers.

Finally, if the product is not serviceable, someone must pay for the hours which a service representative spends isolating a malfunction or for the cost of returned units.

Manufacturability relates to good design practices. In particular, attention to the worst-case timing relationships of delays, setup, and hold times. The design must be tolerant of a specified range of ac parameters for each part in the system. In complex systems, the worst-case timing analysis can become quite complex as timing variations accumulate through several stages of the circuit. Manufacturability also relates to good mechanical design in the PC board and in equipment racks.

Reliability relates most often to the quality of the components and their ability to resist environmental pressures. A reliable design must be tolerant of a specified range of temperature, supply voltage, vibration, pollution and electromagnetic noise. Again this goes back to good design practice. Fortunately, as digital equipment has become more complex, the issues of manufacturability and reliability have become more important in the initial design process.

Serviceability also relates to good design practice. Test engineers generally have the responsibility of seeing that designs can be easily tested in the factory and serviced in the field. Troubleshooting is much easier when testability has been designed into a product. Testability includes accessible test points, built-in testing aids, and self-test features. Field service personnel can also help by documenting common points of failure and weak spots in the design. There must be a continuous flow of information between design engineering, factory and test engineering, and service personnel. And design engineers must respond to problems found in manufacturing and field service by improving designs.

There are two types of troubleshooting: system development and field service. During the developmental stages of a project, the lab technician may be confronted with almost any type of problem, from simple wiring errors to subtle timing violations. When equipment fails in the field, we know at least that the system *did* work but now does not. This knowledge is a great advantage. In addition, field service personnel usually have service literature, including manuals on theory of operation, setup and installation, and maintenance aids. Technicians working in development labs generally do not have service literature. During the developmental phases of a product, there is little or no documentation and the technician must consult with the design engineers to obtain schematics and engineer's notes. In both cases, however, the first step in successful troubleshooting is to learn how the system is put together. *What subsystem blocks or PC boards are there, what are their functions, and how do they fit together?* Without this knowledge, the troubleshooting effort will be random and ineffectual.

After a technician has become familiar with the equipment to be serviced, he or she proceeds with an organized, problem-solving procedure which eliminates properly working sections from further consideration. Troubleshooting is always a process of *elimination*. In a step-by-step fashion, we remove from further consideration any part of the system which cannot be the cause of the failure. We then troubleshoot the part that is left. The process includes the following steps:

1 Inquiry. The purpose of inquiry is to find out what happened to cause the failure (this assumes that the system was working previously). What was the first indication of trouble? Who was using the equipment when it failed? What was being done with the equipment when the failure occurred? What maintenance or engineering changes have been performed recently? Notes should be taken when asking these questions.

2 Visual Inspection. Check fuses, switches, and supply voltages. Note the state of LEDs or other visual indicators. Look for broken wires, discolored components, loose or improperly installed connectors, and oxidation of PC board connections. Use your fingers to check for hot spots.

3 Fault Isolation. With the aid of flowcharts, diagnostic software, or general front-to-back techniques, eliminate the most probable causes of failure from further consideration and isolate the problem to a specific PC board or component.

4 Repair. In the field, a PC board may be swapped with a working board. In the lab or factory, the fault will be isolated further until the exact source of failure is determined. The results of this investigation may signal weak points in the design and will make future servicing easier if it is properly recorded. It may even result in a design change.

5 Test. Make sure that the corrected fault is the only fault. Other sources of malfunction may appear after the obvious source was eliminated. Underlying faults may, in fact, cause other failures in the area just repaired. Some *burn-in time* is required before you can be sure that the problem is fixed.

6 Documentation. This is an important, but often neglected, step. Be sure to document in the appropriate maintenance log everything relevant to the failure. In the lab, this may result in immediate design changes. In the field or factory it may help to expose weak or faulty design practices.

Inquiry

Troubleshooting is a thoughtful and logical process of elimination. Many probable sources of failure may be eliminated with careful questioning of the persons using the system. You cannot count on getting all the facts with general questions. It may be necessary to probe. Very often the person questioned will make assumptions about what is important and omit key pieces of information. It is important to know if anything unusual occurred which may be connected with the failure. Was there a storm? Did the air conditioning fail, or was it turned off? Are there any new sources of noise on the power lines such as motors or other noisy machinery? (Events of this nature should not, in themselves, cause failure and therefore may not be mentioned unless you ask. Marginal designs may, however, may be susceptible to any of these changes.) Were any changes made to the equipment by authorized or unauthorized service personnel? Is there a history of sporadic failure with this piece of equipment? In the lab or factory, has anyone changed any part of the system? Has the equipment been moved? Has new carpeting been installed which may increase static? *Good service technicians become adept at asking questions much like a good diagnostic physician.* The goal is to extract as much preliminary information as possible before starting any detailed troubleshooting.

Visual Inspection

The second level of troubleshooting is visual inspection. This requires an association between equipment and schematic. In particular, it is important to note the following:

1. Are there any fuses, and where are they located?
2. How many voltage supplies are there, what are their values, and where can they be measured?
3. Are DIP-switches or other mechanical switching devices used (relays, Berg pins)?
4. Are there any loose connectors, oxidized contacts, or broken wires? Touch the tops of ICs. Some ICs normally run warm, but a hot chip is a sure sign of trouble.
5. If there are visual indicators, what do they indicate?
6. Are there any analog adjustments such as trim-pots or trim-caps?

Power Supplies

The first and most obvious places to check are the fuses and supply voltages. The fuses may be inspected directly, and the supply voltages may be inspected with the aid of a voltmeter or oscilloscope. An oscilloscope is preferred because it will more readily expose the existance of low-frequency hum or high-frequency switching noise on a supply. Low-frequency hum (60 or 120 Hz) riding on top of a supply voltage most likely indicates an aging filter capacitor. High-frequency noise may be due to poor bypassing between power and ground, or to a failure in a switching power supply. Switching power supplies are often found in digital equipment racks because of their efficiency, size, and weight. However, they can generate their own switching noise (between 20 and 200 kHz) and may be a source of switching spikes on supply lines. When excessive switching spikes are found, the ground connections should be checked. Switching supplies require very good low-impedance ground connections. Given a good ground, the electrolytic capacitors should remove the spikes. These caps must have a low *equivalent series resistance* (ESR). Previous failure and capacitor replacement could be the source of the problem if the replacement caps were not of the low-ESR type.

Improper voltage levels may indicate a loss of regulation in the power supply or excessive loading on a PC board. When linear power supplies are used, all the PC boards should be removed and the output voltage measured again. If the voltage returns to its proper value, then PC boards should be replaced, one at a time, until the voltage sags again. This isolates the board causing the excessive load. Switching supplies, however, may require some load for proper regulation and may break into oscillation when subjected to sudden load changes. A sagging V_{CC} may be caused by an internal short circuit on one of the chips or by a defective diode in the power supply. A shorted chip will feel hot to the touch.

Connections

Switches must always be viewed with some suspicion. Seldom-used switch contacts may become dirty and appear as high-resistance points in the circuit. Often simply exercising a switch will restore it to proper operation, but spray-can contact cleaners provide a better solution. High-impedance points may also be caused by loose or oxidized connectors. Mechanical connectors represent the single most common source of failure in digital systems, and it seldom pays to use low-grade components in this area. Oxidized printed circuit board contacts may prevent signals from reaching the board, or they may just make the logic thresholds marginal. PC board contacts can be examined visually and cleaned when necessary. The most commonly used tool for cleaning oxidized PC board contacts is a pencil eraser. However, an eraser may make the contact more susceptible to further oxidation. The use of a cleaning solvent is preferred.

Broken or discolored wires may indicate a short or a device failure causing a short. Sometimes the insulation on wire-wrap wires can be cut if pulled too tightly around wire-wrap posts. This may cause a short between two signal lines. If some rework had been done on a wire-wrapped board previous to the failure, the new wires should be inspected.

Indicators

LEDs can be very helpful in visual inspections. But it is important to understand what they really mean. This information may be obtained from the service literature or from the design engineer, but the best approach is to study the schematic to determine what signals actually drive the LEDs and what they mean. The same advice applies to other visual indicators such as lamps and displays.

Analog Adjustments

It is generally undesirable to include analog adjustments in digital systems, and their presence should be viewed as a potential source of failure. Any circuit which can be adjusted can also become unadjusted. It is important to determine the exact function of any analog adjustments and their effect on system performance. Change of adjustments should be made only as a last step when it is clear that they are the source of failure.

Fault Isolation

When visual inspection fails to turn up any obvious faults, a more detailed procedure of fault isolation must be applied. This should be undertaken in an organized manner. One of the most common problems in field service is the swapping of boards without knowing exactly what they do and what their effect is on the system. Changing boards haphazardly is equivalent to destroying evidence. A board should be swapped only when it has been isolated as a faulty component. Then, the fact that it has been swapped should be documented along with the effect of the change on the system.

Often entire systems may be too complex to troubleshoot all at once but may be broken down into logical subsystems. This is often possible because many systems are designed using the "black-box" approach. The *black-box*, or *modular*, approach to logic design is a very powerful but sometimes misunderstood concept. Because a digital system may be too complex to grasp all at once, designers will subdivide the system into modules with specific functions. The overall design may be developed by temporarily ignoring the details of each box and assuming that it can satisfy the requirements. At this level, we can represent the system as a block diagram with each block representing black-box circuitry with defined inputs and outputs. This does not mean that we will never be concerned with the circuitry inside the boxes. When system design considerations are satisfied, we can attack each box individually and subdivide it further into more detailed boxes. Ultimately this permits us to focus on digestible blocks of logic without sacrificing the functionality of the entire system. At any given point, we can ignore the rest of the system and attack a given box as an independent design problem. In this way we keep the problem scaled down to a level compatible with our mental abilities.

This same approach may be applied to troubleshooting. Once we determine what each block is supposed to do, we can examine its inputs and outputs independently to see if it works. If it does not work, we can look inside the block to examine the subblocks. In this manner we isolate the portion of the circuitry which is at fault. This modular approach to troubleshooting may be coupled with back-to-front methodology to provide an orderly effort. *Back-to-front troubleshooting* means starting with the output section and progressing toward the input until the proper signal is found. The movement from back to front need not be incremental. After each probe, the checking may proceed to the middle of the remaining stages. This *half-splitting* technique helps to reduce debugging time when there are many stages.

Some types of subsystem are specific to a particular product and must be dealt with individually. Manufacturers sometimes include servicing flowcharts in their service manuals for such circuitry. The flowcharts provide a logical order of checks which should lead to the isolation of the problem to a specific component. Some other subsystems are common to many digital products and may be treated in a general way. The more obvious types of subsystem are:

1. Clock subsystems
2. Reset circuitry
3. Counters and state machines
4. Interface circuitry
5. RAM and ROM

Clock Subsystems

One of the first steps in troubleshooting a digital system is to check the clock. A PC board may receive its clock from an external source via the card edge, it may generate a local clock, or it may *phase-lock* its local oscillator to an external master clock. Local clock sources can usually be identified by the presence of "tin cans" on the board. The tin can may be the crystal for an oscillator, a complete crystal oscillator, or a voltage-controlled oscillator. In each case the tin can includes a crystal wafer which vibrates mechanically at a specified rate to supply a stable time base. The leads of a simple crystal cannot be probed without loading the crystal and disturbing normal operation. When viewing local clock sources, the probe must be applied to the clock output of the oscillator. In microprocessor systems, this may be a pin on the processor chip.

When an external master clock is used, the first step is to probe the point where it enters the system. Problems observable here are no clock or a defective clock. A clock may be defective because of noise, jitter, or reflection. When probing the clock, be sure to use a minimum-length grounding clip with the scope probe. Poor grounding between the scope and the PC board, or between the PC board and the master clock subsystem, may distort the appearance of the clock signal on the scope. It is important to know whether the problem is really in the system or in the test interface. Also remember that the clock is coming into the PC board and into the scope via a transmission line. A 10:1 probe should be used for frequencies below 10 MHz, and an active probe should be used for frequencies above 10 MHz.

When a master clock is distributed to several boards, *phase-locked loops* are sometimes used. A local phase-locked loop circuit provides a stable clock with a fixed phase relationship to the master clock. High-frequency noise or glitches on the incoming clock can be removed by the loop and it is possible to derive fractional multiples of the master clock rate using counters. The loop may appear as in Fig. 12-16*a* or with counters as in

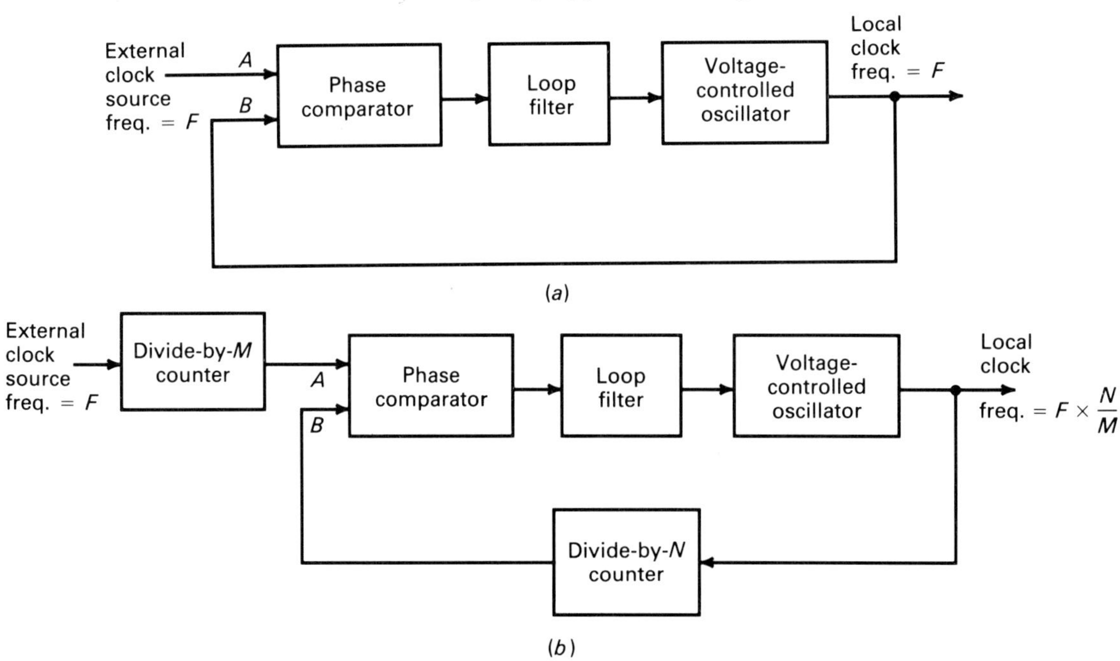

Fig. 12-16 Local clock derived from external source using phase-locked loop. (*a*) Without frequency division; (*b*) with frequency division.

Fig. 12-16*b*. In either case, inputs *A* and *B* are compared, and their phase difference is converted into a slowly changing or dc voltage. The dc voltage is filtered to remove noise and rapid changes before being applied to the input of a *voltage-controlled oscillator* (*VCO*). If the phase of the master clock were to change quickly, as with a noise spike, no change would appear at the VCO output. But any drift would cause the dc error voltage to adjust the VCO until it has the same frequency and a fixed phase relationship with the master clock.

If a phase-locked loop is found on a board and no clock signal is present, then the VCO output should be examined with the scope. The tin can on the circuit board may be the entire VCO, or it may be just the crystal which provides the reference for the VCO. In either case the schematic will indicate where the *VCO output* may be measured. There should be a clock output from the VCO even if no master clock input is being received. The absence of a clock output indicates a dead crystal or defective VCO. Crystals are susceptible to breakage due to excessive vibration.

If the VCO clock output is present, then compare the phase relationship between the two inputs on the phase comparator. At this point you should be able to trigger on one input while the other appears stable with a fixed phase difference. If one input to the phase comparator rolls across the screen and cannot be stabilized or if the frequencies differ, then the loop must be broken to isolate the fault. Figure 12-17 shows how the loop may

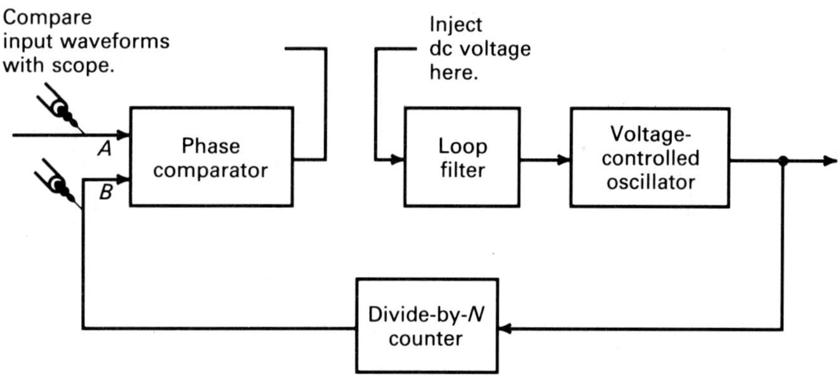

Fig. 12-17 Breaking the loop to test the VCO.

be broken to inject a dc voltage. Changing this voltage within the range specified for a given VCO should cause the VCO output to change over its range of frequencies. It should be possible to find a dc voltage which nearly stabilizes the phase relationship between the phase comparator inputs. As the dc voltage is varied, the unstable waveform should appear to roll across the screen more slowly and then reverse its direction of roll. This indicates that the VCO has enough range to follow the master clock and that the feedback counter (if present) is working properly. The problem, then, will likely be in the phase comparator. The feedback counter can be checked by comparing its input and output to see that it divides the VCO clock by the proper modulus. If the two inputs to the phase comparator are locked but excessive jitter is present on the output of the VCO, the problem may be in the loop filter capacitor.

Reset Circuitry

Among the easiest faults to isolate are initialization problems relating to reset. Power-on reset and push-button reset pulses may be used to initialize the system to its proper state. The reset button should be pushed while observing the reset signal at various nodes in the circuit. In microprocessor-based systems, the reset line provides an asychronous input to the processor which causes it to jump to a specific address in memory. The data at that address is itself an address which directs the processor to initialization programs. These programs must be executed before normal operation can begin. Another asynchronous input, called the *interrupt line* (sometimes labeled IRQ or INT), causes the processor to jump to another specified part of memory where interrupt programs are stored. Addresses for interrupt programs are set up during initialization. Thus if an interrupt pulse occurs during the initialization process, the processor may go off randomly to an unspecified part of memory. The interrupt lines should be checked to be sure they are in the correct state and remain quiet during initialization. Another common problem occurs when reset pulses are too short during power-up or make their transitions too slowly. These can be checked on the scope. If the system does not power-up properly but does reset properly with the manual reset, then the problem is most likely in the power-on reset circuitry.

Sometimes one-shots are used either in the reset circuitry or as missing clock detectors. The *one-shot* is a circuit which produces a single output pulse after receiving a transition on its input. It may be used in the reset circuitry to produce a pulse of known duration after receiving a transition on its input. *Retriggerable* one-shots maintain their output pulse for as long as new transitions occur within the pulsewidth duration. They are used to monitor clock or control lines for activity, and they act like watchdog timers. If a clock or control signal transition does not occur within the pulsewidth duration of the one-shot, the one-shot will return to its unasserted state, notifying other circuitry of the failure. The output pulsewidth of a one-shot is determined by an external resistor and capacitor combination. Figure 12-18 illustrates these two modes of one-shot operation.

The analog nature of one-shot timing makes it particularly susceptible to malfunction. Pulsewidth duration can change and may cause a false alarm. One-shots are also known to false trigger on power supply switching noise. If a one-shot is found in a system, it should be one of the first circuits checked. But be careful—when a system is run at slow speed for testing purposes, a properly functioning one-shot may time-out and cause confusion. Or a one-shot which does not operate properly at normal speed may appear normal at slow speeds. It is important to remember that the one-shot is an asynchronous clock source in an otherwise synchronous system.

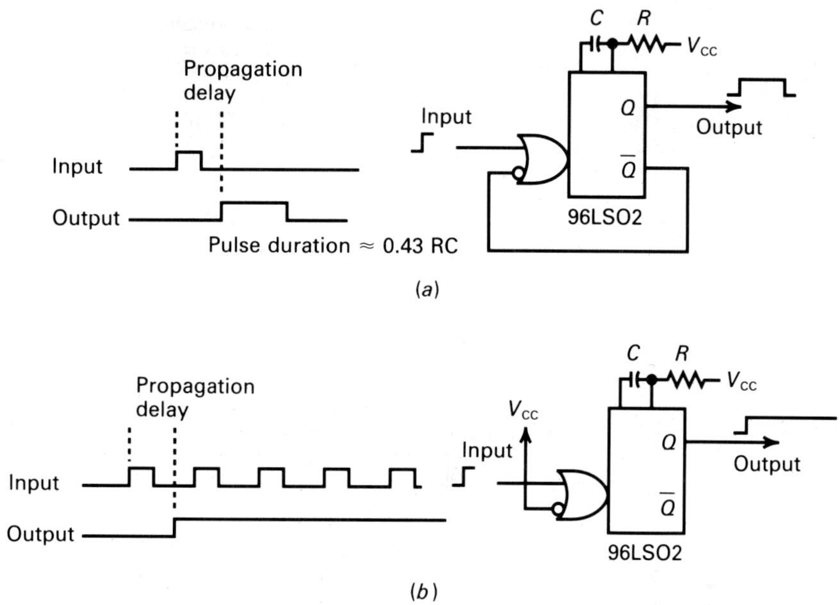

Fig. 12-18(*a*) In the nonretriggerable mode, the one-shot responds to an edge by producing a pulse. (*b*) In the retriggerable mode, the output remains asserted for as long as edges continue to trigger the input (the edges must repeat within a time ≈ 0.43 RC).

Counters

Binary counters can be easily checked with a scope. Start with the front end and check the outputs, two at a time, while triggering on either the slower changing waveform or on the terminal count. When triggering on the terminal count, the trigger level should be set so that it will not respond to a glitch. Adjacent outputs of a straight binary counter should have *half-period* or *double-period* relationships to each other. If unusual patterns are detected, consult the service literature or other documentation to determine whether this is intentional. Unusual counting sequences should always be documented. The period of the waveforms of each counter should also be checked and compared with the service literature. The waveforms of a counter stage may appear normal, but the enable decode circuitry preceding it may cause it to increment too often.

Sequencer and control outputs should be viewed on a timing analyzer and compared with the expected pattern. A scope may also be used if a good reference trigger voltage can be found and applied to the EXT trigger input. This requires some time-consuming pencil-and-paper work since only two or four signals can be displayed on a scope at one time. It is worth noting, however, that a sketch of the output pattern of a control sequencer can be useful when documenting the failure. If the control signals appear normal but cause abnormal behavior in the system, then their signal integrity should be checked with the scope. Sometimes *cold-solder joints* can cause confusion in this area. It is possible to probe a solder joint with a scope and see the signal on the scope without the signal actually reaching the intended input. A bad solder joint may act like a broken wire, or it may simply provide a high resistance point so that an input is unable to reach proper voltage levels. Probing on the actual pin lead will indicate whether a device is seeing the signal properly. If not, the solder joint and points further back may be examined until the point of failure is discovered.

I/O Interface Circuitry

Input/output interfaces to external devices are common points of failure. Many external devices, such as printers and plotters, have self-checking modes. They can be taken offline and run through a test pattern to eliminate their internal circuitry as a point of failure. If the external device checks out, then signal integrity between the device and the control interface should be examined. Often the problem is in the connectors. Grounding problems are the second most common source of trouble in these cases. Repeated accesses to the device should be made while observing the interface waveforms on a timing analyzer or scope. Sometimes the combination of logic analyzer and scope provides the best indication of signal integrity problems between equipment. The logic analyzer can be set up to trigger the scope on reads or writes to the interface circuitry, and the scope can expose marginal voltage levels, poor grounding conditions, or excessive noise or crosstalk. A simple problem, often overlooked, is due to a stuck interrupt line from an external device. If an interrupt line is hung in the active state, it will prevent the processor from initializing properly.

RAM and ROM Tests

Most manufacturers provide diagnostic software which can be loaded into a system on site by the service personnel. This software provides a functional check of the various parts of the system and indicates where faults were encountered. Some diagnostic software is resident in the system. Resident software usually includes RAM and ROM tests. Other diagnostic programs provide test patterns to peripheral devices such as printers, plotters, or magnetic disk drives. These programs exercise the peripheral continuously so that the field service technician can monitor relevant test points for improper operation or signal integrity problems.

Often the lab technician is trained to write simple routines to check various parts of a system. RAM problems, however, will not always be detected with simple software. Simply writing 1s and 0s to every cell and verifying that the same values are read back does not adequately test a RAM. This type of test does not expose problems due to defective address decoding, sense-amplifier sensitivity, memory loss, or pattern sensitivity. A defective address decoder, for instance, could cause every write and read to occur at the same location regardless of the address supplied in the program. In this case, writing and then reading every cell location merely proves that there is at least one location working properly in the RAM.

In order to be certain that every cell is accessible and capable of storing a 0 and a 1, we must use the *marching 1s and 0s RAM test* or some similar technique. In this test, 0s are written into all address locations and each cell is read and then written to a 1. When the next cell is read to be a 0, we know that a new cell is being addressed because we have left a trail of 1s behind. This is done in both ascending and descending address order and repeated with the complementary pattern of marching 0s. The marching 1s and 0s test detects most address decoder problems.

Another test, called the *shifted-diagonal test,* exposes slow recovery in sense amplifiers. When a cell has been written to the same value several times in a row, switching to the opposite value may be delayed because of faults in the sense amplifiers. A test for this problem consists of writing all 0s into memory except for a diagonal pattern of all 1s. The 1s diagonal pattern is then shifted, one position at a time, across the entire memory space. Every write cycle reenforces the 0s until the 1s diagonal pattern sweeps through. The test is repeated with a 0 diagonal.

Memory loss sometimes occurs in dynamic RAM because the individual cells cannot hold onto their values long enough between refresh cycles. Most RAM tests sequence through RAM too quickly to expose this type of problem. This can be checked by loading patterns into memory, inhibiting the refresh for the maximum allowable time, and then reading the entire memory.

Pattern sensitivity means that the reading and writing of some cells affects electronically adjacent cells. Exposing interaction between cells involves testing cells which share common addressing and refresh circuitry. This requires knowledge of the organization of the rows and columns in the RAM being tested. Pattern sensitive faults may be detected by writing 0s and 1s into each cell and checking the cells which share common circuitry to make sure they were not affected.

ROM tests are much simpler. They usually involve either EXCLUSIVE ORing the columns of a ROM or dividing the entire contents of ROM by a binary number. When a ROM is programmed, it is common to set aside one location to hold a checkword. In the simplest case, a *checkword* consists of the bits needed to force an odd or even parity in each of the data columns. As the ROM is programmed, the number of 1s in each column is determined to be either odd or even, and a checkbit is added at the end to ensure that it is *always* one way or the other. Figure 12-19 illustrates this method with odd-1s parity. More sophisticated checksums may be computed by dividing the entire ROM contents serially with a specially chosen binary number. This is the same technique used in signature analyzers. In either case, the checkword provides a *signature* which may be used to determine not only that a ROM is working correctly but also that it contains the latest software.

Repair and Test

In field service, *repair* usually means replacing a defective PC board. Back at the factory or board repair center, it may mean further fault isolation and replacement of a component or connection. In product

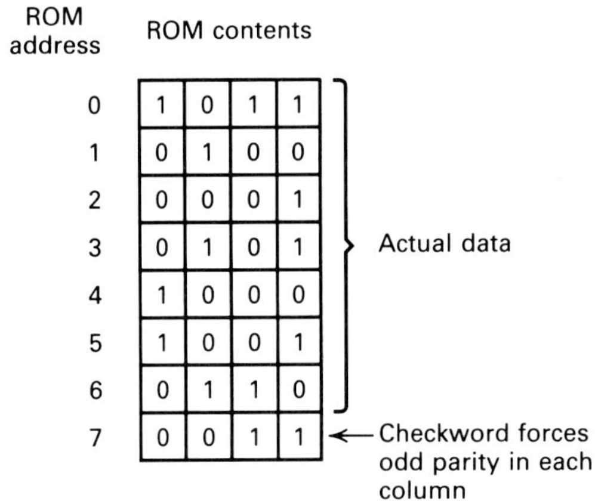

Fig. 12-19 Parity checkwords may be written into a ROM to guarantee that each column contains an odd number of 1s (odd parity) or an even number of 1s (even parity). A diagnostic program will EXCLUSIVE OR each column and produce a *checksum*. The checksum will be 1111 in this example if the ROM is good.

development, *repair* means finding and fixing the problem at the component level. In every phase of a product's life, however, problems need to be found and fixed and documented so that the engineers can be made aware of shortcomings in their design and manufacturing processes. The Japanese have shown us that to compete successfully, we must produce robust designs. Their own success is largely due to the constant supervision of product development, manufacturing, and field service. When yields fall during manufacturing, or when too many boards are being replaced in the field, they stop production long enough to find out why. Patterns pointing to marginal designs become apparent only when faults are isolated to their most basic level and documented. If marginal designs are not corrected, then poor yields will result and cause a loss in profitability.

Testing is always part of the repair process. It is common for faults to mask one another, and fixing an obvious problem may only expose another problem just one layer below it. Futhermore, the same problem may reoccur after repair because of a weakness in the design. The system must be watched carefully for some time after repair to be certain that all is well. During this burn-in time, any changes in system behavior should be documented.

Documentation

There are four very obvious benefits to good record keeping:

1. Some troubleshooting jobs may take hours or even days. Sometimes you will go off on a tangent or lose track of your initial objectives. Experienced troubleshooters make written notes along the way. When they get lost or confused, good written notes can help them get back on track.

2. If reinforcements are called in to help, they will want to know what has happened so far. At this point you may not be able to recall all of the details without notes.

3. Over the long run, good records will help establish patterns in your own mind for specific pieces of equipment. These patterns often point to common sources of failure and make finding them easier next time.
4. Good documentation can expose bad design practices and help to produce more robust products.

Notes should be taken throughout a troubleshooting job. It is important to document each change you make to the system, each test you conduct, and every result you obtain. Changes made to the system mean board changes, wire changes, connector changes, or any other modification. Remember that any change you make has the potential of destroying valuable evidence relating to the cause of failure. In the developmental phases of a product, design or circuit changes should be noted on a copy of the schematic.

It is a good practice to keep a logbook of some kind and to use it daily. Discussions with engineers or technicians pertaining to a piece of equipment should be noted in the logbook along with dates. Not only does this practice help establish a background on a particular system but it helps to straighten out misunderstandings that may arise between staff personnel. In general, a good piece of advice is to make notes on everything and throw away nothing.

Another important aspect of documentation is detailed records of pin connections, test points, system configurations, procedures, etc. There always seems to be someone in the department who knows all the details and can provide them at short notice. Often this is a technician. This person may not be praised for his or her efforts but he or she is appreciated. Such a person is usually considered a key employee whose loss would not go unnoticed. It is difficult to take the time to keep track of small details, but when they are needed, the time will seem well spent.

When you become familiar with a piece of equipment, start a list of questions which may be used to eliminate probable sources of failure by careful questioning of the persons using the system. The goal is to extract as much preliminary information as possible before starting any detailed troubleshooting. This list may be extended to include actual troubleshooting steps which will eliminate the most common points of failure from consideration. Successful television repairpeople limit their service to two or three brands so that they can learn the weak spots which account for most of the service calls. When you know where to look first you can save valuable time.

REVIEW 12-2

1. **Q.** What troubleshooting aids do field service technicians have which are not available to lab technicians in research and product development? **A.** They have service literature, troubleshooting flowcharts, and the knowledge that the system did work and now does not.

2. **Q.** What must always be done after repair of a system? **A.** testing and documentation

3. **Q.** What are common sources of failure in power supplies? **A.** filter capacitors and diodes

4. **Q.** What is a common mechanical source of failure? **A.** connectors

5. **Q.** Define back-to-front methodology. **A.** Back-to-front methodology means starting with the final output and working your way toward the front until the proper signals are found.

6. **Q.** How can an interrupt line interfere with microprocessor start-up? **A.** If an interrupt pulse occurs during initialization, the processor may go off to an undefined part of memory.

7. **Q.** What are some common problems with one-shots? **A.** One shots may false trigger, time-out too soon, or operate differently during testing than during normal system operation.

8. **Q.** What is a ROM checkword? **A.** A ROM checkword is a binary number added to the contents of a ROM to force either odd or even parity in all of the columns.

9. **Q.** What can a logic probe tell us about a counter? Why is a scope better? **A.** The logic probe indicates that the various stages are active but does not tell us if the count sequence and period are correct. The scope allows us to examine the sequence and determine whether the counter is clocking at the proper rate.

10. **Q.** Name some common faults in I/O interfaces. **A.** bad connectors, grounding problems, signal integrity problems, and stuck handshaking lines.

11. **Q.** Does writing and reading ones and zeroes into each cell adequately test the RAM? **A.** No, a defective decoder inside the RAM could cause every read and write to occur at one good location.

12. **Q.** What is the purpose of a ROM checksum? **A.** It provides an indication of the integrity of a programmed ROM.

Single-Ended Drivers

TYPE NUMBER	CIRCUITS/PACKAGE	DESCRIPTION	APPLICATION
SN55/75121	2	Emitter-follower outputs to 75 mA	General purpose
SN75361A	2	Totem-pole outputs to 100 mA	General purpose
SN75451B	2	Open-collector to 300 mA	General purpose
DS78/8831	4	3-state diode clamped outputs	General purpose
DS78/8832	4	3-state outputs to 40 mA	General purpose
SN75150	2	Totem-pole dual polarity	RS-232-C
SN75156	2	Totem-pole dual polarity	RS-232-C/423-A
uA9636	2	Totem-pole dual polarity	RS-232-C/423-A
SN75186	4	Totem-pole dual polarity	RS-232-C
SN75187	4	Totem-pole dual polarity	RS-423-A
SN75123	2	Emitter-follower output to 60 mA	IBM 360
SN75126	4	Emitter-follower output to 60 mA	IBM 360/370
SN75130	4	Emitter-follower to 60 mA with fault flag	IBM 360/370

Single-Ended Receivers

TYPE NUMBER	CIRCUITS/PACKAGE	DESCRIPTION	APPLICATION
SN55/75122	3	Input hysteresis	General purpose
SN55/75140	2	100 mV sensitivity, common reference	General purpose
SN55/75141	2	100 mV sensitivity, common reference, diode protected	General purpose
SN55/75142	2	100 mV sensitivity, internal reference	General purpose
SN55/75143	2	100 mV sensitivity, internal reference, diode protected	General purpose
SN75152	2	Adjustable hysteresis	General purpose MIL-STD 188 & RS-232-C
SN75154	4	Adjustable hysteresis	RS-232-C
SN75189	4	Response control	RS-232-C
SN75189A	4	Wide hysteresis, response control	RS-232-C
SN75157	2	Fixed hysteresis	RS-232-C/423-A
uA9637AC	2	Fixed hysteresis	RS-232-C/423-A
AM26LS32AC	4	±7 V CMR with ±200 mV sensitivity	RS-422-A/423-A
AM26LS33AC	4	±15 V CMR with ±500 mV sensitivity	RS-422-A/423-A
MC3486	4	±200 mV sensitivity ±7 V CMIV	RS-422-A/423-A
SN75173	4	±200 mV sensitivity 12 kΩ min R_{IN}	RS-422-A/423-A
SN75175	4	±200 mV sensitivity 12 kΩ min R_{IN}	RS-422-A/423-A
SN75124	3	Hysteresis, independent strobes	IBM 360
SN75125	7	7 kΩ min input resistance	IBM 360/370
SN75127	7	7 kΩ min input resistance	IBM 360/370
SN75128	8	Strobe active high	IBM 360/370
SN75129	8	Strobe active low	IBM 360/370

Courtesy of Texas Instruments Incorporated

ANSWERS TO ODD-NUMBERED EXERCISES

CHAPTER ONE

1.

$$I = \frac{Q}{T} = \frac{18.9 \times 10^{18}\ \text{electrons}}{2\ \text{s}} = \frac{9.45 \times 10^{18}\ \text{electrons}}{\text{s}}$$

$$1\ \text{A} = \frac{6.3 \times 10^{18}\ \text{electrons}}{\text{s}}$$

$$I\ (\text{in amps}) = \frac{\dfrac{9.45 \times 10^{18}\ \text{electrons}}{\text{seconds}}}{\dfrac{6.3 \times 10^{18}\ \text{electrons}}{\text{second}}} = 1.5\ \text{A}$$

3.

$$R = \rho \frac{l}{t \times w} = 2.8 \times 10^{-6}\ \Omega - \text{cm} \times \frac{7\ \text{cm}}{1\ \text{cm} \times 1\ \text{cm}}$$

$$= 1.96 \times 10^{-5}\ \Omega$$

5.

$$\text{Power (in watts)} = \frac{\text{Energy (in joules)}}{\text{Time (in seconds)}} = \frac{1\ \text{J}}{10\ \text{s}} = 0.1\ \text{W}$$

$$\text{Power (in watts)} = \text{Current (in amps)} \times \text{Voltage (in volts)}$$

$$\text{Voltage (in volts)} = \frac{\text{Power (in watts)}}{\text{Current (in amps)}} = \frac{0.1\ \text{W}}{6\ \text{A}} = 16.7\ \text{mV}$$

7. Sheet resistance $= 0.03\ \Omega/\square$

$$\frac{5\ \text{in}}{12.5 \times 10^{-3}\ \text{in}} = 400\ \square$$

$$R = 400\ \square \times 0.03\ \Omega/\square = 12\ \Omega$$

9. Half the power is lost in the cables. The lost energy appears as heat.
11. $C = q/v = 1\ \text{C}/1\ \text{V} = 1\ \text{F}$. Since it takes one joule of energy to raise one coulomb of charge to a potential of one volt, one joule will be dissipated when the coulomb of charge is returned to ground state. If we assume that the connecting wires have negligible resistance, one joule of energy is dissipated in the resistor.

13. Discharge:

$$v = V_0 e^{-t/RC}$$

$$0.01V_0 = V_0 e^{-t/RC}$$

$$0.01 = e^{-t/RC}$$

$$\ln(0.01) = -t/RC$$

$$-4.6 = -t/(0.1 \times 10^{-6} \times 1000)$$

$$t = 4.6 \times 10^{-4}\text{ s} = 460\ \mu\text{S}$$

this is 4.6 time constants since one time constant $= 10^{-4}$ s.

Charge:

$$v = V_0(1 - e^{-t/RC})$$

$$0.99V_0 = V_0(1 - e^{-t/RC})$$

$$0.99 = 1 - e^{-t/RC}$$

$$\ln(1 - 0.99) = -t/RC$$

$$-4.6 = -t/(0.1 \times 10^{-6} \times 10000)$$

$$t = 4.6 \times 10^{-3}\text{ s} = 4.6\text{ mS}$$

this is 4.6 time constants since one time constant $= 10^{-3}$ s.

15. Discharge:

The capacitor will not discharge to 0 V but to 0.45 V. Thus, $V_0 = 4.55$ V instead of 5 V, but since V_0 cancels out of the equation, $0.01 = e^{-t/RC}$ as before. However, $R = 10\ \text{K}\Omega // 1\ \text{K}\Omega = 909\ \Omega$, and $t = 418\ \mu\text{S}$ (4.6 time constants).

Charge:

Likewise, since the charging cycle begins at 0.45 V, $V_0 = 4.55$ V. However, since the capacitor charges only through the 10 KΩ resistor, the time constant is the same as before, and $t = 4.6$ mS (4.6 times constants).

17. One input load of 0.4 mA across 16 Ω produces an output voltage of 6.4 mV. In addition, there is an unloaded output voltage of 78.7 mV, raising the total output voltage to 85 mV. At 0.4 V, the output can sink 0.4 V/16 Ω = 25 mA. However, it is already sinking approximately 5 mA unloaded. This leaves 20 mA for fanout. Each input requires 0.4 mA of sink current and the fanout (low) is 20 mA/0.4 mA = 50.

CHAPTER TWO

1. ϵ is the permittivity constant: ϵ_{SI} for silicon, ϵ_{OX} for oxide. D, the separation between plates: w is the separation produced by the depletion region, t_{OX} is the separation produced by oxide.

3. $L = 5\mu m$, $W = 10\mu m$, $W/L = 2$.

$$C_{OX}' = \epsilon_{OX}/t_{OX} = (3.5 \times 10^{-13}\ F/cm)/(9.0 \times 10^{-6}\ cm) = 3.89 \times 10^{-8}$$

$V_{DS} \approx 0$ V since capacitive load is being driven.

$$R_{ON} = 1/(800\ cm^2/V - s)(3.89 \times 10^{-8}\ F/cm^2)(2)(5\ V - 0.8\ V)$$
$$= 1/(3.1 \times 10^{-5}\ F/V - s)(2)(4.2\ V) = 1/(2.6 \times 10^{-4}\ F/s)$$
$$= 3.8 \times 10^3\ s/(C/V) = 3.8\ K\Omega$$

5. $R_{ON} = 0.4\ V/1.6\ mA = 250\ \Omega$
Solve for W/L with $V_{DS} = 0.4$ V.

$$W/L = 1/(3.1 \times 10^{-5}\ F/V - s)(250\ \Omega)(5\ V - 0.8\ V - 0.4\ V)$$
$$= 1/(0.029) = 34.5$$
$$W = 3 \times 34.5 \approx 103\mu m$$

7. Data rate$_{max}$ $= 0.318/t_r \approx 763$ KHz.
9. The charging time (rise time) will be 3 times the discharge time (fall time).

CHAPTER THREE

1.

x	y	f
0	0	1
0	1	0
1	0	0
1	1	1

3. (*b*) is incorrect because adjacent squares differ in more than one variable.

5.

$$\bar{f} = z + xy$$
$$f = \overline{z + xy}$$

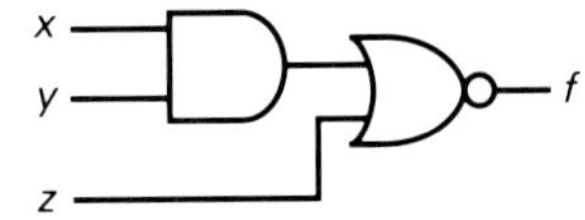

7.

a) $f = \overline{A} + \overline{B}$
b) $f = \overline{AB}$

The truth tables for (a) and (b) are the same. This verifies DeMorgan's theorem.

A	B	f
0	0	1
0	1	1
1	0	1
1	1	0

9.

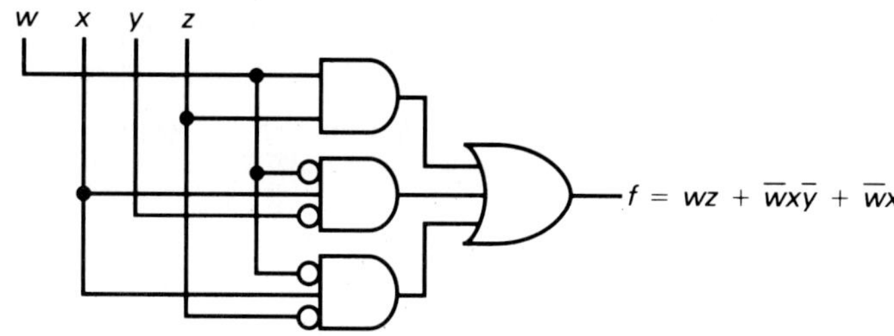

11.

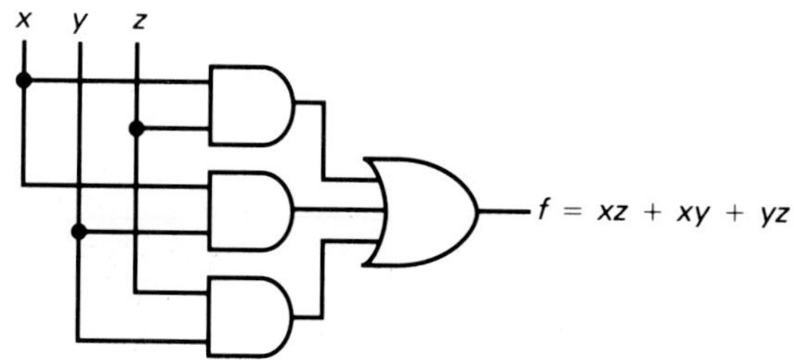

13.

$$f_{(A>B)} = A_1\overline{B}_1 + A_0\overline{B}_0(A_1 + \overline{B}_1)$$

$$f_{(B>A)} = \overline{A}_1B_1 + \overline{A}_0B_0(\overline{A}_1 + B_1)$$

$$f_{(A=B)} = \overline{f_{(A>B)} + f_{(B>A)}}$$

15. Stick diagrams were introduced in Chapter 2 (Fig. 2-23). The stick diagram for the Exclusive NOR looks identical to the stick diagram for the Exclusive OR in Prob. 14, except flipped vertically (V_{DD} and V_{SS} swapped, and diffusion types swapped).

CHAPTER FOUR

1.

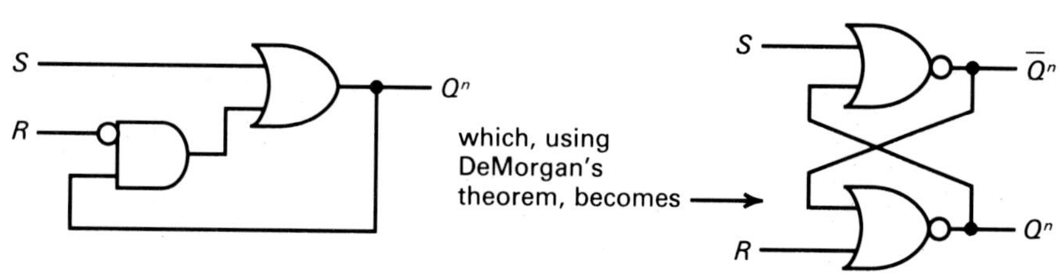

3.

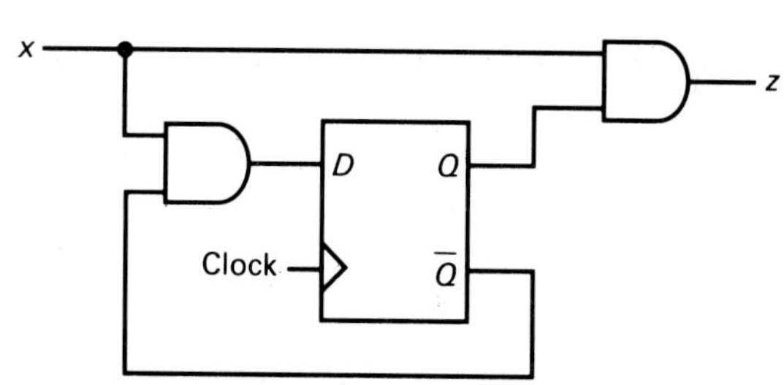

5.

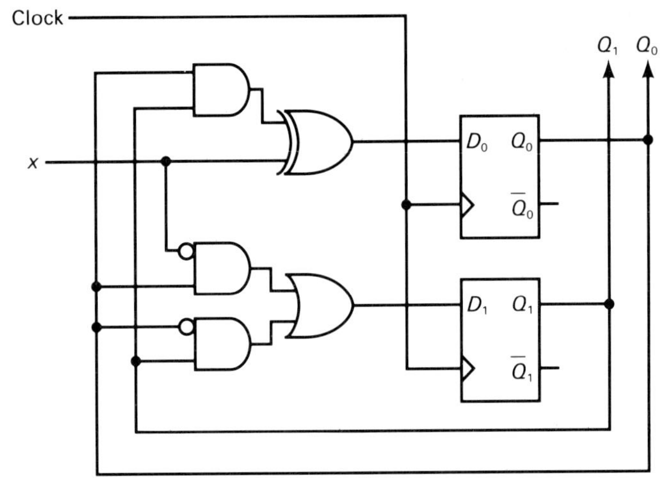

Note: $Q_0^{n+1} = x\bar{Q}_0^n + x\bar{Q}_1^n + \bar{x}Q_1^nQ_0^n = x(\bar{Q}_0^n + \bar{Q}_1^n) + \bar{x}Q_1^nQ_0^n$

$= x(\overline{Q_1^nQ_0^n}) + \bar{x}(Q_1^nQ_0^n)$ Which is an Exclusive Or function.

7. There are four states, two of which must decode to Green. If 00 and 01 are Green, 10 is Red, and 11 is yellow, then the following circuit will satisfy the requirements of Problem 7:

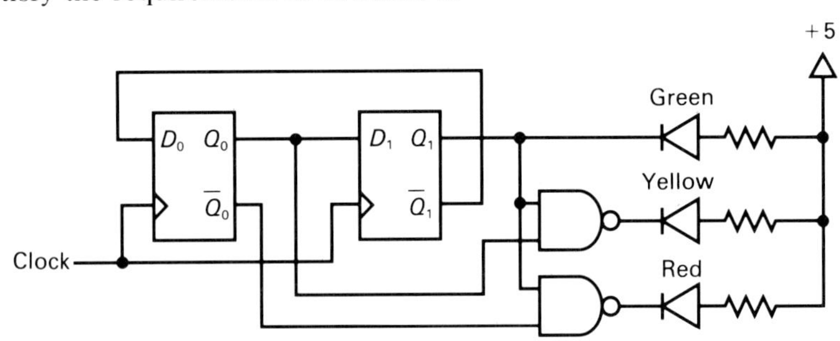

9.

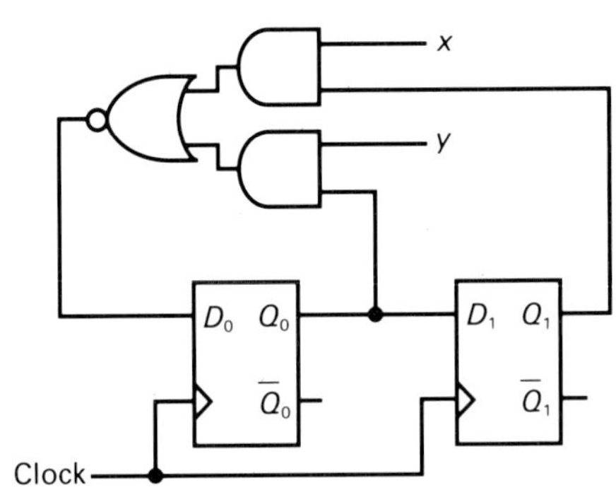

11.

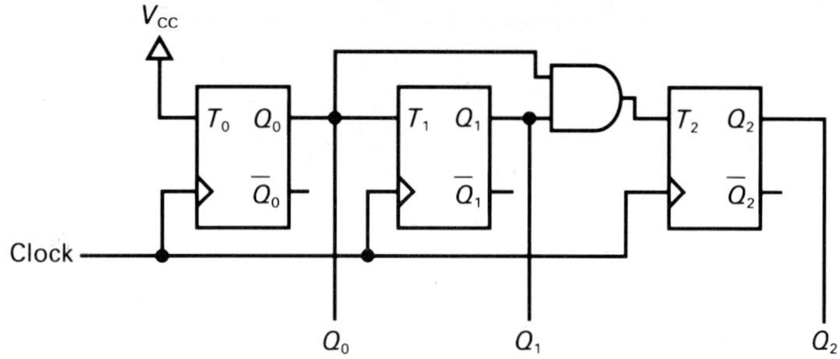

The T flip-flops may be implemented with either D flip-flops or JK flip-flops.

13.

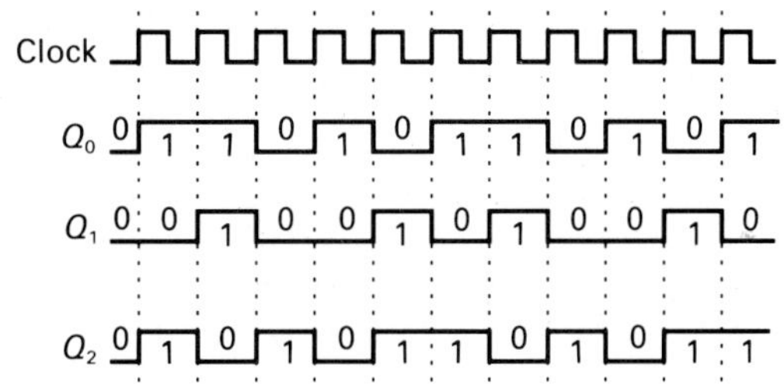

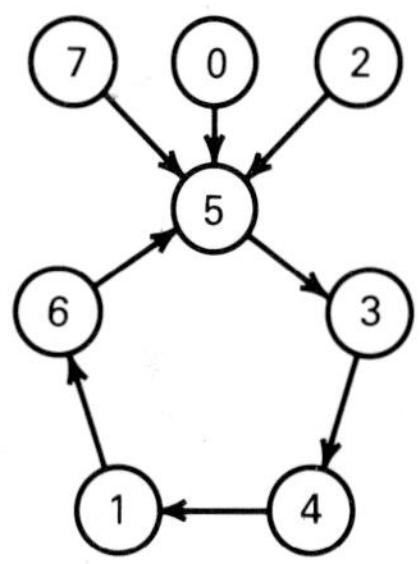

CHAPTER FIVE

1. 00101001, 01000000, 01111111.

3. 60_{10}, 107_{10}, 249_{10}.

5. 1001, 0001 0000, 1001 1001, 0001 0000 0000.

7. The following circuits, if reset prior to receiving data, will indicate whether an odd or even number number of ls is received.

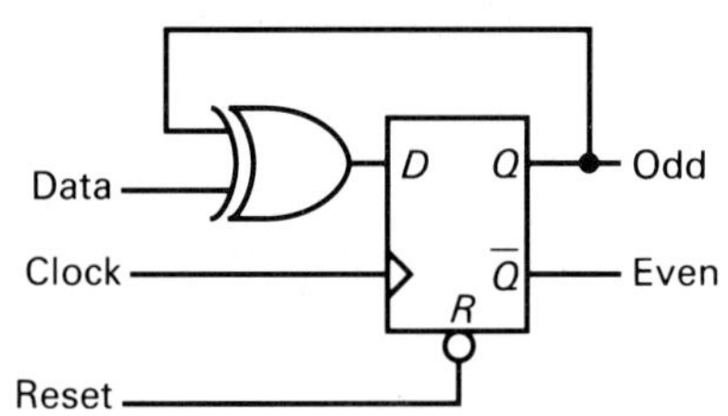

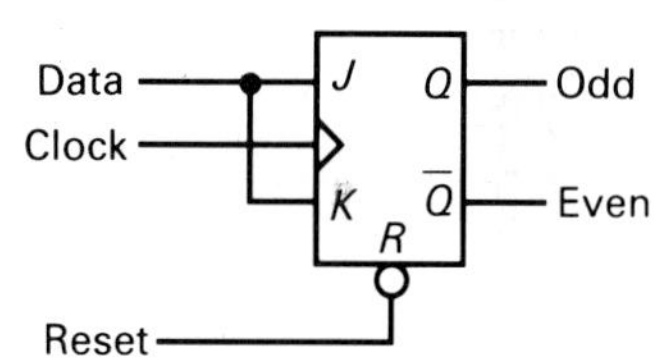

9.

E	Input	Q^n	Q^{n+1}
0	0	0	0
0	0	1	1
0	1	0	0
0	1	1	1
1	0	0	0
1	0	1	0
1	1	0	1
1	1	1	1

$$Q^{n+1} = \overline{E}Q^n + E \text{ (Input)}$$

11. Propagation delay of the AC device may not be long enough to meet the LS data hold time requirement.
13. 66.7 MHz.
15. Only one stage of a shift register needs to be designed since the rest are identical. One possible function code assignment yields the following:

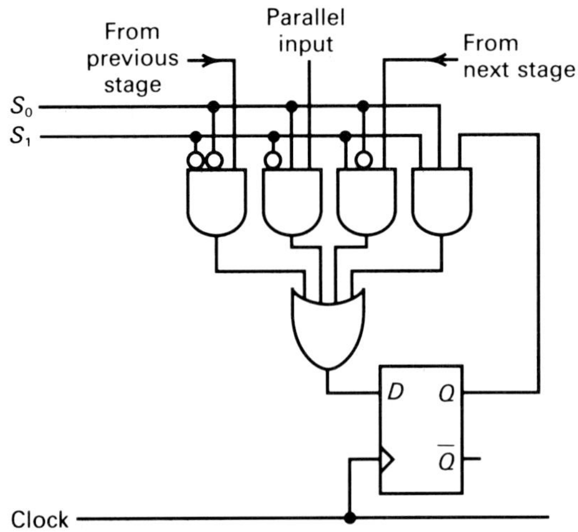

S_1	S_0	Response
0	0	Shift Right
0	1	Parallel Load
1	0	Shift Left
1	1	Hold

17.

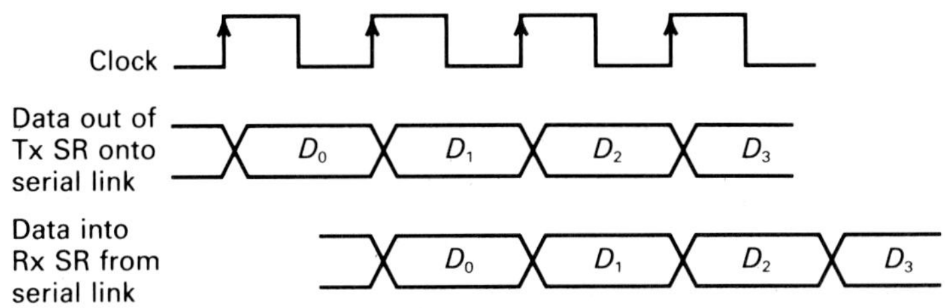

19.

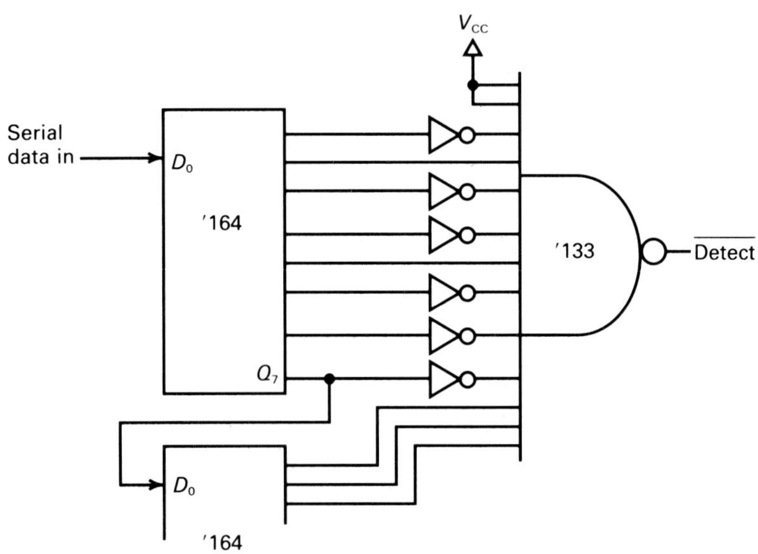

CHAPTER SIX

1. Maximum delay for flip-flops = 21 ns, minimum set-up time = 20 ns.

$$f_{max} = \frac{1}{21\text{ ns} + 20\text{ ns}} = \frac{1}{41\text{ ns}} \approx 24.4\text{ MHz}$$

For all of the scalers except the ÷2 and the ÷7. The ÷7 has an additional gate delay.

$$f_{max} = \frac{1}{21\text{ ns} + 20\text{ ns} + 20\text{ ns}} = \frac{1}{61\text{ ns}} \approx 16.4\text{ MHz}$$

The ÷2 scaler can operate at the maximum clock rate of the flip-flop which is 30 MHz as specified by the table in Fig. 6-2.

3.

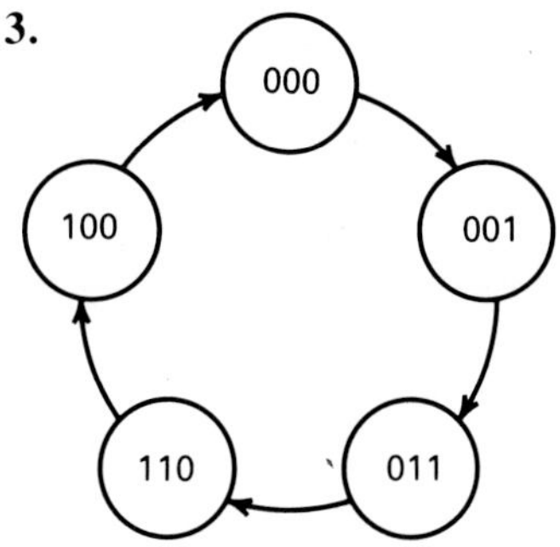

Q_2^n	Q_1^n	Q_0^n	Q_2^{n+1}	Q_1^{n+1}	Q_0^{n+1}
0	0	0	0	0	1
0	0	1	0	1	1
0	1	0	x	x	x
0	1	1	1	1	0
1	0	0	0	0	0
1	0	1	x	x	x
1	1	0	1	0	0
1	1	1	x	x	x

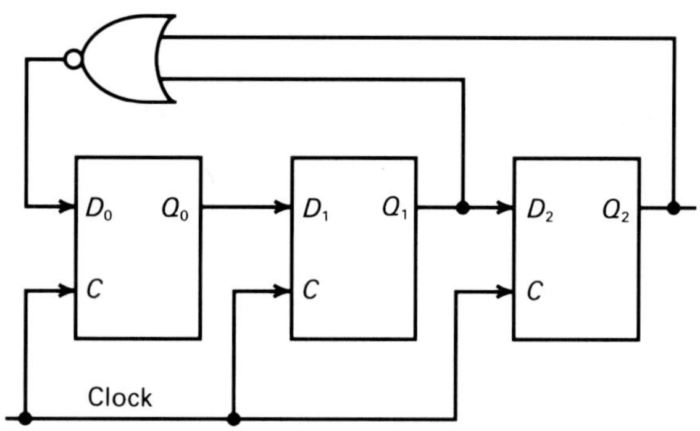

Choose the don't cares so that:

$Q_2^{n+1} = Q_1^n$

$Q_1^{n+1} = Q_0^n$

Use Karnaugh map to find Q_0^{n+1}

$Q_0^{n+1} = \overline{Q_2^n}\,\overline{Q_1^n} = \overline{Q_2^n + Q_1^n}$

Verify that there are no hang-up states.

5.

present state				next state				JK Inputs			
Q_3^n	Q_2^n	Q_1^n	Q_0^n	Q_3^{n+1}	Q_2^{n+1}	Q_1^{n+1}	Q_0^{n+1}	J_3K_3	J_2K_2	J_1K_1	J_0K_0
0	0	0	0	0	1	0	1	0 x	1 x	0 x	1 x
0	0	0	1	0	1	1	0	0 x	1 x	1 x	x 1
0	0	1	0	0	1	0	1	0 x	1 x	x 1	1 x
0	0	1	1	0	1	0	0	0 x	1 x	x 1	x 1
0	1	0	0	1	0	0	0	1 x	x 1	0 x	0 x
0	1	0	1	1	0	1	0	1 x	x 1	1 x	x 1
0	1	1	0	1	1	0	0	1 x	x 0	x 1	0 x
0	1	1	1	1	1	0	0	1 x	x 0	x 1	x 1
1	0	0	0	0	0	0	1	x 1	0 x	0 x	1 x
1	0	0	1	0	0	1	0	x 1	0 x	1 x	x 1
1	0	1	0	0	0	1	1	x 1	0 x	x 0	1 x
1	0	1	1	0	0	1	0	x 1	0 x	x 0	x 1
1	1	0	0	0	0	0	0	x 1	x 1	0 x	0 x
1	1	0	1	0	0	1	0	x 1	x 1	1 x	x 1
1	1	1	0	0	1	1	0	x 1	x 0	x 0	0 x
1	1	1	1	0	1	1	0	x 1	x 0	x 0	x 1

$J_3 = Q_2^n$ $\quad J_2 = \overline{Q}_3^n$ $\quad J_1 = Q_0^n$ $\quad J_0 = \overline{Q}_2^n$

$K_3 = 1$ $\quad K_2 = \overline{Q}_1^n$ $\quad K_1 = \overline{Q}_3^n$ $\quad K_0 = 1$

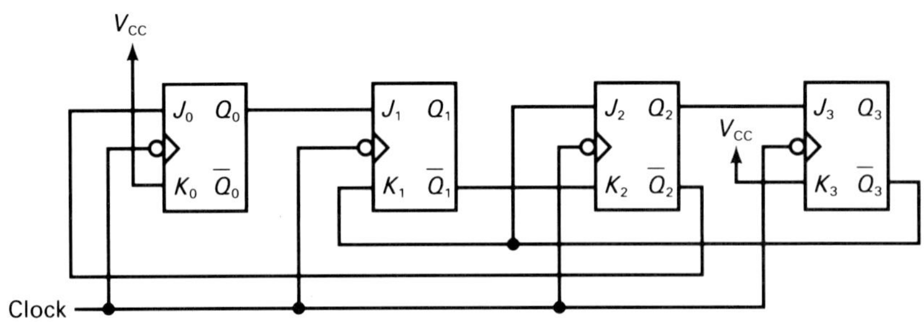

7.

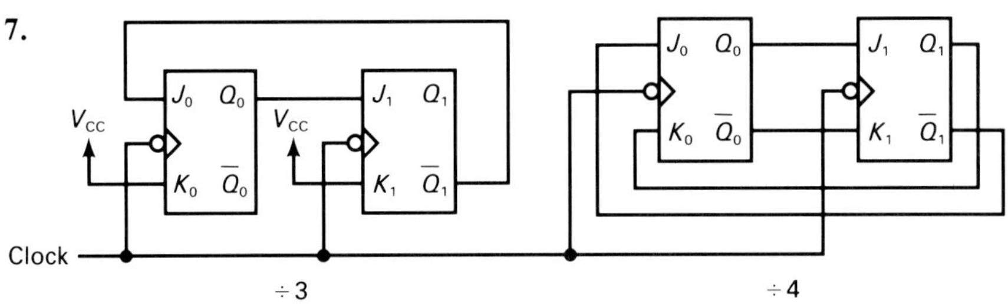

9.

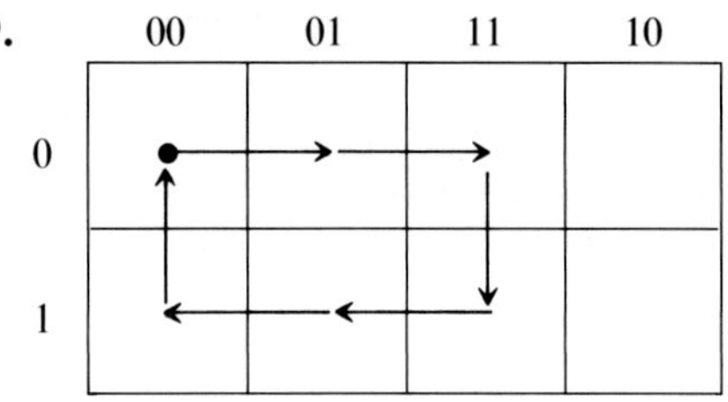

Q_2^n	Q_1^n	Q_0^n	Q_2^{n+1}	Q_1^{n+1}	Q_0^{n+1}
0	0	0	0	0	1
0	0	1	0	1	1
0	1	0	x	x	x
0	1	1	1	1	1
1	0	0	0	0	0
1	0	1	1	0	0
1	1	0	x	x	x
1	1	1	1	0	1

Q_2 \ Q_1Q_0	00	01	11	10
0	0	0	1	x
1	0	1	1	x

$$Q_2^{n+1} = Q_2Q_0 + Q_1$$

Q_2 \ Q_1Q_0	00	01	11	10
0	0	1	1	x
1	0	0	0	x

$$Q_1^{n+1} = \overline{Q}_2Q_0$$

Q_2 \ Q_1Q_0	00	01	11	10
0	1	1	1	x
1	0	0	1	x

$$Q_0^{n+1} = \overline{Q}_2 + Q_1$$

11.

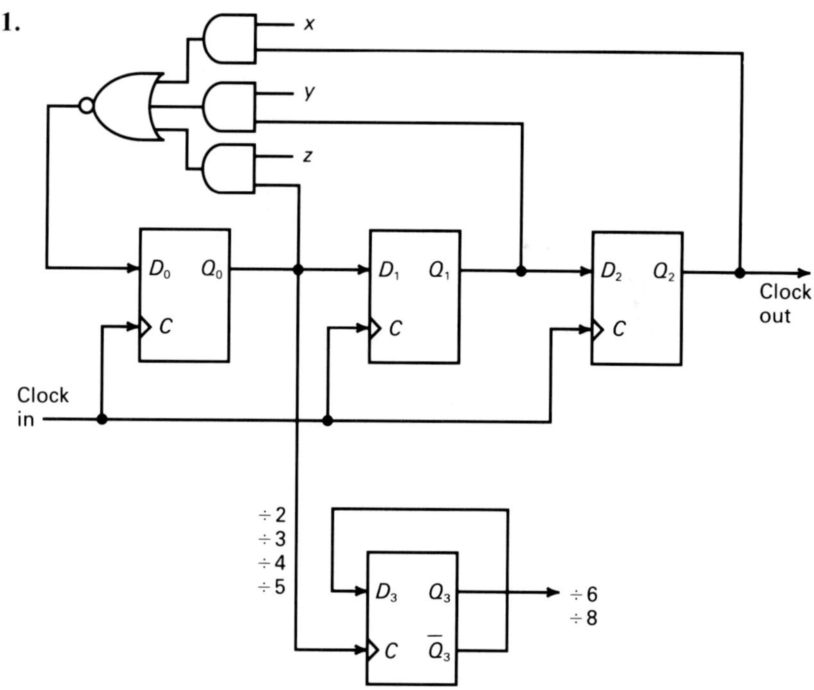

x	y	z	$\div n$
0	0	1	÷2
0	1	0	÷4 or ÷8
0	1	1	÷3 or ÷6
1	1	0	÷5

13.

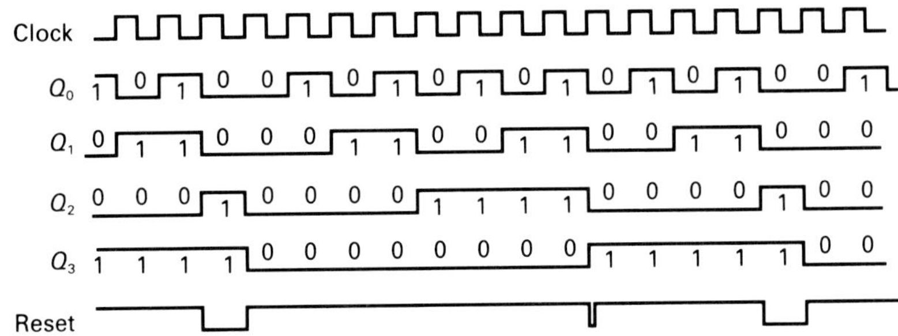
Clock
Q0 1 0 1 0 0 1 0 1 0 1 0 1 0 1 0 1 0 0 1
Q1 0 1 1 0 0 0 1 1 0 0 1 1 0 0 1 1 0 0 0
Q2 0 0 0 1 0 0 0 0 1 1 1 1 0 0 0 0 1 0 0
Q3 1 1 1 1 0 0 0 0 0 0 0 0 1 1 1 1 1 0 0
Reset

15.

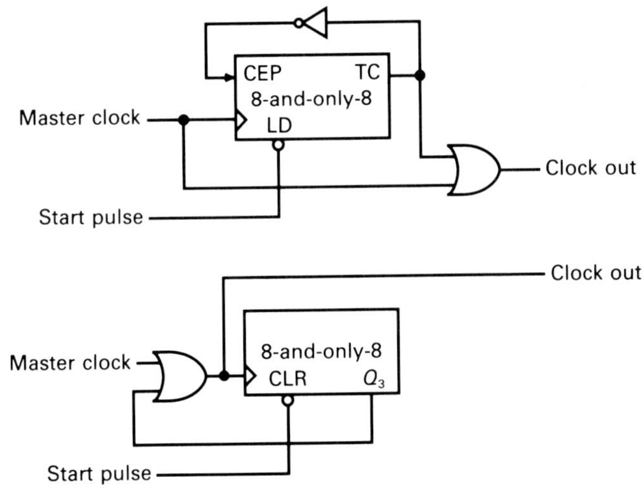
CEP
TC
8-and-only-8
LD
Master clock
Start pulse
Clock out
Clock out
8-and-only-8
CLR
Q3
Master clock
Start pulse

17.

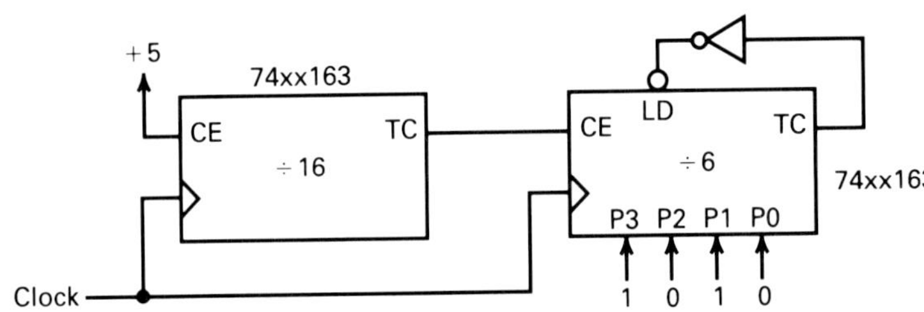
+5
74xx163
CE
TC
÷16
Clock
LD
CE
TC
÷6
P3 P2 P1 P0
1 0 1 0
74xx163

19.

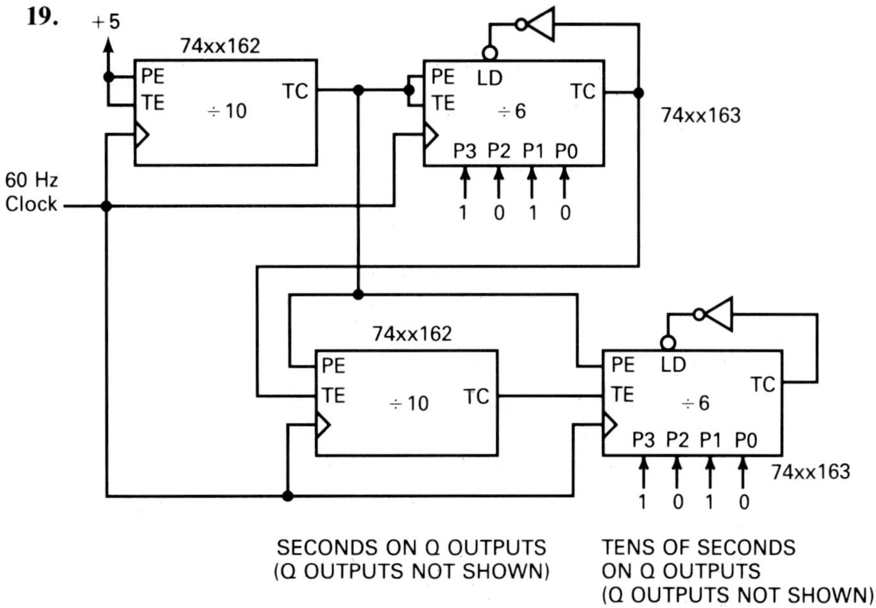

21. Preceed each settable stage with a manual override of the TC enable.

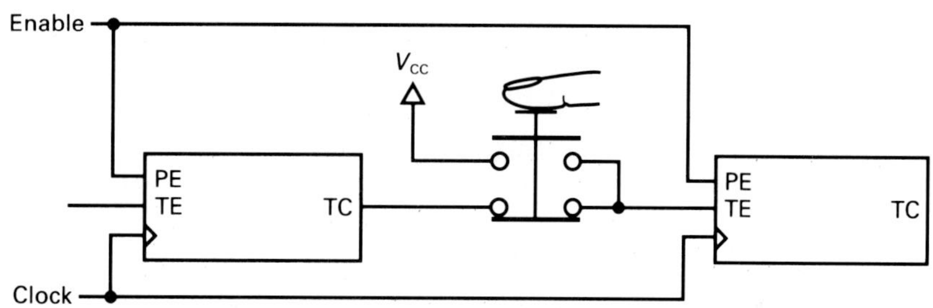

23. The missing circuitry in Fig. 6-28 starts with two ÷16 stages (two '163 counters) to produce 100 Hz, followed by two ÷10 stages (two '162 counters) for hundredths of seconds display. The TC output of the second ÷10 stage enables the remainder of the circuitry every second (1 pulse per second). Seconds and minutes are displayed using the outputs of alternating ÷10 and ÷6 stages as before.

CHAPTER SEVEN

1. (a) 11000001, (b) 11111000, (c) 101111010.
3. The maximum clock rate is 18.2 MHz.
5. Since only 10 ns is required for the two levels of carry logic and 5 ns for the XOR in each full-adder, the maximum clock rate is 40 HMz.
7. (a) 00111001 (+57), (b) 10000001 (−1), (c) 11011111 (−95).
9. (a) 11110011 (−13), (b) 00010011 (+19), (c) 101011111 (overflow).
11. (a)

$$\begin{array}{r} 00001101 \\ +11101111 \\ \hline 11111100 \end{array}$$

(b)

$$\begin{array}{r} 10110010 \\ +11111111 \\ \hline 10110001 \end{array}$$

(c)

$$\begin{array}{r} 01111111 \\ +11111011 \\ \hline 01111010 \end{array}$$

13. (a)

```
     1110
    ×0111
   ------
     1110
    1110
   ------
   101010   Partial Product
    1110
  -------
  1100010   Partial Product
  0000
  -------
  1100010   Product
```

(b)

```
     1010
    ×0101
   ------
     1010
    0000
   ------
    01010   Partial Product
    1010
   ------
   110010   Partial Product
  0000
  -------
  0110010   Product
```

(c)

```
      1111
     ×1111
   -------
      1111
     1111
   -------
    101101   Partial Product
    1111
   -------
   1101001   Partial Product
   1111
  --------
  11100001   Product
```

17. (a) 26.5, (b) −3.375, (c) 1.70898×10^{-3}, (d) −0.0234375

CHAPTER EIGHT

1. Twenty address lines.

3. The hex equivalent of 32000_{10} is 7D00.

5.

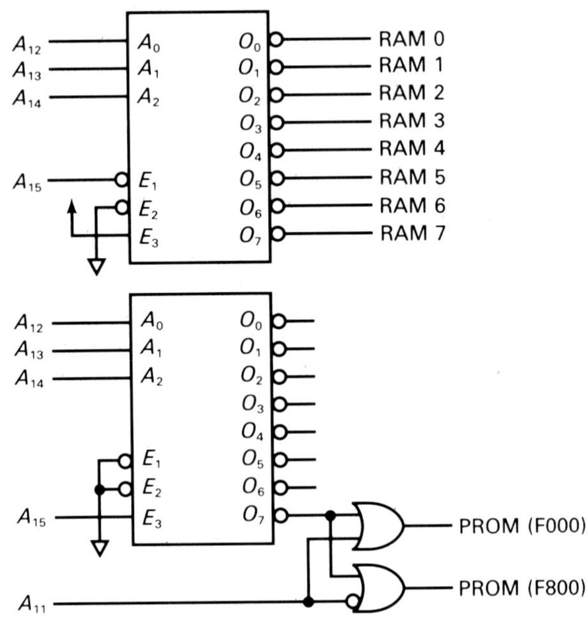

Full Decoding (Active Low)

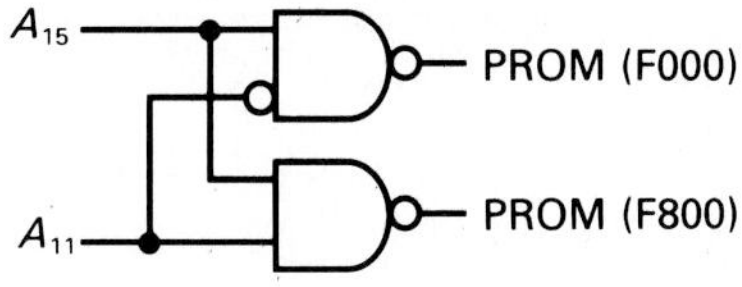

Partial Decoding of PROMs
(Active Low)

When the PROMs are partially decoded, any address containing a 1 in the A_{15} position will enable one of the PROMs. This is called aliasing.

7. Nine '138 ICs would be needed.

9.

	A_{15}	A_{14}	A_{13}	A_{12}	A_{11}	A_{10}
32K	X					
16K	X	X				
8K	X	X	X			
4K	X	X	X	X		
2K	X	X	X	X	X	
1K	X	X	X	X	X	X

CHAPTER NINE

1. Eight.

3.

Address Inputs				Seven Segment Code						
A_3	A_2	A_1	A_0	a	b	c	d	e	f	g
0	0	0	0	0	0	0	0	0	0	1
0	0	0	1	1	0	0	1	1	1	1
0	0	1	0	0	0	1	0	0	1	0
0	0	1	1	0	0	0	0	1	1	0
0	1	0	0	1	0	0	1	1	0	0
0	1	0	1	0	1	0	0	1	0	0
0	1	1	0	0	1	0	0	0	0	0
0	1	1	1	0	0	0	1	1	1	1
1	0	0	0	0	0	0	0	0	0	0
1	0	0	1	0	0	0	1	1	0	0
1	0	1	0	0	0	0	1	0	0	0
1	0	1	1	1	1	0	0	0	0	0
1	1	0	0	0	1	1	0	0	0	1
1	1	0	1	1	0	0	0	0	1	0
1	1	1	0	0	1	1	0	0	0	0
1	1	1	1	0	1	1	1	0	0	0

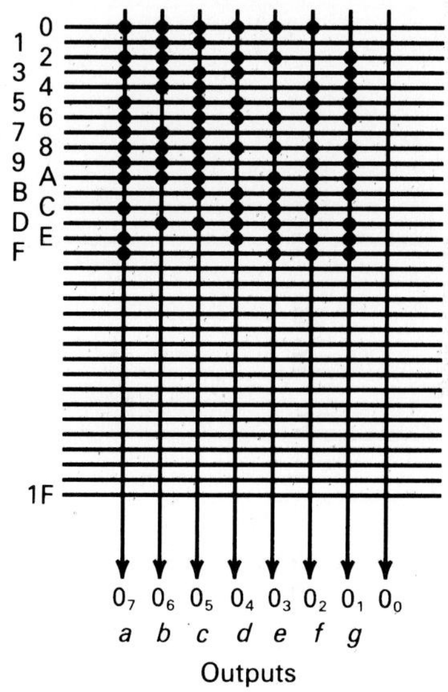

5.

x	y	Q_1^n	Q_0^n	Q_1^{n+1}	Q_0^{n+1}
0	0	0	0	0	1
0	0	0	1	1	1
0	0	1	0	0	1
0	0	1	1	1	1
0	1	0	0	0	1
0	1	0	1	1	0
0	1	1	0	0	1
0	1	1	1	1	0
1	0	0	0	0	1
1	0	0	1	1	1
1	0	1	0	0	0
1	0	1	1	1	0
1	1	0	0	0	1
1	1	0	1	1	0
1	1	1	0	0	0
1	1	1	1	1	0

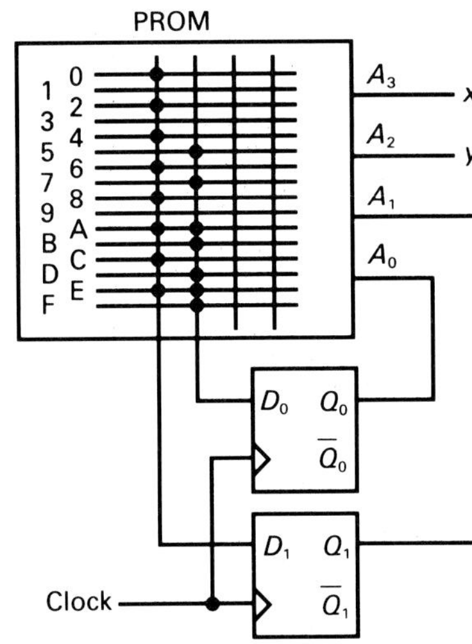

7. The PROM code below represents four possible chase light sequences.

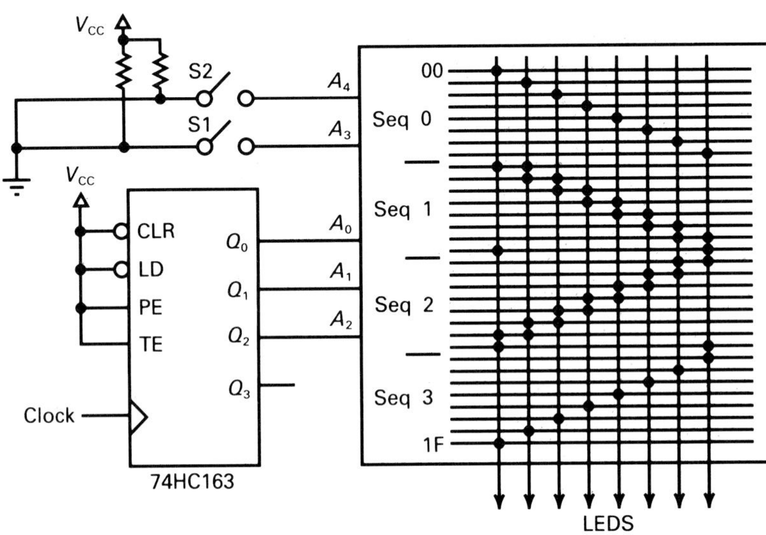

9. Hex Address

40 (7) (9) (12) (14) (15) all low — PC → MAR
(11) (10) high — PC + 1 → ALU latch

41 (4) (8) low — ALU latch → PC

42 (16) (17) low and (21) high — Ext. memory → MDR

43 (18) (9) low (11) (12) (10) (7) high — ACC-MDR → ALU latch

44 (8) (5) low — ALU latch → ACC

45 (2) (3) low and (13) high — Jump back to fetch.

11. Hex Address

Hex Address			
60	(9) (10) (11) low and (7) (12) high	ACC-1 → ALU latch
61	(8) (5) low	ALU latch → ACC
62	(2) (3) low and (13) high	Jump back to fetch.

CHAPTER TEN

1.

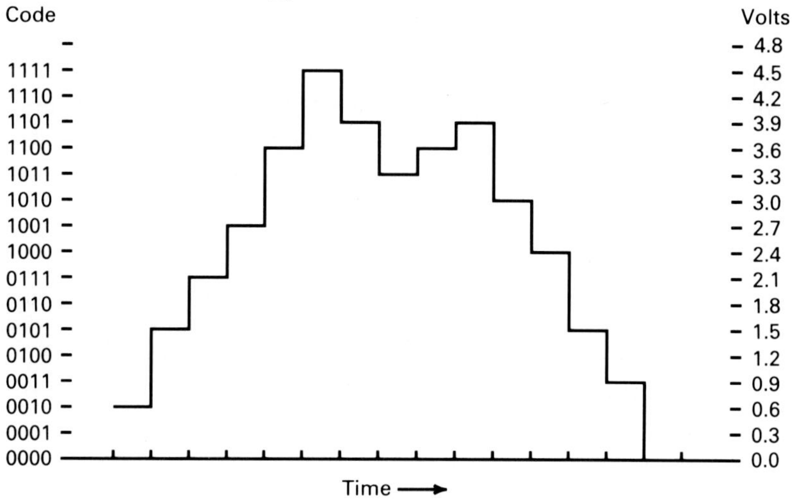

3.

Binary Input				Analog Output
B 1	B 2	B 3	B 4	v_{out} (volts)
0	0	0	0	0.0
0	0	0	1	0.4
0	0	1	0	0.8
0	0	1	1	1.2
0	1	0	0	1.6
0	1	0	1	2.0
0	1	1	0	2.4
0	1	1	1	2.8
1	0	0	0	3.2
1	0	0	1	3.6
1	0	1	0	4.0
1	0	1	1	4.4
1	1	0	0	4.8
1	1	0	1	5.2
1	1	1	0	5.6
1	1	1	1	6.0

5. 11000101_2 represents 197_{10} units steps.

$$\text{One step} = \frac{V_{max}}{2^n - 1} = \frac{6.0\ \text{V}}{255}$$

$$\text{Analog voltage output} = 197 \times \frac{6.0\ \text{V}}{255} \approx 4.64\ \text{V}$$

7.

Code	Ideal Voltage	Actual Voltage	Difference
0000	0.0	0.00	0.00
0001	0.5	0.51	0.01
0010	1.0	1.10	0.10
0011	1.5	1.63	0.13
0100	2.0	2.18	0.18
0101	2.5	2.75	0.25
0110	3.0	3.25	0.25
0111	3.5	3.75	0.25
1000	4.0	4.25	0.25
1001	4.5	4.75	0.25
1010	5.0	5.25	0.25
1011	5.5	5.60	0.10
1100	6.0	6.10	0.10
1101	6.5	6.55	0.05
1110	7.0	7.05	0.05
1111	7.5	7.50	0.00

There is a maximum integral linearity error of 0.25 V, or 1/2 LSB. This implies a maximum differential linearity error of 1/2 LSB.

9. The dynamic range (dB) = 6.02 × 10 = 60.2 db

11. The voltages in the graph represent the V_g input to the comparator in Fig. 10-19*a*. The voltages in the graph are added to $-V_{IN}$ and the sum is compared to ground in Fig. 10-24.

CHAPTER ELEVEN

1. 59 Ω < Z_0 < 116 Ω.

3. 57 Ω the dc resistance is much smaller. 57 Ω represents the resistance the trace presents to high frequency changes in voltage.

5. 24 mA.

7. Between 46 pF–49 pF.

9. Noise margin is the difference between what is required for an input voltage threshold and what is guaranteed as an output voltage for a specified fanout. The noise margin for 74AC driving 74HCT is 0.7 volts LOW and 2.4 volts HIGH (0.7/2.4).

11. The charge stored locally in the bypass capacitor helps to supply extra current needed for fast switching. This reduces the instantaneous current demands on the power rails and thus reduces switching noise. Switching spikes within a package are caused by the lead inductance of the V_{cc} and Gnd pins. Even if the extra current were available from outside the chip, it must flow through the lead inductance and may not get to the internal power bus fast enough to prevent some voltage drop or ground rise. Increased output drive capability and faster switching times result in greater current requirements in less time and therefore cause larger voltage spikes on the internal power bus.

13. Since the drivers are open collector, the switching time is limited to the charging time of the 1 KΩ pull-up resistor and the 50-pF bus capacitance. If the switching time is one time constant, then it will take 50 ns to switch from LOW-to-HIGH. This results in a maximum clock rate of 6 MHz. The HIGH-to-LOW changes would typically be much faster but the slowest transitions limit the effective data rate.

15. A HIGH-to-LOW transition requires 2–3 time constants.

17. a) 0 mA LOW/0 mA HIGH, b) 46 mA LOW/0 mA HIGH for termination to V_{cc}, and 0 mA LOW/24 mA HIGH for the termination to ground, c) ≈20 mA LOW/0 mA HIGH, d) and e) require no dc current.

19. RS-232 specifies +3 V to +15 V for logic 0, and −3 V to −15 V for logic 1.

INDEX